THE ARCHI L FRY

DREW

Ashgate Studies in Architecture Series

SERIES EDITOR: EAMONN CANNIFFE, MANCHESTER SCHOOL OF ARCHITECTURE, MANCHESTER METROPOLITAN UNIVERSITY, UK

The discipline of Architecture is undergoing subtle transformation as design awareness permeates our visually dominated culture. Technological change, the search for sustainability and debates around the value of place and meaning of the architectural gesture are aspects which will affect the cities we inhabit. This series seeks to address such topics, both theoretically and in practice, through the publication of high quality original research, written and visual.

Other titles in this series

The Architectural Capriccio
Memory, Fantasy and Invention
Lucien Steil
ISBN 978 1 4094 3191 6

The Architecture of Pleasure
British Amusement Parks 1900–1939
Josephine Kane
ISBN 978 1 4094 1074 4

No Matter: Theories and Practices of the Ephemeral in Architecture
Anastasia Karandinou
ISBN 978 1 4094 6628 4

The Challenge of Emulation in Art and Architecture
Between Imitation and Invention
David Mayernik
ISBN 978 1 4094 5767 1

Building Transatlantic Italy
Architectural Dialogues with Postwar America
Paolo Scrivano
ISBN 978 1 4724 1483 0

Forthcoming titles in this series

The Architecture of Edwin Maxwell Fry and Jane Drew
Twentieth Century Architecture, Pioneer Modernism and the Tropics
Iain Jackson and Jessica Holland
ISBN 978 1 4094 5198 3

The Architecture of Edwin Maxwell Fry and Jane Drew

Twentieth Century Architecture, Pioneer Modernism and the Tropics

Iain Jackson
Liverpool School of Architecture, University of Liverpool

Jessica Holland
Liverpool School of Architecture, University of Liverpool

Routledge
Taylor & Francis Group
LONDON AND NEW YORK

First published 2014 by Ashgate Publishing

Published 2016 by Routledge
2 Park Square, Milton Park, Abingdon, Oxfordshire OX14 4RN
711 Third Avenue, New York, NY 10017, USA

First issued in paperback 2016

Routledge is an imprint of the Taylor & Francis Group, an informa business

British Library Cataloguing in Publication Data
A catalogue record for this book is available from the British Library.

Library of Congress Cataloging-in-Publication Data
Jackson, Iain, author.
The architecture of Edwin Maxwell Fry and Jane Drew : twentieth century architecture, pioneer modernism and the tropics / by Iain Jackson and Jessica Holland.
pages cm. -- (Ashgate studies in architecture)
Includes bibliographical references and index.
ISBN 978-1-4094-5198-3 (hardback)
1. Fry, Maxwell, 1899-1987--Criticism and interpretation. 2. Drew, Jane, 1911-1996--Criticism and interpretation. 3. Modern movement (Architecture)-- Great Britain. 4. Modern movement (Architecture)--Africa, West. 5. Modern movement (Architecture)--India. 6. Architecture, Tropical--Africa, West--History--20th century. 7. Architecture, Tropical--India--History--20th century. I. Holland, Jessica, author. II. Title.
NA997.F78J33 2014
720.92'2--dc23

2014003011

ISBN 13: 978-1-138-24747-5 (pbk)
ISBN 13: 978-1-4094-5198-3 (hbk)

Contents

List of Illustrations

Note: Every effort has been made to trace all the copyright holders, but if any have been inadvertently overlooked, the publishers will be pleased to make the necessary arrangement at the first opportunity.

Introduction

1 From Classical Beginnings

Abbreviations

AA	The Architectural Association
ABN	*Architect and Building News*
AD	*Architectural Design*
AJ	*The Architects' Journal*
AR	*The Architectural Review*
AYB	The Architects' Year Book
CIAM	Congrès Internationaux d'Architecture Moderne
CIRPAC	Comité International pour la Réalisation des Problèmes d'Architecture
CPRE	The Council for the Preservation of Rural England
DIA	Design and Industries Association
EPIC	Exhibition for Planned Industrial Construction
FDDL	Fry, Drew, Drake and Lasdun
FLL	Francis Loeb Library
GSD	The Harvard Graduate School of Design
HL	Houghton Library
ICA	Institute for Contemporary Art
LCC	London County Council
LPTB	London Passenger Transport Board
MARS Group	Modern Architecture Research Group
MIT	The Massachusetts Institute of Technology
PEP	Political and Economic Planning
PWD	Public Works Department
RIBA	The Royal Institute of British Architects
RIBAJ	*The Royal Institute of British Architects' Journal*
TPR	*The Town Planning Review*
UEA	University of East Anglia
UL	University of Liverpool

Acknowledgements

We have been so kindly assisted by many during the course of researching this book, in particular, we are very grateful to Bruce Aitken, Soumyen Bandyopadhyay, Pradeep Bhagat, Justice Brobbey, Andre Brown, Ann Colin, Richard Bunt, Mark Crinson, Peter Dale, Trevor Dannatt, Elizabeth Darling, Alistair Fair, Robert Home, Grace Horne and family, Julian Holder, Neil Jackson, Rachel Lee, Judi Loach, Sylvester and Nike Okonkwo, Rexford Oppong, Hugh O'Shaughnessy, Jean and Stephen Macfarlane, Shireen Mahdavi, Dusa McDuff, John Morrison, Peter Pace, Simon Pepper, Alan Powers, Peter Richmond, Hannah Le Roux, Yemi Salami, Narrinder Singh, M.N. Sharma, Sangeet Sharma, William Smail, Graham Tipple, Ola Uduku, Steve Ward, Jonathan Woodham.

ARCHIVES:

Arne Hästesko, Alvar Aalto Museum; Edward Bottoms, Architectural Association; Lydia Dean, the Borthwick Institute for Archives, York University; Joanne Burman and staff at BP Archive, Warwick; P.C. Sharma, Chandigarh Architecture Museum; Sangeeta Sharma and Harmani Mann, Chandigarh Library; Justin Hobson at Country Life Picture Library; Charlotte Driessens and Andrew Culshaw, DOW; Inés Zalduendo at Harvard University Graduate School of Design Archives; Staff at the Houghton Library, Harvard University; Office of International Programmes at Ibadan University; Maureen Watry and staff at Liverpool University special collections and archives; David Redfern, Archivist for The London Group; staff at the National Archives, Kew; Nancy Barker, Debra Bennett, Stephen Kelly and Maria White at Pilkington Brothers; staff at McPherson Library and Special Collections; Bob Pullen at the Public Monuments and Sculpture Association; staff at the Public Records and Archives Department, Accra; Valeria Carullo at the RIBA Photographs Collection; RIBA staff at the V&A; Neil Parkinson, Royal College of Art; Diane Backhouse, United Africa Company; the Archivist at Unilever; Bridget Gillies at the University of East Anglia Archives; John Frederick at the McPherson Library and Special Collections, University of Victoria.

The authors are grateful to the editors of *Architectural History* for allowing use of material on Pilkington Brothers' Headquarters in Chapter 7, which was first published in different form, in the 2013 volume of the journal.

The authors wish to express their gratitude to the Leverhulme Trust; Liverpool School of Architecture and Paul Mellon Centre for studies in British Art for their generous grants that enabled this research to take place.

PICTURE CREDITS

Anon: 3.11, 2.1, 2.2, 2.5, 3.14, 3.15, 5.9, 5.15, 5.16, 5.18, 5.19, 5.20, 5.21, 5.22, 5.23
Architectural Association Archive: 3.5
Architectural Press Archive / RIBA Library Photographs Collection: 1.8, 1.9, 2.12, 2.14, 2.15, 2.16, 2.23, 2.24, 2.25, 2.26, 2.29, 2.30, 3.10, 3.11, 3.17, 3.18, 3.19, 3.20, 3.22, 3.23, 7.4
BP Archive: 6.43, 6.49
Chandigarh Art Gallery & Museum: 6.3, 6.8, 6.13, 6.14, 6.22
Churchill Archives Centre 4.9, 4.10, 4.12, 4.13, 4.14, 4.17
Colin Westwood / RIBA Library Photographs Collection: 3.24
Country Life Picture Library, © Country Life: 1.13, 1.14, 7.2, 7.3
Dell & Wainwright / RIBA Library Photographs Collection: 2.11, 2.13, 2.27
John Donat / RIBA Library Photographs Collection: 7.22, 7.23
Estate of Maxwell Fry: 1.2, 1.3, 1.10, 1.11, 1.12, 1.15, 1.16, 2.3, 2.4
Estate of Maxwell Fry, courtesy of University of East Anglia, Pritchard Papers: 2.19, 2.20
Estate of Maxwell Fry, courtesy of University of East Anglia, Pritchard Papers: 1.5, 1.6, 1.7
Courtesy of Grace Horne: 0.1, 0.2, 6.45, 7.7, 7.18
Jessica Holland: 6.25, 6.37, 6.49, 7.16, 7.19, 7.32
Iain Jackson: 4.8, 4.11, 4.12, 4.15, 4.16, 4.19, 4.18, 4.19, 4.20, 4.21, 4.22, 4.24, 4.26, 4.27, 5.2, 5.4, 5.5, 5.6, 5.7, 5.8, 5.10, 5.11, 5.12, 5.13, 5.14, 6.3, 6.4, 6.5, 6.6, 6.7, 6.9, 6.10, 6.11, 6.12, 6.15, 6.16, 6.17, 6.19, 6.20, 6.21, 6.23, 6.24, 6.26, 6.27, 6.28, 6.29, 6.30, 6.31, 6.32, 6.33, 6.34, 6.35, 6.36, 6.37, 6.39, 6.42, 6.43, 6.46, 6.47, 6.48, 7.18
Courtesy of Jean and Stephen Macfarlane: 3.15
John Maltby / RIBA Library Photographs Collection: 3.9, 3.16
Courtesy of John Morrison: 2.9, 3.1, 3.2, 3.3, 3.4, 3.8, 5.1, 6.19, 6.41, 6.42, 7.1
The National Archives: 4.4, 4.5, 4.23, 4.25, 5.3, 5.17
Courtesy of Hugh O'Shaughnessy: 0.3
Pilkington Brothers' Archive, Pilkington Glass Ltd (part of the NSG Group): 7.7, 7.8, 7.9, 7.10, 7.11, 7.12, 7.13, 7.14
Courtesy of Alan Powers: 2.21
Courtesy of Public Monuments and Sculpture Association: 7.22
RIBA Library Drawings Collection: 2.17, 2.18, 7.30, 7.31
RIBA Library Photographs Collection: 2.6, 2.7, 2.22, 2.28, 3.9, 3.16, 3.21, 3.25, 6.39, 7.5, 7.15, 7.20, 7.21, 7.24, 7.25, 7.28, 7.29
Rolls Royce Archive: 7.26, 7.27
University of East Anglia, Pritchard Papers: 2.8, 2.10
University of Liverpool Special Collections and Archives: 1.1, 1.4, 5.1, 6.2

Archives Consulted, Excluding Private Archival Material

Alvar Aalto Museum, Helsinki, Finland
Architectural Association Archive, London
Borthwick Institute for Archives, University of York, York
BP Archives, Warwick University, Warwick
Chandigarh Art Gallery and Museum, Chandigarh, India
Churchill Archives Centre, University of Cambridge
Concordia University, Quebec, Canada
Frances Loeb Library, Harvard University, Cambridge, MA, USA
Gas Archive, Warrington
Houghton Library, Harvard University, Cambridge, MA, USA
Liverpool Record Office, Liverpool
National Archives, Kew
National Sound Archive, British Library, London
Pilkington Brothers' Archive, St. Helens
Public Records and Archives Department, Accra, Ghana
Royal Commonwealth Society, University of Cambridge Library, Cambridge
Royal Institute of British Architects Drawings and Archives Collection, London
Syracuse University Library Archives, NY, USA
Unilever Archives, Port Sunlight
University of East Anglia, Norwich
University of Liverpool Special Collections and Archives, Liverpool
University of Victoria, British Columbia, Canada

To Jo and Pippa

Introduction

This monograph sits between a biography, a select gazetteer, and a social history, using these three historiographical strands to consider the work of architects, Edwin Maxwell Fry (1899–1987) and Jane Beverly Drew (1911–96). Although these two architects are central to the narrative, the intention from the outset has not been to canonise Fry and Drew amongst the saints of a slender and particular reading of Modern architecture. Rather their careers provide a framework through which a distinct set of buildings from the twentieth century are examined. They are at the same time ordinary and remarkable: they both emerged from an aspiring middle-class, taking advantage of university educations to become stalwarts of modernist architecture in twentieth-century Britain, and their marriage in 1942 marked the commencement of a personal and professional relationship that lasted over 45 years.

In his lifetime, Fry enjoyed a reputation as a key figure in the development of the British Modern Movement. He was amongst the first British architects to see a building completed in the concrete, white and geometric manner that became ubiquitous with a certain persuasive type of modern architecture. Discussion of this role perhaps inevitably tends to focus on Fry's involvement in the Modern Architectural Research (MARS) Group, his early social housing schemes and his prestigious partnership with the Bauhaus-founder Walter Gropius, which influenced much of Fry and Drew's later work. This research seeks to provide a well-rounded account of the career of one of Britain's most familiar and yet little known figures.

Drew, meanwhile, is significant for different reasons, not least as one of a handful of well-known female British architects of the twentieth century. Despite her relative celebrity, Drew's contribution to twentieth century architecture remains undervalued; she received five separate nominations for the RIBA Gold Medal, but her work went formally unrecognised. She was highly unusual amongst her contemporaries: firstly, as a woman; secondly, because she chose to step away from design problems related to the 'feminine' sphere of domesticity. Notions of her difficult personality remain (some even speculating it cost her the Gold Medal) and such tales often take precedence over her architecture. Indeed, Drew's strong character did not fit within the cultural conventions of the time and she became a divisive figure. She was apparently viewed as a bad influence for Fry, prompting some of Fry's staff to leave his office following the establishment of the Fry and

Drew partnership. This study seeks to redress the balance by focusing on her architectural achievements.

Despite their considerable contributions to architecture, Stephen Hitchins's book *Fry, Drew, Knight, Creamer: Architecture* (1978) is the sole monograph concerning the work of Fry and Drew. As the title suggests, it focuses on their post-war work undertaken with partners Frank Knight and Norman Creamer. The slim, heavily illustrated volume provides a useful overview of these projects, which this research seeks to complement by providing a comprehensive account of Fry and Drew's respective careers from the 1920s through to the 1970s. Other scholarship has focused on distinct portions of their work, such as Fry's early modernist buildings, their projects throughout West Africa and their collaborative work in India. However, large areas of their careers remain undocumented.

The RIBA Drawings and Archives Collection holds a sizeable collection of Fry and Drew's personal and professional papers, although much was lost before the acquisition of their archive. The couple's offices at 3 King Street were completely destroyed by wartime bombing in 1944,[1] and, later, out-dated records were thrown away in anticipation of the practice's move from their London offices to Sevenoaks in Kent (Figure 0.1).[2] The papers therefore contain very limited material regarding project administration, although there is wealth of personal correspondence between Fry and Drew from 1940 into the 1970s. This correspondence gives details of both home and work, illustrating the way in which these lives were intertwined. It highlights their numerous periods of separation throughout the post-war years: whilst Drew was in India, Fry might be in Africa; and whilst Drew took up a fellowship in America, Fry remained in London. During the 1940s, in the first years of their marriage, the considerable distances – which, of course, could not be navigated with today's ease – were evidently hard felt, but a sense of duty emerges. The letters also give a fascinating view into their respective personalities. Fry used

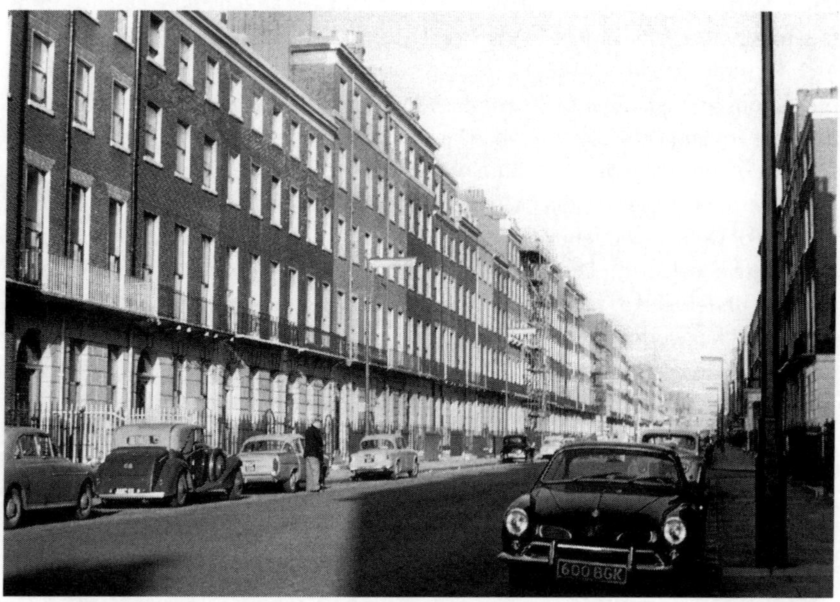

0.1 Fry and Drew's home and office at 63 Gloucester Place, London, c. 1960

writing as a cathartic process, writing in detail of day-to-day frustrations and more significant issues regarding their relationship: 'And this is the last man of yours I am going to entertain. Positively'.[3]

Fry was a talented writer, with a gift for capturing the spirit of personalities and incidences that took place. Although, throughout his lifetime, his writing retained the rhetoric of an interwar modernist striving to promote his cause – leading Marion Coates to refer to him as a 'turgid, if reliable writer'.[4] Fry's enthusiasm for writing began with the self-conscious essays of his student days, he wrote throughout his lifetime to clarify and express his personal and professional life. His letters often contain sketches of his surroundings or the latest project occupying his thoughts. Drew, meanwhile, was more irregular and chaotic in her correspondence, often leading Fry to despair at the lack of news. Her writing style gives the impression that her hand could not keep up with her brain; punctuation was generally not seen as a necessity, thoughts fizz onto the page and, veering from topic to topic, her letters portray a vivacious and generous-spirited character. Fry and Drew's respective writings reveal lives lived in parallel; their memories and observations blur and merge, illustrating the closeness of their relationship.

Although Fry and Drew left abundant notes, diaries and correspondence they are but fragments and disjointed (and not always accurate) recollections and perceptions. Like their modernist collaborators, such as Le Corbusier and Gropius, Fry and Drew were careful to ensure their own posterity, constructing well-worn tropes that appear throughout their writings and interviews with regularity during the 1970s and '80s, but as early as the 1950s Fry's writing, in particular, offers a rather rose-tinted view of his life and work.

As her correspondence shows, Drew was a doer while Fry was a thinker and writer. Her intense practicality was invaluable in (post-)colonial situations where money was scarce but need was great – indeed, this is where Drew flourished. Drew designed a village and school in Kenya in the late 1930s, before she had met Fry, and she brokered much of their work in Africa, India and the Middle East. Drew also instigated the *Village Housing in the Tropics*, a pragmatic publication that gave the duo a strong foothold in West Africa. Fry achieved further recognition with his best-selling credo on architecture, *Fine Building*, and, along with extensive media coverage of their work, they were able to help shape architectural discourse in Britain and its colonies. In 1945, Drew established the *Architects' Year Book*, which included articles from key thinkers and designers. The journal ran for 14 editions, and gives a strong sense of the social 'cause' which was always of utmost importance for Drew.

Furthermore, the architect's vision is not always aligned with that of the client, for example, and many other parties contribute to a final design solution. For instance, engineer Ove Arup contributed to nearly all their major projects and must be considered an integral part of their design team. The role of the client and government is revealing when examining their work in West Africa, particularly the records of the Colonial Office. Information and correspondence has been carefully retained and access to previously 'closed' material has revealed the strong design input made by the client body as well as details of the disputes that occur with major building projects. Other sources that have been generously shared include several private and family archives, giving a privileged view into Fry and Drew's family life. Company records, such as those held by Pilkington Glass and BP, have been particularly insightful and offer a significant in-depth accounts of the workings (and failings) of Fry and Drew's design methods.

The buildings themselves have also provided a valuable source of information. These buildings endure, in some cases, near to the original concept and, in others, vastly altered almost beyond recognition. Where possible, all of the buildings discussed have been visited and analyzed, and a substantial collection of contemporary records of the work has been compiled. The urgency of this task for other significant twentieth century building is heightened as (at least) one of Fry and Drew's buildings has been demolished during the course of the writing of this book.

Fry considered himself a modernist and an ardent follower of Bruno Taut, Walter Gropius and Mies van der Rohe, and his fondness for history and tradition persisted. In particular his desire to use brick and his rejection of concrete as a facing material, along with his dismissal of formalist displays supports revisionist theories of Modernism that argue modernists were much closer to traditionalists than previously argued. Fry's departure from his 1930s pristine white boxes was swift, although his desire to compose a façade with crisp, linear and geometric components remained a constant theme and was clearly important to him, regardless of the building's function, geographical location or materiality. His ability to compose a successful façade arrangement from simple repeated elements was one of his major strengths as an architect. Indeed, his contemporaries have been quick to recognise Fry's abilities as a designer. A former employee (who became an Associate of Fry, Drew, Knight and Creamer) Derrick Lees implies a touch of genius in his work. Describing Fry's method of working out a design problem, Lees later wrote:

> *When Max received a project, he might be briefed say on a Thursday, he would shut himself in his office on the Friday. By Monday or Tuesday a scheme would be ready with plans, elevations and perspectives, all carried out in his own hand, mostly freehand, and with crayon colouring. Max's spontainious [sic] perspectives were especially appealing and works of art in their own right. The client would be immensely impressed at so speedy a response, the design would not represent necessarily the final scheme but it would be an immense step in the right direction.*[5]

Drew, meanwhile, has not received recognition for her work as an architect or as a designer. In contrast to Fry, social concerns were to the fore of her work and this often led to compromise in design or aesthetic terms. This she accepted willingly. As she later observed,

> *I have to confess that the jobs I've done in life, I have always been involved in the cause of the job. Whether the Open University or education in Ghana or doing a city for the refugees. It has always mattered to me tremendously that the object should be something very worthwhile.*[6]

While this monograph seeks to trace the lives and careers of 'Max and Jane' as husband and wife, and moreover as architects, it also seeks to provide details of their overlapping careers as part of the work of a successful architectural office that, at one stage, employed over 50 people from around the world.[7] The amount and geographical spread of their work was enough to ensure that often they had little input or control over the day-to-day running of a project and indeed they allowed their employees to take control wherever possible. Drew thought Fry 'had talent, but not outstanding original talent',[8] and the two of them worked separately on their own projects, although Fry would frequently comment on Drew's proposals.

Fry later claimed that the organisation of the office was 'geared to give me maximum uninterrupted time at my drawing board which has been the source of my power'.[9] This detached approach was not always appreciated by their clients, it cost them at least two projects, and on numerous occasions Fry and Drew were summoned to meetings to resolve various problems and disputes. Drew was particularly adept at smoothing over problems and disagreements, but even her negotiation skills could not placate the strained relationship with the client at Ibadan University, Nigeria. Their projects in Africa were often understaffed, and young inexperienced architects were sent off from London to manage the schemes with very little supervision or effective communication channels. The office also had a high turn-over of staff, which was observed by at least one client. Many young architects, often visiting Britain from overseas wanted to work for Fry and Drew; it was a prestigious firm and even a short stint working with them carried much kudos.

Project management and running an office were not Fry and Drew's strength, yet their genial and friendly manner usually diluted any tension and left the client feeling at ease. Indeed, as a former employee noted, 'It was clear that Max and Jane had magnetic personalities that had impressive effect with clients, who

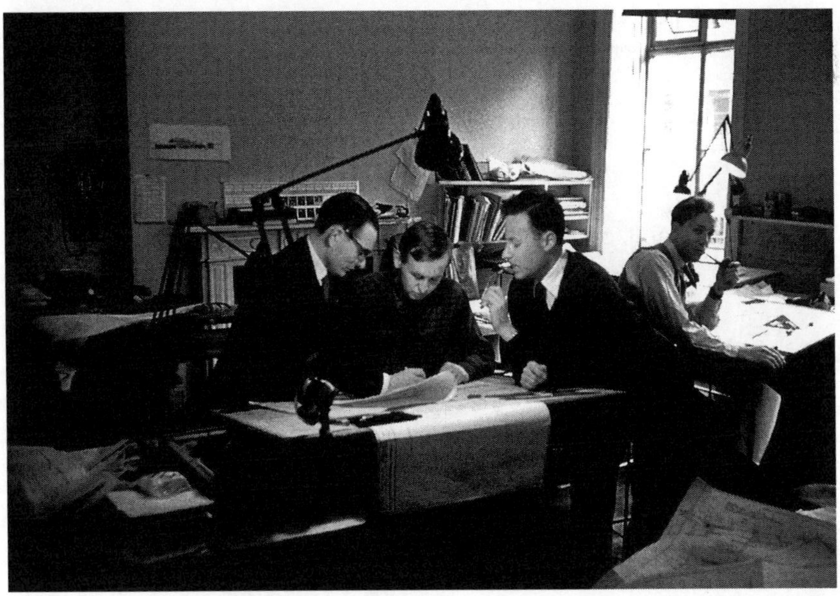

0.2 Draughting Office, 63 Gloucester Place, London, c. 1960

then became personal friends'.[10] This friendship group was diverse and included numerous artists, writers, publishers – many of whom were frequently sent homemade cards and extended letters (Figure 0.3). It seems Fry and Drew were not really concerned with running their practice as an efficient commercial business. Their letters frequently discuss financial 'feast and famine', with concern over where the next job would come from, or not having enough space for the extra drawing boards following unexpected commissions and sudden demands for construction drawings.

Fry and Drew are perhaps most well known for their advancement of 'Tropical Architecture' in former British colonies.[11] This work, spread over at least 16 different countries raises the fundamental questions of this research, namely, why were Fry and Drew designing buildings in the remote places of empire and what kind of architecture were they producing? Without a doubt they were a vector of the late imperial efforts of Great Britain, operating at the tail end of colonial decline and transfer of power. Fry and Drew, particularly in West Africa occupied a central position in this regard, designing schools, colleges, universities and numerous other projects as a result of the Development and Welfare Acts. This pre-eminence resulted in further commercial commissions for banks and offices, and a large collaborative project for the newly independent India. A specific type of architecture was developed for these territories that incorporated perforated concrete screens, rainwater harvesting, window hoods and projecting roof eaves, attempting to passively modify the climate and to respond to the 'local' conditions. It was not a fixed or static style however, and Fry and Drew's work reveals an evolutionary approach that was influenced by the Colonial Public Works Department as well as by the indigenous conditions and their extensive consultations. Consultation and community engagement was an important part of their design process. They did not just fly into a place and start designing, but would seek to collaborate with the 'end users' as well as with the fee-paying client. This is significant, as often the occupants, particularly in Development and Welfare projects were considered anonymous, and there was very little data available that could usefully inform the planning and design. Through lengthy discussions and meetings Fry and Drew sought to understand the desires and needs of the users and did not view

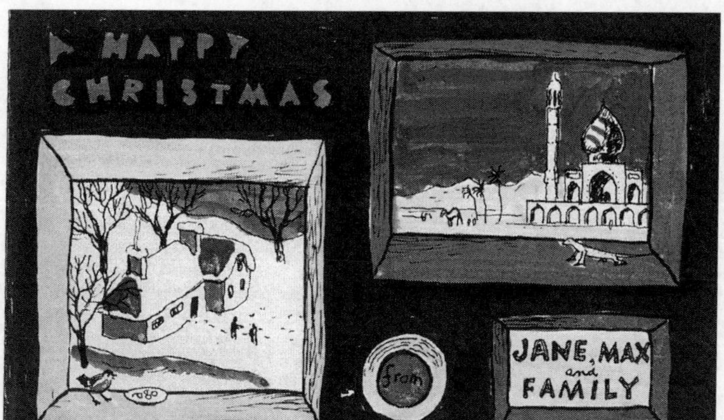

0.3 Christmas Card from the Fry family, c. 1955

the architect as the 'expert' dispensing solutions. Travelling exhibitions were also organized throughout West Africa including one for Kumasi, which required police intervention following the crowds of visitors.

Their design recommendations were often simple and 'hands-on', such as the pragmatic placing of road junctions and designing to follow contours along with methods of lightening the burden of domestic chores and sanitation. With regards to housing, their work investigated various solutions but concluded that the 'traditional' compound arrangement, if built soundly with adequate cross-ventilation, was the most appropriate type, and that the best people to design the housing were the people living in the villages themselves.

As a result of their African experience they were deemed ideal candidates to design in India. Along with Le Corbusier, Pierre Jeanneret and a large team of Indian architects they designed the new Punjab Capital, Chandigarh. Fry and Drew lived in India for three years during the design stages of Chandigarh, leaving Drew's children in Britain under the care of their long-term nanny Maud Hatmil. Their practice, meanwhile, was managed by Denys Lasdun and Lindsey Drake in their absence with the firm renamed Fry, Drew, Drake and Lasdun. The Chandigarh project attracted worldwide interest, but the focus of attention then and afterwards has been firmly fixated on Le Corbusier's contribution. Fry and Drew designed numerous projects focusing on housing but also included a number of other projects such as schools colleges, hospitals and clinics. Despite the interest and opportunity, Fry found this to be a difficult time, not least because of the friendship that emerged between Drew and Le Corbusier. He also felt that he was missing out on opportunities elsewhere, and in the UK in particular. When he returned to London in 1954 he struggled for some years to build up the practice and to position his work within the emerging debates surrounding modernity and monumentality.

Despite Fry's struggle to reintegrate, substantial projects continued throughout West Africa and India largely dominated by the oil industry and associated bank buildings that followed in its wake. Whilst some of this work attempted to respond to climate it was really the introduction of air-conditioning and the equation of height with prestige that influenced the architectural proposals. The non-place specificity of this approach was not lost on Fry and Drew, who attempted to embed a locale root through the introduction of artwork, frequently integrated into the building fabric in the form of mosaics, murals and carved screens. However, they too succumbed to the temptations and possibilities of air conditioning including detailed guidance on its use in their later publications on tropical architecture.[12]

Their final works in Britain include several company headquarters as well as education buildings for the burgeoning university sector. Two of these buildings were in Fry's home town, although the demolition of the Veterinary Science building in Liverpool in 2012 is demonstrative of the current 'value' of Fry and Drew's later British work. This work sits outside of the established and orthodox stories of twentieth century architectural history and modernism in general. It was this 'steady approach' and reticence to depart from the functionalist (plus sculpture) method that rendered their work, by some, as outmoded, despite the quality of the materials, detailing, construction and substantial investment into public art that is found in these works. General attitudes towards 1960s architecture as well as shifts in architectural fashions and the ferocity of a historical narrative that focused on the rise of 'New Brutalism' has been disastrous for this type of twentieth century architecture.[13]

NOTES

1 HL, Harvard University, Walter Gropius Papers in the Bauhaus Archiv, MS Ger 208 (746). Letter Fry to Gropius, 16 March 1944.

2 UL Archive, Peter Dale Papers [uncatalogued]. Jane Drew interview with Peter Dale, 1992. They moved to 63 Gloucester Place on a twenty-year lease funded by Drew's friend Peter Gregory.

3 RIBA Archive, F&D/18/4. Letter Fry to Drew, 1 October 1944.

4 UEA Archive, Pritchard Papers, PP/28/5/41. Marion Coates, Notes on 'Design for To-Day', undated.

5 UL Archive, Peter Dale Papers [uncatalogued]. Letter Derrick Lees to Peter Dale, 23 June 1993.

6 Jane Drew interview with Karen Parker, 19 June 1988. p. 70.

7 UL Archive, Peter Dale Papers [uncatalogued]. Jane Drew interview with Peter Dale, 1992.

8 John Morrison Archive. *Jane Drew Biography*, unpublished manuscript, p. 32.

9 RIBA Archive, F&D/14/4. Maxwell Fry's Memoirs, p. 38.

10 UL Archive, Peter Dale Papers [uncatalogued]. Letter Derrick Lees to Peter Dale, 23 June 1993.

11 See, inter alia: Rhodri Windsor Liscombe, 'Modernism in Later Imperial British West Africa: The Work of Maxwell Fry and Jane Drew, 1946–56', *Journal of the Society of Architectural Historians*, 65 (June 2006): pp. 188–215; Ola Uduku, 'Modernist Architecture and "the tropical" in West Africa: The Tropical Architecture Movement in West Africa, 1948–70', *Habitat International*, 30 (2006): pp. 396–411; Hannah Le Roux, 'Modern Architecture in Post-Colonial Ghana and Nigeria', *Architectural History*, 47 (2004): pp. 361–92.

12 E. Maxwell Fry and Jane Drew, *Tropical Architecture in the Humid Zone* (London, 1956), pp. 228–37.

13 Reyner Banham, 'The New Brutalism', *Architectural Review*, 118 (December 1955): pp. 354–61.

1

From Classical Beginnings

Much has been written of the development of the Modern Movement in interwar Britain and Edwin Maxwell Fry (1899–1987) – or Max Fry as he came to be known – occupies a central role in the 'Grand Narrative' that depicts the movement as a continuous, homogeneous process.[1] Fry's friends and MARS Group colleagues, Alison and Peter Smithson, for example, allied Fry's work with their chronology of the 'Heroic Period of Modern Architecture', presenting Kensal House (1937) as a final British flourish.[2] Fry's work is thus presented as part of a Modern inheritance, part of 'the rock on which we [architects] stand' wrote the Smithsons.[3] Fry, too, contributed to the establishment of his position as one of a handful of home-grown modernists amongst an interwar scene dominated by European ideals and influences. In 1957, he wrote of the difficulties in building his first Modern house, 'it was not before we had been turned off three sites and found one sufficiently remote that I was able to build ... with the limitation imposed by the council that traditional materials should be employed'.[4] These post-war recollections of struggles against an insular architectural establishment – and the flowering of British Modernism nonetheless – reinforce Fry's place at the heart of the heroic canon.

Recent revisionist interpretations of the period also recognise Fry as a key figure in the founding and development of a Modern Movement in Britain.[5] Such interpretations consider the British milieu to be more positively inclined toward Modern architecture and the domestic growth of the movement, with Fry portrayed as a central figure in the 'narratives of modernity' that came together to help Modernism achieve its 'ultimate hegemony' post World War Two.[6] This consensus of Fry's position as the leading light of British Modernism encourages a closer examination of his work. Now, as during his lengthy career, Fry's work generally receives a positive reception; as Anthony Jackson observes, 'Fry was almost unique in his acceptance by the various shades of architectural opinion'.[7] Indeed, there are interesting contrasts present throughout his career: Fry combined a privileged position of associate membership on the RIBA Council (1933–37) and his role as a key figure of the MARS Group with no apparent conflict of interest or agenda; he was a member of the exclusive Athenaeum Club in London – of which the ardent anti-modernist Reginald Blomfield was also a member – while also holding membership to the leftist Political and Economic Planning (PEP) Club. Fry is a perfect example of the 'permissive transgression' between supposed modernists and traditionalists in British interwar architectural culture, highlighted by Hélène Lipstadt.[8]

Looking beyond these social connections, this chapter investigates the roots of Fry's career. Fry's autobiography describes his 'conversion' to Modernism in the mid-1920s, which prompted him to tear up drawings of his earlier Classical work and begin again with a fresh sheet of paper.[9] Yet Fry's architectural training in the international cities of Liverpool and New York suggest that his education and formative influences were more progressive than Fry would care to admit. This chapter reconsiders Fry's route to Modernism, looking beyond his built work to examine a broader scope of his life, to illustrate the continuity of his architectural training and career. It focuses on his formative years in Liverpool, his education, and his early professional collaborations up to the end of the 1920s: firstly, with the town planners Thomas Adams (1871–1940) and Francis Longstreth Thompson; and, secondly, with his fellow modernist pioneers Jack Pritchard (1899–1992) and Wells Coates (1895–1958).

Fry grew up on Merseyside. He was not only a graduate of the Liverpool School of Architecture, but he was a product of the city itself. His childhood instilled in him values that remained throughout his lifetime and directly influenced his approach to the built environment. In his memoirs, written in the 1970s, the long shadow of his childhood experience is evident, played out against the backdrop of a city full of vitality. At the beginning of the twentieth century, the time of Fry's birth, Liverpool was one of the most prosperous ports in the world. Established initially for commerce with Ireland in the thirteenth century, the port had grown dramatically due to the lucrative cotton trade of the late-eighteenth century to create a city of physical and intellectual modernity that was unusual in Britain.[10] This dynamic shipping trade – rather than the manufacturing tradition of industrial centres like Manchester – gave Liverpool an international outlook that thrived on technological advances borrowed from other cultures to ensure its continued wealth. It was the perfect setting for an aspiring architect.

The frequent passenger ships transporting emigrants to the new world facilitated the city's assimilation of innovative ideas. Liverpool's capitalist and cultural ventures that operated through the port thus looked outward to America – rather than into Britain's 'island culture' or to nearby Europe – to such an extent that news from the US was reported in the local newspapers.[11] The port was the centre of life in the city, as Fry later recalled, evocatively describing the cargo and its associated wealth:

> *There was no escaping it. A stone's throw from the politest shopping street were narrow alleys lined with warehouses reeking of cloves and pepper. The smell of molasses came in pungent waves up the slopes from the docks into the financial centre where men in top hats and hands deep in trouser pockets talked money as they walked from one set of chambers to another.*[12]

As a centre of flourishing industry, Liverpool inevitably fostered an under-class of low-paid workers; Christopher Crouch notes, 'Enormous wealth and absolute squalor existed side by side'.[13] The transient emigrants swelled the city's already considerable population of casual dock labourers, keeping wages low and facilitating growth. This extreme inequality led to social tensions and Liverpool was an unusually politicised city that looked to protect the rights of its disenfranchised casual workforce. During Fry's period of architectural training in the 1920s, Liverpool 'was characterised chiefly by the struggles against unemployment'.[14] These struggles were manifest in a series of strikes and riots, beginning almost a decade earlier with the dock strike of 1911, which sparked subsequent action by various workers' groups and garnered lasting cohesion amongst the unions.[15] The young Fry would have been aware of these tensions and the plight of the working classes – and the squalid living conditions of the prevalent slums – and he sought to address these issues in his architecture from an early stage.

The Fry family were typical of the aspirational lower-middle classes that hoped to find opportunity in the city. Fry's entrepreneurial father, Ambrose Owen Fry (b. 1869), had an itinerant childhood; born in Montreal, he was raised first in Greenock, Scotland, and then a few miles from Liverpool city centre, in Walton-on-the-Hill.[16] Ambrose Fry is recorded as a commercial traveller and later a chemical manufacturer, and he travelled on business every few years on passenger liners from Liverpool to Montreal and, more frequently, to New York.[17] These enterprises would have illustrated the nearness of the British Empire and beyond to his eldest son, although Fry's memoirs show a degree of ambivalence towards his father's work and he describes Ambrose Fry as 'nothing if not a businessman with all sorts of irons in the fire – chemicals, electricals, old property, reversions, house conversions'.[18] Instead, Fry chose to align himself with his artistic, piano-playing mother, known as Lily (b. 1869), whom he later portrayed as a refuge from the professional ambitions held for him by his father.

The Fry family moved frequently, according to the success or failure of Ambrose Fry's business ventures. After a spell in Wallasey, across the River Mersey on the Wirral peninsula, the young Fry spent his formative years in 'an undistinguished house in an undistinguished street' in the southern suburbs of Liverpool.[19] He shared the ten-room house with his two elder sisters, Muriel (b. 1895) and Nora (b. 1897), and his younger brother, Sydney (b. 1900), and a domestic servant employed by the family, Annie Blanchard.[20] Straddling the divide between rich and poor, his home was situated close to the large merchant villas that surround the genteel Victorian grounds of Sefton Park but also backing onto an adjacent workhouse. A model of respectability, the Fry family attended the nearby Unitarian church and Fry was schooled at the prestigious Liverpool Institute, a grammar school close to the city centre. Fry was an undistinguished student, his early education coming instead from his friendship with a wealthy local ship-owner, described by Fry as his 'patron'.[21] 'Old Hall', as he later asked Fry to call him,[22] introduced his protégé to English lyrical poetry (an interest that Fry acknowledged lasted a lifetime) and provided an opportunity to experience first-hand the engineering that had brought the city its wealth; as Fry later wrote, 'climbing up and down the cavernous engine-rooms of his ship lying in dock I recognised the purposeful elegance of machinery fortified by a boyish worship of the great steam locomotives'.[23]

Fry's entry into the architectural profession took a convoluted route. Prior to his training he had worked in his father's factory as an office boy, tried his hand as a sales representative and, most significantly, fought as a Second Lieutenant of the First King's Liverpool Regiment in the First World War. Fry writes little of his wartime experiences, suggesting only that his postings were characterised by inactivity: 'to France as fighting petered out into armistice; to Germany in the occupying army; to a period of stagnation by a canal in Belgium'.[24] Yet this uneventful service provided Fry with a much-needed opportunity to study at university via an ex-serviceman's grant. The grants brought numerous students in to the Liverpool School of Architecture, including many from Australia and New Zealand,[25] and Fry was amongst the first post-war cohort, beginning his studies in February 1920 with his matriculation back-dated to December 1919.[26] He accepted his place without his parents' blessing, although an agreement was reached whereby he would receive ten shillings per week and remain at the family home.[27] In addition, Fry's three-year grant covered his tuition fees and provided a maintenance sum of £100 per year.[28] Fry undertook a period of self-preparation in an attempt to compensate for his lack of formal qualifications, which consisted, in his words, of 'reading, observation and day-dreaming' but failed to venture beyond the juvenile 'world of castles and manor houses'.[29] After his first term, in the summer of 1920, he travelled to Oxford with his friend, Barker Jones, and recorded their journey in a nostalgic text. He wrote, 'The architecture of the [Cotswold] buildings large or small seemed to be particularly suited to its purpose ... They seemed to spring from the soil and fit themselves into a beautiful landscape'.[30] Fry's writing demonstrates that his self-education looked beyond the Classical curriculum at Liverpool to include Arts and Crafts theory. His familiarity with the work of Ruskin, Morris and Lethaby is evident, and this would come to have a lasting effect on his work.

Fry lived at this time at the family home, in a ground floor flat of a Victorian villa at 1 Cavendish Gardens. The wide streets skirting around the mid-nineteenth century gardens of Prince's Park had been laid out by wealthy Liverpool merchants and Fry evidently enjoyed living in the generously planned area. He wrote poetry and essays about the park, his nascent interest in the Romantic poets evident in his rather self-conscious prose:

> *I have watched summer creep through the gates at the nod of spring, and how she spreads herself, with the smile of a beneficient [sic] hostess, over everything. The banks are a glory; great waves of nodding flowers, lupins, foxgloves, sweet peas and all lovely garden blooms glow in masses of colour – and the rose hangs over the path.*[31]

Fry writes of the curative properties of the park, remarking on how the fresh air and sunlight it supplied might revolutionise the health of the city's working classes – if only they would realise and visit for ten minutes each day. Fry's musings, tinged with paternalism, reveal his early desire to educate and reform, and the health benefits of the Victorian parks no doubt influenced his later thinking on garden suburbs and how they might transform the lives of the working classes.

C.H. REILLY'S LIVERPOOL SCHOOL OF ARCHITECTURE

By the time of Fry's admission in 1920, the Liverpool School of Architecture had undergone a period of transformation under the direction of Charles Herbert Reilly (1874–1948). Appointed as the Head of School in 1904, Reilly had made an immediate impact through his desire to professionalise what was then the Liverpool School of Architecture and Applied Art. The school had been established in 1895 as the first of its kind, seeking to put into practice the ideas of the architect Thomas Graham Jackson and responding firmly in favour of the Arts and Crafts Movement regarding the 'Profession or Art' controversy of 1891.[32] Addressing students and instructors at the school's inauguration in 1895, Jackson said:

> *You, at Liverpool, are now about to try a novel method, and to apply to the training of architects, and craftsmen who are in touch with architecture, the methods which are being applied throughout the country to the training of handycraftsmen … the result will be waited with anxiety by all who have the progress of our art at heart.*[33]

The school's head, Fred Simpson, was an advocate of Jackson's ideas on architectural education and he sought to implement this blend of arts (including architecture) and handicraft by hiring well-known artists and craftsmen as 'instructors', rather than teachers.[34] Anning Bell, Herbert Jackson, Llewelyn Rathbone and Augustus John,[35] were quickly joined by two of the 'Glasgow Four', Herbert McNair and his wife Frances Macdonald.[36] Classes were given in painting and drawing, modelling and sculpture, wood-carving and a variety of metalwork, enamelling, furniture construction, embroidery, architecture and more besides. The practical application of these skills was encouraged and, out of term-time, students went to work in joiners' workshops, builders' yards or architectural practices.[37]

Quentin Hughes observes the significant similarities in architectural education between the Liverpool School of Architecture and Applied Art and Walter Gropius's subsequent Bauhaus school. However, Maxwell Fry's education was not to come via the Bauhaus prototype of Simpson's school of art and architecture, but rather from Reilly's progressive pedagogy that followed. Fry later wrote that Reilly's pedagogic approach 'consisted of little more than an infectious enthusiasm for architecture'.[38] Yet Reilly's appointment signalled a calculated change – namely, to ensure RIBA exemption – and by 1904 the Liverpool Arts and Crafts 'experiment' had fizzled out. The Fine Art department had been taken over by the City Council and Reilly responded strategically to contemporary debates, particularly in connection to the planning of towns and cities.[39] Fry's own debt to tradition requires investigation particularly in connection to his Liverpool education as his 'rejection' of Classicism in the mid-1920s and the alleged amateurism of Charles Reilly downplays its significance in the development of his architectural agenda.

Reilly was quick to recognise Liverpool's strong philanthropic tradition of non-conformist merchant families, such as the Rathbone and the Holt families,[40] and the opportunities that this might bring for the school's development.[41] Unitarian civic-mindedness became the cornerstone of his pedagogy and he developed a close relationship with the wealthy industrialist William Hesketh Lever (1851–1925).[42] As the co-head of Lever Brothers, a thriving soap manufacturing company situated across the River Mersey, William Lever was a perfect philanthropist for a school of architecture. He held a keen interest in art, architecture and town planning, and he had financed the building of Port Sunlight, a garden suburb complete with art gallery and museum to inspire his workforce.[43] Lever's patronage was instrumental in ensuring the success of the school. In 1908, he funded the establishment of the world's first department dedicated to civic design, 'as it was wisely called', wrote Reilly, 'to indicate that more than mere planning was included'.[44] Reilly's approach is crystallized in his place on the Executive Committee of the 'City Beautiful' Society, established in 1907, and he wrote to the French planner Charles Bonnier: 'In England the corporations of our towns have no artists or architects of any sort to advise them only surveyors and drain men to say how the sewers are going to go'.[45] This interconnected approach to architecture and town planning adopted at Liverpool was unique at the time; it sought to provide a new generation of planners able to create large-scale civic work of 'balanced, symmetrical and dignified buildings'.[46]

Under Reilly's headship, the school adopted an architectural outlook of ambitious internationalism combined with English tradition. Teaching methods were based upon the *Ecole des Beaux Arts* and the *Beaux Arts* methods used in American schools to give students confidence in 'monumental planning', with a decidedly English flavour for detailing, materials and construction.[47] Students were required to undertake a range of conventional classes in architectural design, construction, professional practice and a combined module in architecture and civic design. Although feted as part of the Classical tradition, by the time of Fry's full session, from 1920–21, the staff roster had become more eclectic. It included the Gothicist Giles Gilbert Scott, as Reader in Ecclesiastical Architecture, who was at this time working on his masterpiece of Liverpool Cathedral; the Lever Professor of Civic Design, Patrick Abercrombie (1879–1957); Lever's friend Thomas Mawson as Special Lecturer in Landscape Design;[48] and would-be modernist, Lionel Budden (1877–1956), as Senior Independent Lecturer and Studio Instructor.[49]

Fry quickly assimilated into the artistic life of the school. Located on Ashton Street, a half-hour walk from his family home, the architecture school comprised three large studios each accommodating 50 students (Figure 1.1), a library, a lecture hall and a series of rooms for the Department of Civic Design.[50] Fry's Unitarian upbringing would have provided him with a fundamental understanding of the school's pedagogy that linked civilized, clearly planned towns to commerce and culture. Indeed, this notion recurs throughout his career although it was later dressed in a modernist polemic. Fry typically received very good marks for architectural design and construction projects, although he found it difficult to maintain this standard in secondary courses such as professional practice and sanitation and hygiene.[51] In 1921, he was joint recipient of a Lever second prize for architecture.[52] This evident ability in architectural design ensured that he was favoured by Reilly, a friendship secured when the student's cartoon of the professor as a cherub (heralding a new architecture on a bugle) was spotted by Reilly who 'recognised himself with glee'.[53]

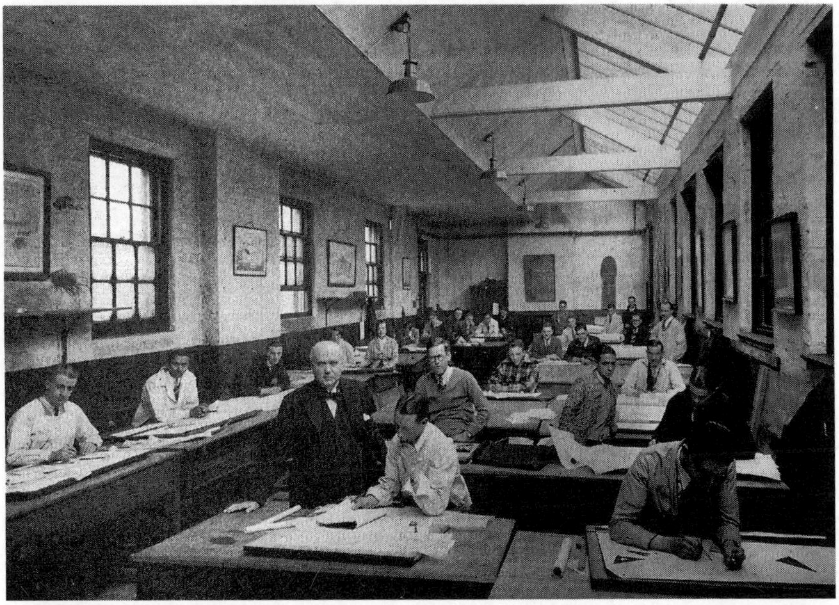

1.1 Architecture Studio (Charles Reilly to the foreground), Ashton Street, Liverpool University, 1930

Although Reilly's appointment had also signalled the end of the fine and applied arts alliance at Liverpool University, Simpson's ideas lived on in the bohemian environment of the Sandon Studios Society. Founded in 1905 by former students of the Art School, the Sandon Studios were organised on the Paris system.[54] Studios remained open at all hours and the resident artists Gerard Chowne and Herbert MacNair provided critique for the students. The society maintained close links to the School of Architecture and – under the auspices of Lever and orchestrated by Reilly – for a time shared the same premises, at the Blue Coat Chambers in the centre of town.[55] Fry joined the club, possibly at the invitation of Reilly, who often elicited membership for his favourite students, with Bernard Miller joining in 1921, and Francis X. Verlade and Herbert Thearle joining in 1925.[56] Described as the Bloomsbury Group of Liverpool,[57] the society was the centre of cultural life in the city and hosted a programme of dances, dinners, exhibitions and lectures, such as Walter Sickert's talk on 'Magnasco and the Baroque' in 1925.[58]

The Spring Exhibition of 1922, which Fry must have visited, was opened to much acclaim by the local press. Paintings by local artists, such as Anning Bell and Augustus John, featured heavily and the architecture school was well represented with drawings and photographs of design work by Professor Reilly, Lionel Budden and Patrick Abercrombie, amongst others.[59] The *Morning Post* reported:

> *The exuberance of French Art since Cézanne ... [is] little in evidence, for workers in Liverpool are out of shouting distance of Mr. Clive Bell, and the voice of Mr. Roger Fry only reaches them in a whisper. Clearly the English tradition of watercolour still holds its own in Liverpool.*[60]

While a significant exhibition of Post-Impressionist art had, in fact, been held at the Sandon Studios in April 1911, there had perhaps been some regression to more traditional work since the heyday of the society's formative years.[61] In 1922, the exhibition review suggests an informed provincialism existed amongst the society, and it nursed young talent such as Fry and his friend Christopher Wood (1901–30) – a fellow architecture student – in the early days of their careers. Yet Wood's time in Liverpool was short-lived. He resigned from the university in April 1920,[62] just two months after Fry's enrolment; the budding artist quickly left for London, after realising that architecture was not for him and, the following year, moved to Paris to absorb the ideas of Picasso and Braque.[63] Wood was evidently influential in Fry's development, and he later described his friend as 'an elegant young man with a Byronic limp' and his 'first educator at the school'.[64] His comment underscores the importance of the Sandon Studios milieu in Fry's development.

The Sandon Studios also gave Fry an opportunity to socialise with fashionable society, which no doubt helped to cultivate his knowledge of Oscar Wilde and Baudelaire, which he later wrote was 'a necessary acquaintance' as a member of the Studios.[65] Fry's family, although notably not his father, also hoped to become part of this society: Fry's mother unsuccessfully applied for membership in April and again in June of 1923; the following year Fry's sister, Muriel, a fashion designer, applied but she too was declined.[66] Thus they would have missed the opportunity to attend an afternoon tea held in 1924 in honour of the well-known actress and honorary member of the society, Sybil Thorndike. The Sandon Studios remained significant for Fry after his graduation and he continued to attend social events, such as the annual summer dance, after his move to London.[67]

At a gateway to the British Empire and the new world, Reilly saw his role as producing architects and planners who would play their part in modern development. Often funded by Lever, Reilly fostered connections in America, visiting universities to seek out exchanges of teachers and summer placements for students in the offices of suitable American architects. Fry was selected for this honour and set sail to New York in the summer of 1922. In keeping with his Classical training at Liverpool, Fry worked for one of the leading Beaux-arts firms in America and a favoured practice for Reilly's students, Carrère and Hastings. Both John Merven Carrère (1858–1911) and Thomas Hastings (1860–1929) had studied at the Ecole Nationale Supérieure des Beaux-arts in Paris, and worked for the pre-eminent Classicist firm in America, McKim, Mead and White.[68] Following Carrère's untimely death in 1911, Hastings continued the practice and, in the same year as Fry's work-experience, was awarded the Royal Gold Medal for his work on Devonshire House in London, a project with Reilly as advisor. Although known for continuing the tradition of refined American Classicism, Carrère and Hastings' former staff included a formidable list of would-be modernists, including Emery Roth, Ely Jacques Kahn, and Richmond Shreve and William Lamb.[69] At the time of Fry's placement, the firm was working on several Manhattan skyscrapers, such as a complete redesign of the Standard Oil Building (1921–28), although Fry was set to work on a Long Island mansion.[70]

Fry's pencil sketches of this first spell in New York reveal his wonder at the city's dramatic skyline; his illustration of Warren and Wetmore's recently completed Heckscher Building (1920–21), for example, gazes upwards at the skyscraper from the pavement, suggesting his awe at the towering modern structures (Figure 1.2).

The building's step-backs respond to the 1916 zoning law, which set massing limits to ensure adequate light and air at street level and Fry's sketch is perhaps a comment on this innovation, rather than any perceived architectural merit in the somewhat bland building. Likewise, his drawing of the Queensborough Bridge (completed 1909), linking the boroughs of Queens and Manhattan, records the innovative double deck construction, originally to convey two pedestrian walkways and two railway lines at the top deck and four lanes of traffic beneath (Figure 1.3). Here in New York, Fry witnessed a city of the future.

Fry's shortening of his name to 'Max Fry' can be traced back to his signature on these pencil sketches of 1922. The nickname perhaps first occurred during his work placement in America and it is telling that his employer, Thomas Hastings, was known to almost everyone as 'Tom' or 'Tommy'.[71] Fry's creation of an architectural persona – hinting at modernity and informality – is the first indication that he had understood the lessons of Reilly in the impact that self-promotion and hyperbole could have on his career. Hastings, too, must have underscored the lesson for Fry in his own portrayal of the charming 'eccentric artiste'.[72]

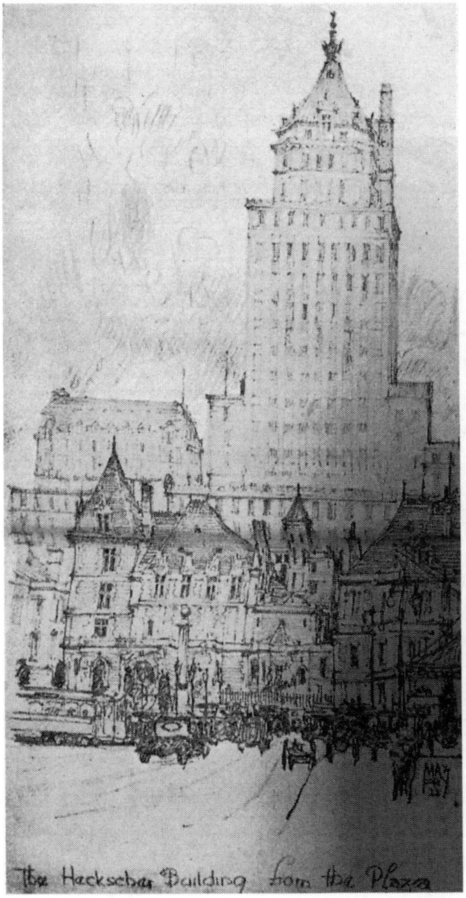

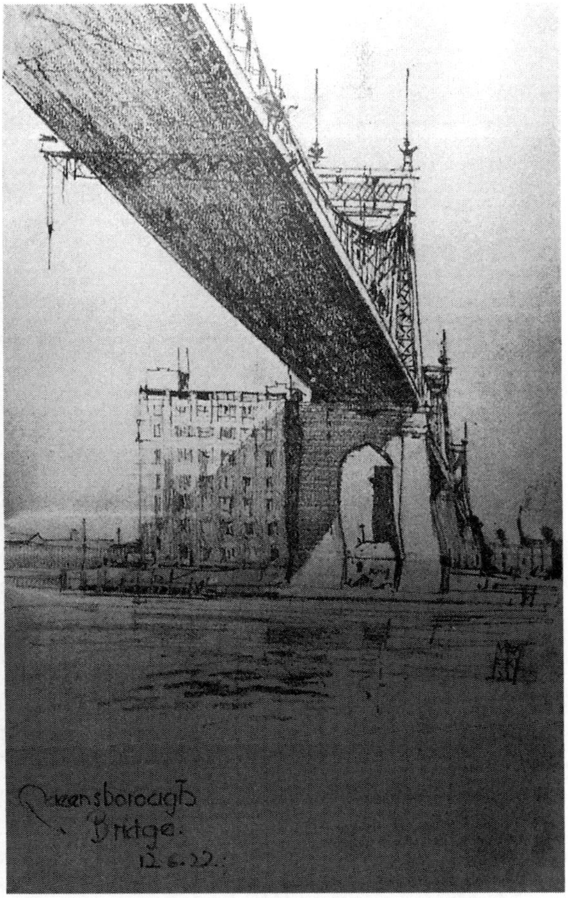

1.2 Sketch of Heckscher Building, Maxwell Fry, 1922 1.3 Sketch of Queensborough Bridge, Maxwell Fry, 1922

In June 1922, Fry's ex-serviceman's grant was withdrawn and he was forced to supplement his wages and the small income from his parents with draughting work during his spare time in New York.[73] Fry returned to Liverpool to complete his studies and was amongst a group of five students to be 'groomed and exercised like racehorses' in preparation for the Rome Prize in 1923.[74] Reilly publicised the success of the Liverpool finalists and a photograph in the local newspaper the *Daily Courier* shows Fry looking every inch a young architect (Figure 1.4). For the preliminary round of the prestigious competition the entrants designed a rail and road bridge over a gorge and Fry's entry, although fairly uninspiring, featured in the school's prospectus. His final design, for a Recreation Centre, was an example of the 'monumental' planning that Reilly had encouraged at the school (Figures 1.5 and 1.6). The *Builder* was unconvinced by Fry's 'Italian scheme', although it praised the rational plan, it dismissed entirely the over-composed elevations: 'Like a collection of travellers' samples, or a selection of plates from an architectural exemplar, the masses jostle one another, here long and low, there unduly tall, here rotund and enciente. The circular loggia is especially bad, although excellently drawn'.[75] Like the other Liverpool entrants, Fry was unplaced in the Rome Prize. However, he received a Bachelor of Architecture with Honours in Architectural Design in July 1923 after passing both the fourth- and fifth-year examinations in his final term.[76]

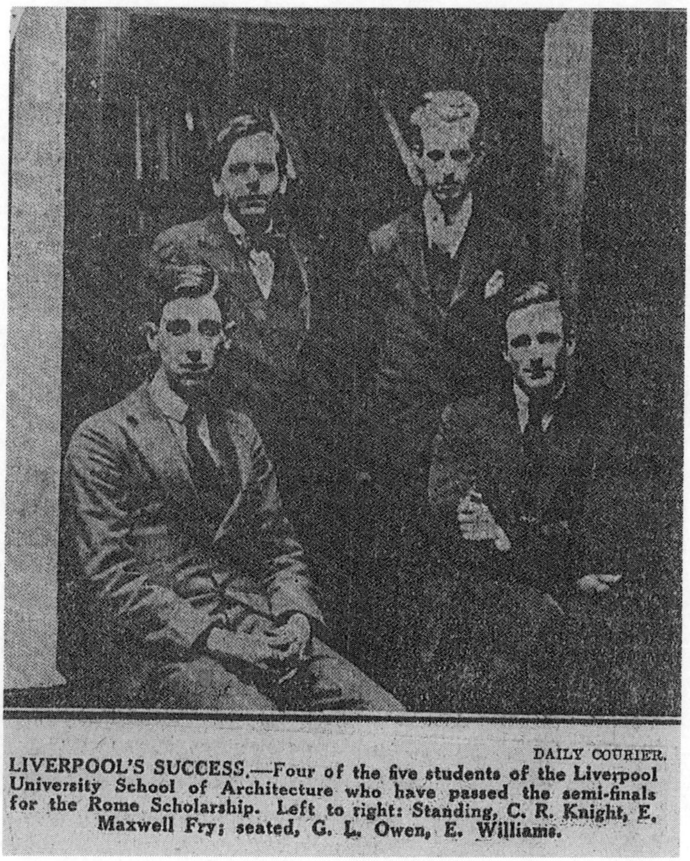

1.4 Liverpool University's Rome Prize Finalists, 1923

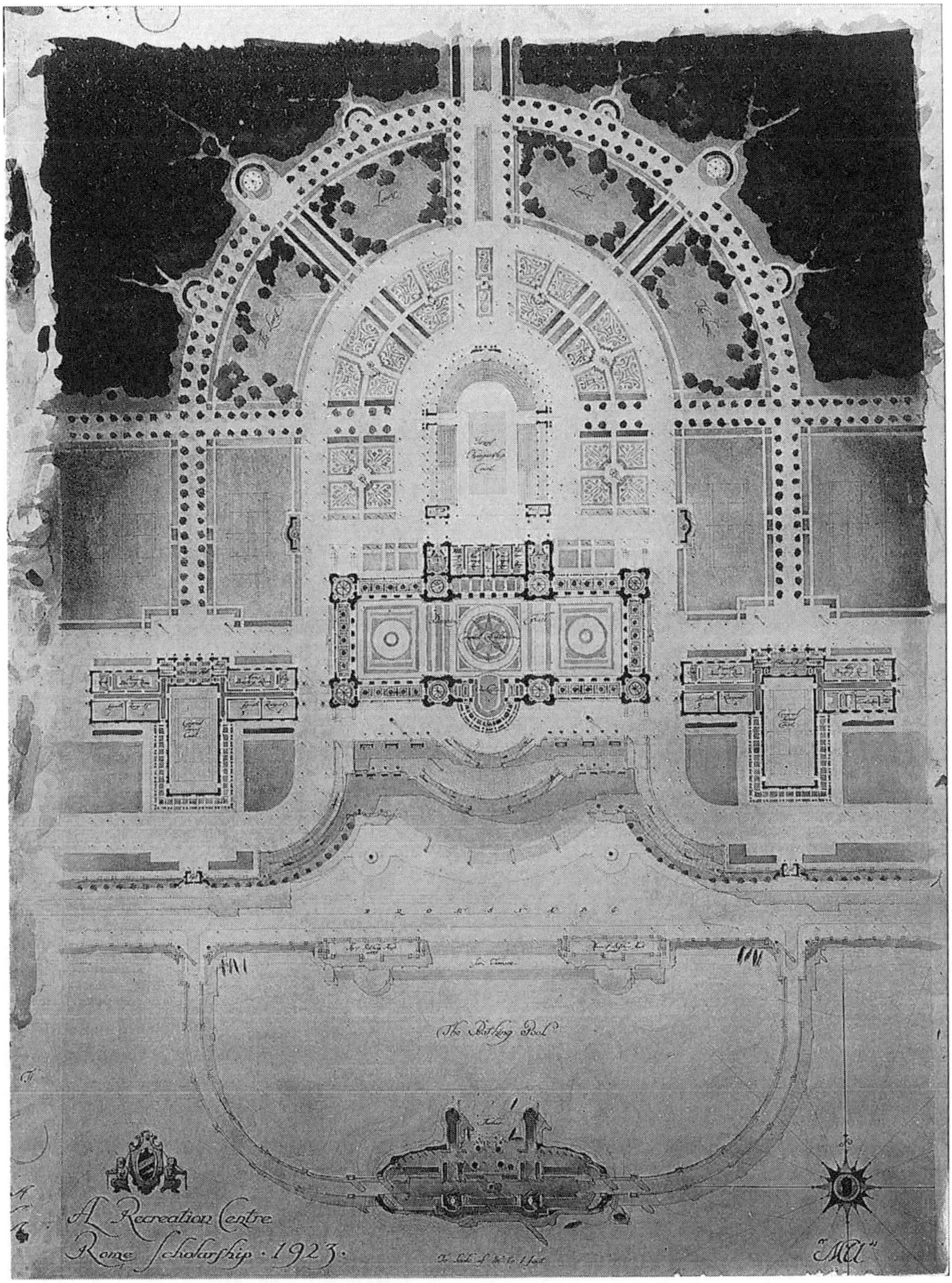

1.5 Site Plan of a Recreation Centre, Maxwell Fry, 1923

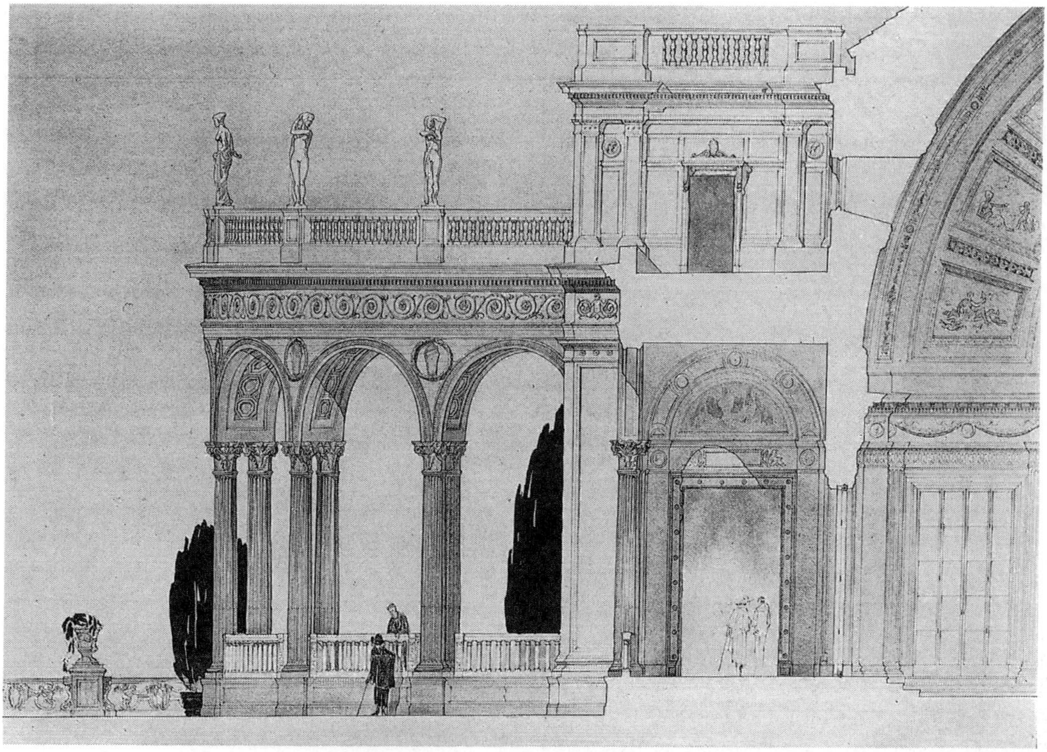

1.6 Part Elevation of a Recreation Centre, Maxwell Fry, 1923

His education complete, in 1924 Fry moved to London equipped with two letters of introduction written by Professor Reilly. The synthesis of architecture and planning in Fry's outlook is evident in these early letters: the first was addressed to the highly respected Classicist, Stanley Adshead; and the second to the Chief Architect at the Office of Works.[77] Stanley Adshead had been appointed as the 'first professor of town planning in England' and the first editor of the *Town Planning Review*, the newly established journal of Liverpool's Department of Civic Design, by his old friend Reilly and had held the post from 1909 to 1914.[78] Adshead's work integrated architecture with town planning, and during the mid-'twenties was involved in the design of numerous ambitious masterplans.[79] This approach had been encouraged during Fry's Liverpool education and was espoused by Fry himself throughout his career. Adshead was known for his excellent draughtsmanship,[80] and Fry described him as 'disarmingly unconventional', admiringly noting the 'delicate Greek Revival style' used in the Students' Union building in Liverpool, which Adshead had co-designed with Charles Reilly.[81]

Fry's other choice of the Office of Works was a far cry from Adshead's established private practice. The Office of Works was supervised by Frank Baines (1877–1933) and, guided by the 1919 Tudor Walters Committee report and Town and Country Planning Act of the same year, producing large estates of workers' housing in the Garden Suburb manner for deprived London boroughs such as Eltham and Camberwell. The choice reveals Fry's early interest in municipal housing and demonstrates that these concerns were based on a desire to tackle social problems, rather than any aesthetic or stylistic programme. Yet, according to Fry's memoirs, he took an immediate dislike to the relaxed work ethic at the Office of Works. Adshead, meanwhile, had insufficient work to take on another draughtsman, and suggested that Fry instead pay a visit to Thomas Adams, of the planning firm Adams and Thompson.[82]

LONDON AND THE FIRST FORAYS INTO PRACTICE

Thomas Adams was an important figure in the planning profession. With numerous large-scale projects underway in North America, he was an excellent match for Fry's early interests. Adams held influential roles as Secretary of the Garden City Association, was a founder of the American Planning Institute in 1917 and of the Canadian equivalent in 1919. His partner, Francis Longstreth Thompson, published the influential *Site Planning in Practice* (1923), a volume mainly concerned with the layout of housing estates based on the Garden City model. Crucially for Fry's development, this approach responded to existing topographical conditions rather than imposing a geometric solution onto a site.[83] During Fry's employment at the practice, Thompson was based in New York due to his involvement in the New York Regional plan. The results of the plan were published in two large volumes supplemented by seven rigorous appendices concerning all aspects of modern city life. The recommendations included proposals for building 'set backs', plazas and transportation for Manhattan.[84] Fry contributed a number of drawings to the research, such as the Fritz-Langian 'Future City of Towers', 'New York's artificial mountain range' as well as illustrations depicting ancient urban models (Figure 1.7). Described as 'the only one to envision radically modern solutions for the city',[85] his drawings were indebted to the American perspectivist Hugh Ferriss and Reilly's six-hour sketching tasks.

Despite these futuristic ambitions, the plan was criticised by the Regional Planning Association of America for its lack of radical urban reform. Lewis Mumford also attacked the scheme, as he 'found little of value' in an exercise that accepted a 'continuation of the status quo as inevitable, and failed in its goal of providing a real vision'.[86] But, as Frank Jackson notes, this criticism was unjust as the plan was 'dependent on the formation of a powerful planning committee' to carry out the proposals, without which the plan remained 'impotent'.[87] Fry viewed this as an important lesson; he became acutely aware of the limits of what could be delivered without a centralised authoritarian approach and, in a publication co-authored with Adams, he argued that 'democratic communities are too timid in getting rid of obsolete buildings'.[88]

In contrast to the New York Plan, Fry's first project for Adams and Thompson was the planning of a new village, Kemsley (1924) in Kent, for a local paper mill owner, Frank Lloyd. In the tradition of other philanthropic industrialists, Lloyd intended to develop a suitable 'garden village' for his workforce and a site was selected close to the majority of workers' existing homes in Sittingbourne. Situated between a light-railway track, the main railway line and Ridham Dock, the setting provided convenient transportation links across the country and further afield, via the English Channel. The amoeba-shaped plan was composed of two concentric roads in a horseshoe

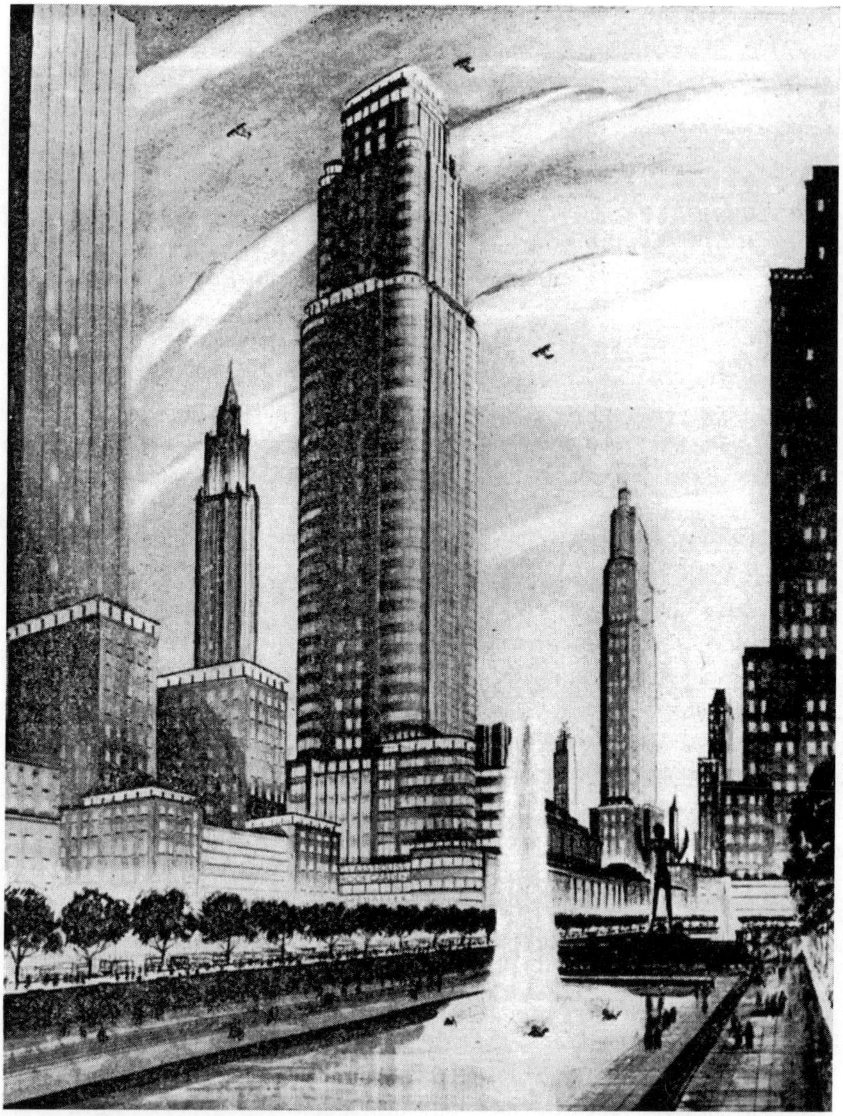

1.7 City of the Future, Maxwell Fry, 1931

arrangement with a main central road that intersected a square planned for future construction of community buildings and shops. Sites were also designated for a school, other public buildings and 'ample open areas' for recreation.[89] The layout was of a type developed by Raymond Unwin, by whom Adams was heavily influenced. The practice considered it a 'great advantage' for the 'planning, land development, and building' to be under the control of a single practice stating that:

> *no satisfactory design can be made for an estate, still less for a town, if it is confined to laying out streets and does not include a study of the prospective buildings. The street system should be planned with a view to serving the types and character of buildings likely to be erected, and should be adjusted to the density and arrangement of the building that it is designed to encourage.*[90]

Adams and Thompson strongly believed that there should be a connection between a site plan and its associated architecture, a view wholeheartedly supported by Fry. Members of the practice wrote a substantial book, *Recent Advances in Town Planning* (1932), that discussed their findings, ideas and suggestions for future plans as well as promoting their favoured Garden Village approach over the profiteering 'bye-law' street. Densities were to be kept low, each house was to have its own garden and the re-creation of rural English village life was paramount. At Kemsley, formal planning devices are limited to a central road leading from the mill, although the road abruptly terminates at the existing railway line where one would expect an architectural finale to the axis, or at least for the road to lead somewhere. The houses are mainly semi-detached cottages repeated across the village; in a direct manner, houses were laid out at 12 dwellings per acre, with little apparent concern for the ornament or variety found in comparable philanthropic villages. However, in the face of the economic slump, demand for the houses was slow, with only 180 houses built by 1932 against the expected total of 750. Of the project Fry later wrote:

> *Designing ... a gentle little garden village in the shadow of the great paper mills of Kemsley above the marshes of Sittingbourne, Kent, finished my interest in village forms as fit material for city building for in the meantime I had fallen in love with the beautifully civilised texture of classical London, so much of it, including most of Regent Street, still extant and usable.*[91]

Fry's mention of John Nash's Regent Street referenced the long-standing concerns over the rebuilding the Regent Street Quadrant, which had been redesigned by Richard Norman Shaw following his commission in 1904.[92] Only part of Shaw's acclaimed scheme was implemented and in 1912, aged 81 years, he resigned from the project after ongoing wrangling with the Quadrant's tenants over desired amendments to the design failed to reach an agreement. A revised scheme was eventually drawn up, principally by Reginald Blomfield, and the five-year rebuilding of the Quadrant began in April 1923, causing considerable outcry amongst the architectural community. Fry was 'incensed' by its 'wanton destruction' and wrote of joining with his Liverpool friend, an architect and journalist, Christian Barman (1898–1980), in appeal of the decision.[93] Fry suggests this was his route into the Design and Industries Association (DIA), and thus the beginnings of his life as a modernist. Yet Fry's education at Liverpool had prepared the ground for this transitional phase; Regent Street was exactly the type of grand civic plan promoted by Reilly. Indeed, Fry's Liverpool professors Charles Reilly and Arthur Trystan Edwards (1884–1973) had made vociferous protestations against alterations to Regent Street and it seems Fry was following in their outrage. Fry's reference to this, the 'noblest of shopping streets in Europe' clearly echoes Reilly's views expressed in two lengthy articles on Regent Street in *Country Life* magazine.[94] Trystan Edwards also repeated this view in *Good and Bad Manners in Architecture* (1924), which includes considerable discussion of 'the most beautiful street in the world'.[95] Edwards's views on manners in architecture may be traced back to his writing for the *Town Planning Review*. Indeed, Edwards wrote a series of articles for the *TPR* which deeply influenced Fry's own viewpoint; 'On Monotony in Street Architecture', for example, illustrates the undesirability of long, monotonous streets of housing and Edwards's writing suggests the root of Fry's rejection of Liverpool's bye-law streets.[96] Moreover Edwards's notion of architectural good and bad manners, of 'taste-that-begets-style',[97] was indebted to Geoffrey Scott's *Architecture of Humanism* (1914) – a book on the Liverpool School of Architecture's reading list.[98] This attempt to establish a critical framework for aesthetic critique would continue throughout the interwar period and was taken up by Fry and his modernist colleagues, as we will see.

Fry's dissatisfaction with existing methods of planning and building during the mid-1920s became increasingly evident. He later wrote that 'neither the garden village nor the revived classic was capable of dealing with the range of building needs of a highly industrialised country'.[99] Fry's questioning of established styles mirrors a more widespread lack of architectural agenda and this dissatisfaction was given form in an enduring acquaintance with his friend and colleague, Wells Coates. In 1923 or early in 1924, Coates entered the office of Adams and Thompson at 121 Victoria Street in Westminster.[100] He sought a job to supplement his wages as a journalist for the *Daily Express* and, although there were no positions for a draughtsman, Coates was taken on as Adams's secretary. Fry and Coates became firm friends. The two 'met frequently outside office hours and spent long hours discussing poetry and other forms of literature'.[101] With his scientific background via Japan and Canada, Coates must have widened Fry's horizons, although it would be several more years before their relationship resulted in any tangible architectural outcome. Coates was amongst a bohemian network of friends established by Fry during the mid-1920s and he spent his evenings frequenting cafes in Chelsea patronised by the local artistic scene. The group included Ethel Leese (b. 1888) and, in June 1926, she and Fry married at Chelsea Registry Office, with Julian Huxley acting as his best man.[102] The couple set up home in Chelsea and, while Fry may not have enjoyed his professional life during the mid-'twenties, the period marks a critical period in Fry's architectural development.

SOUTHERN RAILWAYS

Fry's employment at Adams and Thompson was short-lived and, in 1924, he took a position in the Engineering Department of Southern Railway with the prospect of more work and 'quite a lot more money, to furnish out my taste for books and new company'.[103] Fry was appointed as an assistant to the Chief Architect, James Rob Scott (1882–1965), designer of Waterloo Station, which also housed the Southern Railway architects' offices. According to Fry, he was running the office within a matter of days of starting work and he later wrote, 'the old architect [Scott] fawned on me … it was all too easy'.[104] In keeping with his heroic position, Fry's memoirs downplay his work at the office and he does not claim ownership of any specific projects designed during this period; instead he writes of producing schemes with 'a careless competence' that relied on his Liverpool training.[105] Clearly, Fry was eager to minimize the significance of his work at Southern Railways.

In his *Architectural Review* (*AR*) 'criticism' column of October 1935, Charles Reilly praised Maxwell Fry and Herbert Barton, 'two old Liverpool men',[106] for 'standing up to the railway engineers in their designs for Ramsgate and Broadstairs'.[107] Although Reilly had a reputation for promoting his own students, his reference to Broadstairs Station is curious for it lacks any architectural significance and he perhaps confused it with Margate Station (Figures 1.8 and 1.9). Indeed, in an earlier article, the *Builder* wrote in praise of Margate and Ramsgate stations, claiming that 'architecturally they are excellent, and show a greater advance, both in design and planning, upon anything of a similar type and scale produced in this country within recent years'.[108] Reference to specific designers is not given and Fry's role in these station designs is unclear. The projects were a collaborative process and, therefore, draughtsmen did not initial drawings, although both Margate and Ramsgate stations have subsequently been attributed to Fry.[109] There are however some distinctive Fry motifs, such as the cantilevered canopy above the main entrance of Margate, which hint at his involvement.[110]

Fry's experience at Southern Railway provided him with the opportunity to work on large-scale projects within a collaborative environment of architects and engineers. While Ramsgate and Margate Stations illustrate Fry's debt to his Classical training at Liverpool, the Southern Railway office may have been far more progressive than Fry has suggested with his colleagues including Rodney Thomas and Guy Morgan.[111] The post also gave Fry sufficient freedom to work on architectural competitions in his spare time, which he saw as a means of making a name for himself.

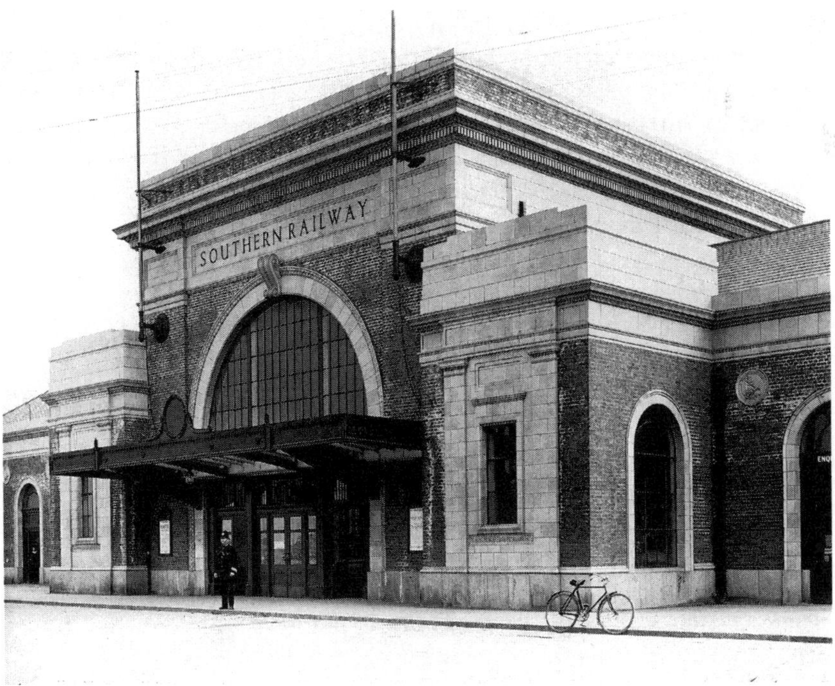

1.8 Railway Station, Margate, Kent, c. 1930

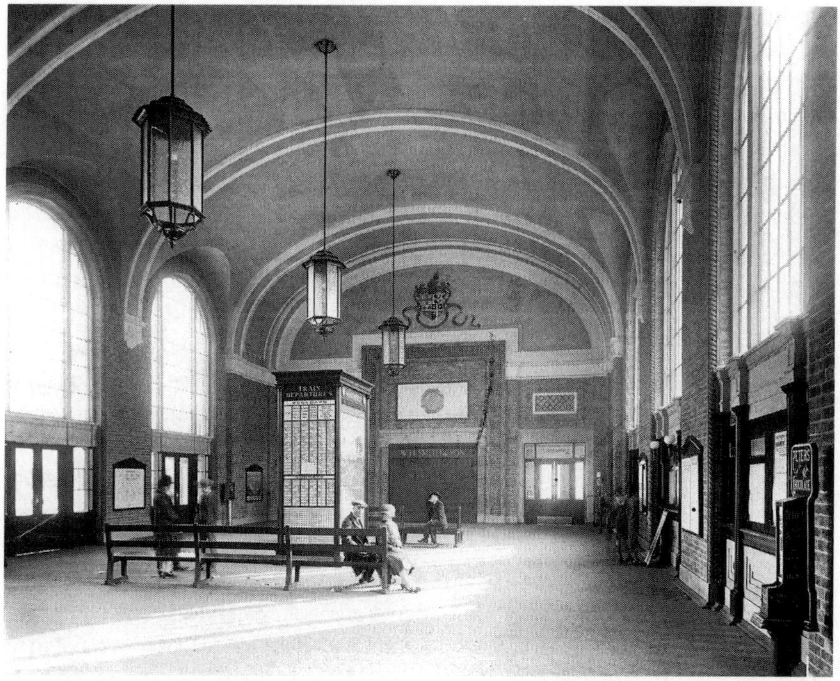

1.9 Booking Hall, Railway Station, Margate, Kent, c. 1930

In 1925 he entered the Manchester Art Gallery Competition with Geoffrey Owen, a university friend who was also a finalist in the 1923 Rome Prize (Figure 1.4).[112] Their entry was well received and awarded 'second premiated design', an outstanding success for the pair of newly qualified architects.[113] Fry and Owen's design was wholeheartedly Classical, with Corinthian tetrastyle porticos and a top-heavy attic story. The ground floor was deeply rusticated with a blank, windowless piano noble to the front façade creating a confident, if detached, façade (Figures 1.10 and 1.11). The scheme filled the available site adopting a trapezoidal plan and the projecting semicircular, apse-like gallery is indebted to the north façade of St. Georges's Hall in Liverpool. For Fry and Owen, this brief was familiar to the monumental student projects set by Reilly. The *Builder* considered the façades to be 'scholarly', but they might have been improved further 'if an attempt had been made to secure even greater simplicity'.[114]

In 1927, Adams offered Fry the opportunity to return to Adams and Thompson with a partnership in the firm, which the 28 year-old accepted. Later in life Fry suspected that the motive behind the offer was to 'bolster the low fees paid for town planning with the higher one paid for architecture'.[115] If this was Adams's intention, it was not to pay off.

ADAMS, THOMPSON AND FRY

Adams, Thompson and Fry was duly formed and the practice entered its first competition as a new partnership: to design a new civic centre for Birmingham.

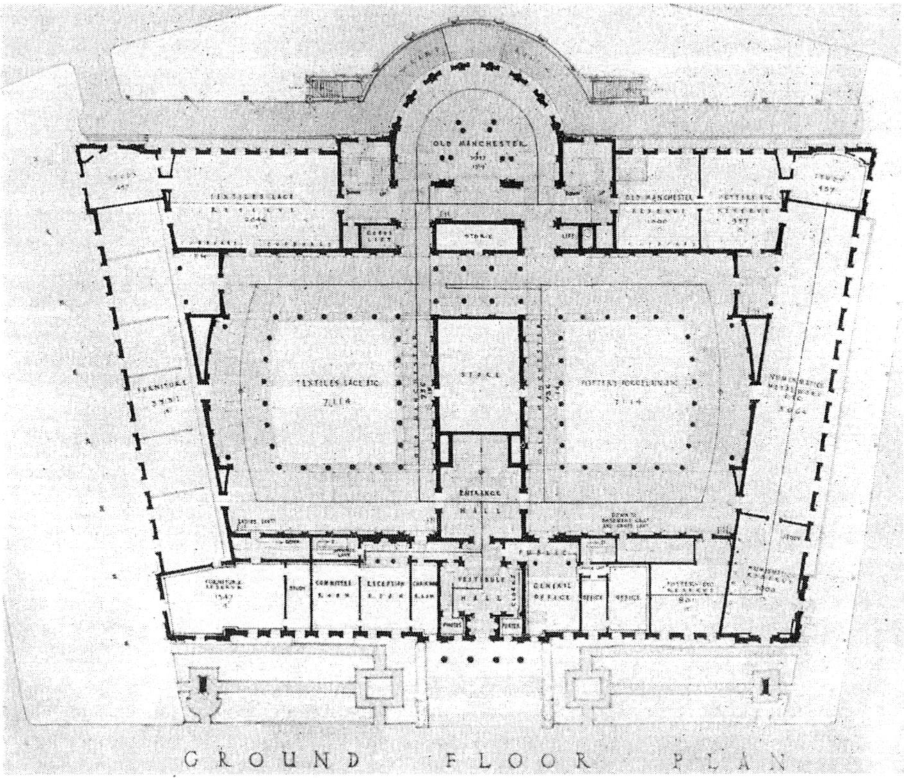

1.10 Ground Floor Plan, Manchester Art Gallery, 1925

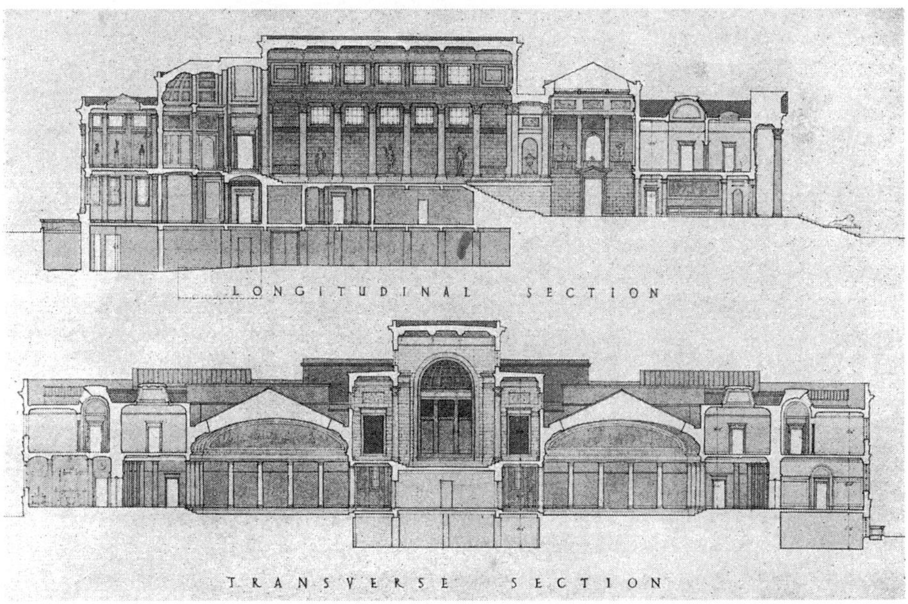

1.11 Transverse and Longitudinal Sections, Manchester Art Gallery, 1925

The extensive brief called for the provision of a City Hall, Mansion House, Municipal and Public offices, Natural History Museum and Public Library, as well as the re-planning of the city centre up to a radius of 900 feet from the existing War Memorial. It was a bold and ambitious competition, echoing the vast Edwardian civic centres such as Cardiff. The prestigious competition attracted international attention, with prizes awarded to designs originating from New York, Zurich, Oslo and Sweden.[116] Like the Manchester Art Gallery, this kind of civic-minded brief was familiar to Fry and was of a suitable scale to test his perspective rendering and composition skills. The competition was judged by Henry Vaughan Lanchester, an architect familiar with civic programmes and known for his rational planning in schemes such as the elaborate Baroque Cardiff City Hall (1897–1906), co-designed with Edwin A. Rickards, which set an obvious precedent for the Birmingham competition.

Adams, Thompson and Fry's master plan was a symmetrical Neo-Classical composition (as were all the other entries) taking inspiration from the Beaux-Arts manner and no doubt appealing to Lanchester's sensibilities. It was, however, certainly more restrained than Lanchester and Rickard's Cardiff scheme, and was perhaps viewed as lacking a playful, human edge to counter the grand gesture. Their design was described as 'an attractive scheme' with the city hall placed to the north of the main square, but was criticised due to the duplication of the war memorial in order to retain a balanced composition (Figure 1.12). As noted in the *Architect & Building News* (*ABN*), such duplication would deprive the original monument of its status.[117] Despite these reservations, the judges awarded Adams, Thompson and Fry third place, securing a £100 prize.

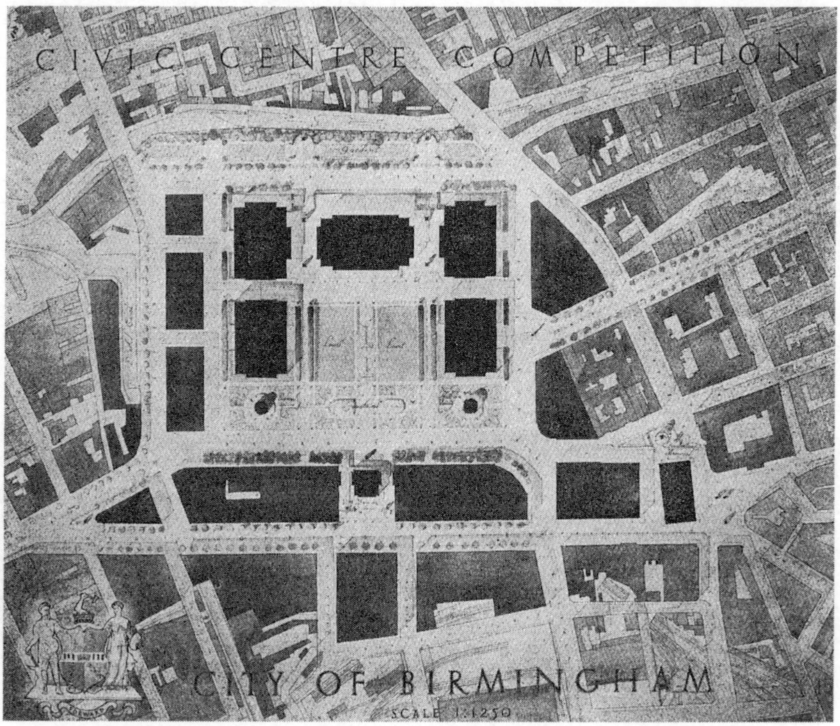

1.12 Site Plan, Birmingham Civic Centre, 1927

Competition successes generated both recognition and some financial revenue that as a partner Fry was expected to attain for the practice. The relatively low winnings were not enough, however, and putting his drafting skills to good use, Fry produced perspective drawings for Adams, who was developing a considerable reputation in North America. A masterplan for Windsor, Ontario, for example, is presented through seductive pencil perspective drawings – it was the vision and ambition that was being presented, rather than any detailed proposals of layout. In addition to the grand civic centres, smaller plans were prepared by the practice, such as a scheme for the sea front at Cramond, Edinburgh, which incorporates a promenade and bandstand area leading to a pavilion placed on a central road axis and flanked on either side by bowling greens and tennis courts.[118]

Adams, Thompson and Fry also gained commissions for several regional plans,[119] with the aim of delivering 'feasible' broad-brush schemes with a 'general outline capable of variation'.[120] The advisory plans were not intended for immediate execution but, with characteristic Adams restraint, would highlight sites suitable for development, state proposed residential densities, select roads suitable for widening and provide suggestions for new highways. Unlike the diagrammatic plans produced by Raymond Unwin or Patrick Abercrombie, these suggest actual proposals on scale plans, but remain indebted to the division of function according to simplistic definitions of residential, business, industry and 'open space'. Fry recognised that these advisory plans would not solve urban problems or offer radical suggestions, but pragmatically accepted that basic planning measures that could be introduced, particularly in the countryside, and with regard to traffic flow.

In addition to the plans prepared by the practice, the First Report of the Greater London Regional Planning Committee was published in 1929, again as an advisory document prepared by Unwin with help from Adshead and Longstreth Thompson – all of whom wanted to convince the Ministry of Health to urgently re-plan London. Unwin wanted the plan to lead to a proper organisation of space, the preservation of open land, low-density dwelling and good transportation. Fry would have been aware of this plan due to Thompson's involvement and he later used the findings of this project for his proposed plan of London developed with the MARS Group.

Fry's bohemian Chelsea network led to a friendship with the wealthy interior designer, Ned Groghan. They spent considerable time together and prior to Fry's marriage in 1926 had shared lodgings for a time, although Fry was living way beyond his means.[121] Groghan's connections led to a commission for Fry to design a large house, which is likely to have been Ridge End in the affluent Surrey village of Virginia Water. This was an important job for Fry, as it was the only work he had secured for the practice other than designs for domestic gas fires.[122] Fry worked incessantly on the project, but the extended time spent resulted in very little profit. To his dismay the rewards were reduced further by the 'handsome cut that Ned Groghan expected',[123] which brought a swift end to their friendship. Built in around 1928, the dwelling utilises a cranked plan of asymmetrical wings and an impressive entrance stair at the pivot point resulting in Arts and Crafts meets 'plainest possible Neo-Georgian'.[124] Built of cream-washed brick, with a slate roof, it features double pilasters at either side of the main entrance and to the rear façade two curved bay windows overlook the garden (Figures 1.13 and 1.14). Due to Fry's journalistic contacts, the house received considerable publicity including an article in *Country Life*.[125] Indeed, with its blend of Arts and Crafts and restrained Neo-Georgian features, the house was typical of the type published by the magazine during this period.

Fry also continued his work at Kemsley and designed the Village Club House in 1929, a two storey, Neo-Georgian brick building, with rather grand aspirations for such a small settlement. The club house faces the village square and, as the main entertainment venue for the paper-mill workers, formed the focus of village life. The scale of the building is only slightly greater than the domestic properties of the village, but in order to illustrate its public function Fry introduced more Classical detailing and gained extra height with the addition of a lantern. The building lacks the restraint of Ridge End, with over-designed railings and keystone details to the main façade. A more elegant result is achieved in the gable façades, with a smaller projecting portico above the central door and fanlight.

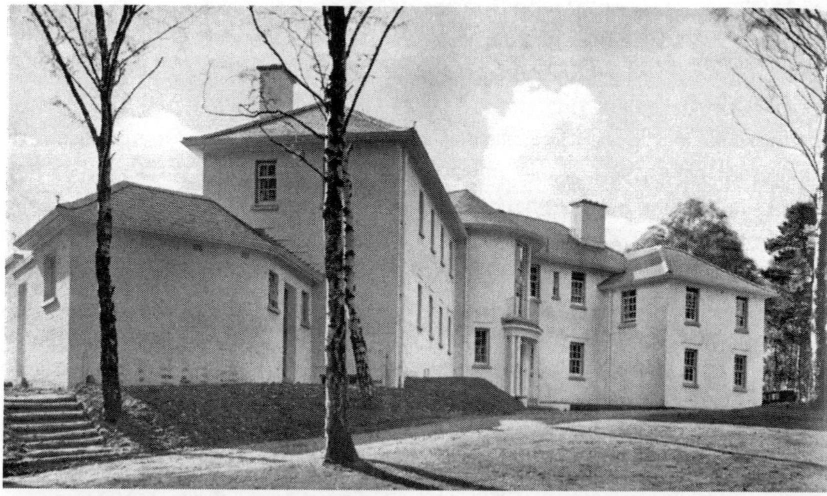

1.13 North Façade, Ridge End, Virginia Water, Surrey, 1932

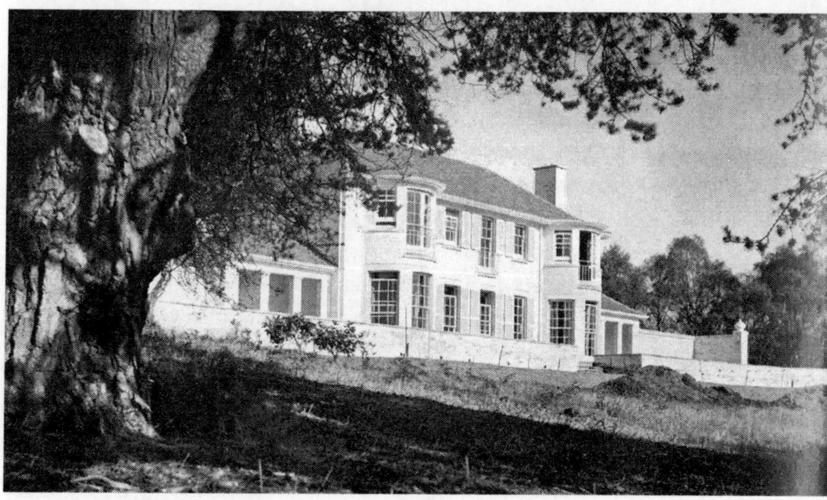

1.14 South Façade, Ridge End, Virginia Water, Surrey, 1932

This was a difficult time for Fry. His failure to attract any substantial work to the practice in the wake of the economic slump must have strained his already 'irksome' relationship with Adams and Thompson.[126] Small-scale, infrequent commissions were not enough to sustain the drawing office and in 1930 he was forced to dismiss his only assistant, Edward 'Bobby' Carter,[127] who went on to a distinguished career as RIBA Librarian and Editor of the *RIBA Journal*.[128] In addition to work pressures in an empty drawing office, Fry's short-lived marriage was also 'in jeopardy' – an issue he later claimed was worsened by his 'growing resolution' in Modernism.[129] Fry's memoirs portray himself as outgrowing his personal and professional lives of the 1920s, his 'conversion' to Modernism marking a fresh new outlook and approach to life. The reality appears less clear-cut, as Fry's professional partnership and developing interest in the Modern Movement were irrevocably intertwined from the late 1920s until his establishment of a new partnership with Walter Gropius in 1936 (see Chapter 2). Indeed, Adams, Thompson and Fry continued to offer opportunities for large-scale masterplanning that enhanced Fry's profile, while he explored questions of style, structure and social agenda.

From the mid-1920s, Fry sought to secure a position of authority amongst the British architectural community: he actively participated in societies; wrote for various architectural and political publications; and taught architectural design to the next generation. His partnership in a well-known firm and a growing list of influential acquaintances gave Fry an unrivalled position amongst Britain's young modernists. As Fry later recalled, 'When I first came to London in the late 20s the RIBA was a learned society and a gentleman's club that gave me a polite welcome. It was not a joke'.[130] Fry's alliances spanned the spectrum of architectural opinion; his friendships were as diverse as they were influential, from Charles Reilly and Robert Atkinson, to Wells Coates and Jack Pritchard. With considerable skill, Fry maintained his respectability while pursuing architectural innovation.

RETHINKING ARCHITECTURE: GROUP WORK AND THE RISE OF FRY'S PROFILE

Fry describes the mid-1920s as his rebirth as a modernist; his autobiography details an exquisite awakening, as he cast off his outdated Beaux-Arts training in favour of an authentic, modern means of expression. Eager to place himself at the forefront of the British Modern Movement, Fry draws a clear distinction between his career as a partner in a well-respected town planning firm and his modernist endeavours, undertaken initially in his spare time. His path towards modernity was aided by his participation in a series of groups, which each gave direction to Fry's shifting architectural agenda. In the late 1920s Fry joined the DIA and, in 1931, the think-tank Political and Economic Planning (PEP). These organisations can be seen as a precursor to Fry's active involvement in the MARS Group and his role in such groups provided useful context for the initial focus of MARS, as he and his co-founders sought to develop projects and ideas first debated at the DIA and PEP.

In *Autobiographical Sketches*, Fry tells of his architectural revelation through his careful observation of the construction of Devonshire House (1924–27) in London.

Designed by Fry's American employer, Thomas Hastings, in collaboration with his old professor, Charles Reilly, the building is the embodiment of Fry's Classical education and training. He describes his visceral reaction to the steelwork frame being clad in stone with Renaissance-inspired detailing, which provoked in him a 'moral revulsion'.[131] This concealment of structure prompted Fry, 'although not without some fondness', to discard his old school drawings in a symbolic act of a new beginning.[132] The following day, Fry writes, he began work on a hypothetical block of working-class flats with a portal truss frame of reinforced concrete. This description suggests a good deal of artistic licence, not out of keeping with Reilly's account of his own career, *Scaffolding in the Sky* (1938). Indeed, this conversion was probably a gradual process rather than the epiphany of Fry's memoirs, a view supported by the conflicting building designs and articles produced by Fry at the time. His groping around for satisfactory answers is evident, as new methods of construction necessitated a reassessment of the accepted architectural practices. His 1928 article for the *AR*, for example, examines the innovative steel structure of the Russell Institute; Fry writes, 'it is difficult to decide whether this framework should influence the building, and … in what way it can do so'. He goes on to say 'the important thing, so far as the building is concerned, is the steel work, the rest is – whatever you like to make it – Renaissance scenery for Piccadilly, eighteenth-century for Regent Street'. Fry recognised that the influence of steel was increasing daily and would 'at last affect the external appearance of all types of buildings'.[133] As he had witnessed in New York (and to a lesser extent in Liverpool), a steel- or iron-frame could support the floors of a building, rendering the walls as non-structural components, thereby enabling greater building heights, thinner wall sections and opportunities to use new materials and techniques, such as glazed façades.

Fry's reassessment of his architectural outlook was significantly aided by his involvement in the DIA, giving 'a vocabulary to his discontent' as Elizabeth Darling notes.[134] Established in 1915 as a much-needed British counterpoint to the *Deutscher Werkbund*, the DIA's founding members included William Lethaby, Frank Pick and Harry Peach. These prominent figures in British art and design sought to improve design and manufacturing standards, following German theories of functional design. The Werkbund's 'machine' interpretation of Arts and Crafts thought could be (and was) re-imported and branded as a continuation of English tradition; thus the in-house publication, *DIA Quarterly*, reported on the latest continental – particularly German – work, such as the *Weissenhofsiedlung* at Stuttgart and Ernst May's 'New Frankfurt'.[135] Fry describes these articles as critical in his move towards Modernism, later claiming that 'the whole thing became quite clear'.[136]

Despite a pioneering focus on architectural Modernism, the DIA retained a conservative edge and was, in Fry's words, rooted in 'a Lethaby world of honest craft'.[137] Following this philosophy, Fry's outlook at this time illustrates a growing interest in the emergent European architecture and a desire to link modernity to tradition. In a review of 'Modern Cottages and Country Houses', he defends national preferences and traditions as being 'rather grand things' that therefore could be 'expected to provide some substantial part of the emotion of architecture'.[138] Fry supports local traditions and the use of regional materials, citing the overhanging eaves of Switzerland and the use of stone slate in Derbyshire as being appropriate, in a manner that recalls his student diary entries on Cotswold villages. Fry argues that such influences and materials must 'continue to affect design wherever they are properly understood and keenly felt', before issuing a warning that 'where they are used as picturesque stock-in-trade, it is time for the work of Le Corbusier'.[139] Fry did not, therefore, view Modernism as a prescriptive solution; but believed that large-scale, contemporary development should reflect the current age. His writing shows a sensitivity to local building traditions that remained throughout his career; at this stage, it manifested itself in a preservationist approach to the countryside.

Around the time of Fry joining the DIA, the architect and then President, Clough Williams-Ellis, had overseen publication of the *Cautionary Guides*, a series of pamphlets that provided case studies of English towns – St. Albans, Oxford, Carlisle – to address the ribbon development that was stretching into the countryside, seemingly without rationale. Yet this paternalistic, cautionary approach to development was viewed as old-fashioned by some of the DIA's more recent recruits. The growing membership had brought in new blood with more progressive views, including Raymond McGrath, Mansfield Forbes,[140] P. Morton Shand and, most significantly for Fry, John Craven 'Jack' Pritchard.[141]

The DIA was thrust into a new phase of modernity from 1930 onwards, mainly through the campaigning and careful strategising of Jack Pritchard. First and foremost a businessman, Pritchard masterminded a 'DIA Plan' for modernisation, which he presented to the committee in late 1931. His brother-in-law, the architectural critic John Gloag, had introduced him to the DIA.[142] A well-respected 'establishment' figure, Gloag was an important ally for Pritchard and, with the window manufacturer W.F. Crittall, they drew up the Plan, later with input from Fry.[143] The Plan marked a significant departure from the DIA's cautionary-guide approach, leaving the newly-formed Council for the Preservation of Rural England (CPRE) to tackle unregulated countryside development. Instead, Pritchard, Gloag and Crittall called for the group's refocusing of its limited resources on 'the design of certain classes of British industrial products of general consumption' to help address the 'prevailing trade depression'.[144] This proposed direct collaboration with manufacturers and designers to manufacture British products that embodied the DIA ideal of 'fitness for purpose'.

Pritchard sent a draft plan to Fry. His reply is insightful, revealing a thorough absorption of DIA doctrine and a solid basis for his development as a modernist. Fry had evidently assimilated elements of German theory and was looking to create a Bauhaus-style school in England. Alongside the plan's aim to 'educate the public', Fry advised that the professionalisation of art through closer ties to industry should begin with a grounded art education, funded by industry or government. He suggested the removal of the amateurish nature of 'profitless and unplanned art-pottering' through the conversion of an existing art school into a central training centre operating 'under the right man' and guided by DIA principles 'to concentrate entirely on work fit for its purpose'. With his inclusion of the training of architects as 'the most important factor of the training centre', Fry removed any distinction between art and architecture – a view that echoed the Arts and Crafts Movement in England, and which also inspired Walter Gropius's teaching at the Bauhaus. Fry also called for an exhibition 'on strictly DIA lines' to crystallise the higher standards of design and 'set the training centres on their feet'.[145] Fry's response for a coherent plan of public education of good design, encompassing journalism, exhibitions and training, illustrates the importance of his involvement with the DIA in the development of his architectural outlook; from these early studies Fry took an active interest in art and design pedagogy that remained throughout his career.

Fry's ideas were formally presented to the DIA Committee alongside the Plan written by Pritchard, Gloag and Crittall. The supplement, signed by Maxwell Fry, appears to be a collaborative document incorporating contributions from Harold Stabler.[146] A DIA founding member and director at Carter, Stabler and Adams pottery (which later became Poole Pottery), Harold Stabler was a talented sculptor, metalworker and ceramicist.[147] Fry and Stabler's proposals aimed to promote the designer in the DIA's refocused programme, as Fry wrote: 'The demand from the manufacturer, stimulated by propaganda, must be met by a supply of first rate designers'.[148] Fry's DIA work, then, was a natural development of ideas undertaken in conjunction with the DIA's founding fathers. This quiet revolution illustrates Fry's ability to promote change whilst remaining loyal to tradition.

In addition to his work at the DIA Fry became an early member of the think-tank Political and Economic Planning. Acknowledging his reliance on memory, he later claimed to be amongst the 40 or 50 founder members of PEP.[149] Although there is no evidence to support this, correspondence shows that by August 1931 – just a few months after its inception – Fry was already deeply involved in the group's activities.[150] Fry recalled his inspiration to join the group followed his reading of an article by the environmentalist (Edward) Max Nicholson. Fry described Nicholson's article – which must have been 'A National Plan for Britain' – as 'a brave new world set out in considerable detail that so fitted my aspirations that I wrote offering help'.[151] Nicholson's plan was published on 14 February 1931 in the *Week-end Review*, a newly-established independent paper edited by Gerald Barry and focused on contemporary political debate and the arts. Emphasizing the cross-fertilization of groups such as the CPRE and the PEP, articles included a series by Barry's (and Fry's) friend Clough Williams-Ellis, providing case studies of ten British cities which asked 'What is Wrong with England?'[152]

The PEP included scientists, sociologists, educators, architects and artists, with many figures who would help to shape Fry's career, such as Gerald Barry, Julian Huxley, Leonard Elmhirst, Ben Nicholson, Henry Moore and Ove Arup. The PEP also led to further collaboration with his DIA colleague, and an early member of the PEP Directorate, Jack Pritchard. A reformer by nature, Pritchard was eager to effect change wherever possible; his involvement with the DIA and PEP often led to frustration due to what he perceived as the cautious approach of many of the senior members.[153] Pritchard was evidently an important ally for Fry and, at his suggestion, Fry became Secretary of the PEP's Town Planning group.[154]

Fry used his involvement in the DIA and PEP to address problems encountered in his professional life. For example, Fry was amongst an influential design collective – also comprising engineers William Muirhead and Sir Murdoch Macdonald, architects W.D. Caröe and Barclay Niven, and Fry's partner, Thomas Adams – who submitted a scheme for a new Charing Cross Bridge and the associated transport routes to both sides of the Thames. The major project had been ongoing for several years and had caused much controversy, prompting the LCC to appoint an advisory committee in June 1930. The committee of 15 eminent figures tasked with the selection of a suitable design, included Frank Pick, Giles Gilbert Scott, Reginald Blomfield, Raymond Unwin and Fry's other partner, Longstreth Thompson; a subsequent call for schemes resulted in 70 proposals from entrants as diverse as Stanley Adshead and Owen Williams.[155]

A shortlist of six schemes was scrutinised, the report by chairman Sir Leslie Scott revealing profound differences amongst the committee members. Fry's collective produced the most far-reaching proposals, with a swathe of re-planned streets south of the river to create a new 'low level' railway station in place of a demolished Charing Cross (Figure 1.15). An extensive network of transport links with underground connections to nearby stations and extended tramways would be complemented by 120 feet wide boulevards, allowing road traffic to a new traffic square at the Strand from an enhanced St. George's Circus in the south, connected to St. Martin's-in-the-Fields church in the north. The station, largely 'open to the sky', was enclosed by 'rich revenue-producing' shops and offices with a towering hotel facing the River Thames.[156] The proposal shows evidence of Fry and Adams's American experience, with large sections of raised roads used to ease vehicles' passage across the river and a skyscraper to contribute a modern landmark to the London skyline. Yet the ambitiousness of the scheme led to

1.15 Competition Entry, Charing Cross Masterplan, London, 1931

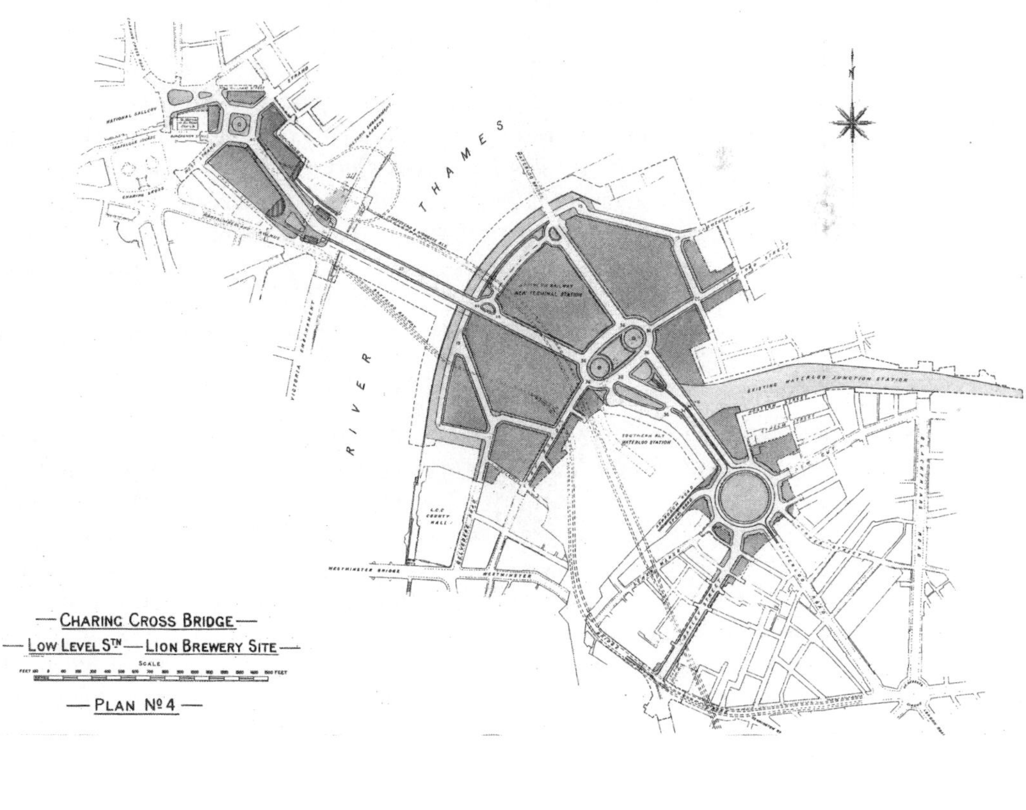

widespread criticism amongst the conservative committee members, who viewed the main disadvantage to be a cost estimate of several million pounds over the £12.5 million budget.[157]

The committee failed to reach a unanimous decision and again, the following year, progress stalled. The chairman's report illustrates the turgid discussions of the committee, with appendices written by members supplying caveat upon caveat to their decisions. Their inability to agree upon a unanimous winner demonstrates the highly conservative nature of town planning at the time and the problems faced by Fry. The Charing Cross planning process shows Fry's rallying of support in an attempt to instigate progress. He used the DIA and PEP agendas to highlight these issues with planning policies, gaining support from like-minded individuals, such as Pritchard, who wrote to Fry in early August 1931:

> *What you told me about Town Planning during our weekend has troubled me considerably and now that the Charing Cross schemes have become fluid again and the Government Housing Bill has gone through, while the Town Planning Bill has not, it is most important that your [planning] group of P.E.P. should work at exceedingly high pressure. ... Wouldn't it be a good thing to make a canvas of Government Ministers. Can't we get Clough and Pick to tackle [the Transport Minister, Herbert] Morrison and [the Prime Minister, Ramsey] MacDonald again?*[158]

Pritchard's letter illustrates the urgency felt at the time by left-wing activists.[159] In response, Fry produced a document on the 'Control of Elevations and the legal system of control' outlining the current problems in legislative control of architecture, responding to recent housing policy and control delegated to local authorities. Fry's manifesto illustrates, unsurprisingly, his support for the role of the architect, which – perhaps drawing on his experience at Southern Railways – he saw as being undermined by local authorities' employment of engineers to carry out architects' work for economic gain: 'The effect of vulgarity – bungaloid growth – chain stores – bad taste – clumsy, barbarous building. This can only be remedied by education. Education of the architect first, then the public. Architecture is slack – mixes its styles – is illogical and un-urbane. It must settle its structural problems and work out its style'.[160]

The text provides a useful summary of Fry's interests and wider ideas of the period. As Anthony Jackson notes, 'many people looked to planning as a panacea' at the end of the 1920s.[161] Town planning and architectural design were seen as interdependent amongst those seeking social reform and Fry's Liverpool education would have provided a strong emphasis on such ideas – albeit with an Edwardian civic-mindedness at its heart. Fry utilised his training to good effect; his polemic shows the continuing influence of his Liverpool professors, with the words of Reilly, Trystan Evans and Abercrombie echoing throughout his work. Both Fry and Pritchard emphasise the importance of organisations such as the CPRE. Recently established in 1926, by then Professor of Civic Design at Liverpool University, Patrick Abercrombie, the CPRE sought to highlight the destructive nature of unplanned ribbon development on the English countryside. Fry and Pritchard's backing of the CPRE is significant, illustrating the interconnectedness of groups, be they 'traditional' or 'modernising', during this period.

THE EXHIBITION FOR PLANNED INDUSTRIAL CONSTRUCTION

Fry's efforts at the DIA and the PEP focused on the modernisation of town planning processes in Britain. Fry occupied perhaps a unique position as a modernist-inclined architect who had a firm grasp of town planning and its current issues. In Wells Coates and Jack Pritchard he had found like-minded individuals, and, together, they were the most progressive unofficial grouping in English architectural culture at the time. While Fry's knowledge and skills focused on planning, Coates's early modern projects experimented with new materials and presentation techniques, for example in his shops for Cresta Silks. Coates excelled at elegant presentation, demonstrated by his competition-winning exhibition stand designed for Venesta, a plywood company for which Pritchard had worked since 1925. Fry, meanwhile, was a respectable runner-up in the same competition, but his early writings on slum clearance and concrete housing show a different emphasis to Coates's interests. Yet Fry and Coates's personalities and interests coalesced to form a strong partnership. Their modernist ideals were further strengthened by Pritchard's marketing skills and, despairing of the reticence of societies such as the DIA and PEP – which were generally perceived as progressive groups at this time – they joined forces in a highly ambitious venture to stage 'EPIC'.

Fry and Coates jointly authored a report for an Exhibition for Planned Industrial Construction, which they sent to Pritchard in late 1931. As architects with 'sole control over the design, plan and selection of exhibits', the project suggests Fry and Coates's primary interest during the early 'thirties was the planning of working class housing. The exhibition intended to demonstrate the meaning of 'Planned Economy', telling a story of three parts of 'a specified section of industrial activity': Un-planning, Planning and Planning in Practice.[162] Firstly, Fry and Coates aimed to portray 'a stagnation and a decay' accumulated around an 'unplanned nucleus' of a factory and a subsequent mass of houses. Countering this misuse of natural assets and amenities, Fry and Coates proposed an illustrative diagram of 'the essential elements of industrial planning', including land use, transport, services, population, availability of labour, and social needs.[163] The exhibition would conclude with detailed proposals of a specific site for a planned factory and adjacent workers' housing. Fry and Coates intended to devise a housing scheme including provision of open spaces, recreation, and shopping and transport facilities to 'show that large-scale housing is not merely a builders' job, but that it requires the brains of sensitive and experienced planners'. The proposal shows Fry and Coates' respective interests at this time, with plans of a workers' housing scheme to be followed by proposals for 'the minimum house'. As a finale, Fry and Coates proposed to enlist 'Professor Unwin's support' – and that of the RIBA, for which Raymond Unwin was then President – by providing a 'small space' for the display of his London and Regional Town Planning Schemes, which Unwin had developed under the aegis of his role as technical adviser for the Greater London regional planning committee. Unwin wrote a series of reports that called for a London green belt and satellite towns, again illustrating the inter-connected issues of planning and preservation.

Fry used his contacts in industry in an attempt to secure funding for EPIC, writing to figures such as Francis Goodenough at the British Commercial Gas Association.[164] Goodenough was unwilling to pledge his support without detailed information, and it seems Fry and Coates were similarly unwilling to undertake a considerable amount of work without a guaranteed income from the project. The exhibition stalled. Yet the work by Fry and Coates was recycled in later projects, particularly in setting the research agenda of the MARS Group and in numerous exhibition designs, as Chapter 2 will discuss. Indeed, in EPIC, Fry and Coates had planned a modern exhibition, combining drawings, photographs, scale-models, bills of quantities and a full-size section of an area of the factory, including furniture and furnishings.

During the same period, Fry and Pritchard were busy working on the first issue of a newly revamped DIA journal. The DIA Plan outlined plans for the journal – renamed *Design in Industry* – with John Gloag assuming the role of General Editor and guest editors appointed for each quarterly issue. The magazine looked beyond the group's membership and aimed to reach a wider audience with its design propaganda. As part of the DIA Plan's campaign of 'constructively criticising' current design, special issues of the journal would consider objects used in the office, the kitchen, the bathroom and the living room. Fry was chosen to edit the first issue on 'The Office', published as the spring issue of 1932. To further influence architectural debate, Fry became a regular contributor to the *Architects' Journal* (*AJ*), which had been co-edited by Christian Barman since 1927.[165] The Charing Cross scheme was supported by the *AJ* – in an article perhaps written by Fry himself – to draw attention to the conservatism of the LCC. Fry's connections with the architectural press ensured his schemes were generally reported in a favourable (if not downright partisan) manner, such as his entry for the new RIBA Headquarters. Launched in 1932, the competition provided an opportunity for Fry to test his new ideas. The *AJ* covered the entries and results in great detail (the radicalised *AR* deliberately did not). An impressive 284 entries were received, but Fry's *AJ* colleague, Frederick Towndrow,[166] noted with some regret that 'only four or five schemes were submitted that were really modern in their treatment. Notable among them is the scheme of E. Maxwell Fry'.[167]

The scheme was not placed, but Towndrow felt that the 'straightforwardness and honesty' of the scheme deserved some kind of honourable mention.[168] Fry's proposal softened the transition from Portland Place to Weymouth Street by gently curving the building around the corner, suggesting that the walls are stretched tight over a modern, rigid structure rather than providing any structural value.[169] Fry attempted to introduce a horizontal emphasis to the building through the fenestration, but an overriding (American) vertical thrust persists, upsetting the balance of the façade. The competition rules stated that the façade must be 'dressed' in Portland Stone (so Fry must have recovered from his aversion to using the material as cladding),[170] which prevented any attempts at using the latest

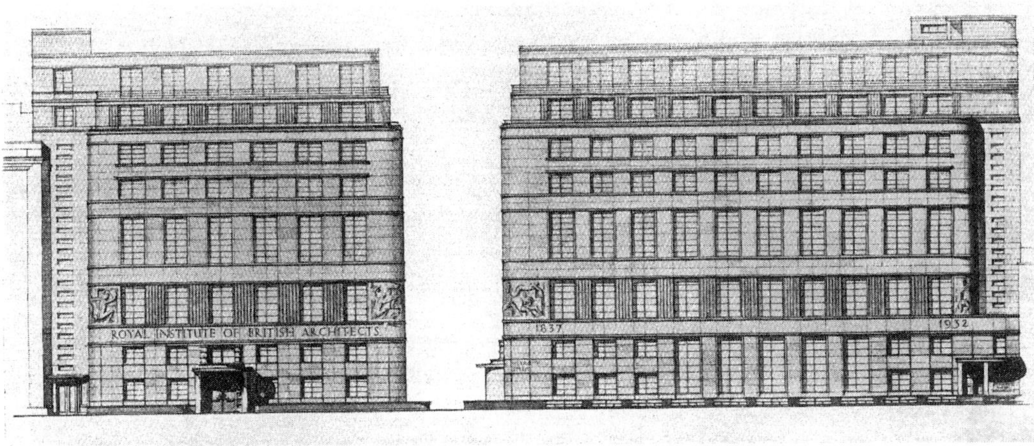

1.16 Elevations to Portland Place and Weymouth Street, RIBA Headquarters, London, 1932

materials and technologies. Judged by Robert Atkinson, the competition was won by Giles Grey Wornum's homage to Swedish Grace.[171] Towndrow used Fry's entry to critique the numerous schemes that had taken influence from Sweden:

> *The elevations of this scheme, in contradistinction to many others, show that the building is properly lighted throughout all its parts. The fenestration, if anything, is all too logical … a design like this seems to be to be a great deal more British than one or two of the other rather emasculated 'Swedish' efforts which were premiated or commended.*[172]

The idea that modern architecture was an imported idea was particularly pertinent given contemporary debate on the use of imported styles. Perhaps what made Fry's design palatable *and* modern was its steady commercial approach reminiscent of the 'Liverpool Manner' with windows still treated as apertures within the wall and a determined vertical thrust to the elevations, which stepped back at the higher levels. It was closer to the work of Reilly and Hastings than Fry would care to admit.

CONCLUSION

It is evident that Fry's Unitarian background and his architectural training at Liverpool and New York instilled in him a progressive outlook in architecture and town planning. At Liverpool School of Architecture Fry was educated by a group of practitioners that were modern in outlook, if not Modern in style. In particular, Reilly, Abercrombie and Trystan Edwards all remained influential figures for Fry's work throughout his career. Echoes of their ideas and work are present in Fry's own work throughout the interwar period and beyond. During the early 1920s, Fry's modernity focused on social and economic forces, and was expressed in the grand civic plans and garden villages produced in collaboration with Adams and Thompson. The legacy of this education and early career remained; his 'conversion' to Modernism might be viewed rather as a shift in outlook to consider recent technical and structural innovation alongside his existing social and economic concerns. His membership in reformist groups helped Fry's development of these ideas and his alliance with figures such as Christian Barman, Wells Coates and Jack Pritchard provided a basis for his work throughout the 1930s.

NOTES

1 John Gold, *The Experience of Modernism: Modern Architects and the Future City, 1928–53* (London, 1997), pp. 1–7.

2 Although they suggest that the movement had all but died out by around 1929. See Peter and Alison Smithson, 'The Heroic Period of Modern Architecture', *Architectural Design*, 35 (December 1965).

3 Peter and Alison Smithson, 'The Heroic Period of Modern Architecture', p. 590.

4 Maxwell Fry, 'English architecture from the "thirties"', *Architects' Year Book*, 8 (1957): pp. 53–6, p. 54.

5 See for example: Elizabeth Darling, *Re-forming Britain: Narratives of Modernity before Reconstruction* (London, 2007); Alan Powers, *Britain: Modern Architectures in History* (London, 2007).

6 Darling, *Re-forming Britain*, p. 3.

7 Anthony Jackson, *The Politics of Architecture: A History of Modern Architecture in Britain* (London, 1970), p. 54.

8 Hélène Lipstadt, 'Polemic and Parody in the Battle for British Modernism', *Oxford Art Journal*, 5/2 (1983): pp. 22–30, p. 27.

9 E. Maxwell Fry, *Autobiographical Sketches* (London, 1975), p. 136.

10 Christopher Crouch, *Design Culture in Liverpool 1880–1914* (Liverpool, 2002), pp. 1–2.

11 Crouch, *Design Culture in Liverpool*, p. 127.

12 Fry, *Autobiographical Sketches*, p. 47.

13 Crouch, *Design Culture in Liverpool*, xiii.

14 Peter Taaffe and Tony Mulhearn, *Liverpool: A City that Dared to Fight* (London, 1988), p. 30.

15 See Taaffe and Mulhearn, *Liverpool*, pp. 15–33 for full discussion of this period, which culminated in the Birkenhead Riots of 1932.

16 Ambrose Fry was a British subject via parentage. John Morrison Archive.

17 In 1919, 1920, 1927 and 1929 to New York, in 1926 to Montreal and Quebec, in 1920 he also travelled from Southampton to Buenos Aires. John Morrison Archive.

18 Fry, *Autobiographical Sketches*, p. 76.

19 Fry lived at Arundel Avenue, see E. Maxwell Fry, *Autobiographical Sketches* (London, 1975), p. 35.

20 John Morrison Archive.

21 RIBA Archive, F&D/13/4. E. Maxwell Fry, 'Maxwell Fry and the Modern Movement', undated.

22 'Can't you think of some better invocation for me than Mr. Hall. Old Hall might be better though I be a bit short of seventy yet'. RIBA Archive, F&D/12/10. Letter Hall to Fry, 31 July 1925.

23 RIBA Archive, F&D/13/4. E. Maxwell Fry, 'Maxwell Fry and the Modern Movement', undated.

24 Fry, *Autobiographical Sketches*, p. 78.

25 Joseph Sharples, 'Reilly and His Students, on Merseyside and Beyond', in *Charles Reilly and the Liverpool School of Architecture 1904–1933*, (eds) Joseph Sharples, Alan Powers and Michael Shippobottom (Liverpool, 1996): pp. 25–42, p. 26.

26 UL Archive, Student Dossier, Edwin Maxwell Fry.

27 Fry, *Autobiographical Sketches*, pp. 84–5.

28 UL Archive, Student Dossier, Edwin Maxwell Fry.

29 Fry, *Autobiographical Sketches*, p. 79, p. 90.

30 RIBA Archive, F&D/12/11. E. Maxwell Fry, 'Some Impressions on a Visit to Oxford', 2 August 1920, p. 5.

31 RIBA Archive, F&D/13/1. Maxwell Fry, 'Prince's Park', [c. 1920].

32 This section draws from the excellent overview provided by: Quentin Hughes, 'Before the Bauhaus: The experiment at the Liverpool School of Architecture and Applied Arts', *Architectural History*, 25 (1982): pp. 102–13.

33 T.G. Jackson cit. in Hughes, 'Before the Bauhaus', p. 105.

34 Hughes, 'Before the Bauhaus', pp. 104–5.

35 Augustus John was appointed in 1901, although his appointment lasted only 18 months.

36 The 'Glasgow Four' also comprised the architect Charles Rennie Mackintosh and his wife, Frances's sister, the painter and glass artist Margaret MacDonald.

37 Hughes, 'Before the Bauhaus', pp. 106–8.

38 RIBA Archive, F&D/13/5. E. Maxwell Fry, 'Thoughts on Architectural Education', 7 October 1960, p. 4. This view is echoed by another of his students, Robert Gardner-Medwin, who described how Reilly 'had no pretentions of being a scholar', but completely 'understood the power of architecture and he loved it as he loved life'. See UL Archive, D255/6/3/12/6. Robert Gardner-Medwin, 'Some notes on Charles Reilly and Patrick Abercrombie', 15 September 1977, p. 1.

39 Peter Richmond, *Marketing Modernisms: The Architecture and Influence of Charles Reilly* (Liverpool, 2001), p. 86.

40 For example, Hugh Rathbone (1862–1940) became the first treasurer of the University of Liverpool in 1903, he was elected as Pro-Chancellor of the University in 1920 and was a Liberal MP, representing Liverpool Wavertree from 1923 to 1924. He was an active member of the Unitarian congregation at the Ancient Chapel of Toxteth.

41 Alan Powers, 'Liverpool and Architectural Education in the Early Twentieth Century', in *Charles Reilly and the Liverpool School of Architecture*, pp. 1–23, p. 4.

42 Lever became Sir William Hesketh Lever in 1911.

43 See for example Edward Hubbard and Michael Shippobottom, *A Guide to Port Sunlight Village* (Liverpool, 2006).

44 C.H. Reilly, *Scaffolding in the Sky* (London, 1938), p. 129.

45 UL Archive, D207/2/2. Letter Reilly to Bonnier, 29 May 1907.

46 Reilly cit. in. Richmond, *Marketing Modernisms*, p. 90.

47 UL Archive, S3163. 'Prospectus of the Liverpool School of Architecture, 1920–21', p. 4.

48 Mawson had worked for William Lever on a number of projects, such as The Hill at Hampstead. Like Reilly, he sought the professionalization of architecture and civic design, and to this end he founded the Institute of Landscape Architects in 1929. See for example Janet Waymark, *Thomas Mawson: Life, Gardens and Landscapes* (London, 2009).

49 UL Archive, S3163. 'Prospectus of the Liverpool School of Architecture, 1920–21', p. 5.

50 UL Archive, S3163. 'Prospectus of the Liverpool School of Architecture, 1920–21', p. 6.

51 Fry's student record shows a typical mark of 800/1000 for architectural design and 450/500 for architectural construction and working drawings, but only 50/100 for professional practice and sanitation & hygiene. UL Archive. Student Dossier, Edwin Maxwell Fry.

52 UL Archive. Student Dossier, Edwin Maxwell Fry, 29 June 1921.

53 Fry, *Autobiographical Sketches*, p. 89.

54 Liverpool Record Office, 367 SAN/4/1/1. Sandon Studios Charter, 1905.

55 Roderick F. Bisson, *The Sandon Studios Society and the Arts* (Liverpool, 1965), p. 54.

56 Bisson, *The Sandon Studios Society and the Arts*, p. 132.

57 John Tiernan, *Liverpool's 'Bloomsberries': the Sandon Studios Society* (Liverpool, 2007)

58 Liverpool Record Office, 367 SAN/1/4/7. Minutes of Sandon Studios Executive Committee, 22 September 1924.

59 UL Archive, S3217 Newspaper Cuttings (22 March 1922 – 26 April 1924). 'Sandon Studios Exhibition. Show of Individual Art', *Liverpool Daily Post & Mercury*, 13 May 1922.

60 UL Archive, S3217 Newspaper Cuttings (22 March 1922 – 26 April 1924). 'A Liverpool Art Exhibition', *Morning Post*, 5 June 1922.

61 Richmond, *Marketing Modernisms*, p. 62.

62 UL Archive. Student Dossier, John Christopher Wood.

63 Richard Ingleby, *Christopher Wood: An English Painter* (London, 1995), p. 13, pp. 16–17; UL Archive. Student Dossier, John Christopher Wood.

64 Fry, *Autobiographical Sketches*, pp. 87–8.

65 Fry, *Autobiographical Sketches*, p. 93.

66 Liverpool Record Office, 367 SAN/1/4/7. Minutes of Sandon Studios Executive Committee.

67 Fry writes of attending the society ball and socialising with William and Marjorie Holford, see UL Archive, D287/10. Letter Fry to Myles Wright, 14 September 1977.

68 Fry later acknowledged the influence of McKim, Mead and White and John Russell 'Jackie' Pope on his early work, see A.E.J. Morris and Cornelius Murphy, 'Max Fry: Inspirations, friendships and achievements of a lifetime in the Modern Movement', *Building*, 229/6907 (31 October 1975): pp. 52–8, p. 58.

69 Laurie Ossman and Heather Ewing, *Carrère and Hastings: The Masterworks* (New York, 2011), p. 8.

70 Fry, *Autobiographical Sketches*, p. 96.

71 Ossman and Ewing, *Carrère and Hastings*, p. 8.

72 Ossman and Ewing, *Carrère and Hastings*, p. 13.

73 Ambrose Fry's unwillingness to provide a statement of his income meant that the Board of Education was 'not prepared' to consider Fry's application for further financial assistance. See UL Archive. Student Dossier, Edwin Maxwell Fry, 20 November 1922.

74 Fry, *Autobiographical Sketches*, p. 98.

75 'Rome Scholarship and Henry Jarvis Studentship, 1923', *The Builder*, 124 (29 June 1923): p. 1046.

76 A note to Fry's student file reads: 'E. Maxwell Fry, in view of exceptional circumstances of his case, viz. his absence on professional practice in America during the summer term of session 1921–22, be permitted to offer the fourth and fifth examinations for the degree of B.Arch at the end of his fifth year'. UL Archive. Student Dossier, Edwin Maxwell Fry, 15 February 1922.

77 Fry describes this as the Ministry of Works in his autobiography, although the office was referred to as the Office of Works at that time. See Fry, *Autobiographical Sketches*, p. 120.

78 C.H. Reilly, *Representative British Architects of the Present Day* (London, 1931), p. 21.

79 For example, for the Thames Valley, West Essex, Scarborough, Teignmouth, and York.

80 Reilly described him as the Cyril Farey of his time, see *Representative British Architects of the Present Day*, p. 16.

81 Fry, *Autobiographical Sketches*, p. 120.

82 Fry, *Autobiographical Sketches*, p. 120.

83 F.L. Thompson, *Site Planning in Practice: An Investigation of the Principles of Housing Estate Development* (London, 1923).

84 T. Adams, L. Orton et al, *Regional Plan of New York and its Environs* (New York, 1931).

85 Jean-Louis Cohen, *Scenes of the World to Come: European Architecture and the American Challenge 1893–1960* (Paris, 1995), p. 106. For other examples of the monumental perspective drawings, see Jack Dunne and Peter Richmond, *The World in One School: The History and Influence of the Liverpool School of Architecture 1894–2008* (Liverpool, 2008), p. 21.

86 R. LeGates and F. Stout (eds), *Early Urban Planning* (London, 1998), xxviii.

87 Frank Jackson, *Sir Raymond Unwin: Architect, Planner and Visionary* (London, 1985), p. 154.

88 T. Adams, F.L. Thompson et al, *Recent Advances in Town Planning* (London, 1932), p. 302.

89 Adams, Thompson et al, *Recent Advances in Town Planning*, p. 292.

90 Adams, Thompson et al, *Recent Advances in Town Planning*, p. 292.

91 RIBA Archive, F&D/13/4. Edwin Maxwell Fry, Text Fragment, undated.

92 This outline draws on 'The Rebuilding of Piccadilly Circus and the Regent Street Quadrant', *Survey of London: Volumes 31 and 32: St James Westminster, Part 2* (1963), pp. 85–100. http://www.british-history.ac.uk/report.aspx?compid=41456 Accessed: 28 April 2013.

93 RIBA Archive, F&D/13/4. Edwin Maxwell Fry, Text Fragment, undated.

94 C.H. Reilly, 'London Streets and their Recent Buildings – I. Waterloo Palace and Lower Regent Street', *Country Life*, 51 (27 May 1922): pp. 691–4; C.H. Reilly, 'London Streets and their Recent Buildings – II. Regent Street', *Country Life*, 51 (10 June 1922): pp. 777–81.

95 Arthur Trystan Edwards, *Good and Bad Manners in Architecture* (London, 1924), p. 47.

96 Fry, *Autobiographical Sketches*, p. 52.

97 Alina Payne, 'Rudolf Wittkower and Architectural Principles in the Age of Modernism', *Journal of the Society of Architectural Historians*, 53 (September 1994): pp. 322–42, p. 332.

98 Scott's book was included on the third-year reading list at Liverpool School of Architecture, see: Liverpool University Special Collections and Archive, S3163. 'Prospectus of the Liverpool School of Architecture, 1920–21', p. 11.

99 RIBA Archive, F&D/13/4. E. Maxwell Fry, 'Maxwell Fry and the Modern Movement', undated.

100 Sherban Cantacuzino, *Wells Coates* (London, 1978), p. 13.

101 Cantacuzino, *Wells Coates*, p. 13.

102 This was Ethel Lilian Speakman's second marriage; she had married Charles Leese, a Lancashire cricketer, in 1911. John Morrison Archive.

103 Fry, *Autobiographical Sketches*, p. 125.

104 Fry, *Autobiographical Sketches*, p. 125.

105 Fry, *Autobiographical Sketches*, p. 125.

106 Herbert Leslie Barton graduated with BArch in 1925 and was a Rome Scholar finalist in 1929, see: *The Book of the Liverpool School of Architecture* (Liverpool, 1932).

107 C.H. Reilly, 'Railway Stations: Terminal and Otherwise', *Architectural Review*, 78 (October 1935): pp. 147–9, p. 147.

108 'New Stations at Ramsgate and Margate', *Builder*, 132 (14 January 1927): p. 87.

109 'Ramsgate Station (British Rail), Ramsgate', *English Heritage Listing* (1988), http://www.britishlistedbuildings.co.uk/en-172019-ramsgate-station-bristish-rail-ramsgate-. Accessed: 15 May 2013.

110 Fry may also have designed a Hotel for Southern Railways in Southampton.

111 Gareth Williams, 'The Return of the Curve: Rodney Thomas, Architecture and Interior Design', *Journal of the Decorative Arts Society 1850 – the Present*, 19 (1995): pp. 41–50, p. 41.

112 Owen may have also worked at Southern Railways. Geoffrey Leyland Owen was recipient of the John Rankin Prize in architecture in 1923 and received his BArch in the same year as Fry; see Lionel Budden (ed.), *The Book of the Liverpool School of Architecture* (Liverpool, 1932), p. 52.

113 The winner was Ernest Berry Webber, who went on to win the Southampton Civic Centre competition.

114 'Manchester Art Gallery Competition', *Builder*, 128 (26 June 1925): pp. 969–75, p. 970.

115 Fry, *Autobiographical Sketches*, p. 126.

116 The winning entry, submitted by a Parisian firm, was described as a 'miracle of neatness and simplicity', see 'Birmingham Civic Centre Competition', *Architect and Building News*, 118 (19 August 1927): pp. 297–309, p. 297.

117 'Birmingham Civic Centre Competition', *Architect and Building News*, p. 298.

118 A similar scheme was also prepared for Newquay in Cornwall for a coastal road and cliff park.

119 Plans were produced between 1930 and 1931 for West Middlesex, South-east Sussex, North-east Kent and Mid-Northamptonshire, amongst others.

120 Adams, Thompson et al, *Recent Advances in Town Planning*, p. 113.

121 Fry, *Autobiographical Sketches*, p. 127.

122 See E. Maxwell Fry, 'Settings for Gas Fires', *Architects' Journal*, 72 (24 December 1930): p. 953.

123 Fry, *Autobiographical Sketches*, p. 127.

124 I. Nairn, N. Pevsner et al, *The Buildings of England: Surrey* (Harmondsworth, 1971), p. 494.

125 It was also published in the *Architectural Review* and *The Book of the Liverpool School of Architecture* in the same year. Randal Phillips, 'A Modern Country House', *Country Life*, 71 (19 March 1932): pp. 332–4.

126 Fry, *Autobiographical Sketches*, p. 136.

127 Stephen Games, 'Bobby Carter: an appreciation', *RIBA Journal*, 89 (August 1982): p. 17. Things evidently picked up, however, as in 1931 Fry employed Liverpool graduate Robert Gardner-Medwin. See UL Archive, D688/1/3/15. Robert Gardner-Medwin, 'Recollections of Sixty Years of Modern Architecture in Britain 1930–1990', 1994, p. 17.

128 Carter had assisted Fry on competitions for Charing Cross Bridge and Accrington Town Hall, see: John Allan, *Berthold Lubetkin: Architecture and the Tradition of Progress* (London, 1992), p. 125 [56].

129 Fry, *Autobiographical Sketches*, p. 136.

130 RIBA Archive, F&D/13/7. E. Maxwell Fry, 'My Life with Architecture', unpublished manuscript, 1982.

131 Fry, *Autobiographical Sketches*, p. 136.

132 Fry, *Autobiographical Sketches*, p. 136.

133 E. Maxwell Fry, 'Modernism on Paisley', *Architects' Journal*, 67 (2 February 1928): pp. 291–9.

134 Darling, *Re-forming Britain*, p. 32.

135 W.J. Bassett-Lowke, 'A Wonderful Experiment at Stuttgart', *DIA Quarterly*, 2 (December 1927): pp. 8–10; W.J. Bassett-Lowke, 'Recent Developments in Germany', *DIA Quarterly*, 6 (December 1928): pp. 9–11.

136 Maxwell Fry, 'How Modern Architecture came to England', *Pidgeon Digital Tape and Slide Set* (London, 1980).

137 Fry, *Autobiographical Sketches*, p. 137.

138 E. Maxwell Fry, 'Modern Cottages and Country Houses', *Architects' Journal*, 74 (8 July 1931): pp. 54–6, p. 54.

139 Fry, 'Modern Cottages and Country Houses', p. 55.

140 McGrath and Forbes both joined in 1929 and the following year the *Quarterly* magazine reported on a visit to Finella, following the invitation of Forbes. 'Visit to "Finella", Cambridge', *DIA Quarterly*, 12 (July 1930): p. 16.

141 Other new members included Howard Robertson (who joined in 1927) and Professor Reilly (joined late in 1929).

142 UEA Archive, Pritchard Papers, PP/8/17/15. Letter Gloag to Pritchard, 12 November 1973.

143 Pritchard's correspondence suggests that he was the main instigator of the manifesto although, as an adept promoter of himself and his work, Pritchard's carefully edited personal papers must be viewed with care.

144 'Annual Report presented to The Annual General Meeting', *Design and Industries Association*, 9 March 1933, p. 1.

145 UEA Archive, Pritchard Papers, PP/28/2/11. Maxwell Fry, 'Comments on Draft DIA Plan', 16 October 1931.

146 Pritchard, apparently commenting on Fry's supplement, wrote to Noel Carrington: 'I like the idea put up by Fry and Stabler that we should attempt to capture one centre where D.I.A. principles of design could be preached'. See UEA Archive, Pritchard Papers, PP/28/2/17. Letter Pritchard to Carrington, 26 October 1931.

147 Stabler's wife, the potter Phoebe McLeish (1879–1955), was educated at the School of Architecture and Applied Art in Liverpool (1901–4). She and her silversmith sisters, Minnie and Annie, were possibly known to Fry through the Sandon Studios. 'Phoebe Gertrude Stabler ARBS', *University of Glasgow History of Art and HATII*, 2011, http://sculpture.gla.ac.uk/view/person.php?id=msib2_1204574221. Accessed: 7 June 2013.

148 UEA Archive, Pritchard Papers, PP/28/2/35. Maxwell Fry, Supplement to DIA Plan, [October 1931].

149 Maxwell Fry, 'The MARS Group Plan of London', *Perspecta*, 13–14 (1979): pp. 165–6, p. 166.

150 See for example UEA Archive, Pritchard Papers, PP/4/1/8 series.

151 Maxwell Fry, 'The MARS Group Plan of London', *Perspecta*, 13–14 (1979): pp. 165–6, p. 166.

152 Suzanne Waters, 'In Search of Sir Gerald Barry, the man behind the Festival of Britain', *Twentieth Century Architecture*, 5 (2001): pp. 39–46, pp. 39–40.

153 See for example UEA Archive, Pritchard Papers, PP/4/1/4/1–4, on Pritchard's exasperation of the lack of impetus amongst the PEP. Pritchard's differing outlook ultimately led to his resignation from the PEP Directorate, although he remained a member of the group and sought to bring about change through a PEP sub-group, Tecplan.

154 UEA Archive, Pritchard Papers, PP/4/1/16/4. Letter Jack Pritchard to Max Nicholson, 6 August 1931.

155 Frederic Leslie Scott, *Proposed Road Bridge at Charing Cross* (London, 1931), p. 5, p. 18.

156 'Charing Cross Bridge', *Architects' Journal*, 73 (11 March 1931): pp. 385–7, p. 386.

157 Scott, *Proposed Road Bridge at Charing Cross*, p. 35.

158 UEA Archive, Pritchard Papers, PP/4/1/8/1. Letter Pritchard to Fry, 5 August 1931.

159 Just a couple of weeks after Pritchard's letter, Ramsey Macdonald's Labour Government had resigned, and a presumably less-sympathetic cabinet for the PEP to lobby. Jon Clark, Margot Heinemann, David Margolis and Carole Snee (eds), *Culture and Crisis in Britain in the Thirties* (London, 1979), p. 9.

160 UEA Archive, Pritchard Papers, PP/4/1/8/3. E. Maxwell Fry, 'Control of Elevations and the legal system of control', PEP paper, 9 August 1931.

161 Jackson, *The Politics of Architecture*, p. 62.

162 UEA Archive, Pritchard Papers, PP/23/2/113. E. Maxwell Fry and Wells Coates, 'Report on project for an exhibition to be held at Olympia in conjunction with the Building Trades Exhibition, September, 1932'.

163 UEA Archive, Pritchard Papers, PP/23/2/113. Fry and Coates, 'Report on project for an exhibition'.

164 UEA Archive, Pritchard Papers, PP/7/2/16. Letter Goodenough to Fry, 21 January 1932.

165 Barman was appointed as the co-editor of the Architectural Review and its weekly sister publication, the *Architects' Journal*, by Hubert de Cronin Hastings.

166 Towndrow (1897–1977) became editor of *Architectural Design* in 1932. He was Senior Architect for the Ministry of Works during World War Two and, in 1947, he left for Australia to become Professor of Architecture at New South Wales University. See F.E. Towndrow and Hazel De Berg, F.E. Towndrow interviewed by Hazel de Berg in the Hazel de Berg collection [sound recording], 1964. http://nla.gov.au/nla.oh-vn1766427. Accessed: 6 July 2012.

167 F.E. Towndrow, 'Some Other Schemes of Note. i: The Design by E. Maxwell Fry', *Architects' Journal*, 75 (11 May 1932): p. 625.

168 Towndrow, 'Some Other Schemes of Note', p. 625.

169 Fry made this argument of Burnet, Tait and Lorne's Mount Royal Flats in London, see E. Maxwell Fry, 'Mass-Produced Shelter: Mount Royal, London: Analysis', *Architectural Review*, 77 (July 1935): pp. 14–16, p. 15.

170 Margaret Richardson, *66 Portland Place: The London Headquarters of the Royal Institute of British Architects* (London, 1984), p. 6.

171 Both of whom sat on an architectural committee for the Gas, Light and Coke Company. Fry would later join the same committee, which led to the Kensal House commission.

172 Towndrow, 'Some Other Schemes of Note', p. 625.

2

Thirties' Development

In reaction to the heroic tales of architectural modernists, recent scholarship has sought to place the movement within the context of contemporaneous, seemingly contradictory, historicist and preservationist movements.[1] This chapter looks closely at these themes to ask what modernity, tradition and preservation meant for Fry, as one of the leading British modernists, and how these ideas were manifest in his work. Fry's professional concerns are contextualised by his continuing group work with the DIA and PEP, in addition to his role as a committee member of the newly-formed MARS Group. Outside of committee life, Fry's social circle will also be investigated to build up a thorough picture of his life and work in interwar Britain.

As Chapter 1 has shown, Fry had a background in both town planning and architecture. For him, the disciplines were two sides of the same coin; one could not be conceived of without the other and he wrote in 1937, 'I cannot see that the art of town planning is in any real sense different from architecture'.[2] Indeed, Fry's route into Modern architecture came via town planning and he was imbued with moralistic conceptions of preservation from the outset. Fry was unusual in this respect; it gave his work an edge, for he tapped into wider debates, bringing him into contact with people from different networks and organisations. His influential acquaintances helped sustain his career during the difficult inter-war period (and post World War Two), whilst many of his contemporaries struggled or failed to find work.

During the 1930s, Fry also benefitted from two successful professional partnerships, first with the housing specialist Elizabeth Denby (1894–1965) and, secondly, with the German architect and founder of the Bauhaus, Walter Gropius (1883–1969). These partnerships with like-minded individuals were central to his development as a modernist, providing opportunity to test and refine his ideas. This chapter charts Fry's collaborations with Denby and with Gropius to investigate his development in architectural style. It examines the key concerns for his work as an architect and planner that led finally to the establishment of his own office in 1937.

With the slow improvement of economic conditions and a series of legislative measures during the early 'thirties to stimulate the building industries,[3] Fry was in a position to benefit and to aid in the development of the Modern Movement. Thus, by the close of the decade, Fry had established himself at the forefront of British

Modernism. His ascendency was recognised by his contemporaries F.R.S. Yorke and Colin Penn,[4] who wrote in *A Key to Modern Architecture* (1939):

> *It is almost impossible to name any one man as the leader in modern architecture to-day, but it is perhaps Maxwell Fry who should be given first place by his colleagues. He has been responsible for a number of houses in which rational planning and delicacy of treatment are combined with a precise selection of materials to produce a fine finish.*[5]

THE MARS GROUP AND EXPERIMENTS IN STANDARDISATION

The formation of the MARS Group is couched in contradictory statements adding to the myth and status of the selective club.[6] John Summerson notes, 'MARS was nothing if not exclusive'.[7] This view was reinforced by Fry, who later acknowledged that its origins were 'obscure' and thought the group might have started as early as 1931.[8] In reality, the British chapter of the *Congrès International d'Architecture Moderne* was officially established on 28 February 1933.[9] Fry's memory suggests that, for him, the MARS Group was the consolidation of his work during the early 1930s, particularly his group work and his early collaborations with Coates and Pritchard. Indeed, the inception of MARS stemmed from Fry and Pritchard's difficulties in implementing their reformist Plan for the DIA and Coates's failed attempts to steer the agenda of Mansfield Forbes's Twentieth Century Group.[10] By establishing a small group, MARS sought to build on existing modernist discourse but do so in a more strategic, efficient manner. In an attempt to instigate real change, a select band of six modernisers was assembled of three 'architects' – Fry, Wells Coates and Coates's then-partner David Pleydell-Bouverie – and three 'non-professional adherents' – P. Morton Shand, H. de Cronin Hastings and John Gloag.[11] Demonstrating the group's plans to implement a forceful media campaign. Each of the adherents was an influential member of the architectural press. Shand was a well-respected writer and frequently contributed articles to the *AJ* and the *AR*, both of which were jointly edited by Cronin Hastings. Gloag, meanwhile, was a well-connected critic (he was Jack Pritchard's brother-in-law) and a prolific writer on contemporary architectural debate.

Fry was a hard-working MARS member and assumed a position of leadership from the outset. He viewed the collective of architects, engineers and allied technicians as a vehicle to promote his research interests, in a similar manner to his earlier involvement in the DIA and PEP. Fry's adopted an apolitical approach to Modernism and believed the position of a Modern architect should be that of a 'neutral (albeit compassionate) doctor'.[12] Fry saw his friend Coates as a politically neutral ally, commenting that, 'Wells and [Berthold] Lubetkin were poles apart. Lubetkin … was constantly trying to drag us into communist politics'.[13] Internal MARS memoranda reveal Fry as the group's town planning and housing specialist, while Coates took the lead on liaison with the other branches of CIAM.[14] Fry's previous work lobbying government ministers at the PEP and his professional knowledge of the town planning system fed into his role as head of the group's 'obstructions' to inappropriate legislation. As an immediate consequence, Fry began to write regular agit-prop articles for the *Architects' Journal*, illustrating the significance of the group's 'non-professional' adherents. The pieces focused on this specialism in town planning and housing – particularly standardisation and mass production – and were one of Fry's major contributions to British architecture and design culture of the period.[15] Developing ideas initially expressed in his and Coates's EPIC report, Fry put forward his case for the urgent need for slum clearance and the subsequent provision of houses for the working classes.[16] These articles demonstrate for the first time Fry's interest in mass-production in housing, utilising manufacturing and component systems to produce standardised housing equipped with well-designed furniture at affordable prices.[17] His theories are heavily indebted to Ernst May's 'New Frankfurt' housing developments, which the DIA had reported on in the mid-1920s. The thoroughness of Fry's argument suggests that, by this time, he was familiar with May's *Das Neue Frankfurt*, which had been published in 1929. The book followed the CIAM Conference on 'The Minimum Dwelling' held at Frankfurt in the same year. Indeed, May's dissemination of material on his Frankfurt dwellings appears to have been significant for Fry and the legacy of this *existenzminimum* research is apparent in Fry's social housing projects of the 1930s.[18]

For Fry, Modernism was defined as the study of a set of criteria, 'the inevitable outcome of cumulating forces, social, economic, scientific and aesthetic, inextricably intertwined but capable of separate study'.[19] Fry contended that architecture should dignify its occupants and, by doing so, fulfil 'the human wish to live a neighbourly and an urban existence', which required 'a new architecture'.[20] Addressing social and economic (rather than aesthetic) considerations, Fry followed May and others, such as Walter Gropius, in looking to Taylorist-Fordian methods to help raise standards in working class housing. Wells Coates was also exploring the possibilities of *existenzminimum* at this time, although he concentrated on its use in homes for more upmarket metropolitan living. Indeed, Coates and Pritchard teamed up to create the Lawn Road Flats (1933–34) in Hampstead, a block of serviced flatlets for busy, professional workers.[21]

However, for Fry, the use of standardised components – bathrooms, kitchens, staircases, roof trusses and partition walls – naturally lent itself to the scientific design of a 'minimum workman's house'. Drawing on the precedent of Grete Schütte-Lihotsky's Frankfurt Kitchen, he recognised the importance of factory production and the economies of scale that could be gained with the standardisation of kitchen and bathroom designs, 'where every action is studied, performed in large measure by machinery, and costed to the penny'.[22] Furthermore, Fry recognised that construction costs must be tightly controlled in order for new dwellings to offer affordable rents for working class incomes.[23] Extending this argument, Fry supported the use of reinforced concrete due to the savings it could offer in the rationalisation of construction, with 'the omission of intricate brickflues, and window lintols, continuous foundations',[24] and suggested that any costs of ornamentation could be better spent on items such as refuse disposal systems.[25]

Fry's involvement with the DIA also helped to foster a culture of modern design methods, thus linking mass-produced objects for the home with prefabricated dwellings. Emphasis was placed on the provision of domestic gas and electrical services, as well as on the design of built-in fixtures, to create ultra-efficient interiors. For Fry, this necessitated a collaborative design process to ensure a unified solution. He wrote, 'the pressure of production costs forces the exact study of function ... there must be such collaboration of technician and manufacturer as will produce a much more highly organised unit that will offer a better standard of design, considered from the standard of function and form of that appearance'.[26] These articles demonstrate Fry's rather blunt attempts to 'convert' English architects to a modernist viewpoint. However, he did recognise the existence of a broad spectrum of architectural opinion and his approach at DIA meetings was rather more nuanced.

Fry's committee roles at the DIA and the RIBA enabled him to introduce continental models of standardisation to England, but he advocated these progressive construction techniques and space-saving measures from a traditionalist standpoint. In the early 1930s, Fry had continued to rise through the ranks of the DIA; he was appointed Vice-Chairman in 1933 and became Chairman in the following year.[27] Fry's position offered him opportunities for legitimising *existenzminimum* in an official capacity, which he attempted in a series of DIA lectures in the early 1930s.[28] Despite his calls for a resolutely mechanised approach, Fry was careful to define what he meant by a new architecture, which he argued had 'its roots in the past – of the last 100 years'.[29] This historicisation, later articulated by Pevsner,[30] was largely the result of work by members of the DIA and the CPRE, who traced 'good design' and 'fitness for purpose' back from Modernism to the Arts and Crafts movement.[31] At a DIA meeting in October 1934, for example, Fry's address asked 'Is Mass-Production Possible in Housing?' He carefully reassured his audience, asserting a tradition of mass-production of housing in England, whilst also affirming the good taste of the eighteenth century:

> *You will find, if you look, that from the eighteenth century onward to today we have been building a minimum workman's house that has been altered only in character, as the trim Georgian cottage became grimmer, harder and uglier, with the advancing years of the nineteenth century. ... it has grown only by the addition of a bath. In other respects it has remained fairly true to type.*[32]

Fry and his DIA colleagues' method of reform – couched as it was in tradition – is neatly summarised in *Design in Modern Life*, a volume edited by John Gloag and including contributions from Frank Pick, Gordon Russell and Elizabeth Denby, amongst others – all of whom were prominent DIA members. Fry also contributed two chapters on ' The Design of Dwellings' and on 'Design in the Countryside and the Town', which encapsulate his architectural preoccupations at this time. Eager to signal a new building age, and rejecting his previous work for Kemsley Village, he writes, 'The Garden City movement was the last despairing effort to escape from the new industrial life: from the control of the machine. It is doomed. We are at the moment of complete reorganisation'.[33] Again, Fry looks back to the 'cultivated, eminently practical' building of the eighteenth century for guidance in standardisation. He suggests a proportion of one to eight for 'working space' to 'living space' (reduced from one to four), compressing the kitchen, bathroom and bedrooms to give a spacious living area 'served by controlled heat, made of beautiful but easily cleaned materials, thrown open to air in fine weather – a flexible, airy dwelling'.[34]

His first attempt at such a dwelling was his entry for 'Concrete House', a competition sponsored by the Cement Marketing Company in 1933. The competition included various categories including high-density terraced houses, so Fry's decision to enter the detached villa category is surprising. The judges thought his design 'a little expensive to build' (it was supposed to be built for under £1000) and that 'the door to the garden seems at present crowded';[35] nevertheless he was awarded second place, with 15 Guineas prize money. It was a hesitant solution, built in rendered blockwork, without the benefits of building in concrete that Fry had promulgated in his writings, illustrating the pragmatism of his ideology.

Yet an opportunity to test out his research came shortly afterwards, with a social housing project in Peckham for the R.E. Sassoon House (1933–34). Fry's friend, Elizabeth Denby, whom he described as 'a specialist sociologist with expertise in housing', recruited him to collaborate on the project and together they produced a small block of 20 flats.[36] Fry's structural and technical experiments joined with Denby's knowledge of working class life to produce a scheme without parallel in Britain at the time. Fry later recalled having met Denby at a party held in Henry Moore's studio, where their common interest in housing reform had quickly become apparent.[37] Contemporary texts by both Fry and Denby discuss blocks of multiple occupancy housing as a catalyst for social reform. In neither case do the formal aspects of architecture form the focus of their publications; rather it is the architectural response to climate, light, access and facilities, resulting in a building 'fit for purpose', that shape the basis of their work.

Denby had worked for a decade in the slums of north Kensington, as an employee of the Kensington Council of Social Service.[38] In the late 'thirties she conducted research throughout Europe, carefully examining housing provision in France, Germany, Italy, and Stockholm and Vienna. Published as *Europe Rehoused* (1938), her research focuses on efficient kitchen design, quality housing for the working classes and the importance of exterior spaces. She was critical of British cities 'for their excessive size, their wasted land, their lack of civic dignity and beauty and of opportunities for enjoyment'. She believed that, 'Beauty with us is too often sacrificed to utilitarian ends and financial gain'.[39] Denby made unfavourable comparisons of the population densities of 'so called overcrowded' British towns with their European counterparts, citing Manchester at 28 persons per acre and Birmingham at just 20 persons. Ultimately Denby believed that poor planning, inefficient road structures and low housing densities must be addressed to improve the lives of the working classes.

Fry's planning work mirrored Denby's and the R.E. Sassoon House enabled them together to design an English prototype for a modern way of living. The project's client was an acquaintance of Denby, Mozelle Sassoon, a wealthy widow who wished to build a block of workers' flats.[40] The project was intended to commemorate her son, Reginald, who had died in a steeplechasing accident in January 1933.[41] The five-storey block of flats comprises 11 two-bedroom and 9 three-bedroom flats, with the scheme costing a total of £7,850 to construct. The building uses a reinforced concrete portal frame with beams cantilevered out over a central span of 16 feet.[42] The south side of Sassoon House is symmetrical, with projecting balconies forming neat bookends to the façade (Figure 2.1). Each flat has a private balcony with a built-in flower box and a trellis for growing plants, and with access to the front doors provided via an access balcony. To the north side, the stair tower projects beyond the parapet, generating a vertical element to the otherwise horizontal composition (Figure 2.2). Externally, the building contributes to the area's 'civic dignity'. Concrete walls were originally finished in 'light tones of yellow, grey and cinnamon' concrete paint,[43] and a mural by Hans Feibusch

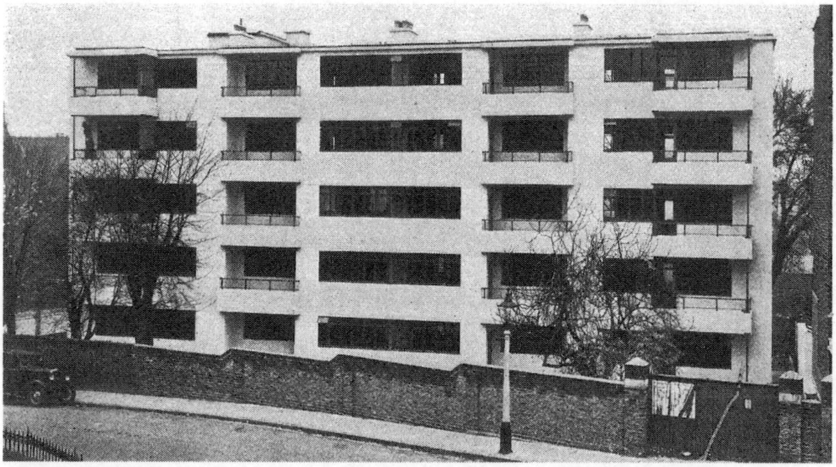

2.1 South façade of Sassoon House, Peckham, London, 1934

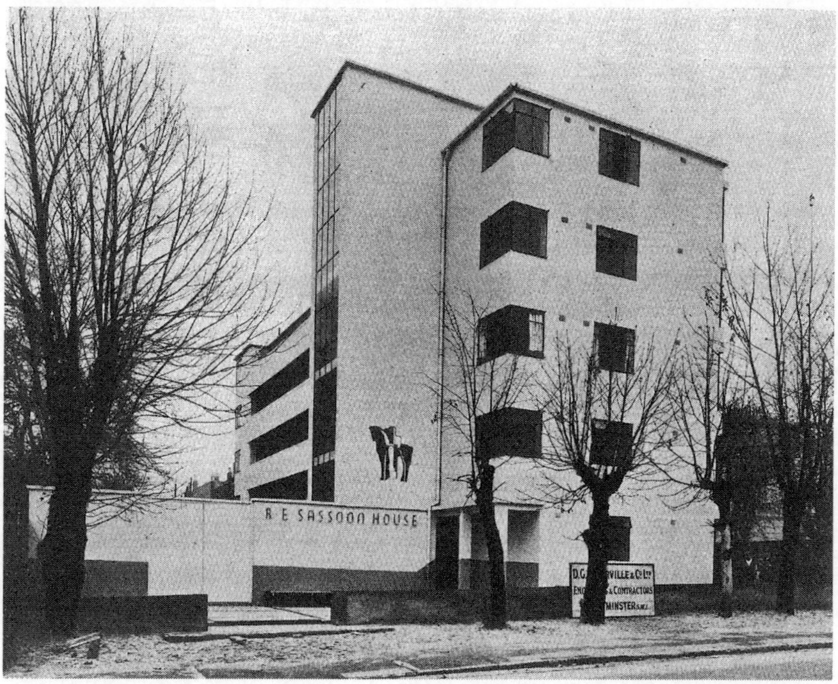

2.2 North façade of Sassoon House, Peckham, London, 1934

(a German émigré, who became a great friend of Fry), depicting a horseman in red, black and white Vitrolite, signals the main entrance. The building is carefully integrated into its surroundings with hard landscaping and, where possible, the retention of existing trees. At night, the illuminated slit window of the stair tower provides a local landmark for the community.

The design drew heavily on the German precedent by Ernst May and Walter Gropius admired by Fry. In both composition and arrangement Sassoon House is reminiscent of the Berlin Siemensstadt housing (1929), designed by Gropius, again illustrating the significance of Fry's DIA membership.[44] The 'very simple' structure, using an alternating grid of 13 feet and 11 feet 3 inches, frees up the maximum available area to provide a generous living space, with an *existenzminimum* kitchen and bathroom on the north side (Figure 2.3).[45] External walls are constructed of four inch, non-loadbearing reinforced concrete shuttered against one inch, cork insulation, with internal breezeblock partitions. For easy cleaning, each flat was originally fitted with linoleum floors and cement skirting throughout. Fry intended the carefully planned kitchen to be 'a machine-room, a room of kindly, helpful machines, designed to simplify and make work enjoyable'.[46] The bathroom, too, is of minimum dimensions, the width determined by the length of a half-size 'sitz' bath.

2.3 Part Plan of Sassoon House, Peckham, London, 1934

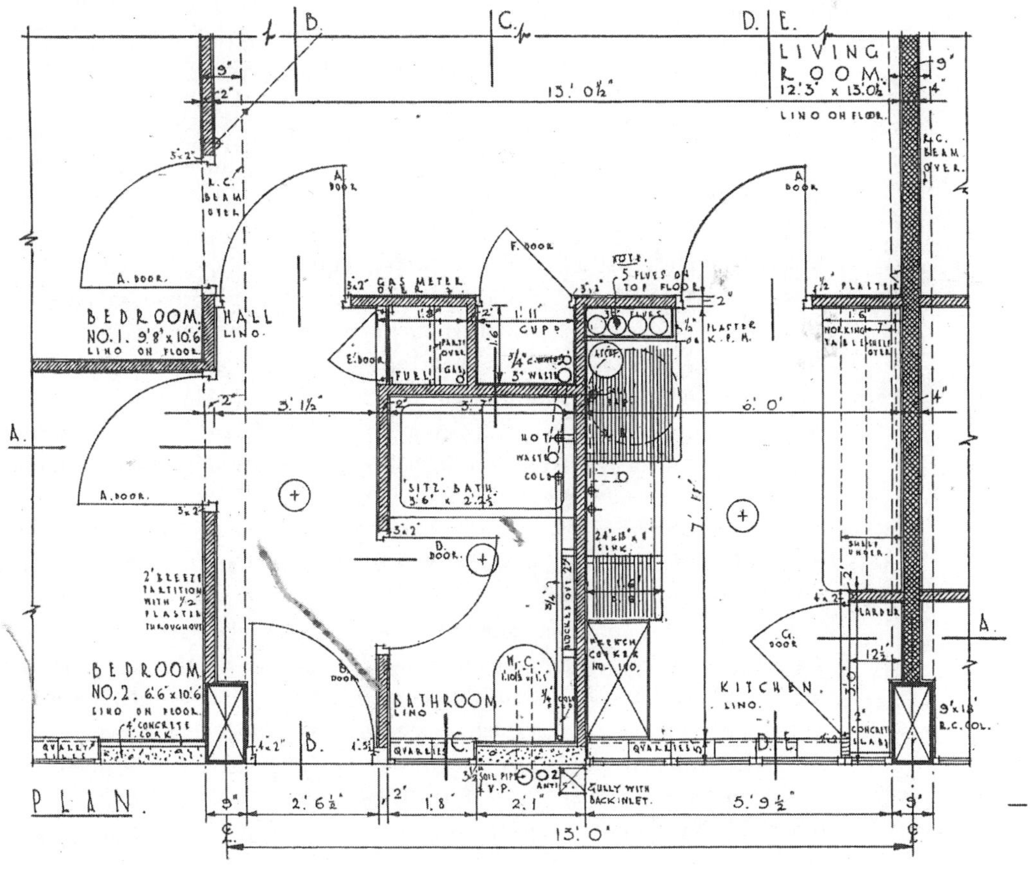

Fry's evident ability as a designer was observed by Anthony Jackson, who noted, 'His work during the few years following his adoption of the modern architectural style has an aesthetic maturity that was rare … [S]olids and voids are disposed as a surface background to a sculptural play of rhapsodic elements. Shadowed in sunlight, his buildings are beautiful to look at'.[47] Contemporary photographs are certainly seductive, yet there was evidently room for improvement upon Fry and Denby's first project. The desired efficiency of the flats left little room for everyday life. The main staircase leading onto open deck access has a meanness of space, despite being adequately illuminated by the opaque glazing; equally, the entrance doors to each flat are of the smallest dimensions possible and, in general, features are trimmed to an almost meagre standard. The careful planning reduced kitchens to such a size that they could not contain an oven, which, instead, was located above a slow-burning coke stove in the living room. As a result, some cooking had to take place away from the kitchen area, something that Denby's subsequent studies revealed was much disliked. As the *AJ* noted, 'The planning of the flats is a matter of rigorous standardisation, a working unit being thought out in great detail, and then repeated throughout the building'.[48] This economy was evidently a step too far for Fry, and in his subsequent work he suggested that all staircases should be heated and enclosed, that balcony access must be avoided where possible, and that private balconies should be large enough to accommodate a table and chairs. Fry's inexperience in designing modernist buildings also became apparent as the ungalvanised steel windows, in poorly detailed openings, suffered from weathering.[49] Nevertheless, Sassoon House opened in November 1934 to considerable acclaim. The project was evidently all-consuming for the designers, as Fry later wrote, 'Elizabeth and I … fell in love with each other as the ideas drew us together'.[50]

Fry and Denby's personal and professional relationship led to another collaboration at Kensal House (1933–37) in Kensington, West London, a development that enabled them to refine their ideas at a larger scale. Commissioned in November 1933, the flats were designed to illustrate the benefits and conveniences that gas could provide for low-income families. The commission was a result of what Fry later described as his 'respectable life with the Establishment',[51] which led to a position within a team of architects designing buildings for the Gas, Light and Coke Company so as to deliver its ambition for gas-fuelled municipal housing throughout London. These architects each designed a scheme for Kensal House and then decided amongst themselves which scheme should be taken forward; Fry's design emerged as the favourite and a collaborative development stage followed.[52] Fry acted as the Executant Architect and Elizabeth Denby was appointed as a Specialist Consultant, with a Consulting Committee of Robert Atkinson, C.H. James, Michael Tapper and G. Grey Wornum. The project was carried out with the co-operation of the Ministry of Health and Kensington Borough Council, who selected tenants for re-housing from the borough's slum clearance areas.

The Governor of the Gas, Light and Coke Company, Sir David Milne-Watson, described the scheme as 'a real urban village with tenants' clubs and nursery school'.[53] This emphasis on tenants' social needs marks Kensal House as a highpoint in the realisation of Fry's interwar modernist agenda and shows the significance of Elizabeth Denby's focus on social reform; Denby's previous employment by Kensal Housing Trust had given her a real understanding of local housing conditions and of the day-to-day issues for the Kensal Rise residents.[54] In a promotional film made by the gas company, Fry noted that 'these flats … were to show that gas can bring to the tenants of such schemes comforts and conveniences that might be thought to be out of their reach'.[55] Denby saw the project as an 'experiment in rehousing families from slum areas', with the development committee ensuring that tenants' outgoings on fuel and light were as reasonable as the weekly rent.[56] The resultant scheme fulfilled Fry and Denby's criteria, combining economic, scientific, aesthetic and, most importantly, social concerns. As John Summerson later commented, 'Kensal House is a humane and cheerful building'.[57]

Situated on a one-and-a-half acre site, on the corner of the old Kensal Green Gas Works in North Kensington, the two housing blocks run from north to south. The 68 dwellings originally housed 380 residents, including 244 children, in two- and three-bedroom flats (Figures 2.4 and 2.5). The semi-circular form of the nursery follows the perimeter of the gas-holder that had previously occupied the site. Dwellings are accessed via internal staircases, which 'largely governed' the plan layout; as Fry explained, 'With internal staircases there is greater privacy, a nice feeling of going up your own staircase, the staircase is undercover, has its own front door and is more civilised for these things'.[58] The bedrooms face east, to take advantage of the morning sun, while the living quarters face west, leading onto balconies with built-in flowerboxes and space to sit out and enjoy the sun. Separate drying balconies are accessed from the kitchen. The kitchens are small and, evidently derived from Fry and Denby's studies of Frankfurt kitchens, intended to be 'really workable without being cramped' (Figure 2.6).[59] They were also equipped with the latest gas-fuelled paraphernalia – an Ascot Water Heater, a gas iron and a copper providing constant hot water – to reduce housewives' drudgery. The largest room was the living room, complete with an enamelled, coke-burning fire to replace the traditional coal ranges that were widely disliked due to the dirt they generated and constant attention they required. The smokeless fuel also ensured a cleaner, healthier environment for the tenants. The coke-burning fires were intended – perhaps slightly optimistically – to spread 'heat from the living-room to the adjoining bedroom and the hall'.[60]

To encourage social interaction between the tenants of the block and the wider community, an existing local club, 'The First Feathers', was invited to occupy part of the new space. This club offered practical classes in dressmaking, carpentry and shoe repairing, alongside social activities such as darts and whist drives; a youth club was also established there. In the nursery one third of the children were Kensal House tenants, with others coming from the surrounding neighbourhoods. Denby described the transformation that took place in the nursery children, rather dramatically, but probably without too much exaggeration, 'It is like a conjuring trick to see how infants entering at two years of age with the expression of men who have been through Borstal and Wormwood Scrubs, are in a couple of months transformed into carefree happy babies'.[61] The nursery cost 1s. 6d. per week and

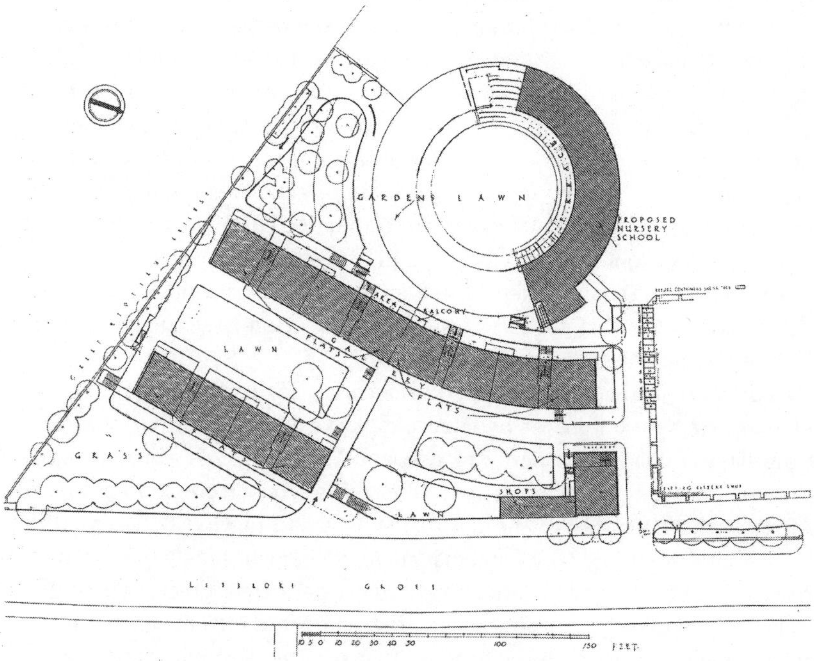

2.4 Site Plan, Kensal House, Kensington, London, 1937

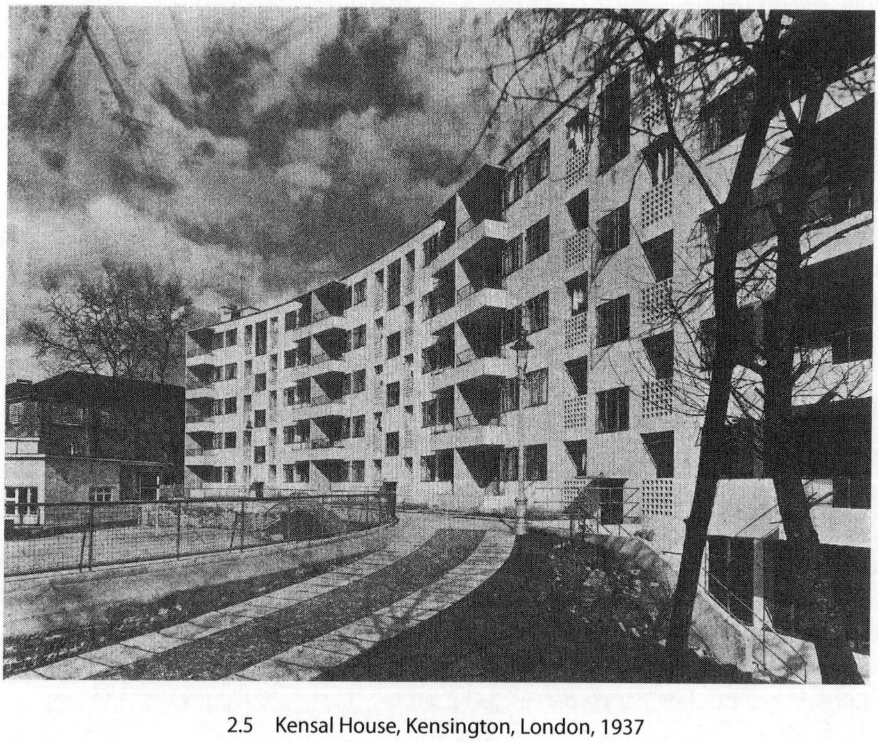

2.5 Kensal House, Kensington, London, 1937

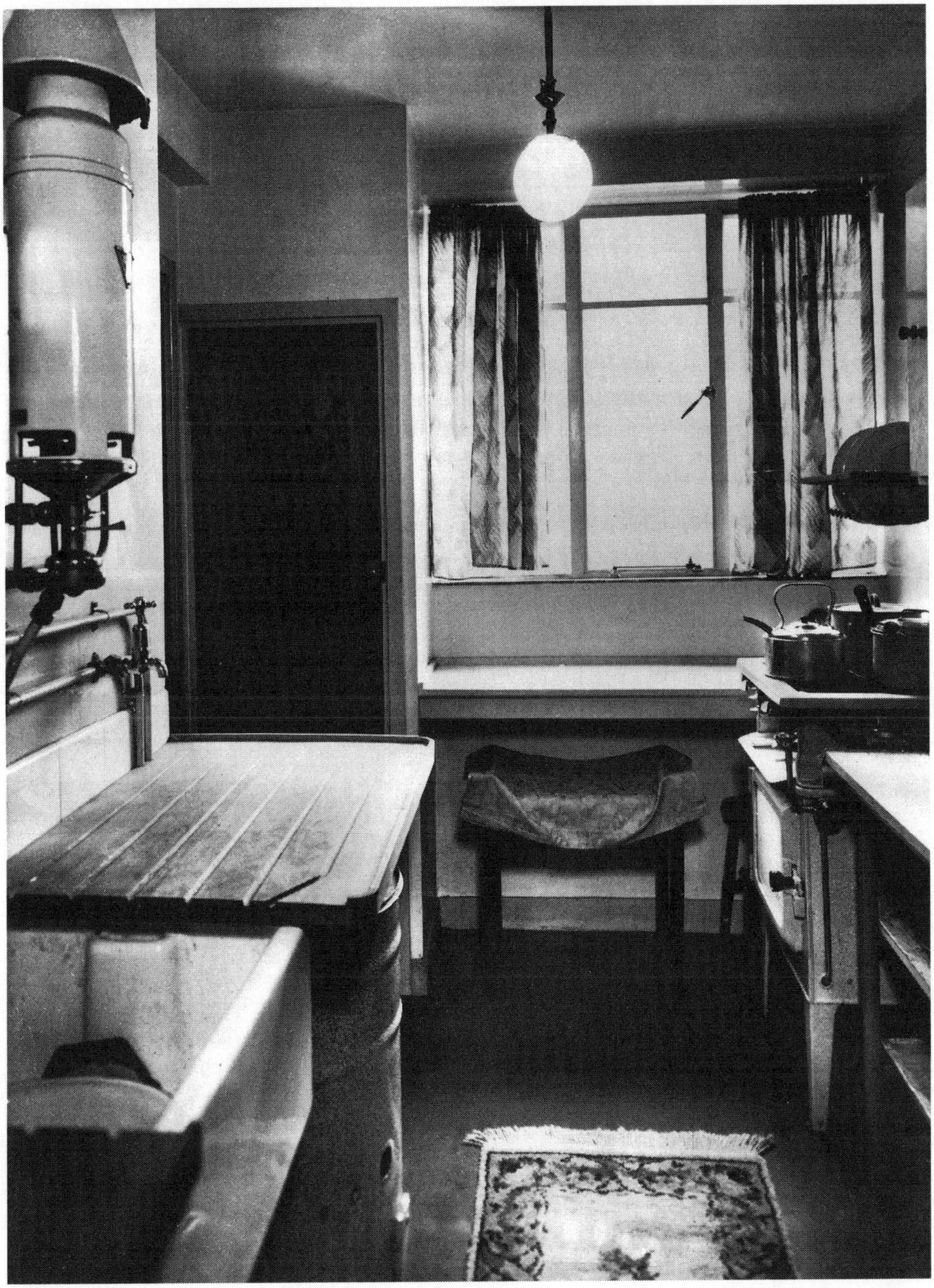

2.6 Kitchen at Kensal House, Kensington, London, 1937

children were cared for from around 8.30am to 4 or 5pm. The nursery looked to Red Vienna for inspiration in the latest methods in childcare, with emphasis placed on Froebel and Montessori principles that promoted a child's independent growth and self-directed learning.

Tenants were re-housed in the modern flats from late in 1936 and Kensal House was officially opened by the Minister for Health, Sir Kingsley Wood, on 15 March 1937.[62] Fry wrote to Gropius of the warm reception received by the development, 'They really do seem to have got hold of people's imagination'.[63] Indeed, Kensal House became an archetype for British Modernism. Appearing on the front cover of J.M. Richards's widely read *An Introduction to Modern Architecture*, Kensal House showed Britain to be capable of producing socially responsible modernist buildings, equal to that of its continental counterparts.[64] As Fry's former employee, John Cordwell, later commented, 'If you read the Penguin book by J.M. Richards on modern architecture, the building on the cover is by Maxwell Fry. Max was known as the leader of the modern movement in England. I mean, that's why I went to his office'.[65]

Fry and Denby's relationship faltered following the completion of Kensal House, as Fry failed to publically acknowledge Denby's contribution to the project.[66] Fry's subsequent work moved away from the overtly social agenda of Sassoon House and Kensal House associated with Denby. Instead, a series of one-off houses followed, despite Fry's distaste for uninformed private clients. He made his feelings apparent in a parody of a property-developer client, who, in setting out his desires for a new house, apparently requested, 'first the garridge. Rolls for me and a runabout for Miss Margaret (his pimply daughter), shuffer above. The rest of the 'ouse to scale and don't stint it Mr Fry. The best is good enough for Jerry Brown'.[67] Fry's caricature is typical of the 'mixture of condescension, hyperbole and irony' used by architecture and design reformers of the 1930s to distinguish themselves from an aspiring nouveau riche that could not distinguish 'good taste' from bad, but had the means to buy it.[68] This stereotyping displays a tendency amongst taste-makers, as Stephen Hayward notes, 'to appropriate the cultural high ground, to draw on a quasi-eighteenth-century notion of "good taste" in promoting the cause of good design in the face of what the *AR* called "genteelisms"'.[69] As we have seen, this notion of 'good design' came to the fore during the 1930s, with the Modern Movement dressed up by reformers in an act of historicising legitimation. It was a polemic that Fry and his partner Gropius would frequently come to employ in their own work of the period.

The issue of good design, and its association with good and bad taste, was also brought to the fore by the widespread difficulties in obtaining planning permission for modernist buildings. As Charles Reilly noted, a successful struggle to gain the necessary consents was fast 'becoming a sign of merit' amongst the architectural community,[70] viewed as a triumph of good taste and common sense over ignorant philistinism. Fry's design for Little Winch (1934–36), a house at Chipperfield Common in Hertfordshire, was typical of the situation during the mid-'thirties. Subject to a planning appeal, the scheme was initially rejected by the local authority due to its modernist aesthetic. The *AJ* reported crossly, 'The Council had no fault to find with its structure, sanitation or any of the aspects covered by their bye-laws, but insisted that a pitched roof was absolutely necessary before they would pass it'.[71] Fry attempted to circumvent the authorities by proposing five different locations for the building, taking the view that a different site would result in a different planning decision.[72] It did not.

Various councillors gave their opinion on the 'Chipperfield Case', as it became known in the architectural press; most felt that the scheme would be improved with the substitution of a tiled pitched roof for the proposed flat roof, which gives an idea of the mainstream attitude to Modern architecture. The entire debate was reported in the *AJ* and suggestions for suitable modifications ranged from tiled roofs to one council member's 'battlements or buttments'. Another member was concerned that the client may use the flat roof as a 'naked sun parlour' and cautioned that there were 'several about'.[73] Such comments fuelled the reaction of modernists and, in response to the council's arguments, the *AJ* again sought to legitimise Modernism with historical precedent. The magazine printed photographs of houses from the immediate area that, like Fry's proposal, employed a strong horizontal emphasis to their roof-lines, most of these being in the Georgian tradition.

However, following rejection by Watford Urban District Council, Fry was obliged to redesign the reinforced concrete house as a timber-frame, clad in oak weatherboarding and local bricks (Figure 2.7). The original plan remained intact, focusing on the impressive open-plan dining and living space. An impressive 45-foot-long window wraps around the south and east corner of the living room, and two of the square windows slide open to connect the interior with the garden. The opening is formed in reinforced concrete and expressed as a projecting frame, with the deep internal sill lined with polished slate.[74] At first floor level, there are

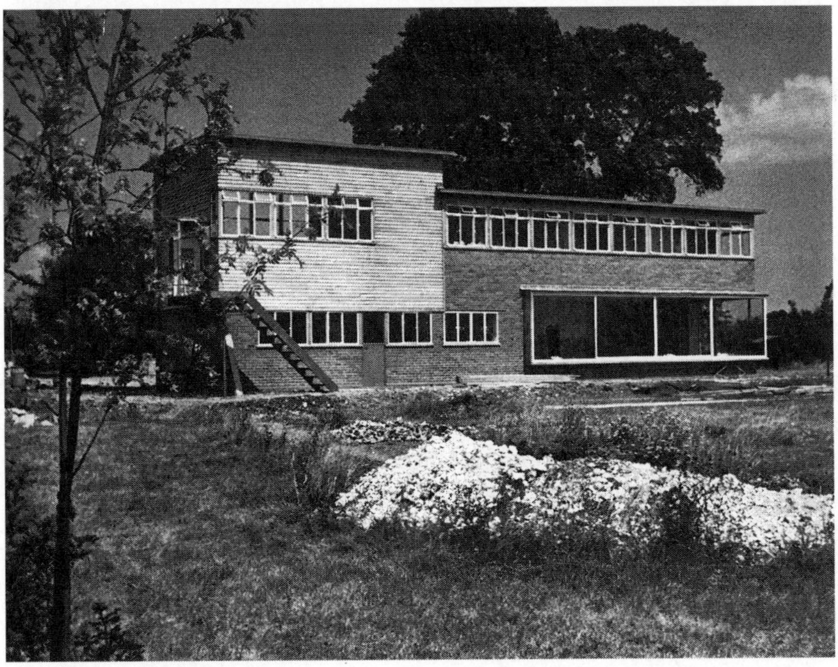

2.7 Garden Façade, Little Winch, Chipperfield, Hertfordshire, 1936

three bedrooms plus a maid's room, a bathroom and a studio for the artist owner. For flexibility, the studio is accessible internally from the first floor or externally, via stairs leading to the garden. The first-floor metal windows give the impression of a continuous ribbon window, a device that Fry would use on later domestic projects. The studio roofline projects beyond the main roofline, offering some variation and forming part of a larger composition, diagonally set against the large windows at ground floor. Little Winch demonstrates that by the mid-'thirties Fry had begun to employ natural materials, such as timber and brick, in conjunction with 'modern' materials; a palette that he refined throughout the interwar period.[75]

'[H]E'S CORNUCOPIUS, GROPIUS'[76]

The arrival of the Bauhaus master in England in November 1934 was celebrated by many of the modernist community and the architectural press wrote eagerly of his impending arrival. The *AJ* (perhaps Fry himself), for example, wrote in hyperbolic fashion, 'Gropius is more than an architect. He is one of the few architectural figures who become prime movers and educationalists through an overflowing of the spirit of creation. He is in this sense a prophet, though unlike Le Corbusier he speaks through his works'.[77] The arrival of 'Pius' the Prophet allowed English modernists,[78] especially Fry as his new partner, to aim for a greater impetus for the movement. A few months before their collaboration officially began, Fry wrote to Gropius that he felt 'honoured and glad' to work with him.[79] Following in the 'historical sensibility' established by the DIA, Fry and Gropius frequently stressed the continuity of architectural tradition in their work and highlighted the similarities present in English and German architecture. The partnership of Fry and Gropius was characterised by that of quiet revolution, rather than of the sweeping reform anticipated by the architectural press. At a time when commissions for modernist buildings were still thin on the ground, this was a sensible strategy securing work with a non-threatening approach to Modernism. It was also a means of ensuring the Gropius's long-term future in England, given their status as temporary residents.

Walter and Ise Gropius were readily absorbed into Fry's network of friends and acquaintances. Gropius soon received a series of invitations to attend dinners, give lectures and write papers. He became part of the DIA network and, at the DIA Annual Dinner in December 1934, he and Ise were invited to share the table of the Chairman, Frank Pick, alongside Maxwell and Ethel Fry, Jack and Molly Pritchard, Amabel Williams-Ellis, Herbert Read and Anthony Bertram.[80] Indeed, in being so welcomed into the DIA fold, the influence of the group on the German architect's architectural approach in England is evident throughout his writings during this period. This was in notable contrast to his more impartial role in the MARS Group, for – although he attended meetings – Gropius chose to remain outside of MARS, despite frequent requests to take an active role and even head the group.

Fry and Gropius were introduced to each other during one of the German architect's early visits to England, probably in May of 1934.[81] The RIBA had organised an exhibition of Gropius's work, which subsequently toured the country, stopping at Manchester, Liverpool, Leeds and Birmingham.[82] Gropius visited London in connection with the exhibition, staying at Fry's Hammersmith home, and gave his first lecture in England – and in English.[83] Fry later recalled this lecture, organised by the DIA, which he chaired at the School of Hygiene and Tropical Medicine, 'I can remember exactly the overcrowded room and he standing among us, speaking with the utmost clarity in broken English of how we could mend the disunity of our machine civilisation, and what moved us was the mixture of humility and authority with which he addressed us'.[84] Fry's account illustrates its significance for the architectural milieu at the time, and marked Fry's transition to an architect of international standing.

Walter and Ise Gropius's journey to England, via Italy, and the work of Jack Pritchard, P. Morton Shand and Maxwell Fry in securing work and accommodation for the German couple is well-known, established in modernist folklore by the writings of Pritchard and Fry.[85] In November 1934 Gropius joined the firm of Adams, Thompson and Fry. This was not a partnership;[86] rather Gropius worked autonomously, although officially under the employ of the firm, as all émigré architects were required to work with a British counterpart, who would act as a guarantor.[87] Their buildings were therefore publicized in the contemporary press as the designs of 'co-architects' Fry and Gropius, although they have since been categorised as designs by Fry or by Gropius. Indeed, the duo generally worked as lead designers on separate projects, coming together to pool skills and ideas. Yet Fry later described his partnership with Gropius as a 'perfect collaboration',[88] and project correspondence reinforces the collaborative nature of their work. Indeed, William Jordy notes, 'Gropius had made collaboration a central tenet of his architectural philosophy, in the sense that he not only believed in the "team" as a method for success in design and in teaching, but welcomed the competition of points of view'.[89] Fry and Gropius's work is, therefore, considered here as a joint effort, taking into account the full building process from design to construction; for example, although Gropius is generally credited with the design of Impington College, Fry was obliged to undertake a redesign following Gropius's departure to America in order to substantially reduce costs, and therefore rightly considered a co-designer.

PRESERVING 'ENGLAND'S GREEN AND PLEASANT LAND'[90]

The partnership's first commission came from Jack Pritchard, a project that he created to ensure Gropius had sufficient work to allow him to travel to England. The scheme for Isokon 2 in Manchester was to be a sequel to the Isokon flats situated at Lawn Road, in the artistic community of Hampstead. Pritchard's long term objective was to establish Isokon buildings throughout the country, providing serviced accommodation for busy professionals.[91] Walter and Ise Gropius experienced this vision firsthand as they spent their three years in England occupying a Lawn Road flat as guests of Jack and Molly Pritchard. The Isokon 2 project intended to address the dearth of flats in Manchester, providing mainly two- or three-bedroom flats for families, again in a professional milieu. The development was to cater for employees of the nearby university, the grammar school and 'large commercial undertakings'.[92]

Again highlighting the significance of the DIA network, the President of the association's Manchester branch, A.P. Simon, had offered the land to Pritchard. A Liberal MP and later the author of *Manchester Made Over*, Simon was keen to promote the DIA's ideals of town planning in the city, by utilising underdeveloped land to provide modern housing. The mature site, Lyndale, at Didsbury featured a detached property at high level with sloping grounds leading down to the river and a concentrated modern development was envisioned, in this case for around 50 dwellings, leaving the remainder of the site untouched and available for recreational purposes.

Although plans for Isokon 2 were shelved, a similar project for Isokon 3 (1934–35) at St Leonard's Hill, near the historic town of Windsor, quickly gained impetus. Illustrating Fry and Gropius's approach of careful innovation, the scheme was promoted as a synthesis of continental experience and an English outlook. Pritchard wrote, 'The combination of Gropius and Fry should be important – Fry's firm being town planning experts – and Fry's own very English point of view combined with Gropius' experience should produce a fine scheme'.[93] The proposal aimed, as Pritchard put it, 'to make a profit from building in the country without spoiling the countryside'.[94] Pritchard's statement shows the influence of the recently established CPRE's calls to curb suburban sprawl. To this end, Isokon 3 was to be built on a 'secluded and historic site' of existing parkland on the outskirts of Windsor.[95] Presented as the answer to speculative and unregulated development, the *AR* publicised the scheme in an alarmist article entitled 'Cry Stop to Havoc', described as 'preservation by development'.[96] The Isokon 3 flats are shown to retain 32 of the 33 acres of existing parkland as open space, with views across Windsor Forest to the south and Windsor Castle to the north-east, 'This view belongs to everybody', commented the *AR*.[97] The company prospectus for Isokon 3 noted, 'The terraces of the former mansion, plantations, and trees will be retained, as it is a fundamental part of the Company's policy to preserve the beauty of the landscape'.[98] The beautiful ruins of the country house and the established gardens (including some particularly photogenic camellia bushes) were well documented by Fry and Gropius, and featured heavily in the promotional literature alongside idyllic rural images of the River Thames and Windsor Park (Figure 2.8).

Fry did not write overtly of Englishness or nationhood, but he did possess a fundamental connection to the English landscape. During his wartime posting to Accra, he wrote of his desire to return home and spend time to 're-absorb the atmosphere of the Ridgeway in Wiltshire, see the lichens on the oaks in Aberglaslyn and sniff the long horizons at Blakeney'.[99] From his earliest student writings regarding the health-giving benefits of Prince's Park in Liverpool to his family's walking holidays in Scotland, Fry enjoyed nature in contrast to the civic dignity of large towns (Figure 2.9). Furthermore, Fry's background in town planning gave him a moralising view of preservation; like his good friend Clough Williams-Ellis, he believed that '"Development" even of a place of great natural beauty ... should be an enhancement and not a desecration'.[100]

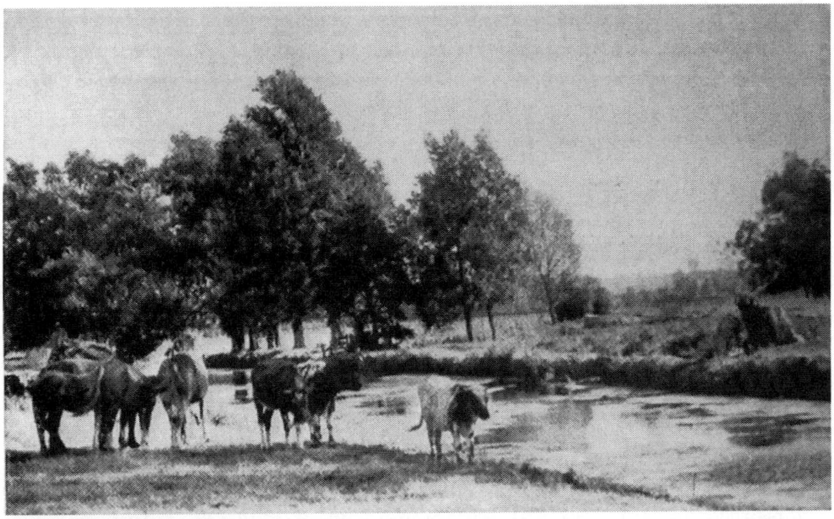

2.8 Existing Gardens, St Leonard's Hill, Windsor, 1934

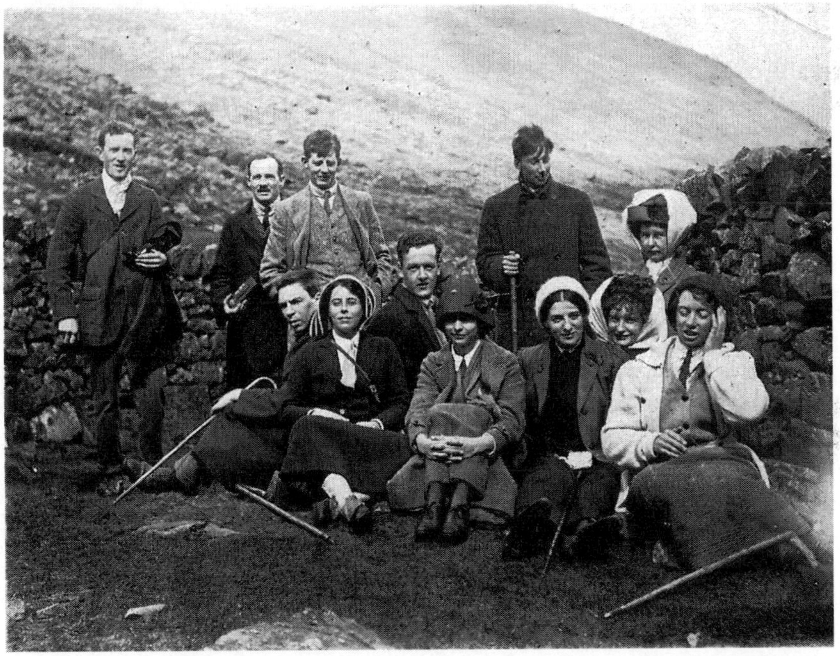

2.9 Maxwell and Ethel Fry walking with family in Scotland, c. 1927

The English landscape also held sway over Gropius. At this time, Gropius was working on several projects for the Dartington Hall estate in Devon, including the alteration of the theatre and advising the resident architect Robert Hening on the design of workers' cottages for the estate's Yelland Farm. Dartington's verdant grounds were much admired by Gropius as he described to his daughter, Manon, during his first visit in 1933, '[E]ngland was a pleasant episode ... a vast park with trees as I have never seen ... the country is very beautiful and green like no other'.[101] The synthesis of this ancient English landscape with Leonard Elmhirst's innovative work in agriculture and education, and new buildings by William Lescaze and Oswald Milne, provided a model of repurposing country estates for twentieth-century living whilst retaining the open countryside. Isokon 3 followed Dartington's example and, although unbuilt, demonstrates the strong ties between the interwar preservationist movement and Modern architecture.

The prestigious scheme comprises 110 flats, split over two parallel slab blocks running north-east to south-west. A block of eight-and-a-half storeys and a smaller seven-and-a-half storey block were to be linked at ground floor by a range of community rooms (Figure 2.10). The reinforced concrete frame construction, with bays of 13 feet 3 inches, was serviced by staircases at every fifth or seventh bay. A range of 11 flat types was planned in various combinations, so that flats could be supplied according to demand without alteration to the building façades. Each flat was intended to cater for different lifestyles, but all were to be finished to a high standard with built-in furniture, central-heating and 'work-saving amenities', such as an integrated vacuum cleaning system. In the larger 'II' block, the smallest 'one room' flat was to comprise a bedroom recess with en-suite bathroom, living spaces and small kitchen; whilst a large 'four room' flat was to include two generous living spaces, a balcony, two bedrooms and no kitchen, with occupants instead able to avail themselves of the full hotel service.[102] The development was to provide extensive recreational facilities, including a restaurant, ballroom, barber's shop, Turkish bath, reading room, swimming pool and tennis courts to cater for an affluent clientele 'who want some approach to a country gentleman's life near London … who are tired of expensive and squalid living in fashionable parts of the West End', as the *AR* reported.[103] As Gropius wrote in an initial description for the project, the flats were designed for 'a more pretentious class of people' (this was subsequently altered to 'a more exacting class') and was certainly far removed from the social housing on which Fry and Gropius had forged their respective reputations.[104]

Fry was later quoted as saying, 'The scheme was nearly all Gropius'.[105] However, his planning expertise was called upon in a study showing the contrast between the proposed Isokon scheme and any speculative building development on the site, to show the 'devastation' that would otherwise be caused.[106] Indeed, project correspondence shows that Fry, plus their employees Hazen Sise and Albrecht Proskauer, contributed to the scheme.[107] Albrecht Proskauer, a German architect, had previously worked in the office of Wells Coates for six months and joined Gropius in November 1934 at Coates's suggestion.[108] The Jewish-Canadian architect, Sise, had moved to London in 1933 and, he later recalled, 'almost immediately got a job with Max Fry'; he became a close ally of Fry during the mid-1930s.[109] Thomas Adams was also involved in the Isokon project, providing expert advice on the potentially contentious planning issues and, in February 1935, Adams, Fry, Gropius and Pritchard all successfully defended the scheme at a meeting at the Home Office, also attended by representatives from Windsor Borough Council, Berkshire's Planning Office and Windsor Castle.[110]

Isokon 3 was formed as a limited company on 1 July 1935, with Jack Pritchard, PEP member Frederick Graham Maw, Jack's brother, Fleetwood Craven Pritchard,

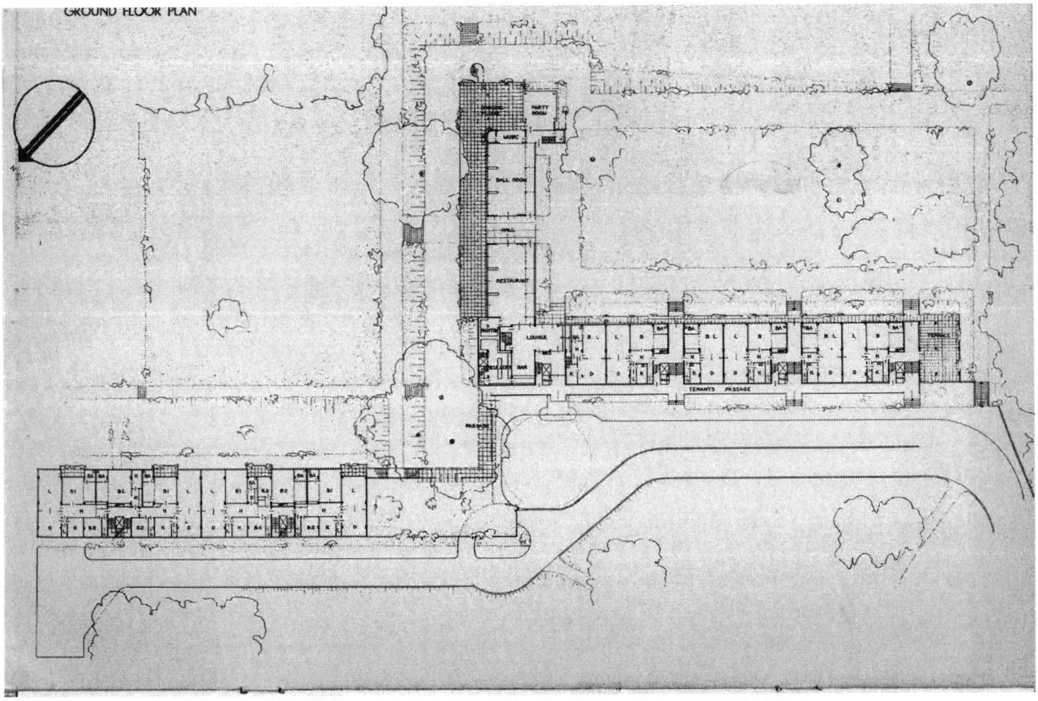

2.10 Ground Floor Plan, St Leonard's Hill, Windsor, 1934

and Fry's former professor, Charles Reilly, appointed as directors.[111] In a bid to secure funding for the first phase of works, the scheme was well-publicised in June and July 1935, which Reilly helped to arrange, including a luncheon at the Café Royal on London's Regent Street to launch the project.[112] But, by the end of July, the Isokon 3 project had been indefinitely postponed due to a lack of investment.[113] The ambitious scheme (as well as proposals for an Isokon development in Birmingham) was not built and with it a viable modernist alternative to the Garden Suburb was deferred, at least until the post-war rebuilding programme commenced.

THE NEW ARCHITECTURE AND A NEW OFFICE

In 1936 Fry and Gropius made an amicable split from Adams and Thompson to form their own practice. In July of the same year, Gropius applied to the Aliens Department for permanent residency in England.[114] They moved just a short distance to a new office at 171 Victoria Street, but it was a bold statement by Fry and Gropius illustrating the optimism they felt at this time. Situated close to some of Fry's favoured haunts, such as the Tate Gallery and the PEP Club, the practice became a centre for nurturing young modernists in Britain. Indeed, it was the practice of choice for many recent graduates, and employees included Jack Howe,[115] Eric Lyons,[116] Edward Mills, Ralph Tubbs, Arthur Baldwinson,[117] Bronek Katz, Reginald Vaughan, Arthur Ling, Freddie Charles, and Hazen Sise.[118] Arthur Korn also worked out of the Victoria Street office and was later described by Fry as 'half on, half off' the staff roster.[119] The formative influence of the vibrant offices is apparent in a comparison of the post-war work of some of the young architects. As Alistair Fair points out, the Indian Students' Hostel in London (1952) by Ralph Tubbs and Arthur Ling's Belgrade Theatre in Coventry (1948–58) both utilise a brick volume with a projecting bay window, raised on pilotis.[120] This expression of the concrete frame bay window was a device regularly used by Fry, suggesting the development of an office aesthetic during the late 1930s that continued post-war.

Many of the young employees came to the practice through recommendations from Charles Reilly of promising Liverpool students or through Fry's involvement with the Regent Street Polytechnic.[121] Fry had been a lecturer at the polytechnic since 1932 and possibly secured the post due to his acquaintance with Frederick Towndrow, an *AJ* colleague who was also amongst the school's teaching staff. In 1932, as the polytechnic's magazine reported, the architectural staff had been 'recast and augmented' by new members to keep 'pace with the most recent developments in architectural education in Europe'.[122] Fry and other modernists, including Serge Chermayeff, were appointed to help steer this change in direction. The post enabled Fry to keep up to date with new ideas in architecture and education, as did his involvement in the MARS Group and CIAM, ensuring the Fry and Gropius office was well connected to continental developments.

At this time Fry began to take a more active role in CIAM. In 1936, Fry attended the congress of CIAM's executive committee, *Le Comité International pour la Résolution des Problèmes de l'Architecture Contemporaine* (CIRPAC), at La Sarraz in Switzerland. Alongside William Tatton Brown, Fry represented the MARS Group as an official delegate,[123] and the meeting inspired him to promote new ventures in Britain. This new-found energy and altered perspective of Britain from that of his continental colleagues is evident in a letter to Leslie Martin and Sadie Speight. Written from La Sarraz, Fry observed, 'above all we feel the sense of danger that surrounds this struggle to carry the clear ideas of urbanism through the difficult years that lie ahead of us all'.[124] Fry was convinced that 'England' could contribute to European developments as it was 'in the throes of self-analysis, with complicated problems such as the rehabilitation of South Wales and the South-east region to consider with vast new extensions in the South growing before our eyes'.[125] Here Fry found a framework for his work in town planning; he saw how British issues, tied to social and economic problems, and might be addressed in a modern manner. Thus, full of enthusiasm, Fry asked Leslie Martin if he was willing to form a small working cell of MARS in Hull. Fry also wrote of his intentions to approach William Holford in Liverpool with a similar proposal. It seems these bold plans came to little, but Fry's vigour suggests a renewed impetus for MARS Group activities during the late 'thirties; this enthusiasm became manifest in the exhibition of 1938.

Despite this desire for reform, Fry and Gropius constantly contextualized their work within historical progress and innovation. This is exemplified by Gropius's *The New Architecture and the Bauhaus*, published in 1935. As Louise Campbell notes, the book stresses the 'continuity not rupture' of the New Architecture.[126] Gropius wrote, 'I belong to Prussian family of architects in which the tradition of Schinkel – the contemporary as well as the opposite number to your own Soane – was part of our heritage'.[127] With a foreword written by DIA Chairman, Frank Pick, the book was given a public endorsement by the English design establishment. This approach of quiet reform may well have been encouraged by the art critic and MARS member, Herbert Read. Read had first suggested the possibility of publishing Gropius's lecture to the DIA meeting in modified form, and subsequently arranged for publication with Faber & Faber, after re-translation by P. Morton Shand, all whilst Gropius was still in Germany.[128]

Gropius's book followed Read's *Art and Industry*, which had been designed by the German's Bauhaus colleague, Herbert Bayer, and included photographs sourced by another Bauhaus instructor, Lazlo Moholy-Nagy (1895–1946). Using these publications, Read sought to familiarise English readers with Modern architecture and design, promoting an approach that combined tradition and modernity, art and industry. As Fry floridly wrote in a review of Read's book, 'Not until the union of the two [art and industry] is felt to be part of a re-creation of human values, and necessary to a new way of living, can the movement which it represents turn, perhaps suddenly, into a channel of positive creation, deep and full enough to irrigate the vast, receptive plains of toiling humanity'.[129] The similarity of Gropius and Read's views was evidently significant for Fry, providing a rationale for the integration of contemporary art, design and architecture to humanise the machine.

Projects such as Sassoon House, Sun House and Miramonte highlight Fry's early integration of artwork and contemporary design into his buildings. These domestic projects were central to Fry's post-war reputation, demonstrating his skill in composition and an understanding of hygiene and climate that pre-dates his work in tropical architecture. Sun House and Miramonte are constructed of four-inch structural walls of reinforced concrete that permitted wide structural openings, expressed in the long runs of windows that dominate in both projects. Fry later, still keen to stress the historical inevitability of Modernism, wrote of the structural possibilities that led him to 'a fascinating set of new proportions that were governed nevertheless by rules of contrast, integration and balance as old as time: Sir John Soane, in the same circumstances, would have responded with equal joy'.[130]

Situated in the affluent and artistic neighbourhood of Hampstead, Sun House (1935–36) is one of Fry's best known houses of the 'thirties. Fry later described the project, his last under the partnership of Adams, Thompson and Fry, and the ease of working with his clients:

> *The client for the Sun House was a working sporting Italian tailor with an Italian wife and a shop in Hanover Square ... he was that exceptional thing, an architectural addict, and I managed to hook him on a shopping round around the Mars Group architects. He set me a financial target and two pages of typescript saying exactly how they lived, working hard by day and entertaining friends in the evening, the whole outfit seemed centred upon an old Scottish housekeeper.*[131]

Situated on a steeply sloping site, to maximise views across the city, the four-storey building includes a garage and small plant room at ground floor level and a full-size roof terrace (Figure 2.11). To facilitate the owners' evening entertaining, the sun terrace is installed with a dumb waiter to send up drinks from the kitchen.

Befitting its name, the living spaces and principal bedrooms face due south to provide sunlight throughout the day, with the self-contained service areas, to the north.[132] The house combines structural innovation with careful consideration of climate. Constructed in 12-foot bays of reinforced concrete, the carefully balanced façade combines two floors of continuous sliding windows with projecting balconies and sheltered sun terraces.[133] Slender steel columns, supporting the balconies and terraces, provide an open plan. The concrete was cast against 'wall board' shuttering and finished with a 'mechanical rubber' before the final application of concrete paint.[134] The south-facing rooms are fitted with external roller blinds for summer cooling and, for the winter, underfloor heating and ceiling heating panels. This technical innovation was tempered by murals to the interior walls by Hans Feibusch and by sculpture by Henry Ellison (Figure 2.12).[135]

At Miramonte (1936), near Kingston in Surrey, a generous site allowed Fry to plan an expansive house complete with a gatekeeper's lodge, garages and landscaped grounds with a kitchen garden, a tennis court and a swimming pool. Situated on high ground to the north of the site, the principal rooms all face south overlooking a garden of mature trees (Figure 2.13). Fry's awareness of the problems of heating concrete structures with very limited insulation is demonstrated by his use of different types of heating system: ceiling panels in the bedrooms; adjustable radiant panels in the bathrooms; tubular heaters throughout; and 'specially

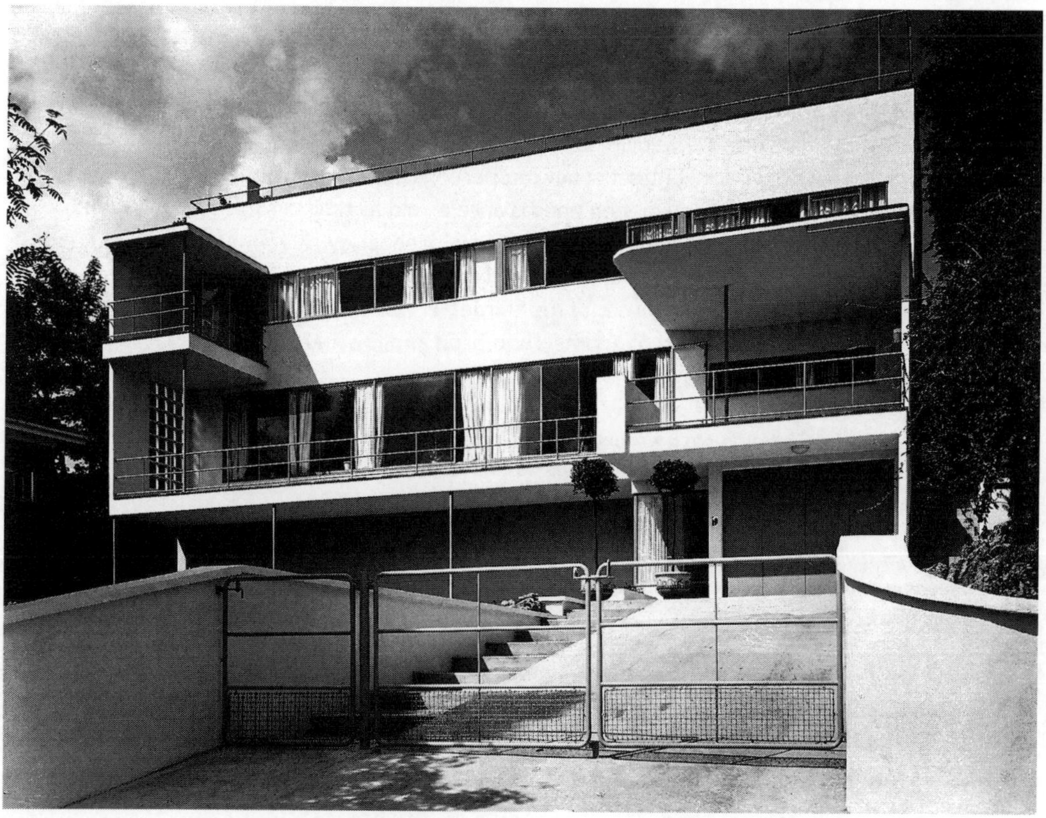

2.11 South Façade, Sun House, Hampstead, London, 1936

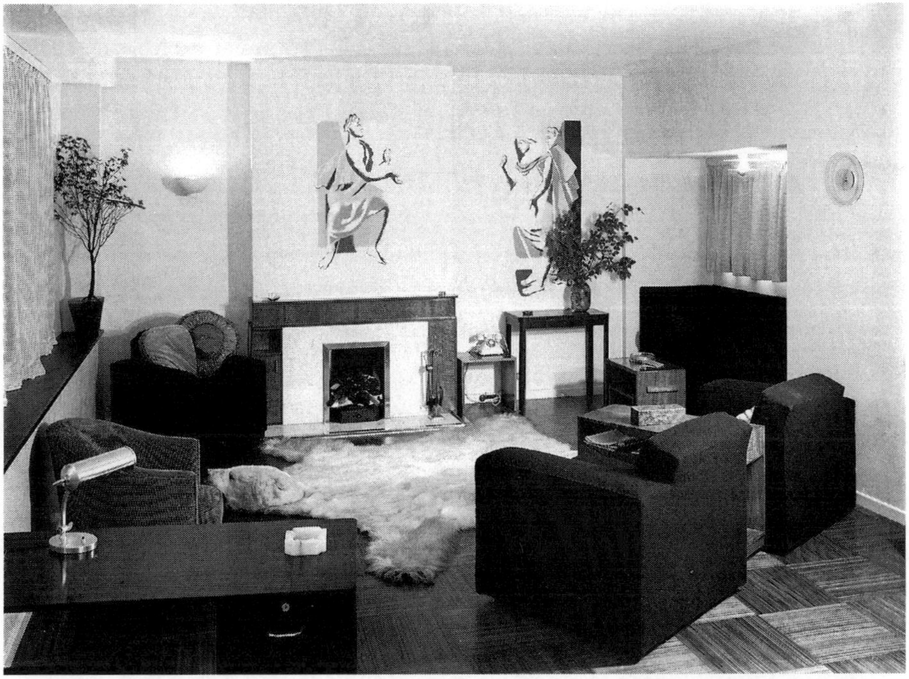

2.12 Living Room with Murals by Hans Feibusch, Sun House, Hampstead, London, 1936

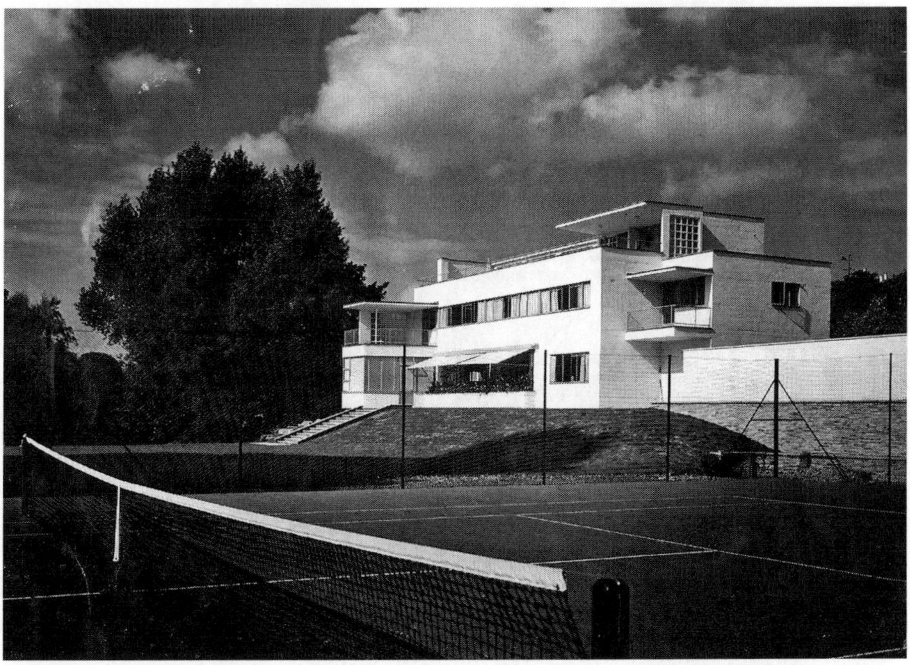

2.13 South Façade, Miramonte, Surrey, 1936

designed convection heaters in the living room and sun-room'.[136] A generous budget allowed Fry to design all of the built-in and much of the free-standing furniture, yet the interiors lack the innovation of the structural solution. Perhaps catering to the client's taste, the dining room has a Deco-Moderne feel; walls are covered in wood-veneer paper, and the centre-piece is a pear-wood dining table and chairs covered in yellow hide (Figure 2.14).[137]

During this period, Gropius acted as lead designer for two further modernist houses: 66 Old Church Street (1936) in Chelsea and the Wood House (1936–38)

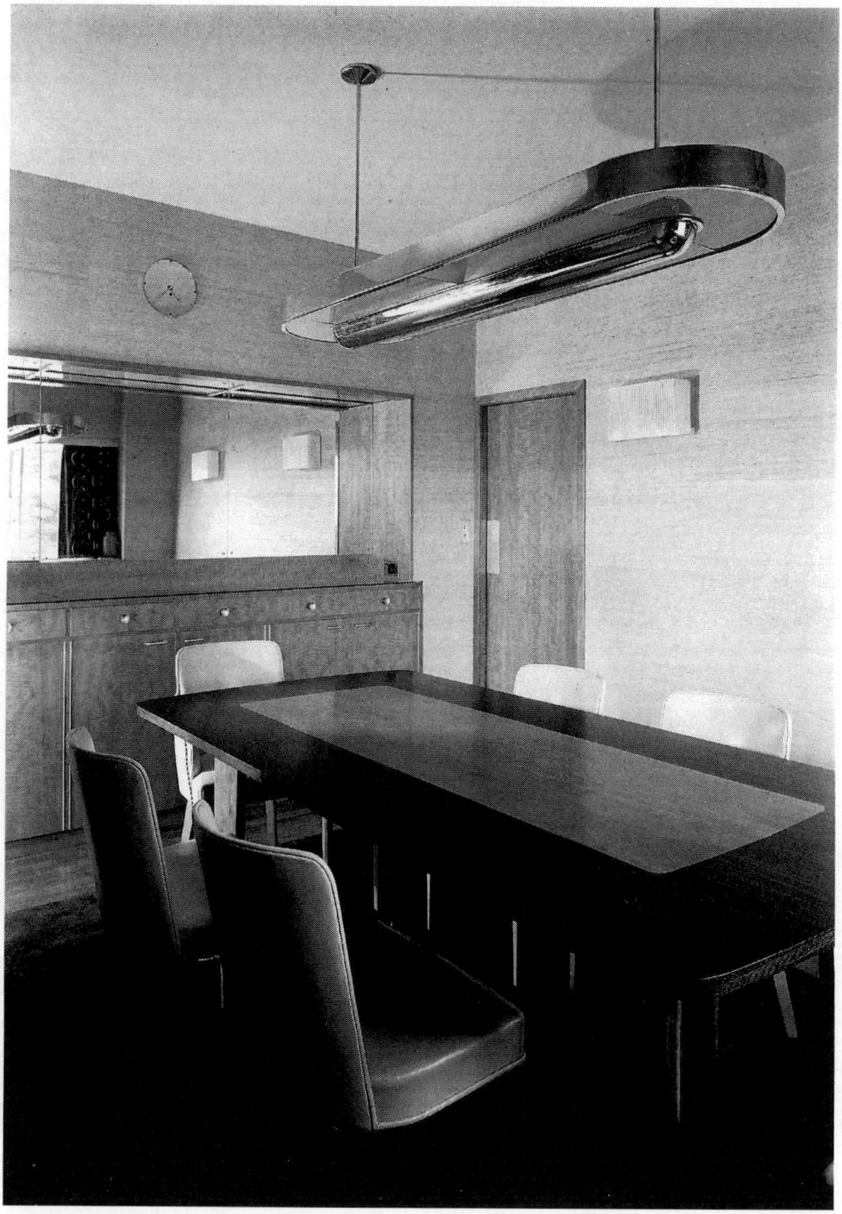

2.14 Dining Room, Miramonte, Surrey, 1936

at Shipbourne in Kent.[138] 66 Old Church Street was designed for the playwright Ben Levy and his wife, the American actress Constance Cummings, who requested a dazzling white, Californian-style villa. Fry and Gropius were keen to stress the continuity of tradition at Old Church Street and the building responds to its Georgian neighbours. Fry and Gropius's scheme is situated next to a similarly respectful house by Eric Mendelsohn and Serge Chermayeff, and the two houses form a neat composition in contextual Modernism (Figure 2.15).[139] Yet the building was dogged by complications from the outset. The 'carbo' render specified by Gropius stained badly in the dirty London environment. Gropius was quick to point out that this was Levy's choice and wrote to Fry, 'there is no doubt that the London soot will always collect on any surfaces. I told this to Levy at a very early date, before we had given any orders. If you remember, I was in favour of a brick building but Levy wanted to have a white house'.[140] The building subsequently suffered from water ingress and underwent two conversions, first by Jane Drew and then by Theo Crosby, resulting in a building that today bears little resemblance to its original concept.

Like the Chelsea house, Wood House responds to its setting; here the result is a timber pavilion rather than an urbane villa. As the name suggests, Wood House is an oak timber-frame building, infilled with concrete blocks and woodwool insulation between softwood joists.[141] Externally, it is clad with cedar weatherboarding (Figure 2.16). As Jeremy Gould notes, Wood House bears similarities to Bernard le Mare and

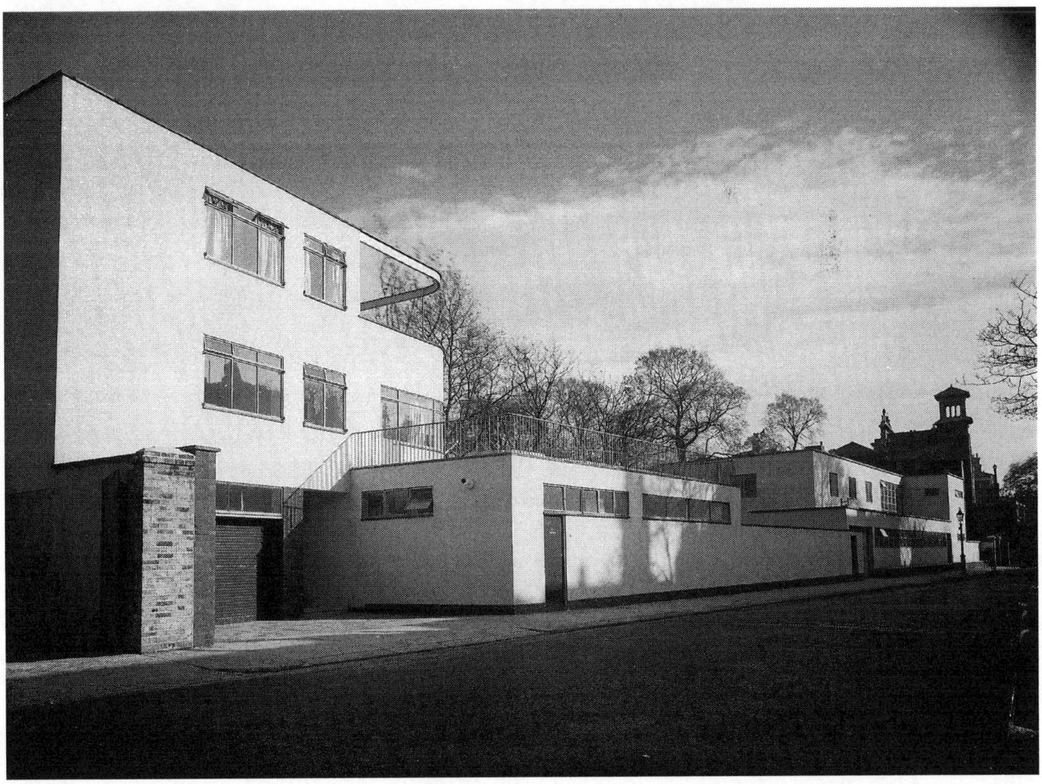

2.15 64 and 66 (to the foreground) Old Church Street, Chelsea, 1936

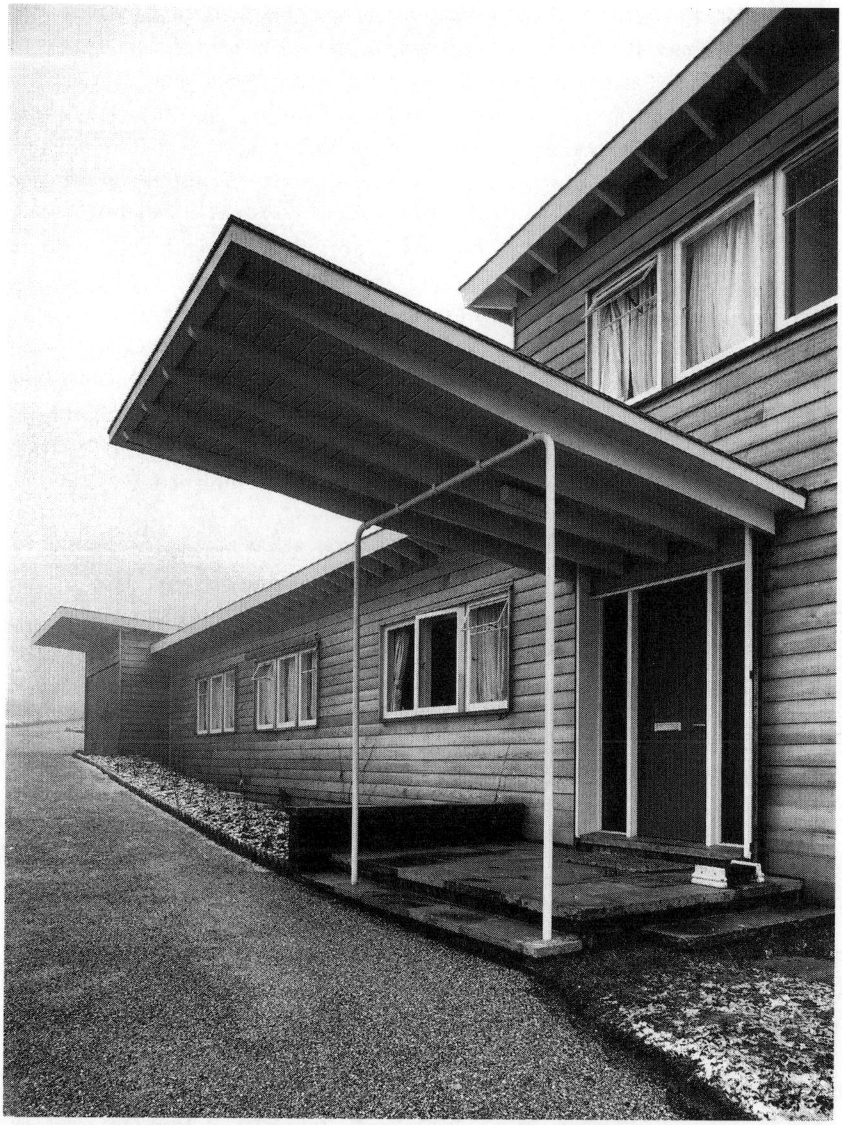

2.16 Wood House, Shipbourne, Kent, 1937

Albrecht Proskauer's competition-winning scheme for a timber house.[142] The house was built at Woodford Green (1935–36) and, given Proskauer's employment by Fry and Gropius at the time, a transference of ideas seems likely.[143] The use of timber-frame construction may also have developed out of Gropius's association with the Resident Architect at Dartington Estate, Robert Hening. The construction of Wood House is identical to Hening's flat-roofed, timber-frame workers' bungalows, which Gropius had critiqued early in 1935. He had praised for their rational planning and form, noting the effectiveness of using elements of differing height and later used this device at Wood House.[144] The house was only in the early stages of construction when Gropius left for America and its execution was overseen by Fry, enabling him to study the benefits of using natural materials.

HENRY MORRIS'S HUMANISM AND THE VILLAGE COLLEGE MOVEMENT

Impington College (1936–39) in Cambridgeshire is the most well known of Fry and Gropius's projects. Built under the aegis of Cambridgeshire's long-standing Chief Education Officer, Henry Morris (1889–1961), the college was one in a series of successful Village Colleges built throughout the county.[145] Morris developed the Village College Movement in the mid-1920s in an effort to 'abolish the duality of education and ordinary life'.[146] He sought to provide further education to the population of rural areas and help stem the exodus of residents to urban areas. Morris intended the education centres to not only 'provide the training ground for the art of living, but the place in which life is lived'.[147] His statement echoes the tagline used throughout the Isokon 3 brochure, which promoted the Windsor development as a place 'where life is living'. Indeed, Fry, Gropius and Jack Pritchard regularly met with Henry Morris from 1934 onwards, suggesting that ideas for alleviating rural decline were thoroughly discussed by the group; at Isokon 3 these ideas were applied as a means to generate new rural, albeit rather exclusive, communities which contrasted with Morris's centres for honest country folk.

Morris had overseen the construction of colleges at Sawston (1930), Bottisham (1937) and Linton (1937), before the Impington project began. Sawston was designed in a Neo-Georgian manner and Morris's philosophy was indebted to Scott's writing on the *Architecture of Humanism*. Like Scott, Morris endorsed the humane values of Classicism and later noted, 'As Geoffrey Scott has said, we transcribe ourselves into terms of architecture: also, we transcribe architecture into terms of ourselves'.[148] In his Village Colleges, Morris sought to provide buildings that embodied modern rural life, while also tracing a history of human endeavour and achievement in the countryside. Fry had 'imbibed' Scott's text during his architectural training and used these ideas to historicise his Modern approach;[149] in Morris he found a like-minded individual and, in the company of Gropius and Pritchard, Fry was able to demonstrate to Morris that Modernism was a continuation of Scott's humane tradition.

Gropius assumed the role of lead designer for Impington College, although Fry undertook a substantial redesign following Gropius's departure for America. The college is often criticised for its functional design and disjointed wings that meet in a somewhat awkward manner.[150] However, as Andrew Saint points out, 'what matters is not a precise solution to any educational or architectural problem, but the sense of congruity between form and social intention: the relaxed grouping of classrooms, community space and shared hall'.[151] The building centres on a fan-shaped assembly hall, with one- and two-storey wings simultaneously spreading into the landscape and framing exterior space as a series of courtyards (Figures 2.17 and 2.18). Built in rough-textured bricks with steel roof trusses, the college has a dedicated adult education wing comprising a library, a common room, a committee room and further rooms for lectures, billiards and table tennis.

Late in 1937 Jack Howe took over as the Project Architect and, with the scheme over budget, he and Fry began a redesign in order to cut costs.[152] Fry, growing tired of cost-cutting, later wrote to Gropius, 'The main struggle has been to maintain a decent standard inspite of the awful cuts we have had to make, coming back again and again to the Committee to tell them we could save some more'.[153] With a vastly reduced project, including the omission of the gymnasium, Fry appears reluctant to send on the revised plans to Gropius, who wrote, 'for six months I have written, in every letter, asking you for this set of drawings – why do you hide them from me?'[154]

Impington College led to subsequent educational projects, such as an unbuilt scheme for a school in Papworth (1935–37), also in Cambridgeshire.[155] Intended for children with tuberculosis, the plan is centred on a wedge-shaped hall with wing for children up to 14 years of age, and a smaller nursery wing for infants. All

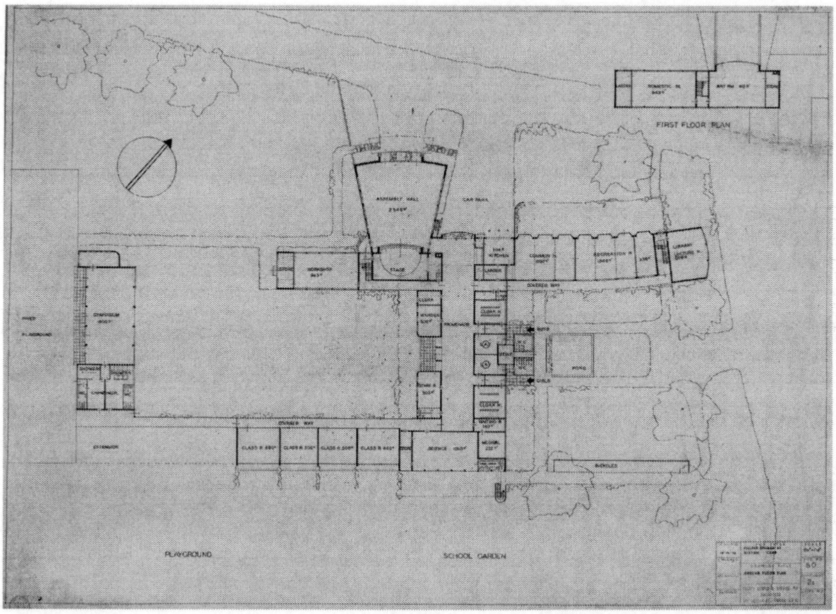

2.17 Ground Floor Plan, Impington College, Cambridgeshire, 1936

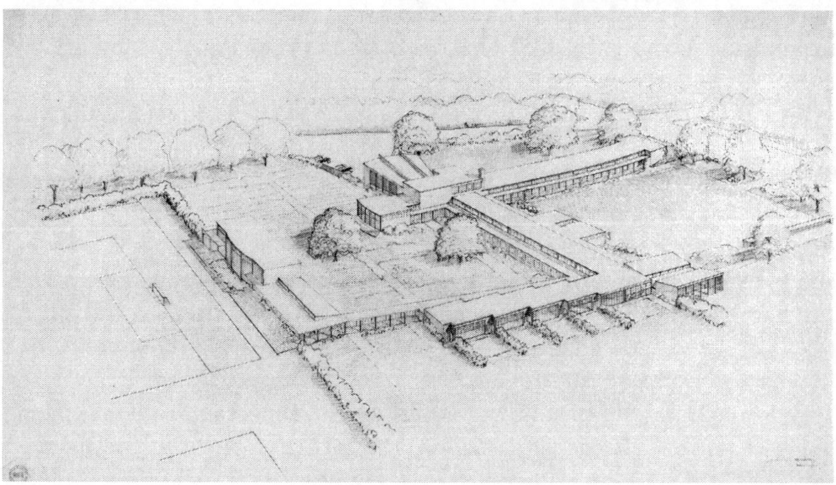

2.18 Aerial Perspective, Impington College, Cambridgeshire, 1936

teaching spaces were to face south, with full glazing to allow open-air teaching to take place. When war broke out in September 1939, Fry was working on a modern school at Altofts in the West Riding of Yorkshire.[156] The school was to be constructed predominantly of timber, showing Fry's continuing revisionism of his modernist palette of materials. Designs for two Lincolnshire schools at Market Deeping and Stamford were also undertaken around this period,[157] as Fry later wrote, for 'one of Henry Morris's protégés'.[158] Like Impington College, his design for a mixed secondary school at Market Deeping is composed of low banks of classrooms reaching into the landscape and the L-shaped plan is centred on a multi-purpose hall.[159] Despite some tentative post-war correspondence between Fry and the West Riding Education Authority, Altofts School (and the Lincolnshire schools) remained unbuilt.[160]

Henry Morris's acquaintances at Cambridge University led to Fry and Gropius undertaking the design of a building for Christ's College (1936–37). Gropius had courted Conrad Hal Waddington, a Fellow at Christ's College and modernist sympathiser, who would later marry Drew's good friend, Justin Blanco White. In an explanatory statement of the design, Gropius presents himself as a scholar drawing on English architectural tradition: 'Looking at the masterpieces of architecture in the past I found out that their architects never limited former styles, but using the new techniques of their time founded the beauty of their buildings on the way it solves the problem. That is what modern architects try to emphasize again'.[161] The sentiment is characteristic of the historical deference portrayed during his spell in England, yet it was ineffectual in convincing the university fellows. Despite Waddington's optimism and lack of concern regarding Gropius's imminent departure – Fry was considered 'perfectly capable of looking after anything which may arise' – the scheme was considered too modern and, ultimately, rejected.[162]

The steel-frame building is frankly modernist in design, responding to its context in formal rather than structural terms (Figures 2.19 and 2.20). Aesthetically, the design mixes old and new. It was to be clad in Ketton stone, like St John's College, in contrast to the glass blocks at ground floor level and continuous glazing to the stair tower.[163] The five-storey building was to provide a shopping parade at ground level and three floors of student flats, each comprising a bedroom and living space. The study flats were to be accessed via a central corridor, with a communal living room and bathrooms to each floor. At rooftop level, two three-room Fellows' flats were designed to open onto a large terrace. The unbuilt scheme explored some formal ideas that were present in Fry's contemporaneous scheme for All Souls College in Oxford, which also remained unbuilt. This work finally found expression in Fry's subsequent work at Kensal House flats and the Cecil Residential Club.

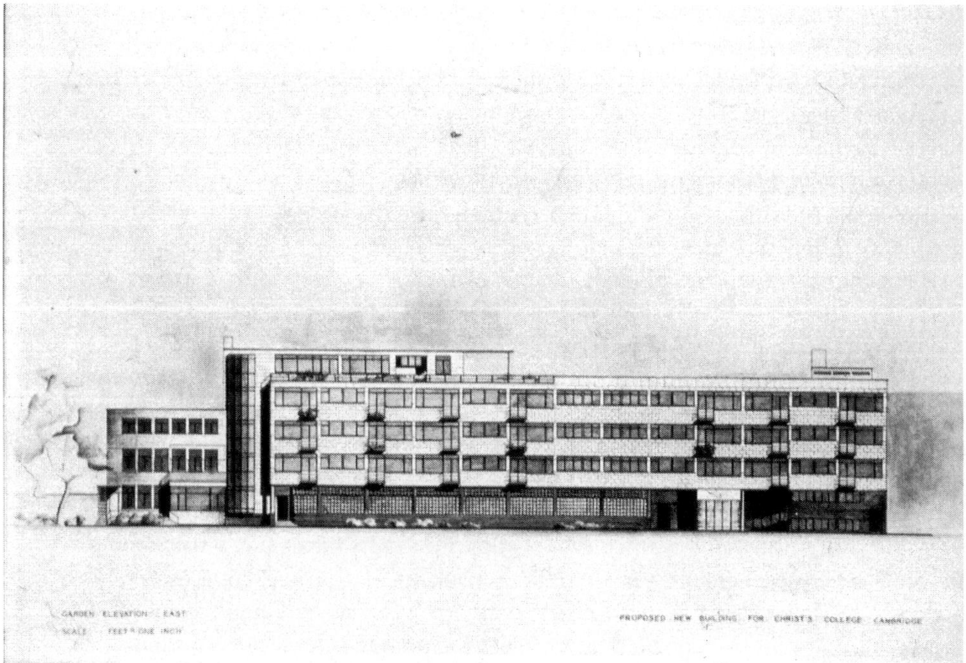

2.19 Garden (East) Elevation, Student Residence, Christ's College, Cambridge, 1936

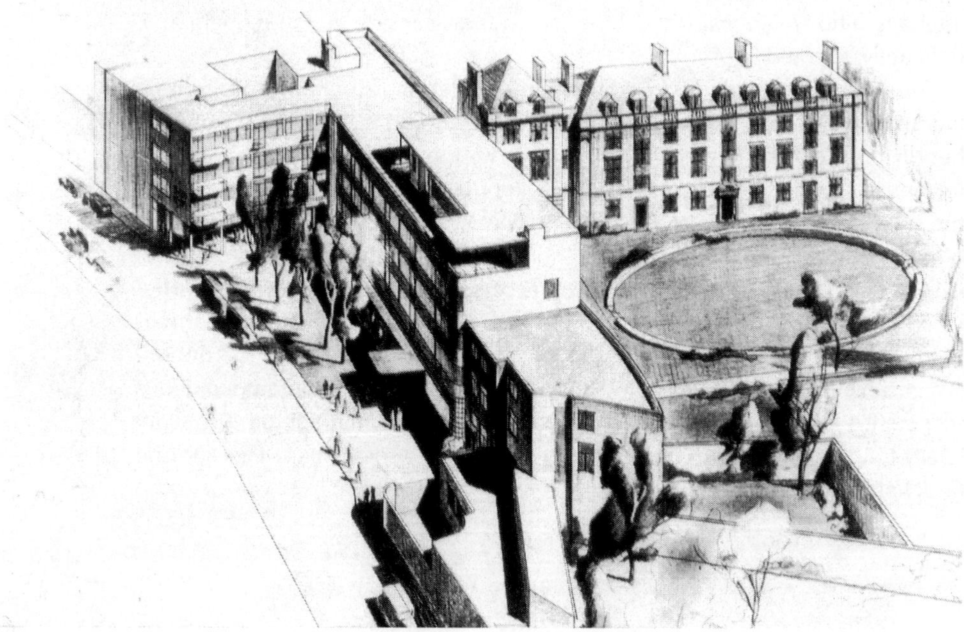

2.20 Aerial Perspective, Student Residence, Christ's College, Cambridge, 1936

NEW BEGINNINGS

After Gropius's departure for Harvard, Fry wrote to his former partner of the difficulties he faced, 'It is rather like starting all over again. Actually our fortunes at the moment are at a low ebb. ... I have been having to reduce the staff to Vaughan, Howe, Charles and Katz, who is staying in without pay, rather than leave.[164] However, just a few months later, the office was busy again with new commissions. In March 1938 Fry travelled to Helsinki to work as a Consultant Architect to the International Nickel Company.[165] He worked on a town plan in the northern region of Petsamo in Finland, presumably to create a new centre for the lucrative nickel-mining trade in the area. Fry later wrote that the commission came from his client for Ridge End, Edgar Pam.[166] Yet the project came to nothing, as the land was officially ceded to the USSR in 1947 following the Continuation War.[167] Whilst in Finland, Fry met Alvar Aalto and visited his Paimio Sanatorium (1928–33), which he later described as 'the most lyrical building I've ever seen in modern architecture'.[168] Indeed, Fry's trip to Finland and particularly his meeting with Aalto perhaps helped him to clarify his shifting ideas regarding organic materials and a modernist aesthetic.

Back in England, Fry was able to test out these ideas with a subsequent commission for a home for the Hudson-Davies family.[169] Warham Ash (1938), in Herefordshire, shares similarities with Fry's design for Little Winch and shows his continued use of natural materials (Figure 2.21). A scheme for a block of flats at 65 Ladbroke Grove (1938) in Kensington was also underway at this time, following a commission from Fry's friend, the property-developer Charles Kearley.[170] The five-storey block (plus a recessed penthouse designed by Raymond Myerscough-Walker) comprises 16 small flats overlooking a communal garden to the rear. The reinforced concrete frame, infilled with flint brick and areas of blue vitreous tile,

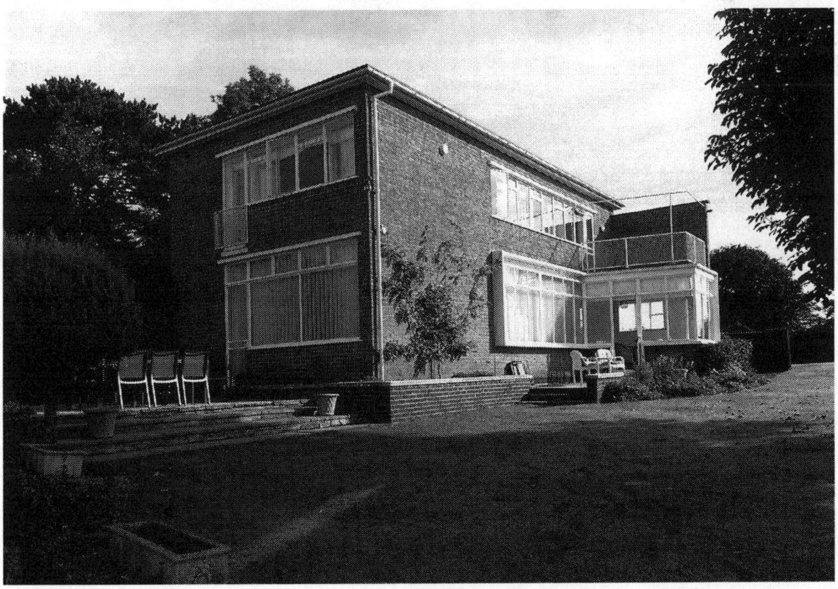

2.21 Warham Ash, Breinton, Herefordshire, 2013

was constructed rapidly; before the year was out, photographs of the completed building had appeared in the *AJ* (Figure 2.22).[171] The building has a similar arrangement to R.E. Sassoon House: flats are accessed via an enclosing stair leading onto north-facing, open galleries. A typical floor contains two one-bedroom and two two-bedroom flats, arranged with tightly planned kitchens and bathrooms that open out to more generous south-facing bedrooms and living spaces with small balconies. The Impington aesthetic is continued, with buff-coloured bricks and blue tiling contrasted against concrete flower boxes, glass blocks and tubular handrails.

The contemporaneous Cecil Residential Club (1938–40) at North Gower Street,[172] London, also uses a palette of brick, blue tiling, glass blocks and sliding metal

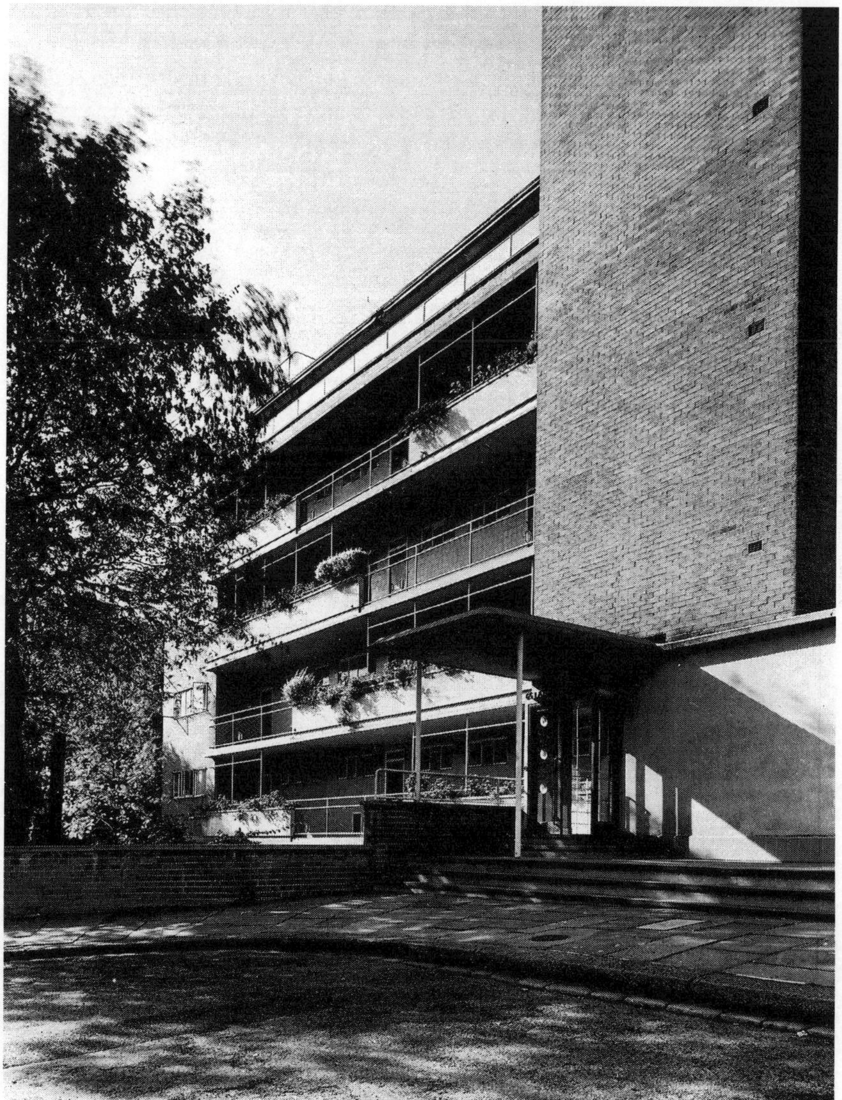

2.22 65 Ladbroke Grove, Kensington, London, 1938

windows within a concrete frame (Figure 2.23). Here Fry perfected his synthesis of what he called 'the metallic and the organic', to give texture and interest to the carefully composed façades.[173] The hostel provided affordable housing to young women working in the capital, away from home. Dormitories for four accommodated 72 women in total, with shared bathrooms and ample communal spaces, including a ground floor restaurant. A mural in the entrance hall depicts a map of London and the location of the club within the city, allowing the residents to situate themselves within their new community (Figure 2.24); this is perhaps the earliest example of Fry using artwork to communicate the notion of an individual contributing to wider human endeavour and this theme reappears throughout his work of the 1950s. Indeed, the residential club is a prototype for Fry's postwar work in its use of instructive artwork, synthesis of natural and machine-made materials, and in the external expression of its structure: the exposed concrete frame wrapping up over a recessed top floor is a precursor to his Veterinary School (1955–60) for Liverpool University.

EXHIBITION EXPERIMENTS

Alongside Fry's architectural work of the 1930s, he also undertook commissions for showrooms, exhibitions and furniture. This work deserves scrutiny in its own right

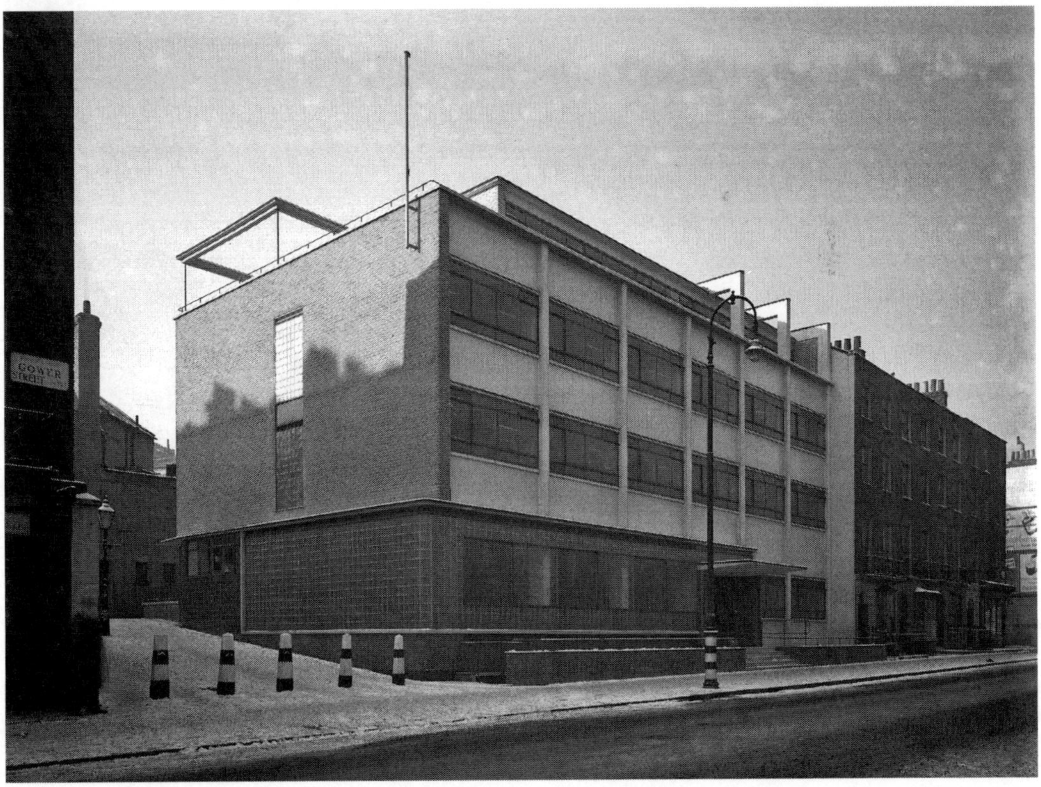

2.23 Street Façade, Cecil Residential Club, North Gower Street, London, 1940

as it provided Fry with an opportunity to work with other designers and artists, and to experiment with new techniques, which, in turn, fed back into his architecture. In 1933, a significant commission came from an emerging market – the supply of electricity and new electrical goods.[174] Fry designed showrooms for Westminster Electric Supply Corporation, situated on Victoria Street, close to the offices of Adams, Thompson and Fry. Fry followed the luxurious example set by Mansfield Forbes and Raymond McGrath's 'Finella', and by Coates's shop designs for Cresta Silks. He experimented with the latest in modern composite materials: the shop-front was a sweeping curve of metal-faced plywood painted in yellow and grey,

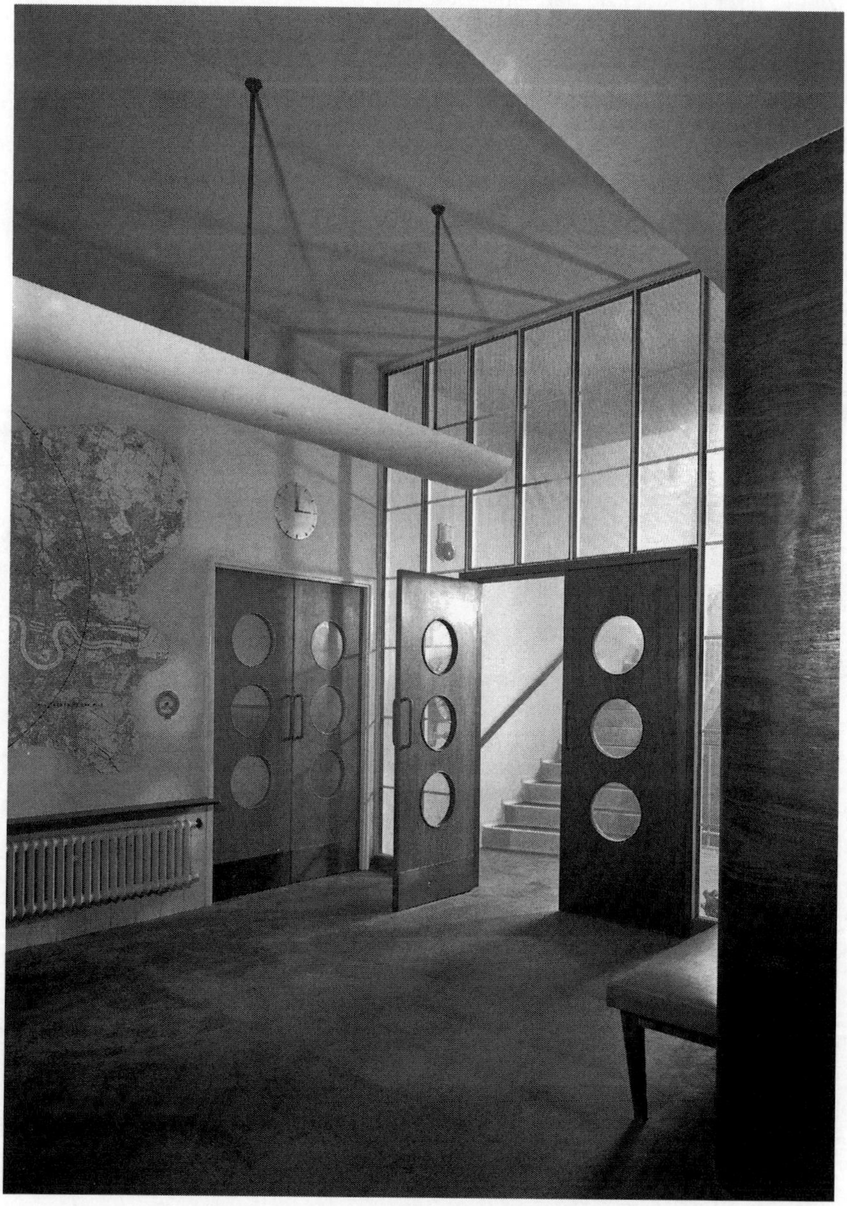

2.24 Entrance Hall, Cecil Residential Club, London, 1940

framed with stainless steel and black marble.[175] The company sign, in neon tubing with a stylised flourish, helped to lead the eye down the shop-front and toward the entrance (Figure 2.25). The interior was decorated with a mural, showing the area served by the electricity corporation, and was fitted with rows of glass cabinets displaying labour-saving electrical goods (Figure 2.26).

A series of small exhibits followed. Fry was commissioned by the Gas, Light and Coke Company to design their display for the 1934 Exhibition of Contemporary Industrial Design in the Home, held at Dorland Hall.[176] Then, in 1935, he designed the Glass Gallery at the Exhibition of British Art in Industry at the Royal Academy

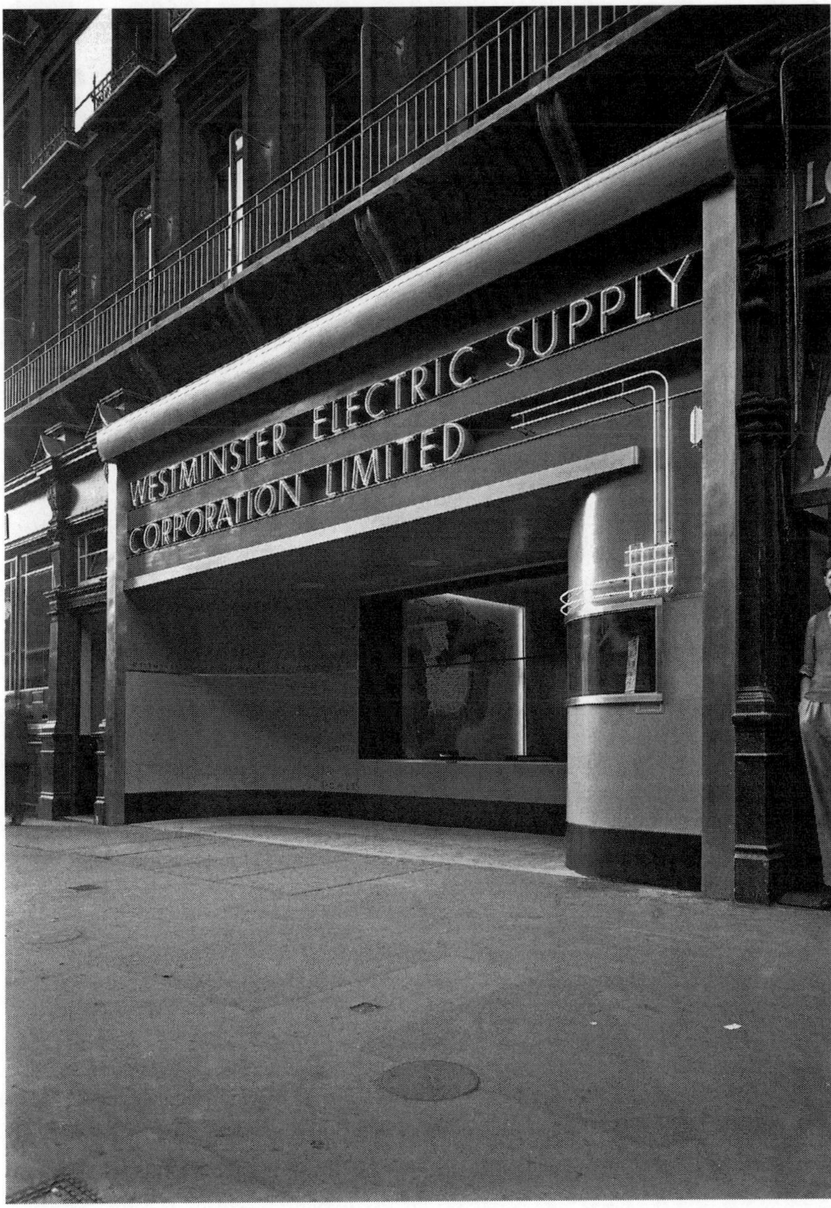

2.25 Westminster Electric Supply Corporation Showrooms, Victoria Street, London, 1933

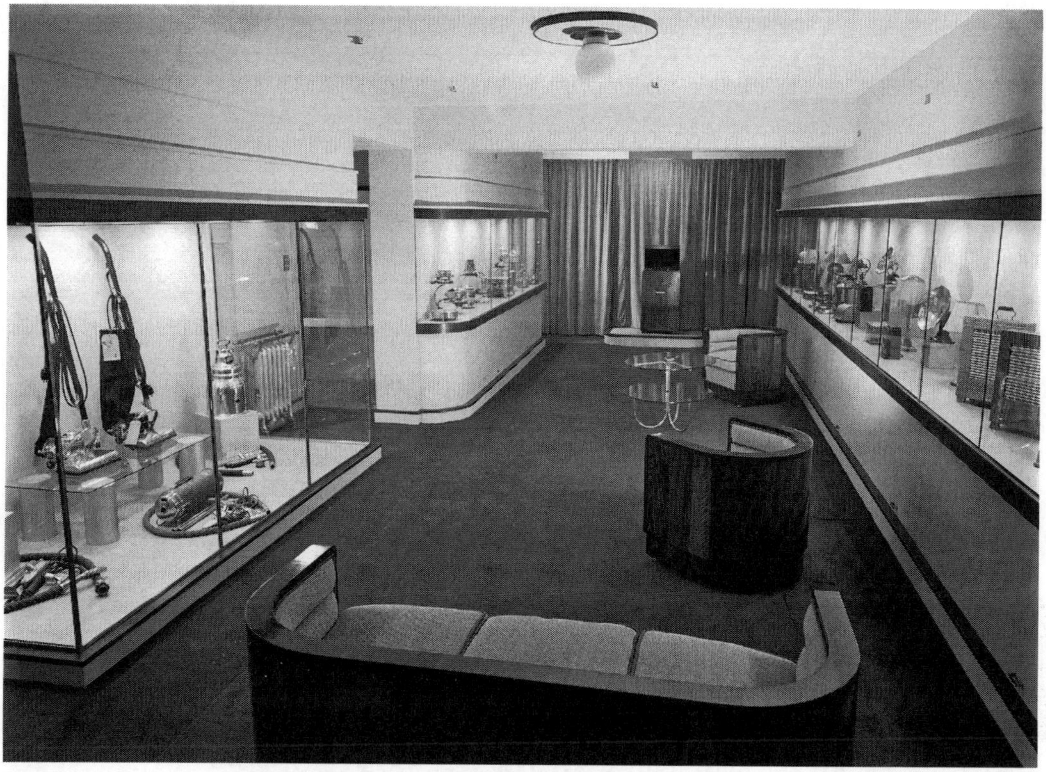

2.26 Interior of the Westminster Electric Supply Corporation Showrooms, Victoria Street, London, 1933

in London. Both make use of photo-murals, the latter including a particularly successful depiction of glass production yet, in comparison to Fry's architecture of the period, there is little innovation in his exhibition work. Later in the same year, he was commissioned by fellow DIA member and designer Ambrose Heal to organise an exhibition of contemporary furniture by a group of architect-designers. The project became known as '7 Architects' as Fry invited contributions from Raymond McGrath, Christopher Nicholson, Brian O'Rorke, Jack Howe, Christopher Heal, Marcel Breuer and 'A member of Tecton'.[177] Their designs were manufactured by Heal's and exhibited at the Mansard Gallery, at the company showrooms on Tottenham Court Road in London, in 1936. Fry and Jack Howe's collaborative designs included a living room furnished with a cocktail cabinet in Indian laurel and easy chairs with ebonised bentwood arms and detachable latex rubber cushions.[178] Despite evident attempts to use new materials in an innovative fashion, their work lacks the grace of Breuer's bent sycamore chairs designed for the same exhibition.

Meanwhile, Fry's work for the MARS Group had begun in April 1934 with the first research programme tackling 'slum clearance', which he coordinated with émigré architect Eugen Kaufmann and his employee, Hazen Sise.[179] The MARS study of Bethnal Green was presented at the 'New Homes for Old' housing Exhibition at Olympia, held in September 1934, at the invitation of the exhibition's chair, Judith Ledeboer and her colleague Elizabeth Denby.[180] Here too, the exhibits were somewhat conventional in design. However, in 1936 Fry was introduced to the Hungarian artist Lazlo Maholy-Nagy, who joined the MARS Group and instigated a dramatic shift in Fry's approach to exhibition design. A great friend of Gropius, Maholy-Nagy and his wife Sybil had moved to England in 1935 and, for a time, lived at the Lawn Road flats. His association with Gropius no doubt helped Moholy-Nagy to assimilate into the close-knit architectural scene. Fry later wrote of Maholy-Nagy and his conception of space:

> *At that time I knew little about him other than he had turned up from Germany, was from the Bauhaus, and in need of work. But we quickly became friends and one day he said to me that I must go with him to the Science Museum in South Kensington to see something that would astonish me ... we came to it and it was a simple enough apparatus consisting of two discs – or were they semi-spheres? – with little chains on them which seen through dual eyepieces created an illusion of limitless space that was breathtaking. I was delighted and Moholy was satisfied.*[181]

As Bauhaus colleagues, Gropius, Moholy-Nagy and Herbert Bayer had collaborated in the design of various exhibitions, such as the Exhibition of Building Workers' Unions of 1931 held in Berlin. The exhibition was devised as an 'interior landscape', with a clearly articulated route taking the visitor on a journey in altered perspectives. A raised platform accessed via stairs and ramps gave visitors an overview of the whole exposition, while the use of New Vision photographic techniques,[182] over-size typography and photomontage were designed to constantly shift perceptions of space. As Mary Anne Staniszewski notes, 'Some exhibited elements could only be seen by leaning over railings ... Peephole constructions and large-scale photographs provided unusual close-up displays of materials'.[183] This shifting perspective was mirrored in Moholy-Nagy's photographic displays, which used bird's eye views, photograms, projections, X-rays, and microscope images to demonstrate the diversity of modern photography.

In England, Gropius and Moholy-Nagy sought to publicise this innovative work with a publication on 'Exhibition Architecture' and Gropius sent a book proposal to the Architectural Press in 1936.[184] Although the publication came to nothing, their novel methods of display are readily apparent in Fry's showrooms and exhibitions of the late 1930s. Fry's later work has an element of surrealism – particularly in the exploitation of different forms of lighting – that reveals Moholy Nagy's influence. Indeed, in the opening speech of an exhibition of Moholy-Nagy's paintings held at the London Galleries in 1936, Walter Gropius said, 'Moholy recognised ... that we can only comprehend space by means of light. His whole work is a mighty battle to prepare the way for a new vision'.[185]

The ideas of Moholy-Nagy and the Bauhaus exhibition techniques coalesced at Fry and Gropius's Electricity Show Rooms at Regent Street (1937) in London. For a restricted site, the trio produced an elegant and innovative solution for this emerging market. The main façade at street level is pulled back behind the neighbouring building line, effectively creating an entrance canopy and additional window display area. To add to the drama, a spiral staircase was put 'front of shop', lifting the eye up through the space to the first floor showrooms (Figure 2.27). The glass cabinets of earlier schemes were dispensed with; instead,

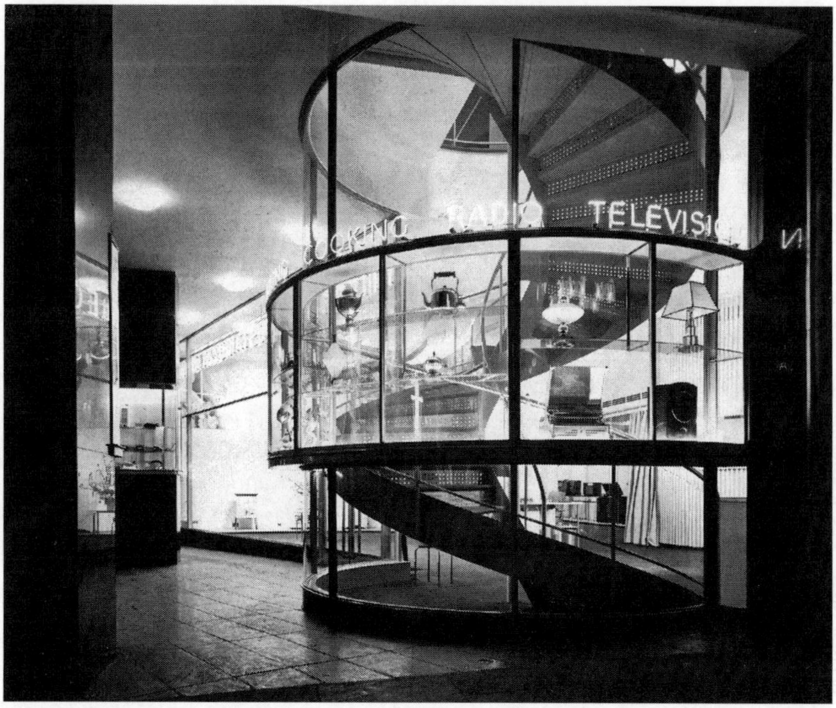

2.27 Electricity Show Rooms, Regent Street, London, 1937

electrical appliances were arranged in unusual juxtapositions to emphasise their sculptural nature. Moholy-Nagy created a photo-mural, including the lights along the Thames Embankment (which, it later emerged, were gas lights) and a light machine that projected 'melting and merging colour images' onto one wall of the showroom (Figure 2.28).[186] Both Gropius and Moholy-Nagy had left for America before completion of the work,[187] but for Fry it remained a perfect project and he would return to ideas first used here throughout his career. The spiral staircase was a particular favourite and he wrote to Gropius, 'The great joy of my life at the moment is Regent Street which is lovely, especially the staircase, a perfection of good metal craftsmanship'.[188]

The Regent Street showroom shares similarities with the MARS Group Exhibition of 1938, held at the New Burlington Galleries also on Regent Street. Like the 1934 Olympia Exhibition, Fry took the lead and headed an exhibition committee that also included Moholy-Nagy, Godfrey Samuel and Serge Chermayeff.[189] The committee was formed early in 1937 and, according to Fry, he and Moholy-Nagy produced the 'lion's share' of the work.[190] Fry reported to Gropius in November 1937 of their progress, 'The Mars exhibition is now in full swing … the whole group, or the active part of the group, is working hard. … I begin to believe that we will do a good exhibition yet'.[191] To Fry's considerable disappointment, Moholy-Nagy left for America in August 1937 and his exhibition duties were assumed by Misha Black.[192]

The legacy of Fry's work with Gropius and Moholy-Nagy is apparent throughout the exhibition. An interactive route, with changes in floor finish – from wood block

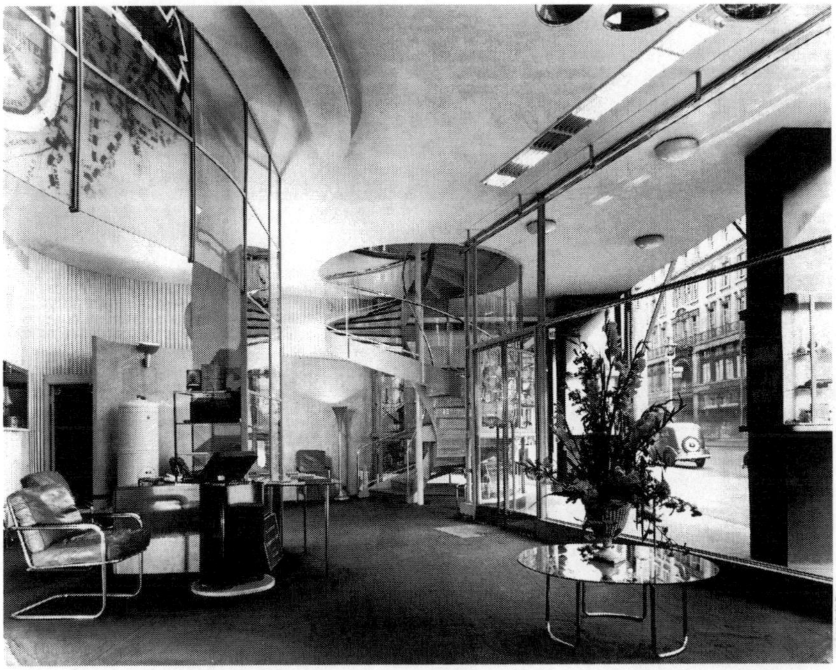

2.28 Electricity Show Rooms Interior, Regent Street, London, 1937

to paving to grass – and overhead canopies, guided visitors through the displays (Figure 2.29). Photomontage, murals, display boxes and full-scale mock-ups of model rooms were all utilised to create a visually stimulating result. As John Gold observes, the organisers 'were at pains to naturalise the new architecture wherever possible', with the exhibition brochure referencing the historic origins of the movement, such as the engineering feats of Telford and Paxton.[193] Giving further historical context, the exhibition was organised under Sir Henry Wotton's words, 'Well building hath three conditions: Commoditie, Firmenes, and Delight' (Figure 2.30). These conditions were interpreted by the MARS Group as a summary of modern building requirements (commodity), the work of scientists and engineers in building techniques (firmness), and the synthesis of these ideas to create the 'new architecture' (delight).[194]

The exhibition was an important stepping-stone for architects and designers. Perhaps due to his involvement in the high-profile exhibition, Fry made it onto the shortlist of architects for the British Pavilion at the 1939 World's Fair in New York.[195] Although he did not receive the commission Fry did contribute an exhibit of a 'model community centre', which continued to develop the ideas first explored at Impington College with Gropius and Henry Morris.[196] As Misha Black later asserted, alongside the 1951 Festival of Britain, the MARS Exhibition was hugely influential for British architecture, as the following chapter will investigate.[197]

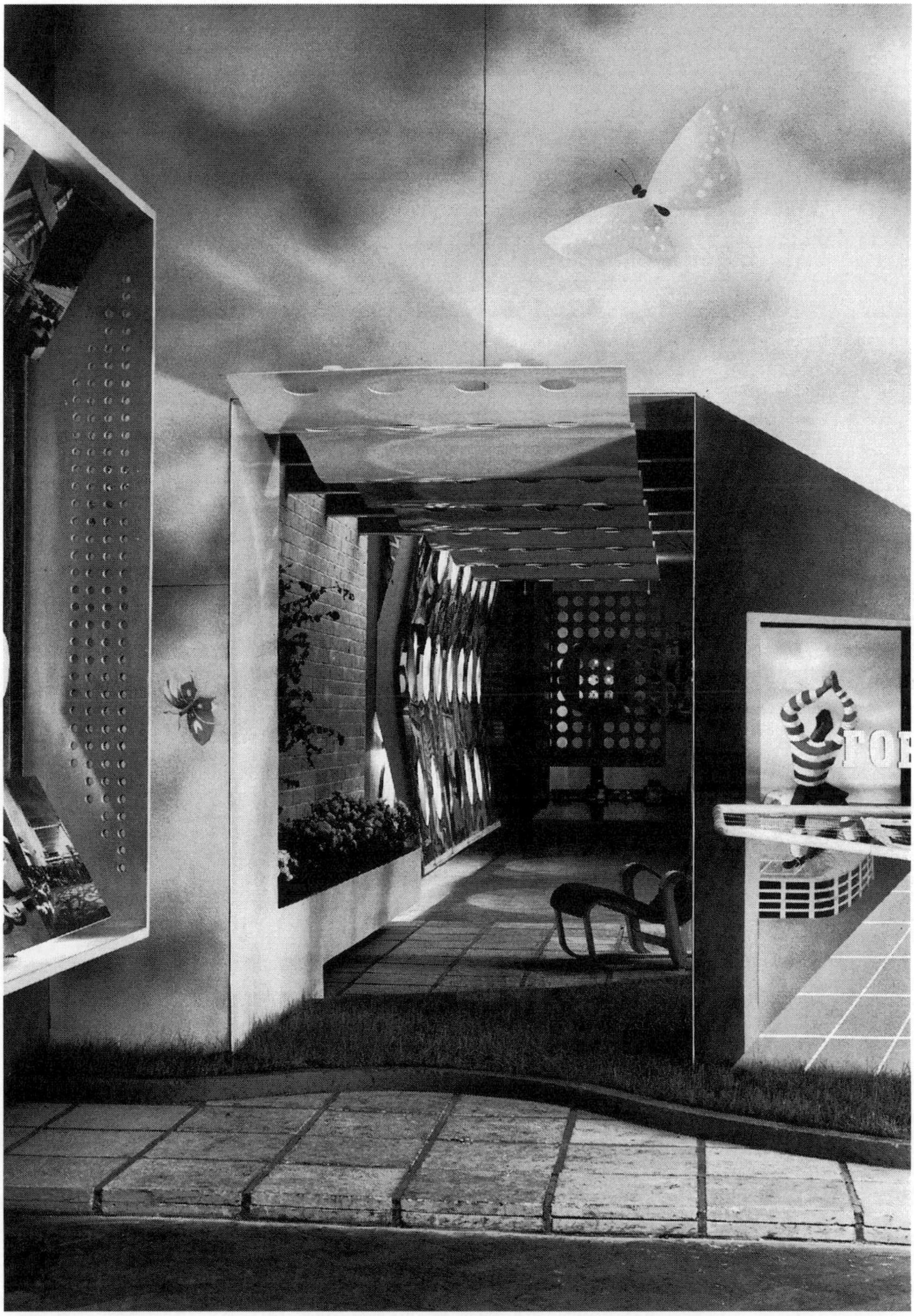

2.29 MARS Exhibition, New Burlington Galleries, London, 1938

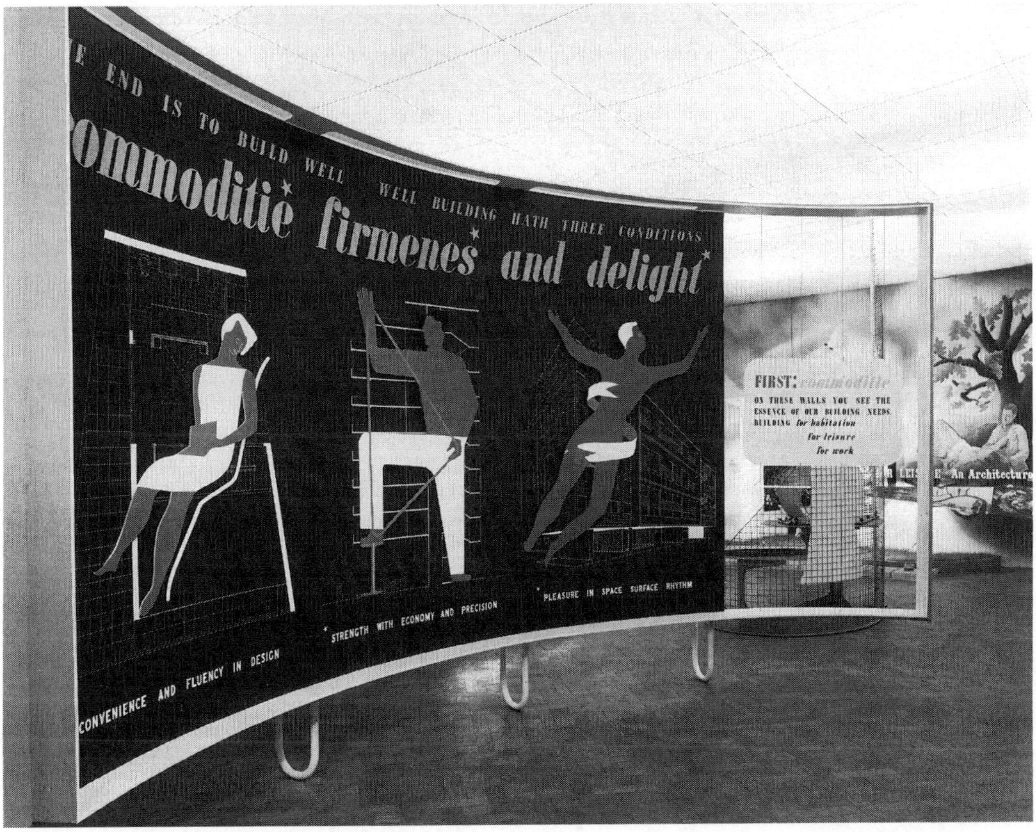

2.30 MARS Exhibition, New Burlington Galleries, London, 1938

CONCLUSION

In the 1930s Fry's efforts took root. His roles as a writer for the *Architects' Journal*, a lecturer at Regent Street Polytechnic and as an active participant of the DIA, the PEP, the MARS Group and the RIBA ensured that, by the close of the decade, Fry was the leading light of British Modernism. He played a crucial role in the evolution of the movement in 1930s' Britain, a development that can be charted in microcosm in his own architecture. Fry's collaboration with Elizabeth Denby produced two of the most successful examples of social housing of the period, combining structural and technical innovation with a social programme. His subsequent work with Walter Gropius helped to establish a pedigree for British Modernism drew from the DIA's desire to base Modern architecture and design in a tradition leading back to Morris and Ruskin. Although eager to spotlight this heritage, and equally the precedent of eighteenth century architecture, Fry took a more reformist approach than many of the DIA members. With his equivocal position on committees and groups that spanned the architectural spectrum, Fry recognised the need for an approach that would unite the architectural community. He therefore developed an outlook that he believed might encompass all classifications of modernity and moved from his early interests in mass-production and standardisation to an architecture that bridged the gap between man – that is, industrialised society – and nature. Fry believed that architecture should resonate with both to define man's position in the universe. It was a view that distinguished Fry from many of his fellow architects, and positioned him with figures such as Herbert Read and Henry Morris.

Furthermore, the Fry and Gropius office was significant in its training of many of the next generation of modernists, such as Howe, Tubbs and Lyons, which helped to advance the status of Modernism in post-war Britain. This modernised architectural milieu was one that Drew would use to her advantage.

NOTES

1 See for example: Powers, *Britain*; Darling, *Re-forming Britain*; David Matless, *Landscape and Englishness* (London, 1998); Michael T. Saler, *The Avant-Garde in Interwar England* (Oxford, 1999); William Whyte, 'The Englishness of English Architecture: Modernism and the Making of a National International Style, 1927–1957', *Journal of British Studies*, 48 (April 2009): pp. 441–65; Peder Anker, *From Bauhaus to Ecohouse: A History of Ecological Design* (Baton Rouge, LA, 2010).

2 E. Maxwell Fry, 'Town Planning', in *Circle: International Survey of Constructive Art*, (eds) J.L. Martin, Ben Nicholson and N. Gabo (London, 1937): pp. 190–2, p. 190.

3 Such as the Housing Act of 1930, which offered subsidies for local authorities for displace tenants, and the Town and Country Planning Act of 1933, which encouraged speculative building.

4 Fry and Yorke first met as journalists working for the *Architects' Journal*. See E. Maxwell Fry, 'F.R.S. Yorke, 1906–1962', *Architectural Review*, 132 (October 1962): pp. 279–80, p. 279.

5 F.R.S. Yorke and Colin Penn, *A Key to Modern Architecture* (London, 1939), p. 45.

6 The early years of the MARS Group have been discussed in some detail and need not be repeated here. See for example: Louise Campbell, 'The MARS Group, 1933–1939', *RIBA Transactions*, 4/2 (1985): pp. 68–79; John R. Gold, '"A Very Serious Responsibility"? The MARS Group, Internationality and Relations with CIAM, 1933–39', *Architectural History*, 56 (2013): pp. 249–75; Darling, *Re-forming Britain*, esp. pp. 42–4.

7 John Summerson, 'Architecture', in Boris Ford (ed.), *Early 20th Century Britain: The Cambridge Cultural History* (Cambridge, 1992): pp. 213–45, p. 240.

8 Fry, *Autobiographical Sketches*, p. 140.

9 CCA Archive, Wells Coates Papers, Box 12/A. Minutes of meeting, 28 February 1933. Sincere thanks to Elizabeth Darling for providing copies of her notes from the Wells Coates Archive at CCA for the authors' consultation; all references within to the Wells Coates Papers are taken from this source.

10 The Twentieth Century Group was established in 1930 by Forbes and Raymond McGrath. As Elizabeth Darling notes, Coates 'hijacked' the group and sought, unsuccessfully, to impose his own agenda. See Darling, *Wells Coates*, p. 106.

11 CCA Archive, Wells Coates Papers, Box 12/A. Minutes of meeting, 28 February 1933.

12 Fry cit. in Allan, *Berthold Lubetkin*, p. 322. He later reinforced this idea, writing: 'I hated, as I still hate, dogma. My friends who toyed with Communism gave up their judgement and sense as though they were paltry things and while the fever lasted they were not worth talking to'. See Fry, *Autobiographical Sketches*, p. 151.

13 Fry cit. in Allan, *Lubetkin*, p. 322.

14 RIBA Archive, Ove Arup Papers, ArO/1/5/13. 'Draft report in policy and programme', MARS Central Executive Committee Memorandum, February 1935.

15 Elizabeth Darling notes the significance of Fry's articles in establishing the 'modernist orthodoxy on social housing' in Britain, see *Re-forming Britain*, p. 116.

16 E. Maxwell Fry, 'Housing Problems in 1933', *Architects' Journal*, 77 (8 February 1933): pp. 208, p. 211.

17 E. Maxwell Fry, 'Mass-Production in Housing', *Architects' Journal*, 80 (1 November 1934): pp. 629–30, p. 630.

18 May's efforts to publicise his work included a series of four films on new building in Frankfurt, the style of which is echoed in the Gas, Light and Coke's promotional film of Fry and Denby's Kensal House. For May's 'ability to extract maximum publicity from his ventures', see Thomas Elsaesser, 'The Camera in the Kitchen: Grete Schütte-Lihotsky and Domestic Modernity', in *Practicing Modernity: Female Creativity in the Weimar Republic*, (ed.) Christiane Schönfeld (Würzburg, 2006): pp. 27–49.

19 E. Maxwell Fry, 'My Province is all Structure', *Architects' Journal*, 77 (12 April 1933): pp. 492–3, p. 492.

20 E. Maxwell Fry, 'De-Slumming', *Architects' Journal*, 77 (15 March 1933): p. 366.

21 '"The Minimum Flat", Lawn Road flats, Hampstead, London; Architect: Wells Coates', *Building*, 9 (August 1934): pp. 310–14.

22 E. Maxwell Fry, 'The Architect's Problem', *Architects' Journal*, 77 (22 June 1933): pp. 844–6, p. 845.

23 E. Maxwell Fry, 'An Urgent Question to the Manufacturers', *Architects' Journal*, 79 (5 April 1934): pp. 499–500, p. 499.

24 'Sassoon House Flats, Peckham', *Architects' Journal*, 79 (26 April 1934): pp. 611–15, p. 611.

25 E. Maxwell Fry, 'London Housing, an Itinerant Survey of Typical Structures', *Architects' Journal*, 79 (17 May 1934): pp. 712–29, p. 713.

26 Fry, 'An Urgent Question to the Manufacturers', p. 499.

27 UEA Archive, Pritchard Papers, PP/28/1/1/2/1. 'The Design and Industry Association Annual Report presented to The Annual General Meeting. March 9th, 1933'; Darling, *Re-forming Britain*, p. 42.

28 See for example: E. Maxwell Fry, 'My Province is all Structure', *Architects' Journal*, 77 (12 April 1933): pp. 492–3; E. Maxwell Fry, 'Mass-Production in Housing', *Architects' Journal*, 80 (1 November 1934): pp. 629–30.

29 Fry, 'My Province is all Structure', p. 492.

30 See Nikolaus Pevsner, *Pioneers of the Modern Movement from William Morris to Walter Gropius* (London, 1936).

31 Matless, *Landscape and Englishness*, p. 51.

32 Fry, 'Mass-Production in Housing', p. 629.

33 E. Maxwell Fry, 'The Design of Dwellings', in *Design in Modern Life*, (ed.) John Gloag (London, 1934): pp. 29–36, p. 34.

34 Fry, 'The Design of Dwellings', p. 36.

35 'Competition News: Cement Houses', *Architects' Journal*, 77 (8 February 1933): pp. 197–8, p. 198.

36 Fry, *Maxwell Fry: How Modern Architecture came to England*, Pidgeon Digital.

37 Fry, *Autobiographical Sketches*, p. 138.

38 For further discussion of Denby, see for example Elizabeth Darling, '"The star in the profession she invented for herself": a brief biography of Elizabeth Denby, housing consultant', *Planning Perspectives*, 20 (July 2005): pp. 271–300, p. 276.

39 Elizabeth Denby, *Europe Re-housed* (London, 1938), p. 260.

40 Darling, "The star in the profession she invented for herself", pp. 277–8.

41 Darling, *Re-forming Britain*, p. 66.

42 'R.E. Sassoon House', *Architects' Journal*, p. 614.

43 'R.E. Sassoon House', *Architects' Journal*, p. 615.

44 See for example Sigfried Giedion, *Walter Gropius: Work and Teamwork* (London, 1954), p. 219.

45 'R.E. Sassoon House', *Architects' Journal*, p. 614.

46 Fry, 'The Design of Dwellings', p. 34.

47 Jackson, *The Politics of Architecture*, p. 55.

48 'R.E. Sassoon House', *Architects' Journal*, p. 611.

49 Jackson, *The Politics of Architecture*, p. 56.

50 RIBA Archive, F&D/13/4. E. Maxwell Fry, 'Maxwell Fry and the Modern Movement', unpublished and undated manuscript.

51 Fry, *Autobiographical Sketches*, p. 143.

52 'Kensal House', *RIBA Journal*, 44 (20 March 1937): pp. 500–5, p. 500.

53 David Milne-Watson, Governor of the Gas, Light and Coke Company, '*Kensal House: A Model Housing Estate*' film, directed by Frank Sainsbury.

54 For further discussion of Kensal House, see: Elizabeth Darling, 'Kensal House: The Housing Consultant and the Housed', *Twentieth Century Architecture*, 8 (2007): pp. 107–15, p. 114.

55 National Gas Archive. 'Kensal House: A Model Housing Estate' film, directed by Frank Sainsbury.

56 Elizabeth Denby, *Europe Re-housed* (London, 1938), p. 61.

57 John Summerson, 'Architecture', in Boris Ford (ed.), *Early 20th Century Britain: The Cambridge Cultural History* (Cambridge, 1992): pp. 213–45, p. 242.

58 E. Maxwell Fry, 'Kensal House', in *Flats: Municipal and Private Enterprise* (London, 1938): pp. 56–60, p. 57.

59 Fry, 'Kensal House', p. 57.

60 Fry, 'Kensal House', p. 59.

61 Elizabeth Denby, 'Kensal House, An Urban Village', in *Flats: Municipal and Private Enterprise* (London, 1938): pp. 61–4, p. 62.

62 'Kensal House: A Gas Undertaking's Model Housing Scheme', *Gas Bulletin*, 26 (April 1937): pp. 54–5, p. 55.

63 RIBA Archive, F&D/23/3. Letter Fry to Gropius, Regis Hotel, New York, 30 March 1937.

64 Published in wartime, the book was sanctioned by the government as suitable propagandist material to illustrate the new society that Britain was fighting for.

65 Betty J. Blum, *Oral History of John Donald Cordwell* (Chicago, 1993), p. 71.

66 Fry, *Autobiographical Sketches*, p. 144.

67 Fry cit. in Esher, *A Broken Wave*, p. 301.

68 Stephen Hayward, '"Good Design is Largely a Matter of Common Sense": Questioning the Meaning and Ownership of a Twentieth-Century Orthodoxy', *Journal of Design History*, 11/3 (1998): pp. 217–33, p. 224.

69 Hayward, '"Good Design is Largely a Matter of Common Sense"', p. 224.

70 C.H. Reilly, 'The Year's Work at Home', *Architects' Journal*, 85 (14 January 1937): pp. 91–102, p. 96.

71 'The Chipperfield Case: Another House Rejected', *Architects' Journal*, 80 (20 September 1934): pp. 406–8, p. 406.

72 Fry, *Autobiographical Sketches*, p. 147.

73 'The Chipperfield Case', *Architects' Journal*, p. 408.

74 'A House at Chipperfield Common, Buckinghamshire', *Journal of the Royal Institute of British Architects*, 43 (7 March 1936): pp. 483–6, p. 483.

75 An idea discussed in more detail in Alan Powers, 'Maxwell Fry and Jane Drew – The Romantic Turn', *The Influence of Fry and Drew*, unpublished conference paper, 2013.

76 UEA Archive, Pritchard Papers, PP/22/6/1/11. Speech by Maxwell Fry at Dinner for Walter Gropius, 12 April 1956.

77 'Notes and Topics', *Architects' Journal*, 79 (3 May 1934): p. 628.

78 Gropius became known amongst his close friends in England as 'Pius', although it is not clear who was responsible for the nickname, which clearly references his dogmatic approach to Modernism.

79 UEA Archive, Pritchard Papers, PP/32/1/1. Letter Fry to Gropius, 15 June 1934.

80 HL, Harvard University, Walter Gropius Papers in the Bauhaus-Archiv MS Ger 208.2 (109). 'Design and Industries Association Annual Dinner List', 7 December 1934.

81 Gropius had made an earlier visit to England in June and July 1933 at the invitation of the philanthropic couple, Leonard and Dorothy Elmhirst. The Elmhirsts asked Gropius to advise on work at their country estate, Dartington Hall in Devon. For full discussion of Gropius's transition to England, see: James Lewis, 'Walter Gropius in England 1934–37: Adaptation, Expectation and Reality', *Docomomo Journal*, 40 (March 2009): pp. 4–7; David Elliott, 'Gropius in England: A Documentation 1934–1937', in *A Different World: Émigré Architects in Britain 1928–1958*, (ed.) Charlotte Benton (London, 1995): pp. 107–23.

82 Elliott, 'Gropius in England', p. 108.

83 UEA Archive, Pritchard Papers, PP/32/1/1. Letter Fry to Gropius, 15 June 1934.

84 E. Maxwell Fry, 'Walter Gropius', *Architectural Review*, 117 (March 1955): pp. 155–7, p. 155.

85 Fry, *Autobiographical Sketches*, pp. 146–9; Pritchard, *View from a Long Chair*, pp. 101–9.

86 As shown by a cheque returned to Jack Pritchard by Fry, asking for it to be made out to 'Adams, Thompson and Fry' rather than the original 'Adams, Thompson and Fry and Professor Gropius'. UEA Archive, Pritchard Papers, /15/3/11/26. Letter Fry to Messrs Isokon Ltd, 6 February 1935.

87 To comply with the Aliens Registration Act of 1914 and the Aliens Order of 1920, émigrés were required to demonstrate their ability to support themselves and their dependents via an offer of work or a British guarantor, see: Benton, *A Different World*, p. 45.

88 Morris and Murphy, 'Max Fry', p. 54.

89 William H. Jordy, 'The Aftermath of the Bauhaus in America: Gropius, Mies, and Breuer', in *'Symbolic Essence' and Other Writings on Modern Architecture and American Culture*, William H. Jordy (New Haven, CT, 2005): pp. 187–224, p. 204.

90 UEA Archive, Pritchard Papers, PP/23/2/73 Letter Pritchard to Coates, 25 November 1931.

91 This aggravated the already strained relationship between Pritchard and Wells Coates. Coates claimed ownership of the Isokon, Lawn Road concept: 'the whole idea of the small flats on that site was mine'. See Cantacuzino, *Wells Coates*, p. 59.

92 HL, Harvard University, Walter Gropius Papers in the Bauhaus-Archiv MS Ger 208.2 (134). 'Notes following inspection of site of A.P. Simon's land "Lyndale", The Beeches, Manchester, on Saturday, 23rd June, 1934'.

93 UEA Archive, Pritchard Papers, PP/15/3/1/27–28. Letter Pritchard to L.K. Elmhurst [sic], 10 May 1935.

94 UEA Archive, Pritchard Papers, PP/15/3/1/6. Letter Pritchard to Ian MacAlister, 12 February 1935.

95 UEA Archive, Pritchard Papers, PP/15/3/2/16. Prospectus of Isokon (Windsor) Limited.

96 'Cry Stop to Havoc', *Architectural Review*, 77 (May 1935): pp. 188–92, p. 189.

97 'Cry Stop to Havoc', *Architectural Review*, p. 191.

98 UEA Archive, Pritchard Papers, PP/15/3/2/16. Prospectus of Isokon (Windsor) Limited.

99 RIBA Archive, F&D/18/4. Letter Fry to Drew, 30 October 1944.

100 Clough Williams-Ellis, *Architect Errant* (London, 1971), p. 138.

101 Gropius to Manon Gropius, 11 September 1933 cit. in James Reidel, 'Walter Gropius: Letters to an Angel, 1927–35', *Journal of the Society of Architectural Historians*, 69/1 (March 2010): pp. 88–107, p. 99.

102 UEA Archive, Pritchard Papers, PP/15/3/3/22. 'Memorandum for Nr. 3', 22 December 1934.

103 'Cry Stop to Havoc', *Architectural Review*, p. 192.

104 HL, Harvard University, Walter Gropius Papers in the Bauhaus-Archiv MS Ger 208.2 (109). 'Memorandum for Nr. 3', [1934]. Gropius's typescript has pencil alterations in what appears to be Fry's handwriting.

105 David Dean, *The Thirties: Recalling the English Architectural Scene* (London, 1983), p. 60.

106 Dean, *The Thirties*, p. 60.

107 UEA Archive, Pritchard Papers, PP/15/3/7/26. Photographs of Windsor Site; PP/15/3/3/14. Letter Fry to Pritchard, 29 November 1934.

108 Coates offered Gropius Proskauer's services, suggesting that Gropius might find it useful to employ a German-speaking architect with experience of English practice; Gropius accepted, initially for a two-month contract. HL, Harvard University, Walter Gropius Papers in the Bauhaus-Archiv MS Ger 208.2 (109). Letter Coates to Gropius, 16 November 1934.

109 Sise had worked for Le Corbusier and for Howe and Lescaze, and was persuaded to attend the 1933 CIAM Conference in Athens by Erno Goldfinger. See *Charles Hill Interview with Hazen Sise*, 1 February 1974, pp. 20–1, http://www.gallery.ca/cybermuse/servlet/imageserver?src=DO918–1000&ext=x.pdf. Accessed: 10 May 2013.

110 UEA Archive, Pritchard Papers, PP/15/3/3/31. Meeting at Home Office, 21 February 1935.

111 John Gloag and Frank Pick were approached by Jack Pritchard to act as directors, but both declined. UEA Archive, Pritchard Papers, PP/15/3/2/16. Prospectus of Isokon (Windsor) Limited.

112 See for example: 'Housing Experiment. Concentration of dwellings amid natural beauty', *Yorkshire Post*, 29 June 1935; 'The Flat in the Park', *Manchester Guardian*, 29 June 1935.

113 UEA Archive, Pritchard Papers, PP/15/3/1/6. Letter Sybil Shand to Messrs. Frigidaire, 3 August 1935.

114 Elliott, 'Gropius in England', p. 122.

115 Howe worked with Fry after Gropius's departure until the office was wound up in September 1939. In 1943 he went to work with Edric Neel, Raglan Squire and Rodney Thomas at Arcon, designing experimental pre-fabricated housing. See UEA Archive, Pritchard Papers, PP/9/19/3–4.

116 Lyons worked for Fry and Gropius from February 1936 to March 1937, and was paid £5 per week. See Neil Bingham, 'The Architect in Society: Eric Lyons, his circle and his values', in *Eric Lyons and Span*, (ed.) Barbara Simms (London, 2006): pp. 1–21.

117 Baldwinson (1908–69) returned home to Australia in 1936.

118 Sise is recorded in the MARS Group minutes as late as January 1936. RIBA Archive, ArO/1/4/5. Minutes of MARS Group Meeting, 29 January 1936.

119 Maxwell Fry, 'The MARS Group Plan of London', *Perspecta*, 13–14 (1979): pp. 165–6, p. 166.

120 Alistair Fair, '"A new image of the living theatre": the Genesis and Design of the Belgrade Theatre, Coventry, 1948–58', *Architectural History*, 54 (2011): pp. 347–82, p. 353.

121 Freddie Charles was a Liverpool graduate; Howe, Lyons, and Mills all studied at the polytechnic. Lyons graduated from the evening school in 1932, just before Fry began his lecturing post, and joined the practice in 1936 following Gropius's arrival.

122 G.A. Mitchell, 'School of Architecture, Surveying and Building', *The Polytechnic Magazine*, 72 (September 1932): p. 184.

123 RIBA Archive, Ove Arup Papers, ArO/1/5/41. Annual Report of [MARS Group] CEC, [1936].

124 RIBA Archive, Leslie Martin Papers, MaL/1. Letter Fry to Leslie Martin and Sadie Speight, 11 September 1936.

125 RIBA Archive, Leslie Martin Papers, MaL/1. Letter Fry to Martin, 11 September 1936.

126 Louise Campbell, 'Gropius in Old Church Street: A "comedy of errors" in Chelsea', *Stylistic Dead Ends?*, unpublished conference paper, 2013.

127 Walter Gropius, *The New Architecture and the Bauhaus* (London, 1935), p. 112.

128 HL, Harvard University, Walter Gropius Papers in the Bauhaus-Archiv, MS Ger 208.2 (650). Letter Read to Gropius, 15 June 1934.

129 E. Maxwell Fry, 'Design and the Machine', *Architects' Journal*, 81 (24 January 1935): pp. 157–8, p. 157.

130 RIBA Archive, F&D/13/4. Maxwell Fry, 'Maxwell Fry and the Modern Movement', undated.

131 Fry cit. in Lionel Esher, *A Broken Wave: the rebuilding of England, 1940–1980* (Harmondsworth, 1981), p. 301.

132 The Analysis section of the *Architects' Journal* gave an in depth review in June 1935.

133 The Project Engineer was Kirkwood Dodds.

134 'Sun House', *Architects' Journal*, 84 (13 August 1936): pp. 210–14.

135 Hans Feibusch, *Mural Painting* (London, 1947), p. 15.

136 'House near Kingston, Surrey', *Architects' Journal*, 86 (18 November 1937): pp. 784–7, p. 787.

137 Syracuse University Library Archive, Marcel Breuer Papers. Letter Fry to Isokon, 8 June 1937.

138 Drawings in the RIBA Collection give Gropius's name as part of 'Adams, Thompson and Fry' suggesting he was the lead designer.

139 For further discussion of these building's relationship, see Campbell, 'Gropius in Old Church Street'.

140 RIBA Archive, F&D/23/3. Letter Gropius to Fry, 25 April 1938.

141 UEA Archive, Pritchard Papers, PP/24/14/60. Letter Roger Thompson to Jeremy Gould, 16 September 1972.

142 Jeremy Gould, *Modern Houses in Britain, 1919–1939* (London, 1977), p. 24.

143 For a short biography of Proskauer, see Benton, *A Different World*, p. 197.

144 HL, Harvard University, Walter Gropius Papers, MS Ger 208 (265). Letter Gropius to Hening, 21 March 1935.

145 Morris held the post from 1922 to 1954, see: Harry Rée, *Educator Extraordinary: The Life and Achievement of Henry Morris* (London, 1973), ix.

146 Henry Morris, *The Village College: Being a Memorandum on the Provision of Educational and Social Facilities for the Countryside, with Special Reference to Cambridgeshire* (Cambridge, 1924), p. 21.

147 Morris, *The Village College*, p. 21.

148 UEA Archive, Pritchard Papers, PP/22/1/1/1. Henry Morris, 'Architecture, Humanism, and the Local Community', uncorrected proof for *RIBA Journal*, [c. 1956].

149 Fry, *Autobiographical Sketches*, p. 94.

150 Jackson, *The Politics of Architecture*, p. 58.

151 Andrew Saint, *Towards a Social Architecture: The Role of School-building in Post-war England* (London, 1987), p. 42.

152 In 1949, Henry Morris arranged for the appointment of Howe to design a Regional College of Further Education in Cambridgeshire. Although unbuilt, Howe's design shows the clear influence of his work for Fry and Gropius at Impington. See Rée, *Educator Extraordinary*, pp. 105–7.

153 RIBA Archive, F&D/23/3. Letter Fry to Gropius, 25 August 1938.

154 RIBA Archive, F&D/23/3. Letter Gropius to Fry, 2 March 1938.

155 Plans and photographs of Fry's Miramonte featured in the same publication. For the Papworth project, see Martin et al (eds), *Circle*, pl. 21–22.

156 RIBA Archive, F&D/12/1. Letter A.L. Binns, Education Officer for County Council of the West Riding of Yorkshire, to Fry, 19 July 1944.

157 RIBA Archive, F&D/12/6. Maxwell Fry, typescript of qualifications and selected work, [c. 1945].

158 Fry, *Autobiographical Sketches*, p. 158.

159 'The "Building Now" Exhibition', *Builder*, 170 (12 April 1946): pp. 351–2.

160 RIBA Archive, F&D/12/1. Letter Fry, Gold Coast, to A.L. Binns, 14 November 1944. Architect's drawings of the Lincolnshire or West Riding schools have not been located to date.

161 HL, Harvard University, Walter Gropius Papers in the Bauhaus-Archiv, MS Ger 208 (8). Christ's College project typescript, [1936].

162 HL, Harvard University, Walter Gropius Papers in the Bauhaus-Archiv, MS Ger 208.2 (763). Letter Waddington to Gropius, 14 December [1936].

163 For further discussion of the project, see Alan Powers, 'Conservative Attitudes: Walter Gropius in Cambridge and Maxwell Fry in Oxford', *Twentieth Century Architecture*, 11 (forthcoming).

164 RIBA Archive, F&D/23/3. Letter Fry to Gropius, Regis Hotel, New York, 30 March 1937.

165 RIBA Archive, F&D/23/3. Letter Fry to Gropius, 12 March 1938.

166 Fry, *Autobiographical Sketches*, p. 127.

167 RIBA Archive, F&D/12/6. E. Maxwell Fry, untitled typescript of qualifications and selected works, [c.1945].

168 Fry cit. in Morris and Murphy, 'Max Fry', p. 57.

169 Alan Hudson-Davies became a good friend of Fry and, in his position on the board of Pilkington Brothers' directors, later worked with him on the new Pilkington Headquarters (1955–65) at St. Helens.

170 Fry writes that the job was commissioned on 11th March 1938. RIBA Archive, F&D/23/3. Letter Fry to Gropius, 12 March 1938.

171 'Flats in Ladbroke Grove', *Architects' Journal*, 88 (29 December 1938): pp. 1067–72.

172 The building is now known as Prankerd House. Refurbished in 2002 by Todd Architects, it provides accommodation for university students.

173 Fry cit. in Powers, *Britain*, p. 71.

174 Fry's friend Christian Barman was an important innovator in this field, designing electric appliances in modern materials and it is possible that Barman recommended Fry to the electricity show room owners.

175 'Working Details: 57–58. Shop-Front. Showrooms in Victoria Street, S.W.1. Adams, Thompson and Fry', *Architects' Journal*, 79 (31 May 1934): pp. 801–2.

176 'The Contemporary Industrial Design Exhibition', *Architects' Journal*, 80 (1 November 1934): pp. 644–5.

177 Syracuse University Library Archive, Marcel Breuer Papers. Letter Fry to Breuer, 3 October 1935.

178 'Exhibition of Contemporary Furniture at the Mansard Gallery, W.1'., *Architects' Journal*, 83 (30 April 1936): pp. 652–3.

179 With seven exhibition sub-committees to co-ordinate, Fry reported to Gropius of the 'sluggish' progress of research in June. See Allan, *Berthold Lubetkin*, p. 316.

180 Darling, *Re-forming Britain*, p. 120.

181 E. Maxwell Fry, 'Introduction', in *Moholy-Nagy: Paintings and Collages 1914–1946* (London, 1961), unpaginated.

182 The New Vision was a term invented by Moholy-Nagy to describe his belief that modern photographic techniques could lead to a new way of seeing. This idea formed the basis of his book *The New Vision, from Material to Architecture* (1938).

183 Mary Anne Staniszewski, *The Power of Display: A History of Exhibition Installations at the Museum of Modern Art* (London and Cambridge, MA, 1998), p. 44.

184 HL, Harvard University, Walter Gropius Papers in the Bauhaus-Archiv, MS Ger 208 (164). Letter Gropius to Architectural Press, 8 November 1936.

185 HL, Harvard University, Walter Gropius Papers in the Bauhaus-Archiv, MS Ger 208 (5). Speech for the Opening of the Moholy Nagy Exhibition at London Galleries, 31 December 1936.

186 E. Maxwell Fry, 'Introduction', *Moholy-Nagy*, unpaginated.

187 Kepes was left to complete the photomural for Moholy-Nagy. RIBA Archive, F&D/23/3. Letter Fry to Gropius, 8 November 1937.

188 RIBA Archive, F&D/23/3. Letter Fry to Gropius, 8 January 1938.

189 RIBA Archive, Ove Arup Papers, ArO/1/2/22. Minutes of MARS Group meeting at 55 Gordon Square, 20 April 1937.

190 RIBA Archive, F&D/23/3. Letter Gropius to Fry, 7 June 1937.

191 RIBA Archive, F&D/23/3. Letter Gropius to Fry, 8 November 1937.

192 Fry wrote to Gropius, 'how much we will miss Moholy! His wife told me today that his appointment was confirmed and my heart sank a lot'. RIBA Archive, F&D/23/3. Letter Fry to Gropius, 17 August 1937.

193 Gold, "A Very Serious Responsibility", p. 260.

194 'The MARS Exhibition', *AA Journal*, 54 (February 1938): pp. 386–8, p. 386.

195 He wrote to Gropius: 'The committee of selection looks pretty ga-ga, but with bright spots. I am preparing a careful portfolio'. RIBA Archive, F&D/23/3. Letter Fry to Gropius, 15 January 1938.

196 'Community Centre', *Focus*, 4 (Summer 1939): pp. 25–9.

197 Harriet Atkinson, *The Festival of Britain: A Land and its People* (London, 2012), p. 208 [25].

3

Jane Drew and the Partnership's Origins

At the opening of the 1930s, Fry was making his name as a great hope of the British Modern Movement. Jane Drew, meanwhile, was just beginning her architectural education. Drew was amongst the student cohort that entered the Architectural Association in 1929, which included 13 or 14 women, and she collaborated with several of her contemporaries in the years that followed. Indeed, Drew's work during the late 1930s and the Second World War was significant in the continuing development of Modernism in Britain. As such, Drew might be grouped with a 'second generation' of modernists who began their careers at the AA between 1926 and 1931.[1] Her marriage to Fry in 1942 invigorated both of their personal and professional lives. The union brought together two generations of British modernists, which had a significant impact on their work, collectively and individually. As Fry later wrote, 'I began my life again at forty with a zest that carried me in the fullest exercise of all my faculties for the next twenty years'.[2]

Later, Drew would often be introduced as the wife of the well-known architect, 'Mrs Maxwell Fry', rather than as an architect in her own right.[3] Such stories reflect the difficulties present for women architects of the period. As Catherine Burke notes, during Drew's lifetime, to succeed in architecture 'women were forced by social and cultural expectations to tread a very fine line with regard to gender identity'.[4] Traditionally feminine attributes, such as gentleness and modesty, seemed to be important. Drew, however, did not adhere to these rules; 'There was nothing narrow about Jane. She was impatient with narrowness. She balked at snobbery. She was intolerant of injustice'.[5] Yet her strong personality led to almost a caricature of this strength amongst her peers. Recollections of Drew constantly reference a lack of architectural ability, her adulterous nature and a desire to shock and upset. Drew's 'value' as a designer is often downplayed or dismissed entirely, and Fry seen as the creative talent of the duo. Drew's former employee and friend Trevor Dannatt, for example, describes her as a 'rainmaker' and 'not a talented architect'.[6] This chapter looks beyond these simple readings of her character, drawing out Drew's work and ideas to illustrate her distinctive role in the partnership.

Recent scholarship has drawn attention to the women architects working in twentieth-century Britain,[7] although there still remains a scarcity of literature. Women architects were (and remain) in the minority, yet their work was vital to the period and this chapter contributes to the reassessment of women architects and designers of the interwar era. For Drew, gender was a non-issue. Frank Knight, an employee at the practice from 1947, later recalled, 'The first thing I learned from Jane was that the Office did not employ men and women, but architects and each was rewarded on merit unrelated to sex'.[8] Following Drew's outlook, this chapter does not seek to highlight any 'female' qualities of her work as a architect, but to highlight her significance in twentieth century British achitecture.

This chapter investigates Drew's formative influences, architectural training and early work, including her professional partnership with her first husband, James Alliston (1908–2000). It considers the origins of the Fry and Drew partnership in the 1940s and the wartime work in Britain, during Fry's absence in Africa, that followed. The wartime period seems pivotal to Drew's career. Her union with Fry brought her into contact with a new network of influential acquaintances in London, which helped to shape her work in post-war Britain.

Joyce Beverly Drew (1911–96) grew up at 8 Parchmore Road in Thornton Heath, a middle-class suburb on the outskirts of London. The family lived comfortably in a 12-room house, with a nurse and two servants. Her father, Harry Guy Radcliffe Drew (1880–1958), was a designer and manufacturer of surgical instruments and her mother, 'Stella', née Emma Spering Jones (1873–1947), was a school teacher and an amateur botanist (Figures 3.1 and 3.2).[9] Drew had an elder sister, Dorothy (1908–88) (Figure 3.3). Drew benefitted from a particularly a close relationship with her beloved father, whom she described as a 'kind and gentle' man.[10] Guy Radcliffe Drew was a humanist. He 'despised the profit motive and abhorred cruelty',[11] and, as such, he was against the patenting of medical instruments for he believed it would be contrary to the public interest. Radcliffe Drew undertook pioneering research, including his development of new uses for stainless steel in the medical profession,[12] and he was the founder of the Institute of British Surgical Technicians. He was a formative influence for both of his daughters: Dorothy became a doctor and Jane's architectural work was imbued with humanitarian concern from the outset. As one friend later wrote, 'She based her work on the principle that architecture should provide a space in which human beings can flourish, both physically and spiritually'.[13]

Drew benefitted from an excellent visual memory and, whilst writing her autobiography in the 1980s, she sketched scenes from her childhood. With remarkable clarity she recalled her surroundings, most notably her parents' bedroom and the nearby Grange Park, where she was taken for walks as an infant. In contrast to the comfortable nature of her home life, the bleakness of the nearby inner-city surroundings made a deep impression on the young Drew. She later wrote at length of shopping trips with her mother to the Caledonian Market, close to St. Pancras Station:

> *A grey world of purple slate-roofed houses without front gardens or trees: inhabited by unspeakable ill-formed women in caps and curl pins, who beat mats or carry newspaper parcels or sordid bulging bags. ... Up the hill, plough the throng, past a grim sausage factory and a public house which seem[ed] to have stepped straight out of Dickens.*[14]

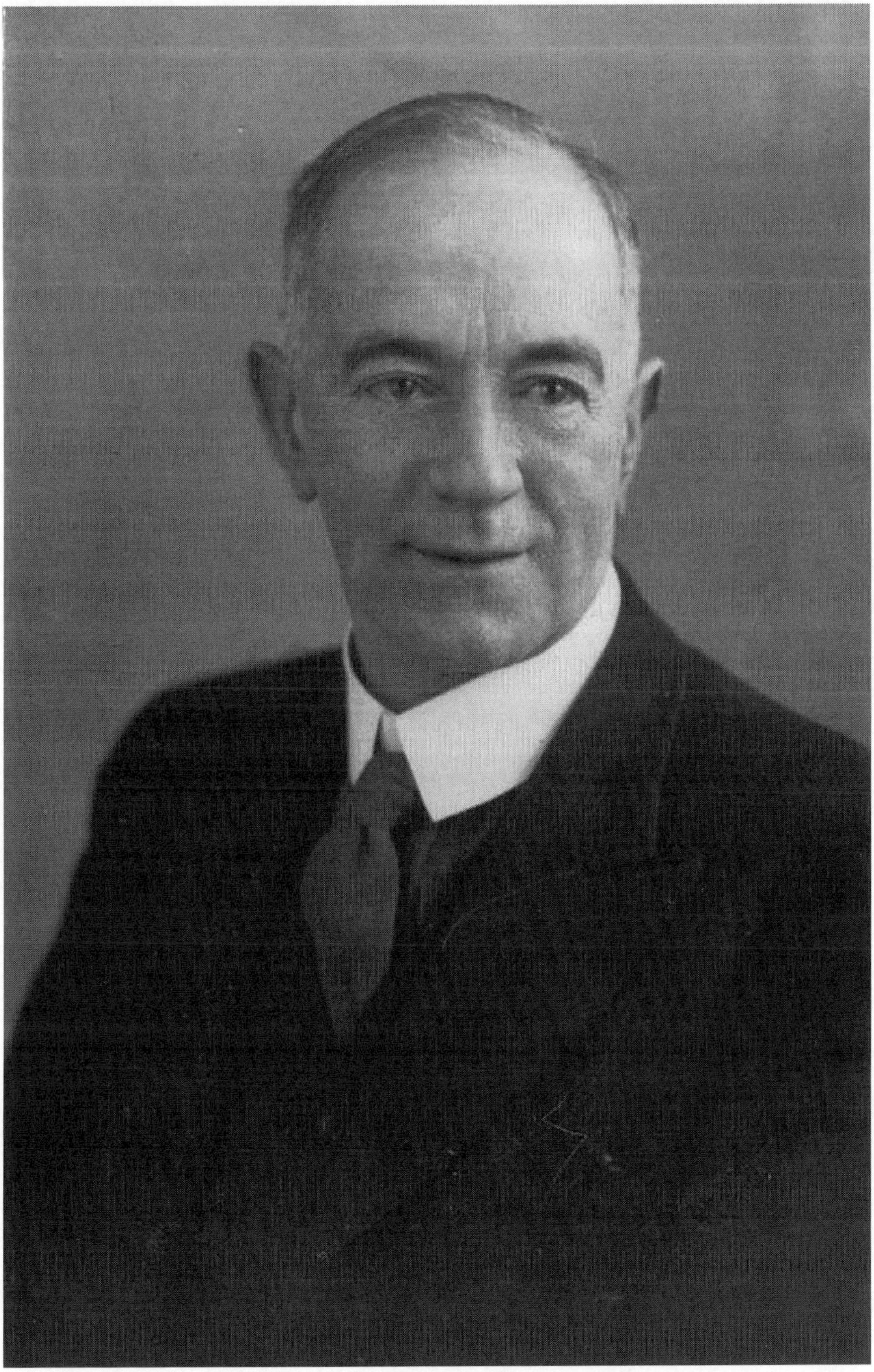

3.1 Harry Guy Radcliffe Drew, 1946

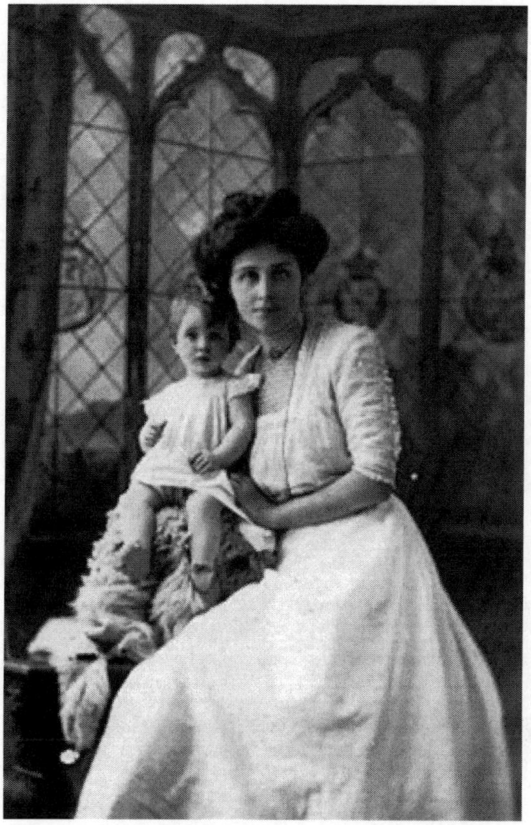

3.2 Emma Spering Drew with daughter Dorothy, 1910

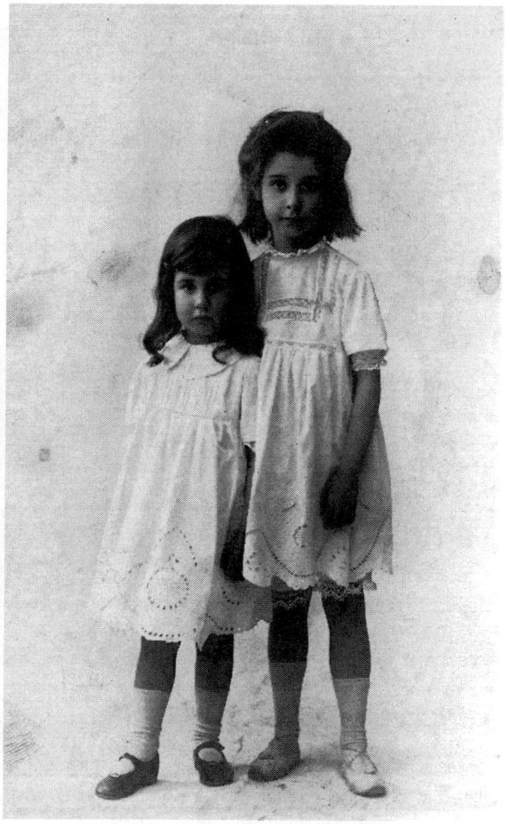

3.3 Jane and Dorothy Drew, c. 1916

Drew's sensitivity to the poor urban conditions sat alongside her appreciation of the vitality of such communities; enthralled by the 'great open bedlam' of the market, she keenly recalled the stallholders's handcarts of vibrant fruit and flowers. Then, 'Lastly we came to the fruit section. This was the liveliest part of the Market. Here was real salesmanship. The showman would stand on the back flap of his van and address the crowd who gathered for the fun. Here was life'.[15] This early exposure to urban life came to have a lasting effect on Drew's architectural outlook.

With her sister, Drew was schooled at the nearby Woodford House School in Croydon. It was here that she befriended Diana Wynyard and Peggy Ashcroft, later both well-known actresses, and the trio made a vow never to adopt a man's surname; a pact that they each preserved.[16] The school for gentlemen's daughters provided opportunity for Drew's independent spirit to flourish, illustrated in a photograph of Drew performing cartwheels on the lawn while her classmates look on in bewilderment (Figure 3.4). Subsequently, at the Croydon High School for Girls, Drew excelled at sport, languages and acting. Drew's leadership qualities were recognised early on and, to her 'great surprise', she became Head Girl until she graduated, aged 18.[17] She was evidently sociable from an early age, taking part in

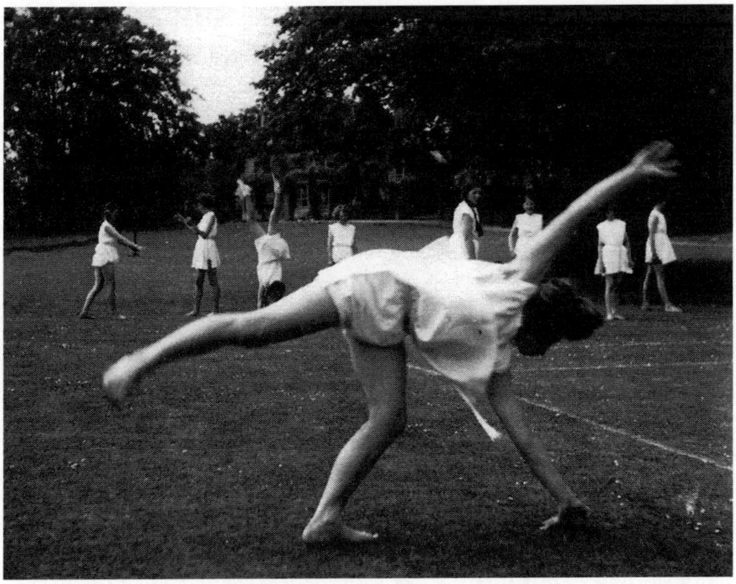

3.4 Jane Drew at Croydon High School for Girls, 1925

amateur dramatics and the school orchestra. Drew finished her schooling placed in the 'Second Division' with qualifications in English, Elementary Mathematics, Latin, Chemistry, French and English History.[18]

THE ARCHITECTURAL ASSOCIATION AND EARLY WORK

In September 1929, Drew enrolled for a five-year architectural course at the Architectural Association (AA). The school, located in the urbane Georgian terraces of 34–36 Bedford Square in Bloomsbury, signalled – in her own words – 'the beginning of a new epoch'.[19] Drew's *Beaux Arts* training came before the modernist transformation in the school's educational programme, which began in earnest in 1933. Yet the students were aware of shifting architectural ideas outside of the rarefied atmosphere of the school and these unofficially filtered through the school. AA summer excursions to Europe were also influential, particularly the 1930 'landmark' trip to Scandinavia,[20] which had an immediate impact on students' work including that of Drew.

Indeed, Drew later spoke of her early influences that reflect the eclecticism of the period: she admired Tengbom and a lot of Swedish architects, and 'the old Gothic things, such as Salisbury Cathedral and Durham [Cathedral]'. Although she commented, 'Where I differed from my fellow students was that it was very fashionable then to like Stockholm Town Hall and all that very thin stuff, which I thought was lacking in vigour and fineness'. Instead, Drew admired 'gutsy' buildings, such as Nicholas Hawksmoor's church St George in the East (1714–29), in Stepney, which she chose as a case study for a measured drawing task.[21] Her distinctive approach was noted by one of the AA tutors, Geoffrey Jellicoe, who later recalled that her work 'was very original, slightly different to the other students ... her personality came out in the drawing'.[22] Drew's reference to 'gutsy' buildings shows the origins of the architectural strong forms that she favoured, in contrast to the more slender proportions of Fry's buildings. Despite her evident passion for the subject, Drew later suggested her architectural studies were rather undistinguished, 'I showed no great talent. ... I pursued my work in a kind of fog not understanding much of what I did'.[23] Her student marks show inconsistency, with some of her best results awarded for two covered market projects, perhaps showing the significance of her trips to Caledonian Market.[24] She struggled with construction modules but, demonstrating a significant imagination and good drawings skills, she enjoyed the *esquisse* tasks.

The close-knit AA community was important to Drew. She later wrote fondly of the comradeship of studio life and remained in contact with many of her classmates after their training had finished.[25] She participated in the school's annual pantomime in 1930 and in 1931. The pantomimes were a vital part of the school's social calender and their production reached almost professional levels; the 1930 performance included choreography and costumes by a young Carmen Dillon, later an Oscar-winning art director. Alongside her friends Humphrey Spender, Bill Edmiston and Camilla Epps, a 'blacked up' Drew featured as 'Amama' in a 1931 sketch of 'British Goods' at 'Jumbo's Home in the Tropics' (Figure 3.5).[26] This busy life at the AA was interrupted by a serious car crash in 1932, which left Drew and her sister, Dorothy, hospitalised. Drew missed the third term of the academic year and graduated later than her classmates in 1934.[27]

An all-woman group of Judith Ledeboer and Jessica Albery (who entered in 1926), Mary Crowley (1927 entry) and Margaret Justin Blanco White (1929 entry) were amongst a new generation of modernists, who began their first year at the AA during the late 'twenties. Following their studies, these women all worked in

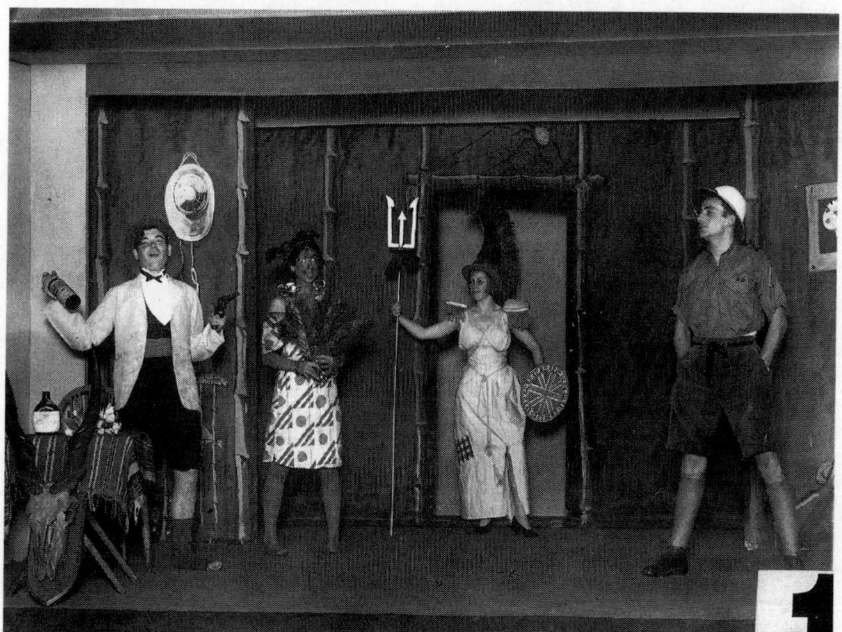

3.5 'British Goods' Sketch, AA Pantomime (Drew, 2nd left), 1931

the voluntary housing sector and, along with Janet Fletcher and Elizabeth Denby, were involved in numerous influential schemes that sought to improve working-class living conditions.[28] Drew may also be added to this significant group. She entered the AA in the same year as Justin Blanco White (1911–2001) and they became firm friends, although Drew did not match the impressively high marks of her classmate.[29]

Justin Blanco White commented that the only women to have successful careers in architecture were those who married architects.[30] Perhaps heeding her friend's words, Drew met and married James Alliston, a fellow AA student, whom she described as 'a big athletic man and very keen on games'.[31] They married on 27 December 1933 at St. Paul's Church in Thornton Heath. Drew's account of early married life reveals a much-desired liberation from what she described as the 'rotten suburb' of her childhood.[32] She travelled in Europe with her mother and sister, and spent time in Cambridge with Alliston's family (Figure 3.6). Most significantly, Drew and her new husband moved into a small, top floor flat at 28 Brunswick Square in Bloomsbury. She later wrote of their bohemian existence:

> *The glamour hung around us. We gave parties and we went to parties: it was a carefree life. ... It was all rather young nineteen-twenty-ish and Bloomsbury. At one time our lives became inextricably mixed with those of a young lawyer and his Russian wife, our passions volatile and surface, especially mine.*[33]

They worked hard to preserve their new-found independence. Alliston was employed by a nearby architect and, whilst finishing her studies at the AA, Drew worked part-time in the offices of two well-known architects: Charles Holden and, later, George Grey Wornum. Both Holden and Wornum were working on high profile projects during the early 30's. Holden was undertaking his series of underground and bus stations for Frank Pick's London Passenger Transport Board. Inspired by the moderate Modernism of Northern Europe, Holden used a palette

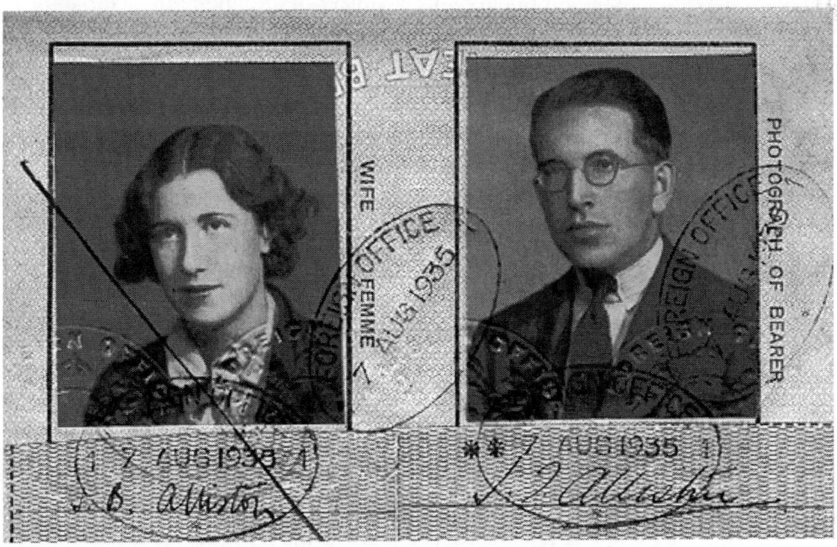

3.6 Passport photographs of Drew and Alliston, 1935

of brick, reinforced concrete and glass combined with artwork and good design to educate the public. Grey Wornum, meanwhile, was developing his competition-winning RIBA Headquarters at Portland Place that also looked to the new Swedish work for inspiration.

Despite her impressive work experience, after graduation Drew experienced considerable difficulty in securing a full-time post. Few architectural practices would appoint women architects at this time, although in 1934 she secured employment with the architect Joseph Hill (1888–1947). Work in Hill's practice, like that of the AA studios, reflected the eclecticism of the 'thirties; he specialised in Neo-Georgian public houses, Art Deco theatres, and Modern factories and apartment buildings. Projects undertaken during Drew's period of employment would have included the Art Deco Odeon Cinema in Surbiton (1934) with stylish interiors by Mollo and Egan (Figures 3.7 and 3.8).

While Drew found working on Hill's Art Deco cinemas and 'sham' Georgian pubs demoralising,[34] the post brought her into contact with members of bohemian London who would have a lasting impact on her work. Just a ten-minute walk from Drew and Alliston's flat, Hill's architectural practice at 34 Gordon Square (now demolished) was better known as the address of the renowned actors, Elsa Lanchester and Charles Laughton. Drew later recalled socialising with Lanchester and Laughton; following difficult evenings, struggling with her work, the couple would invite her for drinks at their home above Hill's offices.[35] In 1934, Lanchester had bought the top three floors of the Georgian terraced house and commissioned Wells Coates to modernize the interior as a backdrop fitting to their informal, modern lifestyle (Figure 3.9).[36] Spread over three floors, Coates's accommodation provided the couple with spacious living and dining rooms, two servant's rooms, individual bedrooms and an office.

The main living spaces were linked via a series of sliding doors to give flexible, functional spaces to suit different events. As Lanchester noted, 'the flat was rather unfurnished, but not quite as unfurnished as most people thought'.[37] Laughton and Lanchester's model of modern living served as inspiration for Drew's work of the late 1930s. A similar informality and flexibility of arrangement is apparent in her first domestic designs built for similarly progressive clients just a few years later.

During this period, Drew and Alliston worked hard to establish their own practice and, by late 1936, the couple had moved to a nearby flat at 24 Woburn Square. They set up a small office, working in their own time in an effort to secure commissions that might allow them to set up in private practice. They employed some of their AA classmates, including Gordon Tait and Michael Thornley.[38] The latter described the office as 'great fun',[39] but there was hard work too and, in 1937, this resulted in a first prize of £100 in a competition for a new cottage hospital at Dawlish in Devon.[40] The couple also placed third in a competition for Chester Royal Infirmary, entered an open competition for St. George's Hospital (1938) at Hyde Park Corner in London and Drew also worked on Putney Hospital with the architect Halliburton Smith.[41] The influence of Drew's father is evident, for Drew held a lifelong interest in hospital design and she championed the provision of adequate healthcare facilities throughout her career.

Although their scheme for a cottage hospital at Dawlish was not built, it provided a starting point for Alliston and Drew's fledgling practice. A series of domestic

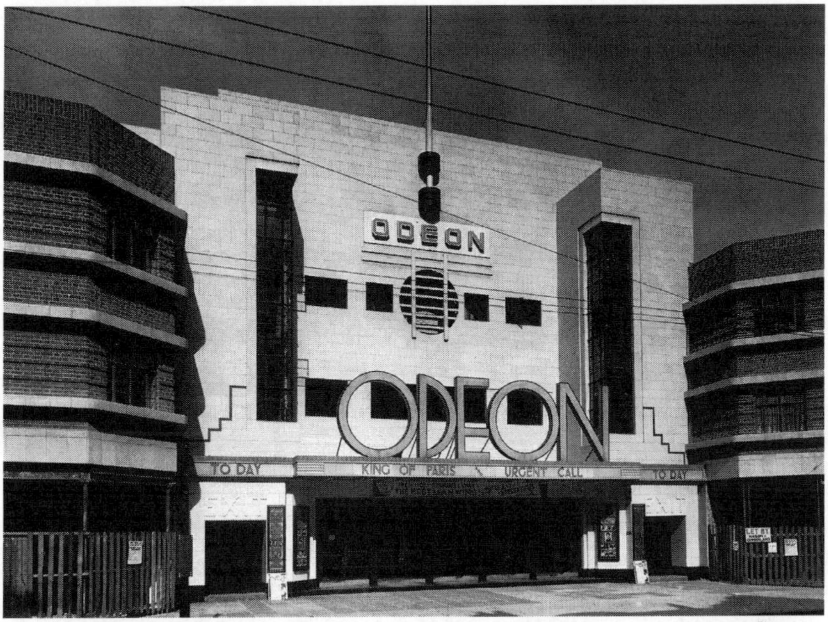

3.7 Odeon Cinema, Claremont Road, Surbiton, London, 1935

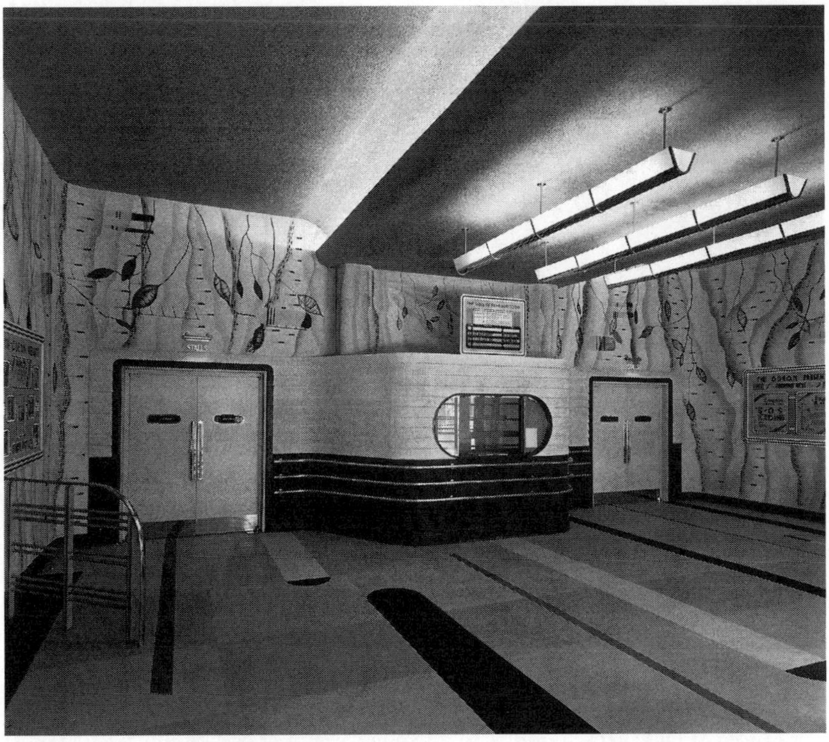

3.8 Entrance Foyer by Mollo and Egan, Odeon Cinema, Surbiton, 1934

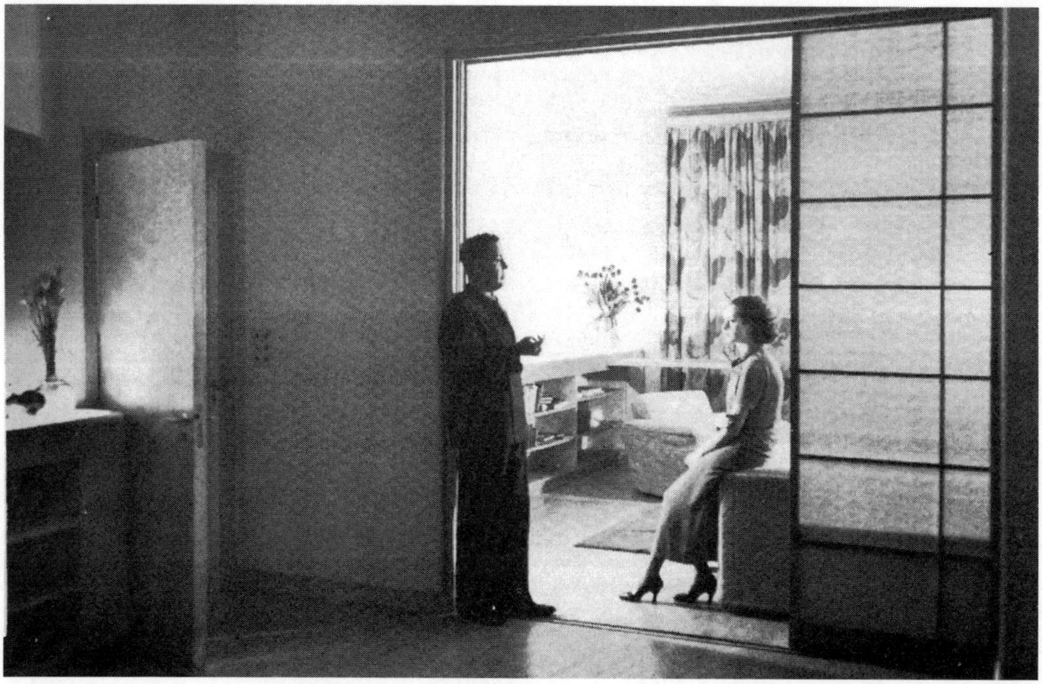

3.9 Lanchester and Laughton at 34 Gordon Square, Bloomsbury, London, 1938

commissions followed, including a house in Cambridge for Alliston's mother. Drew attributes the Cambridge house solely to Alliston in her autobiography, noting how he stayed in the city to supervise its construction.[42] Certainly the building is more traditional than two contemporaneous designs by Drew for houses in Kent and in Hampshire. Both of these houses were designed for healthcare professionals, suggesting that the commissions may have come from contacts of Drew's father. It seems likely, therefore, that the duo worked as lead designers on their own projects under the umbrella of the partnership – perhaps following some early collaboration in the design stages – in a similar manner to Fry and Drew's later working arrangement.

One of Drew's first houses was built at Cliftonville in Kent, constructed of distempered Fletton brick, with a flat roof and metal windows (Figure 3.10). Situated on a corner plot of the newly-laid Gloucester Avenue, the house is distinct from the neighbouring new houses in a typical 30s style. Drew designed a house of brick construction for the client, Dr Stone,[43] which also included a consulting surgery. The scheme shows Drew's experimentation to create a series of interlinked spaces for modern living: at the front of the building at ground floor level, a suite of hall, waiting room, surgery and apparatus store are provided; at the rear of the house, a living room, dining area and waiting room, with a Columbian pine dancefloor, could be thrown open to provide a spacious entertaining area (Figure 3.11).[44] The need for a self-contained suite of hall, waiting room and surgery creates a somewhat awkward layout, but the flexibility of space provided for her modern, professional client suggests that Drew was influenced by the lifestyle enjoyed by her Bloomsbury friends, such as Lanchester and Laughton.

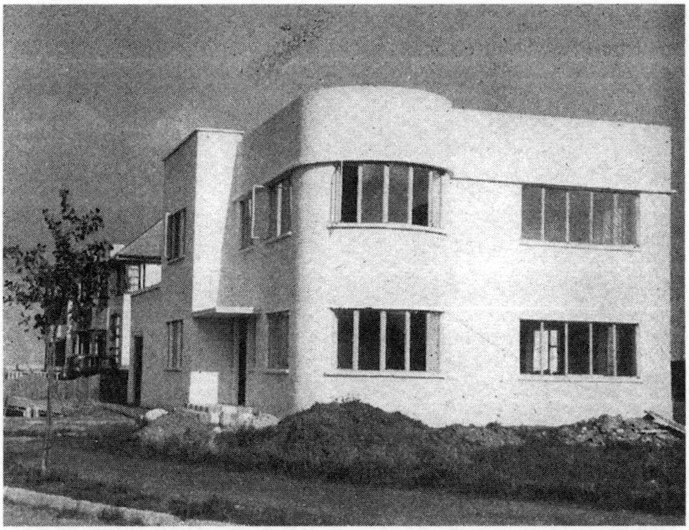

3.10 View from the South-West, Gloucester Avenue, Cliftonville, Kent, 1937

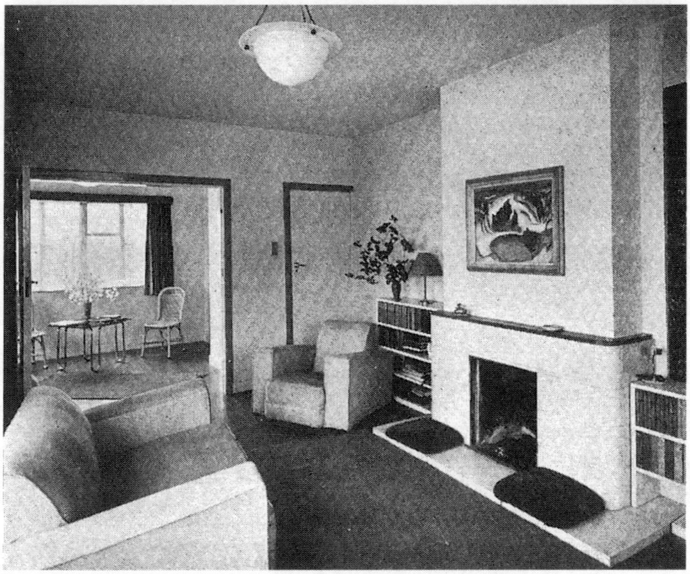

3.11 Living Room towards the Waiting Room, Gloucester Avenue, Cliftonville, Kent, 1937

In the following year, Drew began work on a house at St. Giles's Mount in Winchester. Published in the *Architect & Building News* (*ABN*), the journal attributes the building solely to Drew with the garden layout designed by the landscape architect Richard Sudell (1892–1968).[45] Like the Cliftonville scheme, the house was required to include a consulting room, this time for the female psychologist client. Drew employs a similar plan, with rooms accessed from a generous central hall, although she later acknowledged it was 'not a very good plan. It was not a free plan'.[46] Built of loadbearing-brick with smaller elements clad in red cedar weather-boarding, the house reflects the wider trend for modernist houses constructed of natural materials. In contrast, Suddell's garden layout was traditional in approach and appears to draw nothing from the outdoor living encouraged by the south-facing loggia and pergola, accessed from the dining room (Figures 3.12 and 3.13).

THE BEGINNINGS OF FRY AND DREW

While Drew and Alliston's professional lives were progressing, their personal relationship was not a happy one. In 1939 the couple separated and Drew became

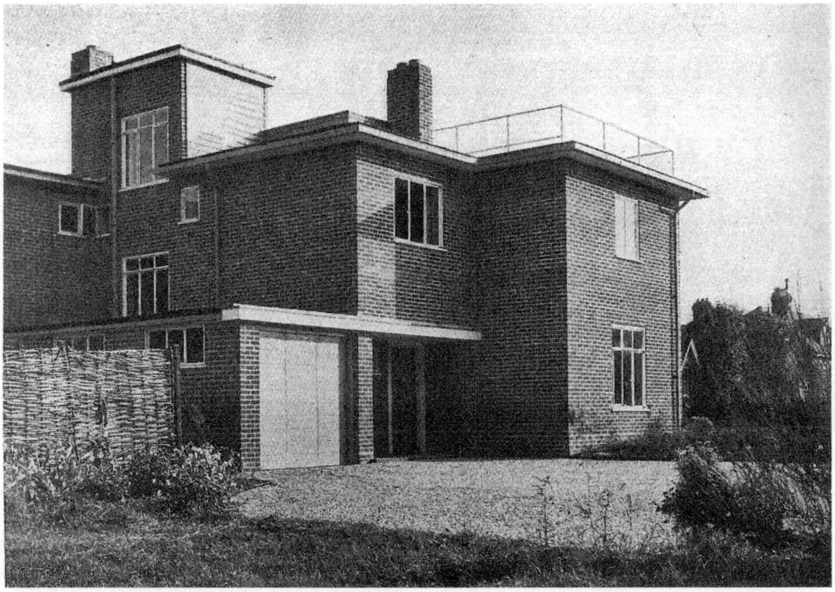

3.12 View from the North-East, St. Giles's Mount, Winchester, 1938

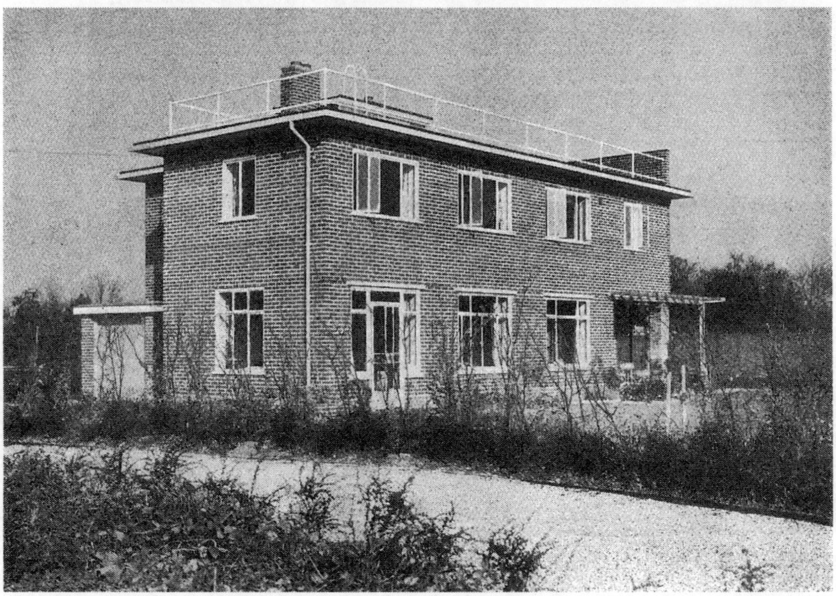

3.13 South Façade, St. Giles's Mount, Winchester, 1938

solely responsible for their twin girls, Georgina and Jenny Alliston (b. 1937). It was clearly a difficult time for Drew and she threw herself into committee life, 'It was a form of outlet. I missed the warm community of studio life – some 60 of us together. Committees added to this and to my sense of self-importance'.[47] At one such RIBA meeting, Drew and Fry met for the first time. Fry recalled his difficulties leading up to their fateful encounter in a letter to Gropius:

> *Having broken every … [relationship] within reach I found myself temporarily without support, and for a short period lived as nearly like a character in a Dostoevsky novel as it is possible to imagine, being on occasion nearly chucked out of the Café Royal … In her own circle my Jane was going through a similar experience, and our own circles met and mingled … by the spring of 1940 we were installed in the delightful little flat in St James.*[48]

Drew joined Fry at his flat at 3 King Street and there, in 1940, the beginnings of the Fry and Drew practice emerged. With Justin Blanco White, they collaborated on the design of a Rodent House for London Zoo, a project which stemmed from Julian Huxley's role as the Secretary of the Zoological Society of London.[49] It was funded by the natural historian Sir John Ellerman, who provided £500 for Fry to spend 'at leisure' on the initial design development. However, this project and a scheme for a 'nursery suite' for Fry's friend, Captain Williams, both came to nothing.[50] These small-scale, faltering projects were typical of the period, as were the numerous, informal collaborations that were formed in an effort to share or create work. Blanco White, for example, had collaborated with Erno Goldfinger and Mary Crowley in the previous year, winning second prize in a competition for prefabricated children's accommodation.[51] Despite these difficulties, the architectural community in London remained active and the MARS Group held meetings well into 1940; in July of that year, John Summerson chaired a session at Goldfinger's house at 2 Willow Road in Hampstead on 'The Influence of Class in the Profession'.[52] Indeed, Fry's correspondence to Gropius during this period articulates a life of privileged freedom amongst the uncertainty of wartime:

> *while our lives are confused within the limits of this horrid structure we have many hours of most delicious liberty to buy books in Charing Cross Road, to listen to string quartets at the National Gallery, to gather as we did last week to talk architecture at a Mars meeting.*[53]

The period was further unsettled by the personal difficulties experienced by Fry and Drew. Fry's acrimonious divorce, and his family's dim view of the matter,[54] was compounded by the couple's enstrangement from their respective children. Drew's daughters, Jenny and Georgina, had been evacuated to North Wales with their nanny, Maud Austin.[55] While Fry had initially intended to send his daughter, Ann, to stay with the Gropius family in Massachusetts, but reconsidered and she instead remained at boarding school.[56] Then, in 1941, Fry was billetted to the Royal Engineers in Derby, lodging at Warwick Avenue with, as he wrote, 'a cockney woman who has a heart for cooking and kindness'.[57] He was 'nowhere near the war and terribly bored',[58] and spent his free time working on the MARS Plan of London and his manuscript for *Fine Building* (1944).[59] The couple had sizeable debts, presumably from the lengthy divorce proceedings, and their correspondence from this period talks of undertaking other work, such as writing and editing publications, to supplement their incomes.[60] Finally, in October 1941, Fry's divorce was finally settled. On 25 April 1942 and a few days before Fry left for a posting in West Africa, the couple married at Caxton Hall in Westminster (Figure 3.14). Julian Huxley was again Fry's best-man and a celebration followed at their favoured haunt, the Café Royal.

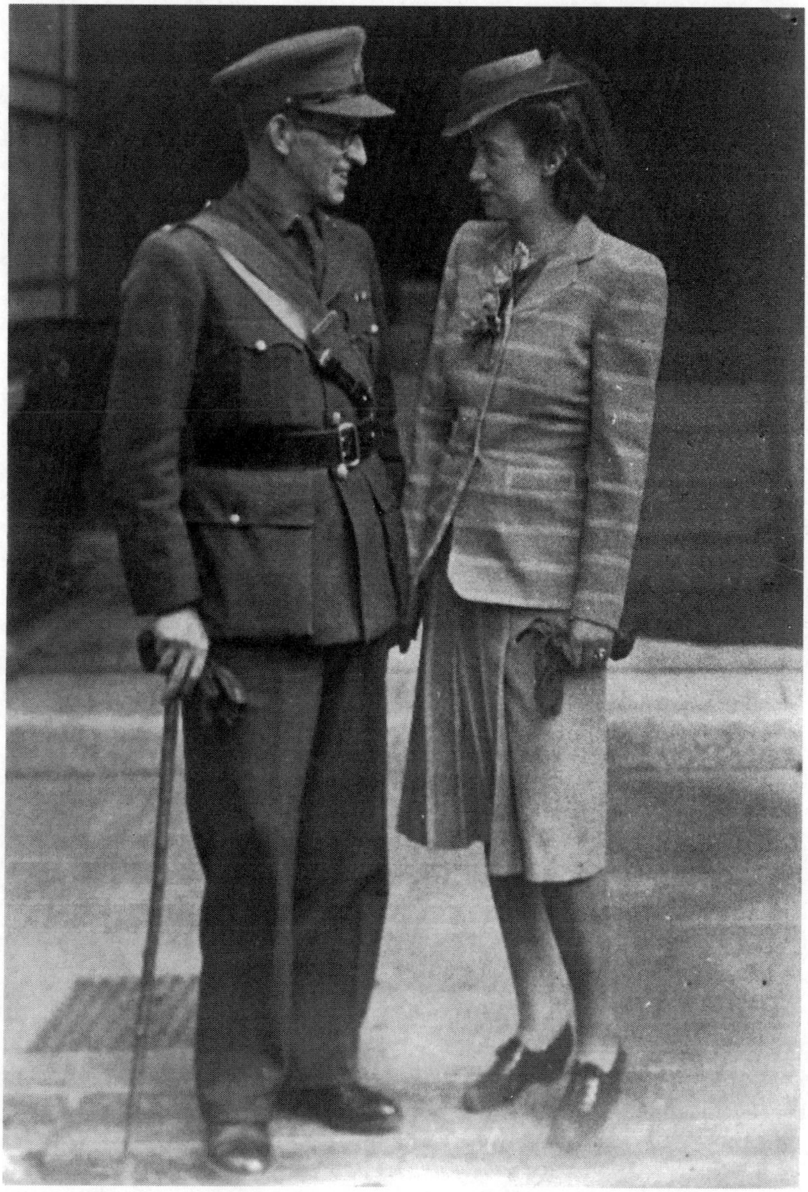

3.14 Fry and Drew on their wedding day, 1942

Meanwhile, during Fry's period in Derby, Drew had set up her own office on the first floor at 3 King Street, below the couple's former flat. The practice was established following a commission for Walton Yacht Works at Walton-on-Thames in Surrey, a 'shadow factory' for the War Office, a project secured for Drew by Fry from his friend Charles Kearley.[61] Drew had initially intended to employ only women staff in reaction to the discrimination she had received as a graduate. However, it seems her intentions were short-lived and amongst her earliest employees were John Terry, Diana Rowntree, Trevor Dannatt, Riehm Marcus, Denis Roberts, Kurt Linden and F.L. Marcus, a German émigré who joined the practice in 1944.[62] Dannatt later recalled his memorable job interview and first meeting with Drew, 'She sat me down on a Marcel Breuer recliner covered with white sheepskin ... and she took me on'.[63] The office was a busy, family affair, with the secretaries Miss Lesser and Miss Barnett, looking after the filing and the Drew's twin girls.[64] Dannatt described it as 'an odd group' swelled by 'an incessant stream of visitors, from all walks of life' when Drew was in the office.[65]

During this period, Drew began to socialise with Fry's circle of friends in London. This influential group was paramount in the development of her ideas and Drew recognised the importance of the cross-fertilisation amongst artists, architects, writers, scientists, psychologists and engineers. Herbert Read, Kenneth Clarke and in particular Peter Gregory (1888–1959) all became close allies for Drew. Gregory was the wealthy Chairman of the art publishers Percy Lund, Humphries & Co and he took an active role in the promotion of contemporary artists. During Fry's absence, Drew came to rely on Gregory and he often provided financial support during this difficult period. Work was in short supply. Dannatt worked on a factory extension in Birmingham for the store fitters, Harris and Sheldon, a project obtained by Drew through her acquaintance with one of the company's directors.[66] Drew also continued with her committee work. In 1941, she became Secretary of the RIBA Public Relations Committee and was, according to Fry, a 'moving force in the Reconstruction Committee' which had been established in March of the same year.[67] The Committee formed groups of specialists to investigate different aspects of reconstruction, and they considered; policy, planning and amenities, housing, building legislation, building technique, architecture and building industry, professional status and qualifications, and public relations.[68] The Committee's findings were presented as part of the 'Rebuilding Britain' exhibition of 1943 to 'help the public clarify its own views', although the exhibition emphasised the need for the British public to decide on the priorities for rebuilding. The exhibition catalogue noted, 'One of the war posters portrays Mr. Churchill saying: "Deserve Victory!" Victory may be achieved by effort, sacrifice, and vision. Reconstruction makes similar demands. The people of Britain will get the reconstruction they deserve'.[69]

The exhibition was curated by Drew and staged at the National Gallery in London, with Rodney Thomas appointed to oversee the project's construction. Fry worked with Drew on the exhibition design and his 1930s', DIA-centred ideas on town planning are still to the fore; the exhibition catalogue described the evolution of the eighteenth century's civilised building tradition into 'great and uncontrolled growth' of late-nineteenth century towns as 'a blight on our landscape', echoing Clough Williams-Ellis's views.[70] Yet this 'twenties and 'thirties polemic was moved forward by Drew, as discussion of the Garden City movement was followed by that of Tony Garnier's Industrial City and Le Corbusier's Ville Radieuse. The catalogue remarked of these schemes, 'They should not be brushed aside as too ambitious, for we have to try and think on this scale',[71] and Baron Haussmann's replanning of nineteenth century Paris was used to illustrate the historic precedent of such large-scale work. Drew later commented on the exhibition:

> *it was probably one of the earliest things on new towns on linear planning, on Tony Garnier's and other theories on industry and so on. It was quite new compared to the Garden City movement and all that had gone before; much influenced by my having been connected with the MARS group and the ideas about planning and CIAM.*[72]

Yet the exhibition was criticised by John Gloag, who wrote to Fry (who was stationed in Accra at the time) of the ambitious scope of the work, 'The fact that I think it's a God awful exhibition doesn't matter ... It certainly looks quite jolly – Rodney Thomas has made a good job of it – but it absolutely and utterly fails to say anything to the ordinary man'.[73] Gloag's view was echoed by Alan Hudson-Davies, who wrote to Fry, 'My highbrow friends think it was too highbrow'.[74] Although critics failed to appreciate the ambitiousness of the project, the exhibition encapsulates Drew's current research into modern town planning. Town planning was a new area of interest for Drew, and one that she quickly grasped. She had been involved in the development of the MARS Plan of London,[75] which Fry had planned in collaboration with the German émigré Arthur Korn (described by Fry as the 'prime mover'), Arthur Ling, William Tatton-Brown, Aleck Low, Felix Samuely, Elizabeth Denby, Bronek Katz, Robert Shaw and Christopher Tunnard.[76] The Plan was heavily indebted to the work Fry had undertaken whilst working with Adams and Thompson. The architectural outcome of individual buildings may not have been the same, and clearly there were shifts in housing density, but the architectural ideas can be traced to *Recent Advances in Town Planning*. In particular, the use of 'Neighbourhood Units' and 'Residential Units' – both of which had been discussed in the New York Regional Plan – were the basis for planning, with populations determined by neighbourhood schools.[77] The MARS Plan adopts the same principle stating that the 'neighbourhood unit' should be 'centred about the elementary school'.[78] Furthermore, Fry acknowledged the use of Le Corbusier's 'grille' system which was overlaid with 'divisions of social activity evolved by the biologist-urbanist Patrick Geddes – work-relation-shelter-communication'.[79] Such ideas formed the basis of Fry and Drew's subsequent work in West Africa and at Chandigarh, as the subsequent chapters will discuss.

Drew also studied for her forthcoming post as an Assistant Town Planner in West Africa, undertaking a course in town planning. She chose Derby, the location of Fry's former army posting, as the subject for her exam and wrote to him, 'I've given Derby a central park, an airport, a Cathedral, a central ring road and removed miles of that horrible narrow fronted housing. It had been nice doing it and thinking of all the delightful walks we had together'.[80] This training was supplemented by her own research, including practical studies with Jacqueline Tyrwhitt on 'activated sludge' and reading of Patrick Geddes's unpublished manuscript on town planning in India. All of which she considered to be 'very relevant' and good preparation for her forthcoming role.[81]

DESIGN AND INDUSTRY

Meanwhile, the war effort was a catalyst for renewed industrial production which brought a range of modern, synthetic materials to the attention of architects. Drew was quick to seize on the possibilities of material innovation for building, perhaps due to her father's pioneering work in stainless steel surgical implements, and she immersed herself in a world of lionide, formica and enamel. While she was interested in the new aesthetic properties of such materials, Drew was more aware of the social impact of new materials. Drew's appointment in 1943 as a Consultant Architect to the Domestic Commercial Heat Services Committee, established by the British Commercial Gas Association, enabled her to investigate these implications in detail and undertake significant research in the planning of modern kitchens for postwar Britain. The commission was typical of those that Drew might expect to gain during this period, as women architects were presumed to pursue work of 'domestic and small scale design as befitted their sex'.[82] For Drew this was not a small matter; she was committed to elevating the lives of ordinary women and commented, 'I feel that every woman agrees that household drudgery must be banished after the war and that's why I'm concentrating on kitchens'.[83]

Fry returned from Accra in September 1943 and in January of the following year they travelled to America. There Drew undertook a six-week research tour to gather information and study the possible uses for new materials, such as chip-proof enamel used as a hygienic coating for stoves. Fry accompanied Drew in order to study the Tennessee Valley Authority development, in connection with his impending post as Town Planner to the Resident Minister in West Africa.[84] The trip was evidently an inspiring one for Drew, and not just regarding possible labour-saving in the home. She and Fry spent a week in Lincoln, Massachusetts, visiting Walter and Ise Gropius at their home. Drew wrote to her new friends shortly afterwards, 'I wish I could tell you about a lot of developements [sic] that will change life after the war. I think apart from you both Thoreau is my biggest discovery – somehow I don't think the Gas Industry will cherish him'.[85]

During the couple's visit, Drew's office at King Street, 'her darling office, the apple of her eye', wrote Fry, was completely destroyed in an air raid.[86] The staff moved into the basement of Peter Gregory's office at 12 Bedford Square and work continued.[87] Drew's research was published as *Kitchen Planning: A Brochure of New Plans and Suggestions for Labour-Saving Kitchens* (1945). The volume is a common-sense, practical guide to the design of kitchens, that also reaches beyond its original aim to include other aspects of the modern home related to health, hygiene and welfare. The kitchen is not considered 'in vacuo', but in relation to the remainder of the house, to ensure good working and living arrangements;[88] the housewife is therefore not working in isolation, but part of the integrated modern household. Drew considered kitchens for all budgets in her research, from a package kitchen to a kitchen in a large house. The most revolutionary of her designs was a kitchen-bathroom unit for flats. She applied the existing idea of a central duct and plumbing, serving a kitchen to one side and a bathroom to the other, to mass-production techniques, creating a hollow wall unit with built-in kitchen appliances and bathroom fittings (Figure 3.15). The wall unit is constructed from a lightweight aluminium frame and faced with anodised aluminium sheets, backed with sound-proofing cork.[89] Drew's research was designed specifically for the anticipated new blocks of flats to be built after the war, and was publicised at a Kitchen Planning Exhibition, held at the Dorland Hall in London in March 1945 whilst the couple were stationed in West Africa.

Upon her return to London towards the end of 1945, and with Fry still in post, Drew set up new offices at 63 Gloucester Place.[90] A letterhead was created for the partnership, which read, 'The Office of Maxwell Fry & Jane Drew. Architects. Town Planners. Industrial Designers'.[91] Their inclusion of 'Industrial Design' reflects the widespread post-war campaign to use design as a means to rejuvenate British industry, and was evidently a movement that Fry and Drew embraced. Following her kitchen research, Drew, in particular, was eager to develop this strand of their work and she became the partnership's specialist regarding modern materials, furniture and components.

This domestic work continued with Fry and Drew's contribution to the 'Britain Can Make It' Exhibition of 1946, held at the Victoria and Albert Museum. The high-profile exhibition displayed work by architects, artists and designers, including

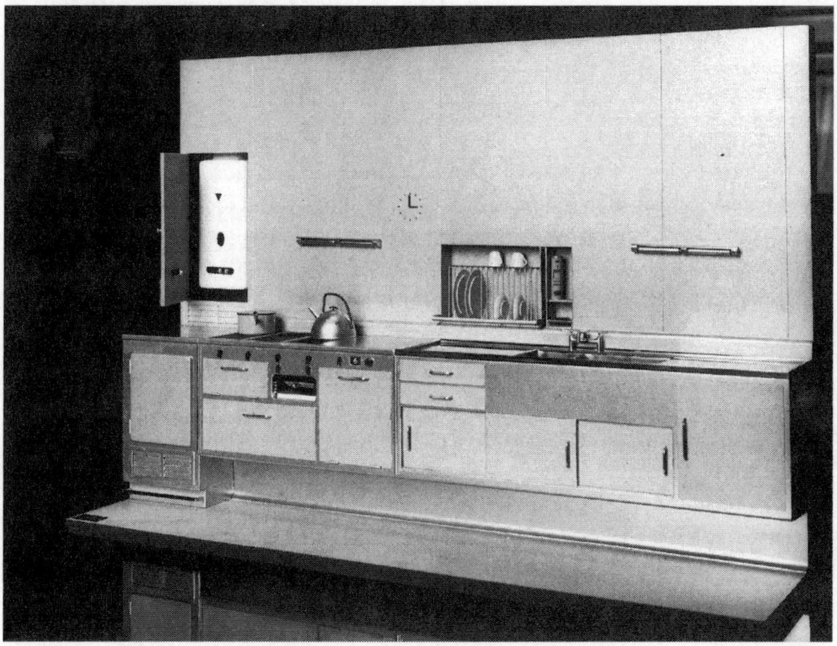

3.15 Kitchen Side of a Kitchen-Bathroom Unit, Kitchen Planning Exhibition, 1945

Christopher Nicholson, F.R.S. Yorke, James Gardner, Laurence Scarfe, Frederick Gibberd, Dorothy Braddell, Hans Feibusch and Richard Noad. Drew's contributions again focused on domestic reform, with an exhibit of well-designed 'Domestic Appliances (Powered and Non-Powered)'. She commented, 'A revolution in our attitude towards domestic appliances has been taking place. … The standard of these fixtures … is still low because of a denigration of the value of women's labour in the home'.[92] Drawing on her work for the Gas Industry, she also designed kitchen exhibits for a range of circumstances, from a large, well-appointed house to a compact kitchen (Figure 3.16). This work ensured Drew was a specialist in the

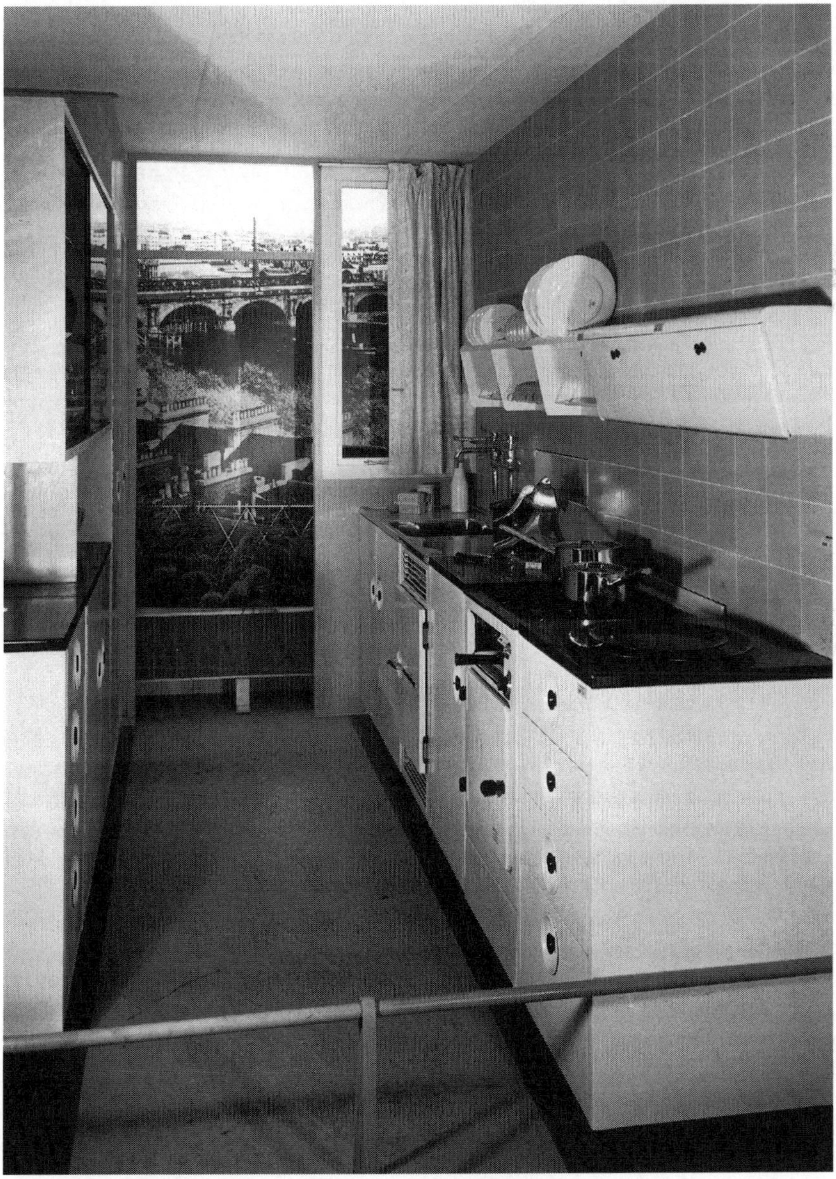

3.16 Compact Kitchen, 'Britain Can Make It' Exhibition, 1946

field and, in the same year, she undertook a lecture tour to Switzerland on behalf of the British Council, speaking on new developments in domestic architecture and design.[93] Indeed, Drew was part of a group of influential women designers and architects in wartime London. She was, for example, a member of the RIBA Housing Group with Elizabeth Denby, Jessica Albery and Judith Ledeboer. She regularly lunched with Elizabeth Denby and was responsible for the repair of Fry and Denby's relationship.[94]

HOUSING, NEW TOWNS AND THE RISE OF MODERNISM IN BRITAIN

In austerity Britain – like many other architectural practices during the late 1940s – Fry and Drew also looked overseas, mostly to the British Empire, for new commissions. As Nicholas Bullock notes, firms were encouraged by senior RIBA figures to work in the Empire, and more specifically in areas such as Africa and the Caribbean, where architectural skills were in short supply.[95] With British building restrictions in force, the prospect of work at home was limited and Fry and Drew therefore focused much of their attention on West African projects. Yet this did not dampen Fry and Drew's hopes for rebuilding Britain. Writing in 1941, Fry set out a broad manifesto stating that 'the new Britain must be planned' and his article, published in the popular *Picture Post*, attempted to highlight the current planning failures. Provocative images of children playing on bomb sites, traffic jams and so on, contrasted with a sketch of a town and country, with clearly-defined boundaries, straight roads, modern flats and clean factories.[96] Restating the views of town planning that he had held since the mid-1920s, Fry again argued that 'the town of the future can well be a place of open spaces diversified and dignified by building and interlaced by traffic ways for vehicles and walkers' but associated with this was 'sacrifice' of land ownership, which he thought 'made nonsense of town planning'.[97] Fry had long viewed land ownership as being the great impasse of planning and believed that 'unless land resources are pooled by one system or another, town planning will come to a dead end and we shall lose our chances of recovery, if not our will to survive'.[98] Fry's desire to participate in this post-war rebuilding is evident, as he wrote to Drew from Accra:

> *I think of the work we must do at home one day. I believe that must be done, and that we, in our spheres, must do it. Give them courage and a belief in their future for I think the English have a grand job to do for the world. Never grander at this time, even if it be the last and final grandeur.*[99]

Writing shortly after the war had ended, Drew hoped to provide 'help and inspiration to war weary architects once more starting the joyous act of creation' through the pages of the *Architects' Year Book* (*AYB*).[100] Established in 1945, Drew co-edited the publication with her employee Trevor Dannatt, supported by an editorial board of Fry, Herbert Read, Charles Reilly and, later, Ove Arup.[101] In an effort to inspire a new generation of architects, Drew modelled the journal's editorial on the short-lived but influential *Circle* (1937).[102] Originally intended as a serialised publication, *Circle* had promoted collaborative work between artists and architects, and was edited by Leslie Martin, Ben Nicholson and Naum Gabo. With contributions from scientists, architects and artists across the world – including Fry and Gropius – *Circle* encapsulates London's artistic milieu of the late 1930s, which Drew joined following her marriage to Fry. She later recalled this influential community:

> *The Café Royal used to be a sort of unofficial club where we all met. The Huxleys were there, that old rogue Augustus John was there with some of his children (those he could recognise), Allen Lane of Penguin … used to be there a lot, a man Justin Blanco White married who was a great biologist [Professor Waddington], … Desmond Bernal.*[103]

Drew was ideally positioned amongst the Café Royal group of artists and architects in post-war London, and she was approached by the publisher Paul Elek to act as the journal's editor.[104] The annual provided opportunity to promote this artistic community, the Fry and Drew practice and to disseminate their own brand of Modernism. True to her empirical outlook, it also included technical reports and discussions on the latest materials, as Drew explained in her inaugural editorial, 'we have balanced the book with technical and aesthetic information and with the sociology necessary for the modern, humanitarian architect'.[105] The *AYB* offered a third way to the more well-known, established journals of the *Architectural Review* and *Architectural Design*. Whilst Pevsner and Richards promoted a picturesque revival at the *AR*, from 1953 Theo Crosby and Monica Pidgeon at the relaunched *AD* presented the views of a younger generation of modernists and 'followed more closely the trends of the times'.[106] The *AYB*, meanwhile, took up the slack between these polemics and sought to galvanise the architectural community in post-war Britain.

Indeed, the *AYB* took an active role in steering the debate on rebuilding Britain. In the first issue Drew contributed an article on 'Housing', which opened with the statement 'we have before us an even bigger job than we had after the last war'.[107] Of the four million houses built following the First World War, a substantial number were designed in a semi-detached format, following existing roads or laid out in large estates with 12 houses per acre. Although the increase in housing was a considerable achievement (particularly as one million of those houses were built by local authorities to alleviate slums and shortages), the architectural qualities of the developments were deemed 'one of our greatest failures … wasteful and ill-designed',[108] and the planning 'haphazard'.[109] The Second World War prompted a reflection, and offered a chance to do things differently, to develop more coherent methods in planning and design and to improve standards of construction, space layout, heating, kitchen equipment and sanitation fittings. Discussions on the matter did not wait for the war to end, reports were published by government, the popular press and media reported almost every exchange of ideas;[110] this propaganda sought to raise morale whilst the British landscape and communities were under threat.

In addition to the houses lost through bombing many more were damaged, unfit for habitation or had been seconded for use in other purposes; an estimated four million homes were needed to be built within just 10 to 12 years.[111] The RIBA set to work on drawing up guidelines for a National Plan in 1943,[112] a move proposed the previous year by Julian Huxley,[113] but most other discussions and recommendations were more focused and pragmatic, content to improve the plight of the urban working classes and to stop the suburban sprawl into the countryside.[114] Regardless of the scale there was an ambition to do things differently, as expressed by Percy Ford who went as far to describe the ideas of the war period as a 'transformation of public attitude' and that 'inhibitions on our social thinking and will were shed'.[115]

The proposed Government-sponsored solutions were in favour of mixed housing types and stated that 'more attention should be paid to the satisfactory groupings of buildings in relation to each other',[116] and that 'unity and character are best achieved in low-density areas by the use of terraces and semi-detached houses in contrast with blocks of flats'.[117] This was a solution that Ralph Tubbs had previously and more poetically described, in his *Living in Cities* (1942) booklet:

> *The solution is surely terraces around open quadrangles of lawns and trees, punctuated with high blocks of flats. How pleasant to walk from one quadrangle to another, to enjoy the sense of seclusion and the peace of the inner courts, with a skyline ever changing with the silhouettes of towering flats.*[118]

Although Tubbs's cloistral vision could be accused of fantasy, the post-war idea for a variety of housing types was also part of the desire to encourage a mix of residents, in terms of age, size of family and class. The government encouraged construction through the continuation of subsidies,[119] and local authority building rose from approximately '25,000 in 1946 to over 190,000 in 1948', after a sluggish start in 1945 when just 3000 dwellings were constructed.[120] Although the existing prejudice remained,[121] it was acknowledged that in the 'most high-density areas the provision of a house for all such families will be impracticable'.[122] As a result, significant studies were undertaken into the design of flats and maisonettes in particular, which were seen to offer the benefits of a house along with higher density occupation.[123] The increased complexity of designing flats, coupled with the desire to improve the appearance of new housing, prompted statements proclaiming that 'design is the function of the architect' and that 'too little use has been made of trained architects in the planning and design of housing estates'.[124] Local authorities in particular were encouraged to 'employ a trained architect in connection with their housing schemes' and, as such, a significant opportunity arose for architects.[125] 'Freda White' writing in the *AYB* overstated the case, but clearly outlined what was required and expected of the architect:

> *the architect of today is fortunate. He works in a moment of time when people see him as one of the great ministers, of the servants who are also masters. He is longed for like the doctor in illness, but with more hope. For the doctor may remedy a bodily disease, but the architect can cure our sick way of life.*[126]

It was in London and the home counties that the housing needs were most acute. In the *County of London* plan of 1943, Henry Forshaw and Patrick Abercrombie recommended that substantially more 'adjacent open space' should be provided for flats, although they too were conscious 'of the needs and wishes of those very large sections of the public who prefer houses', suggesting that a house and garden 'fit the English temperament'.[127] The decision to plan for flats was not proposed for ideological reasons or to preserve density; the planners were in favour of decentralising and suggested flats due to the limited availability of land.

Despite the large numbers of architects employed in local authorities, a significant amount of work was awarded to private firms and new housing estates were quickly constructed. The borough of Lewisham commissioned Fry and Drew to design Passfields (1949–50), a small estate of 101 flats to provide 25 dwellings per acre and with a statutory height limit of five storeys.[128] The variety of dwelling types aligned with the government reports, ranging from a single-room flat up to a five-room maisonette. It was to be a model development achieving 90 persons per acre, with modern facilities such as a lift, refuse compartments, a laundry, a workshop, a children's play area and lawned gardens set between the four housing blocks. Each dwelling would have a view of trees and grass, and the desire was to create a sense of community through the shared overlooked gardens.

The three smaller blocks are set perpendicular to the main road with the larger L-shaped five-storey wing wrapping around the perimeter of the site to create a sense of enclosure and privacy in the communal gardens (Figure 3.17). The larger block offered more scope for architectural intervention, driven by the arrangement of single-storey flats on the ground floor with two sets of maisonettes above. At the time, the maisonette was hailed as a model solution because it required balcony

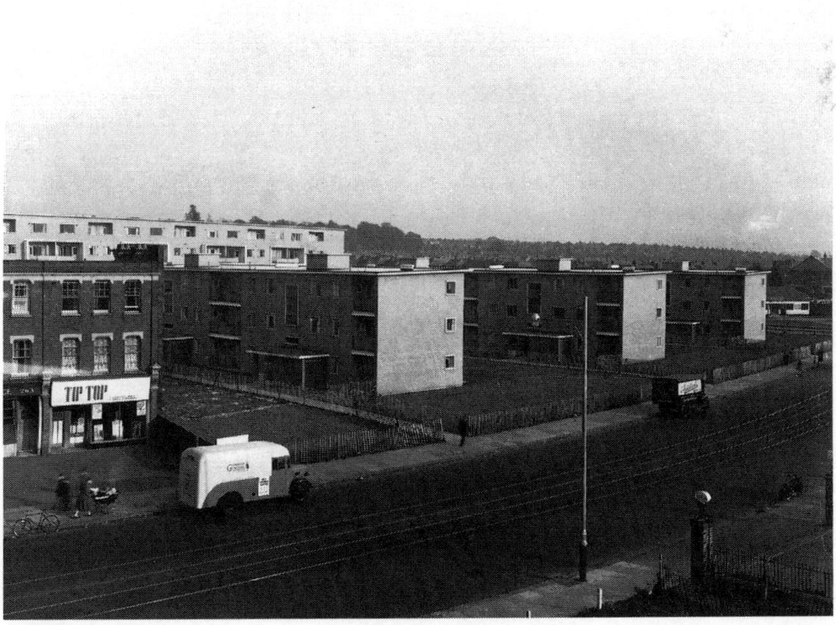

3.17 Passfields, Bromley Road, Lewisham, London, 1951

access at every alternate level and therefore did not overshadow the living rooms to the floor below or result in disturbance in front of the bedrooms.[129] Fry and Drew counter the horizontal walkways with bold concrete rectilinear protrusions, which contrast with the warm buff-coloured bricks and help to punctuate the overall mass of the arrangement. These exaggerated additions to the façade had only the mundane function of housing a refuse store, but they also serve as markers for each front door and associated staircase. Facing into the development the effect is even more pronounced, in a reprise of Fry's interwar dwellings. Three-storey concrete frames with balconies projecting beyond the main building line are interspersed with recessed balconies and windows to form a lively façade (Figure 3.18). The main block also incorporates a gentle curve reminiscent of Kensal House that softens the geometry of the arrangement and opens up the site.

There are other motifs that demonstrate Fry and Drew's formalist approach, such as the projection of the access corridor above the workshop. This simple flourish signifies the end of the block and provides a comfortable transition around the corner of the corridor. Fry and Drew's skill in composing a façade using simple geometries distinguishes their work of the period and invites comparisons with contemporaneous schemes by the Tecton partnership. Showing a continuation of the formalism of Lubetkin and Tecton, Lasdun wrote in the *AYB* that 'the mind and the spirit have to find means of expression in the environment if an integrated life of well-being is to be achieved'.[130] The Hallfield Estate (1946–54) at Bishop's Bridge Road in Paddington, a Tecton project that Lasdun took up following the partnership's dissolution in 1948, attempted to achieve this through a sophisticated elevation treatment. However, the scale of the overall development proved problematic, resulting in rather disjointed and out-of-scale elements, such as the entrance porches. Regardless of the fenestration and balcony patterns, the overall monolithic mass was difficult to reconcile with notions of home, which the smaller units at Passfields were perhaps more successful in achieving.

The rebuilding programme helped to institutionalise Modernism and post-war the movement experienced a cultural dominance that had been long desired by the modernist community.[131] Britain became a focal point for Modernism, with the MARS Group providing the galvanizing force for CIAM's activities. This was recognised by the CIAM President J.L. Sert, who wrote to the group's Secretary, Sigfried Giedion, in 1949, 'I would like to stress that I, like you, believe that the next [CIAM] Congress should take place in England, and we should, by all means, encourage the Mars Group, which, as you say, is the best and most active group in the Congress today'.[132] Much of this activity was led by Fry and Drew, and as their former employee Theo Crosby later noted, they were 'largely instrumental' in the 'triumph of Modernism'.[133] Fry and Drew's efforts focused not only on rebuilding work and the dissemination of CIAM's work in the *AYB*, but also on the education of the next generation of modernists. In 1948, an experimental CIAM Summer School was planned by the MARS Group to be headed by Fry, with assistance from Jacqueline Tyrwhitt.[134] The school was delayed due to difficulties in securing students and volunteer instructors, who were (according to Fry) busy in their attempts to secure employment.[135]

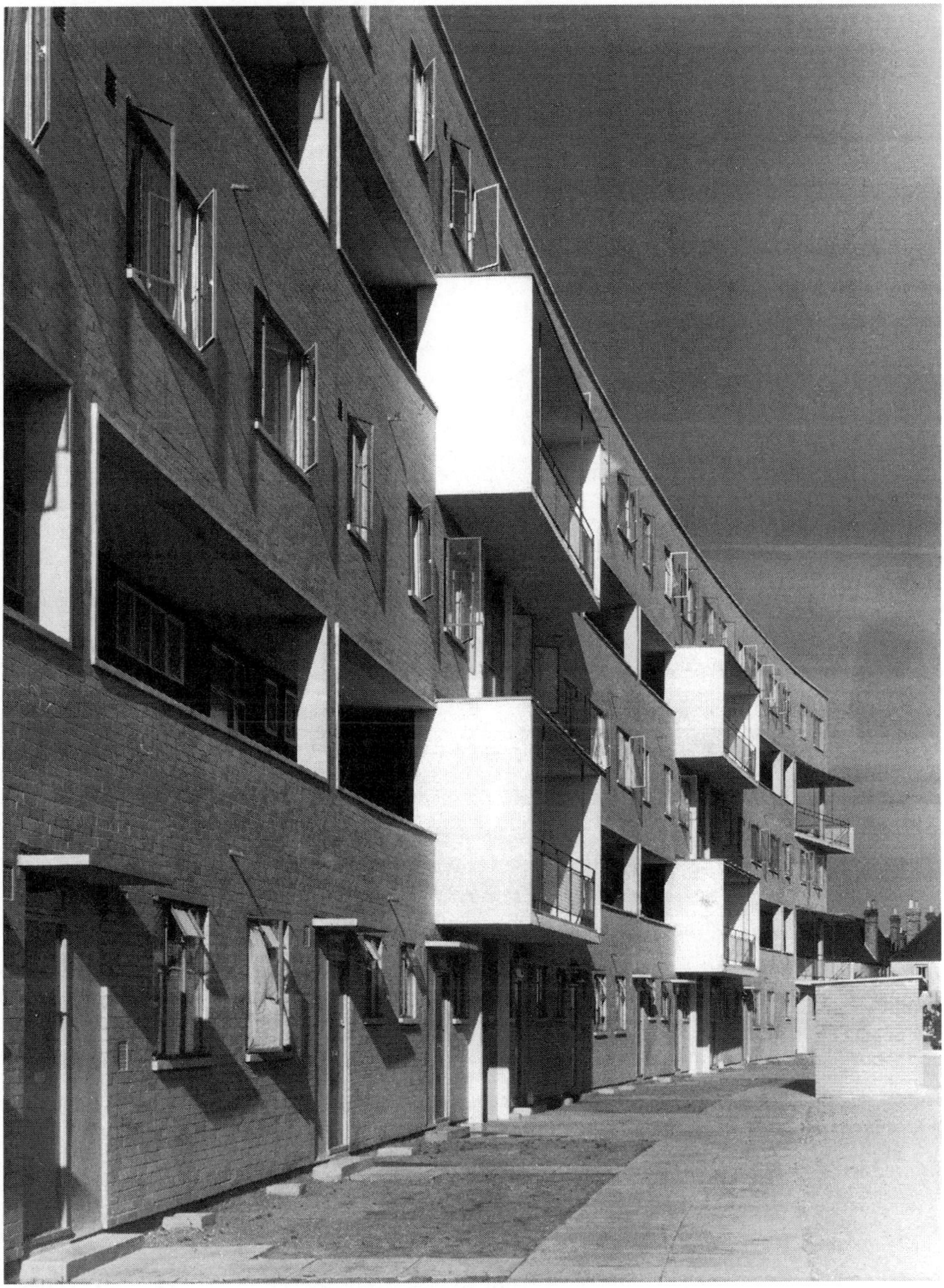

3.18 East Façade, Passfields, Bromley Road, Lewisham, London, 1951

Held instead during the following summer at the Architectural Association, the four-week programme focused on the rebuilding of central urban areas 'whether war-damaged or not', illustrating the desire to use the opportunity for further reform.[136] Four design projects were undertaken: a housing scheme for 3,000 people tutored by one of the New Town architects, Peter Shepheard; a large office building, instructed by Cadbury Brown; a national theatre project led by C.K. Capon; and a large traffic junction, headed by Fry's good friend Arthur Korn, all based on sites in London. The student cohort of young architects recommended by their respective CIAM Groups mostly travelled from Europe to take part, although representatives from Australia, South Africa, Argentina and Colombia were also in attendance.[137] The school ended with an exhibition of 16 projects with critique from Fry, the AA Principal, Robert Jordan, and two CIAM Council members, van Eesteren and Ernesto Rogers. The school was evidently successful and plans to hold a similar programme annually in a different country were circulated amongst CIAM members (although these plans did not materialise).[138]

For Fry and Drew, the work of the summer school brought ideas of the city 'core' to the foreground. The grouping of Fry, Tyrwhitt and Ernesto Rogers illustrates the beginnings of this theory which was subsequently explored at the eighth CIAM Congress held at Hoddesdon in 1951. The conference on 'The Heart of the City' tapped into existing modernist discourse and in England, for Fry and Drew, this was bound up with the building programme for the New Towns. Proposals for extending existing villages and towns with new housing estates had been widely published and discussed in volumes such as the *Housing Manual*. These proposals originally involved a limited number of houses utilising the existing infrastructure and amenities of the host town, but discussion shifted towards larger developments and the introduction of new schools, community buildings and so on, leading to more ambitious plans for entirely new towns, especially in the London area.[139] *The Greater London Plan* prepared by Abercrombie and published in 1944 outlined a number of proposals to deal with the 'overspill' of population requiring accommodation as a result of the planned reduction in congestion, density and decentralization of the population. Of the 1.25 million people estimated to be 'displaced' by the plan, around 500,000 were to be re-housed in ten new towns located approximately 30 miles from the centre. In addition to housing the large population and 'to obviate the necessity for high-density in London',[140] the intention was to prevent the 'spoliation of our countryside by ill-considered building', and to impose a 'limit to the haphazard sprawl of our existing cities'.[141] Furthermore, they were to serve as 'experiments in town planning', helping to consolidate scattered communities and diminishing populations previously employed in agriculture or small industries in the counties surrounding London.

The first New Town to be planned was Stevenage, closely followed by Crawley, Hemel Hempstead and Harlow, all of which were to incorporate existing settlements. The new developments were instructed to be 'wisely sited and skilfully planned, a proper balance between housing and industry' and most importantly, towns in their own right as 'the antithesis of the dormitory suburb'.[142] Harlow, planned by Frederick Gibberd in 1947, was to have a population of 60,000.[143] The plan proposed four main residential quadrants and as Gibberd explained, 'instead of the usual "neighbourhood" planning of six or seven large units, the housing is planned as 13 small compact units, separated from each other by the existing topography'.[144] Secondary schools and sports centres were located within expansive green spaces, isolated from the industrial areas which were planned to the northern periphery of the town and close to the railway. Each residential quarter was equipped with several primary schools and a sub-shopping centre with local amenities, such as libraries, banks and shops. The hospital and main town centre were located in the north-western quarter, some two miles away from the residents in the opposite quadrant. A housing density of 15 dwellings per acre and large swathes of green space running through the town suggested 'sprawl with a lack of general compactness',[145] a criticism that must be tempered with advantages of open space and the close proximity of the countryside.

In terms of dwelling design, the Report of the New Towns Committee was confident that 'the great majority of occupiers will undoubtedly want single-family houses with gardens; the proportion of flats for family occupation must again depend on actual demand'.[146] There was also concern for a variety of occupants, with a separate chapter in the report entitled 'Social structure', the desire was for a 'balanced community' made up of a variety of social groups with 'dwellings of all classes ... built in due proportions'.[147] At Harlow, variety of housing type and style was to be achieved by the appointment of different architects to design the various housing areas; but, despite the agenda for mixed occupancy, the outworkings nearly always correlated class with dwelling size. The only restrictions on the designs related to cost and maintenance, which resulted in over 200 different types of house and flat designs being constructed by a list of eminent pre-war architects including Richard Sheppard and Partners, Cadbury-Brown, YRM, Norman and Dawbarn, as well as the Harlow Design Group and Gibberd.[148]

Urbanity was important to Gibberd's plan, as was a 'greater sense of neighbourliness' in his residential areas, which he hoped to achieve through small, compact units.[149] Equally, he desired picturesque towns where 'the motif of the planning is rather of the romantic than the formal school'.[150] Buildings were positioned amongst mature trees and each area was planned to accommodate the existing field and hedge patterns 'thus the new housing need not obliterate the existing features, and each area can, by reason of the variety in the landscape itself, have its own character'.[151] In contrast to Gibberd's picturesque English towns, Drew looked to fellow CIAM members for models of post-war building. In 1947, she requested details of J.L. Sert's town plan for Motor City in Brazil, which she published in the third edition of *AYB* two years' later. She wrote to Sert of her interest in his Brazilian city:

> *Most of our work is in Africa as building in England is extremely difficult; there is a general gloomy feeling about our financial position and great shortages of material and labour. Our hopes rest with the new towns but the plans so far published do not look very exciting. There is certainly nothing comparable to "Cidade dos Motores".*[152]

'Cidade dos Motores' was conceived by Town Planning Associates, a partnership between Sert and the German-born, New York resident, Paul Lester Wiener. The town plan dates from 1943, following their appointment by the Brazilian Airplane Factory Commission to design a new town for an existing aeroplane engine factory. The town was designed to house a population of 25,000 on 250 acres of reclaimed marshland, to the north of Rio de Janeiro. The centre was to be the heart of the city, consisting of an administrative, amusement and commercial centre, plus cultural and sports areas; most significantly, the centre contained traditional elements of Brazilian towns, a 'praca' or town square, and a 'passeio'. The *AYB* reported, 'Here the social life of the whole town takes place, where one meets, can see and be seen'.[153] The scheme marks an important stage in Sert's development of the multi-layered civic centre, as Valerie Fraser notes, and his work in Latin America formed the basis of his ideas on town planning presented in his Presidential Address at the CIAM Congress of 1951 held at Hoddesdon.[154] Yet it would be several years before Fry and Drew were able to utilise Sert's research in their own work.

Indeed, Drew's letter illustrates the gulf between Fry and Drew's theoretical work and the opportunites available in Britain at this time. Despite Drew's ambivalence towards the New Towns, in 1949 the partnership was brought in to the project by Frederick Gibberd's 'unfailing kindness', as Fry later wrote.[155] Fry and Drew were commissioned to design two areas of housing: The Chantry and Tanys Dell, situated in the Mark Hall North neighbourhood unit to the north-east of Harlow in Hertfordshire. Terraced houses were proposed to both areas in three- and four-bedroom arrangements, as well as a health centre that would eventually be converted into dwellings when a new purpose-built clinic had been constructed. Drew's preference was for terraced housing 'wherein all services and flues are gathered together in one central core', and in the *AYB* she set out her critique of the semi-detached '"Universal" plan' illustrating the extra drainage, roof tiles and so on that this type demanded in contrast to the terrace.[156] She noted that 'there are signs that the ground-floor plan of this house type will at last yield to change; the living room getting bigger, with a dining recess instead of a separate room',[157] and at Harlow the housing interiors were 'opened up' to form a singular living space running from the front to the rear of the property.[158] Equally careful to avoid the criticisms of the 'mechanical and dreary' bye-law street,[159] the runs of terraces are of a limited length and track the contours of the site in shallow curves. The prefabricated solutions proposed by Drew in 1945, especially for kitchens and bathrooms, were slow to be taken up by a construction industry content to remain in its 'habitual state and with a high proportion of site work',[160] yet such kitchens and modern equipment was available and advertised in the *AYB*. The house construction was of standard brick at ground floor level, with render to the first floor (like many of the suburban semi-detached houses), a strong soffit line and a monopitch roof (Figures 3.19 and 3.20). Each individual dwelling was expressed by a projecting party wall, which breaks up the 'larger unit of design', in contrast to the effect achieved by Geoffrey Jellicoe in his terraced house designs that clearly influenced Fry and Drew.[161]

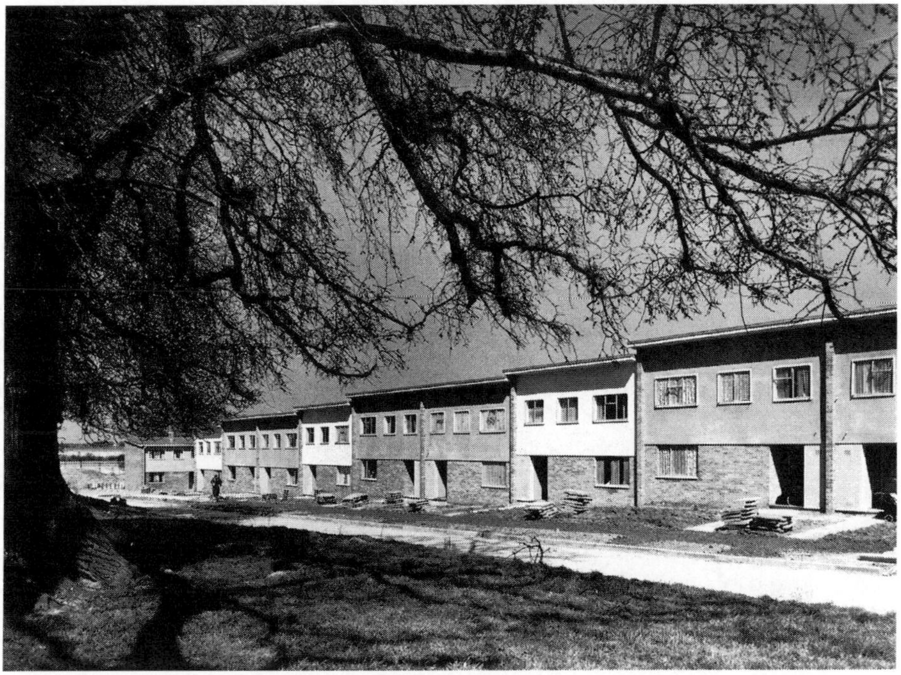

3.19 Mark Hall North, Harlow, Essex, 1952

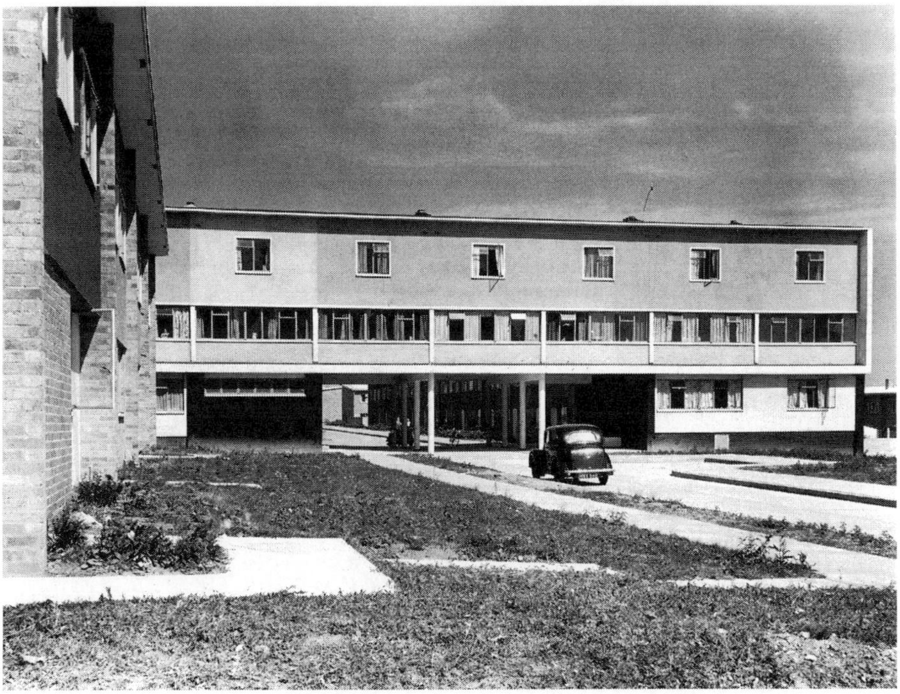

3.20 Mark Hall North, Harlow, Essex, 1952

At the Chantry, to the highest point of the neighbourhood, flats and maisonettes over three- and four-storeys were proposed, adopting a very similar arrangement to that at Passfields. The balcony upstands and railings are arranged to form a pattern rather than for any pragmatic reasons and, like Fry's other flats, the façade is determined by a careful placement of windows, balconies, recessed elements and contrasting materials. Where additional amenities are introduced, these are given their own form to distinguish them from the housing, such as the small community hall which bridges the two rows of flats and subtly cantilevers out over the walkway below.

Fry and Drew's housing at Harlow has been described as 'constrained aesthetics and diluted Modernism',[162] yet this is an unfair critique. The work is a continuation of their pre-war brick work and the houses are not dissimilar to Fry's 'concrete house' competition entry of 1933. The Harlow flats, although lacking the photogenic face of Kensal House, provide generous spaces and are positioned to maximize sunlight and views, as well as considered proportions and façade treatment. The construction process and materials are conventional, but this must be countered with the materials and labour that were available at that time, and the urgent need for housing which did not lend itself to untried techniques and the increased possibility of failure. There is also, of course, a considerable gulf between an architect's desire and the buildings they produce.

More widely, the New Towns came in for significant criticism even before their completion, particularly during the eighth CIAM Congress on 'The Heart of the City'. Held at Hoddesdon in Hertfordshire, the 1951 congress was organised by Fry's good friend, the town planner Jaqueline Tyrwhitt.[163] The delegates travelled to nearby Harlow and the rumblings of discontent are evident in the conference proceedings, published the following year. Philip Johnson commented:

> *I have read Maxwell Fry's explanation of why he left a green space among the houses in Harlow New Town. He says "this corresponds to the place where the church was in old English villages". Well, it may correspond, but does it perform the same function? … my guess is that this will always remain just an open space.*[164]

Johnson's criticism came in light of the conference's main theme of the 'core' of the city in both social and architectural terms, as a means to 'return to the human scale and the assertion of the right of the individual over the tyranny of mechanical tools'.[165] This view was subsequently taken up by J.M. Richards (a CIAM delegate at Hoddesdon) in articles for the *Architectural Review* proclaiming 'The Failure of the New Towns' (1953);[166] an argument reinforced by Gordon Cullen's accompanying photographs that illustrate the sparse, unclaimed public spaces spoken of by Johnson. The low density, Garden City model for the New Towns that had fostered 'prairie planning' received particular criticism by Richards, a view that became the accepted response in subsequent years. John Cordwell, who worked on Harlow New Town for Fry, Drew and Partners, later observed, 'It was a beautiful plan in two dimensions, and in actual fact when I saw it built, it was bloody awful … The beautiful crescents there that we had in the plan were only two stories high. They just disappeared, so you didn't get the feeling of enclosure anywhere'.[167]

However, more recent discussion of the planning of the New Towns has highlighted the compromised position in which Fry, Drew and Gibberd found themselves, as the 'overpowering influence of the low-density Garden City model' took precedence over their modernist agenda.[168] Indeed, Fry himself renounced the low density of Harlow shortly afterwards, writing:

> *I have seen in my lifetime an idea of a garden city embedded into the law of the land with every provision for its promulgation and acceptance; and what came out of the latter end I have seen also. The idea was not a good enough one and no power on earth could protect it.*[169]

Thus Fry and Drew looked to CIAM's discussion on the 'Heart of the City' and work such as Sert's 'Cidade dos Motores' to achieve a humanised urban 'core', thereby creating a social and architectural hub in place of the uninhabited spaces of the New Towns. Indeed, these ideas informed their work in Chandigarh and their subsequent projects in Britain, as Chapter 7 will discuss.

DESIGN AND INDUSTRY AT THE FESTIVAL OF BRITAIN

Alongside their town planning and housing schemes, Fry and Drew's other main source of work in post-war Britain came from exhibitions and shop fit-outs. Fry and Moholy-Nagy's work of the late 1930s set useful precedent for these post-war initiatives, which helped to establish Modernism as the pre-eminent architecture in Britain. Likewise, Drew's wartime industrial work fed into this developing sector. Fry and Drew were therefore ideally placed to contribute to the 'Britain Can Make It' Exhibition (1946) and projects such as a Design Centre for British Rayon (1948) in London. The centre was the first to be incorporated following the formation of the Council of Industrial Design in 1944, which had been established to assist industries in the setting up of their own design centres.[170] As the *ABN* noted, 'Although the main object is to interest buyers and designers, the Centre is a place where architects and students can go for advice and ideas'.[171] Such places were crucial in the post-war development of British design culture.

The following year, Fry and Drew redesigned the foyer to the Ceylon Tea Centre (1949) on Lower Regent Street in London. The lightweight metal staircase and mezzanine is reminiscent of Fry's Regent Street showrooms and the integration of a mural by John Farleigh is typical of their work at this time (Figure 3.21). The integration of art and architecture was also explored in depth in their work for the Festival of Britain (1951). The Festival was the brainchild of the interwar cohort of reformers that came together in groups such as the PEP, the DIA and the CPRE. Following Labour's return to power in 1945, the left-wing design lobbyists again took up their cause, for Clement Atlee's government offered them renewed hope – particularly in sympathetic figures such as Herbert Morrison (1888–1965). Now the Deputy Prime Minister, Morrison had been an ally of Frank Pick in his work for the LPTB and as leader of the LCC in the mid-'thirties, he had directed the development of the city's transport services, health, education and housing. Thus in 1945, John

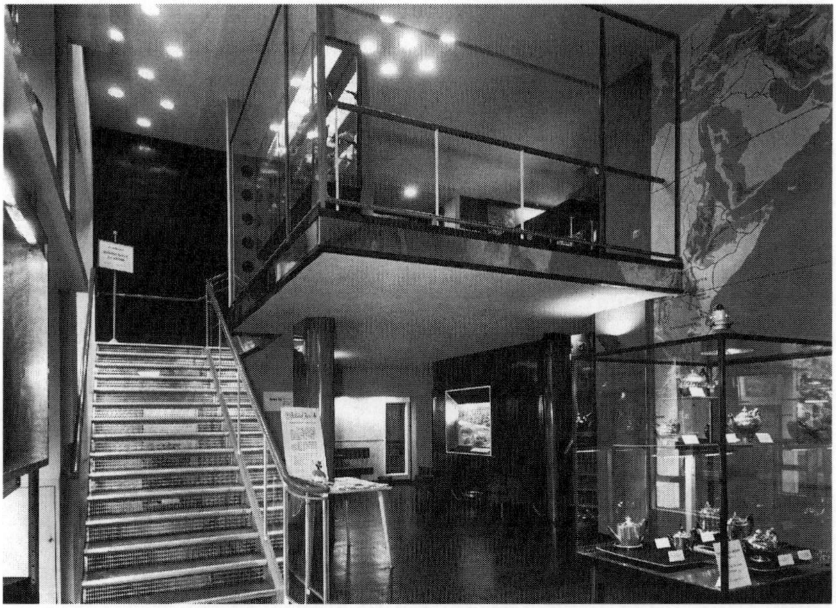

3.21 Ceylon Tea Centre, 22 Lower Regent Street, London, 1951

Gloag wrote a letter to the editor of *The Times* newspaper suggesting the idea of a centenary celebration of the Great Exhibition of 1851. The suggestion had first been voiced by the Royal Society of Arts in 1943 and finally gained recognition with Gerald Barry's open letter to the President of the Board of Trade, Stafford Cripps, which was published in *News Chronicle* on 14 September 1945.[172] The editor of the leftist *News Chronicle*, Drew later noted that Barry held 'terrific charisma' and was a 'friend' to Modern architects.[173] As such, he was a natural choice as Director General for the Festival and took up his post in April 1948; the council members included Leonard Elmhirst, T.S. Eliot and Drew's ally, Kenneth Clark.[174]

Given the involvement of figures such as Barry, Gloag and Morrison – who was appointed as the Government minister responsible for the project – one would assume that Fry would be keen to work amongst this group, to push forward ideas on architecture and planning for which he had campaigned so vigorously since the 1920s. Yet Fry's participation in the Festival was minor; Drew later observed that Fry 'never took to it' for he believed that architecture should not be a temporary conceit. In contrast to Fry's somewhat puritanical outlook, Drew threw herself into the work and was the lead designer for the practice's contributions. Indeed, late in 1950, Fry departed for India to take up a post as Site Architect at Chandigarh before the Festival had opened; Drew remained in London to finish the job. For Drew, the festival captured a spirit of co-operation between artists, designers, engineers and architects that she had experienced in wartime London and she was 'keen' to be involved.[175]

From 4 May to 30 September 1951, the Festival of Britain was staged at venues across the country. Events were focused on a 27-acre site at London's South Bank and the festival is remembered as a light-hearted intermission amongst the austere, immediate post-war years (Figure 3.1). As Drew later commented, the South Bank 'really looked like a piece of Dickens' before the arrival of the festival buildings and beyond the festival grounds this landscape remained.[176] With the exception of the LCC Concert Hall, all of the exhibition buildings were to be temporary structures. An article of 1949 described a modern palette of '[c]oncrete, steel, brick, wood, asbestos, cement, glass, aluminium and fabrics of various kinds'.[177] Yet the buildings needed to be frugal. A lack of materials and skilled labour encouraged the architects to be creative in their designs and pavilions were generally designed to be dismantled and re-used again after the festivities. As Mary Banham notes, the festival was not a homogenous exposition but combined a range of architectural and artistic styles. Different interpretations of Modernism and 'more traditional styles' blended with differing styles of exhibits: 'Some of the artefacts shown inside the pavilions … were hard to swallow, being too "cute" for words. There was much, however, that was forward-looking and tough-minded'.[178] Fry and Drew's work was amongst the latter group, as Drew later described, the opportunity allowed the creation of 'experimental work that [we] wouldn't dare do' for standard building projects. Indeed, their festival designs contributed innovative structures and enjoyable spaces. They collaborated closely with structural engineers and artists to offer a synthesis of an optimistic, post-war British culture. These experiments informed Fry and Drew's other work, particularly the African schools programme, which lent itself to the use of alternative building techniques.

Conceived in part as a post-war morale-boosting exercise, the exhibition was also part of a wider initiative to use design to re-launch British industry. As Chapter 2 has discussed, tangible links between art, design and industry had been called for since the mid-'twenties, and it was a movement that Drew fully endorsed, as her wartime industrial work demonstrates. Drew later spoke of the great excitement that the project aroused amongst the architectural and artistic community in London, commenting that it was 'like lifting the lid off a pressure cooker', providing an opportunity to design well for the whole of life; 'It was the first time since the war that people were asked to be inventive … and they responded. … The spirit was terrific'.[179]

The practice's design team was headed by Drew, and included Norman Starrett and Kathleen Greenwell,[180] J.C. Todd and Christopher Knight.[181] The South Bank site was close enough to the Gloucester Place offices, allowing Drew and her staff to undertake frequent site visits, and attend regular progress meetings with the festival's Director of Architecture, Hugh Casson.[182] Gerald Barry assigned Fry, Drew and Partners a sizeable area at the northern tip of the festival site, and they designed a cluster of buildings with associated access routes to link this group to both the site and the city. Work comprised: the Waterloo Bridge Entrance, to connect Waterloo Bridge Road with the LCC Concert Hall; the Harbour Bar; and the Riverside Restaurant. One of five public entrances, the entrance from Waterloo Bridge brought visitors into the festival site from high level via stairs and ramps (Figure 3.22). The cantilevered concrete bridge was engineered by Ronald Jenkins, an employee of Ove Arup and Partners, whom Drew later described as 'an absolute inspiration'.[183] It was the first time that a pre-stressed concrete bridge had been constructed in England,[184] and the elegant tapered columns of the terrace perfectly balanced the lightweight language of the festival buildings (Figure 3.23). The route into the site was enlivened by Barbara Hepworth's *Turning Form*, a slowly rotating sculpture, commissioned by Drew.

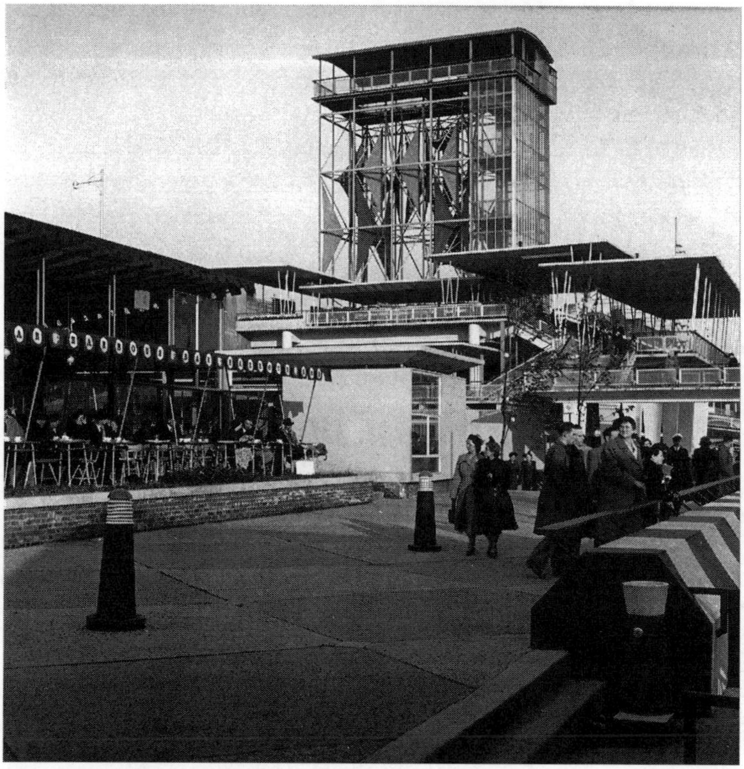

3.22 Waterloo Bridge Entrance and Observation Tower, Festival of Britain, South Bank, London, 1951

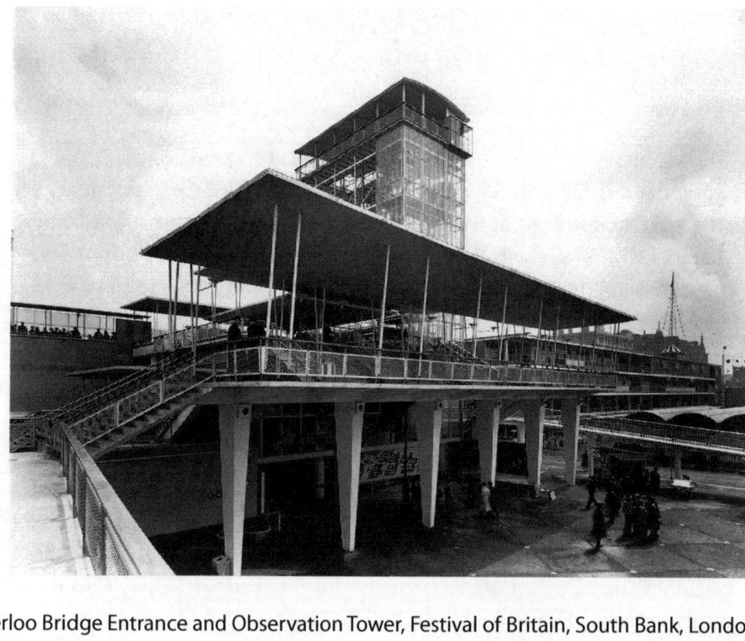

3.23 Waterloo Bridge Entrance and Observation Tower, Festival of Britain, South Bank, London, 1951

Underneath the bridge entrance, the 'New Schools' Pavilion was housed in a prefabricated building and although the practice was responsible for the structure, Drew later noted that they played no part in organisation of the exhibition.[185] The display was instead designed by Neville Conder and Patience Clifford to tell the story of the Hertfordshire Schools building programme. The adjacent observation tower was designed as point of reference due to vastness of site. Constructed from open tubular framework and enclosed to the northern side with glazing, a glass-sided lift conveyed visitors to a viewing platform 80 feet above ground level. Drew commissioned the sculptor Lynn Chadwick to design artwork for the tower. Evolving from his work designing exhibition stands with the architect Rodney Thomas, Chadwick had been experimenting with large-scale metal sculpture following the war. His delicate, kinetic structures complemented the festival's aesthetic and he produced three works in total for the South Bank site. For the top of the viewing tower Chadwick created a hanging *Second Tower Mobile*, which measured seven-and-a-half feet in diameter and was composed of metal and wood. He later described his work:

> *I filled it [the tower] in with coloured triangles of, of ... canvas triangles, and then I had a mobile in the top of it. But the mobile, that's where I found that it is impossible to have a mobile in the top of this tower, you see, and the wind was blowing through, and although the bits were very heavy, it still wasn't very satisfactory.*[186]

The canvas triangles were intended to evoke a ship's sails,[187] echoing the little boats to the dock below the tower. The adjacent Harbour Bar was also clad in canvas, with colourful canvas awnings sheltering a dining terrace that opened out onto the dock area (Figure 3.22). The fringes of the festival site was accommodated the Riverside Restaurant, which skirted the river, running under Waterloo Bridge to give a view of St. Paul's Cathedral. As Drew later noted, 'on visiting the site I perceived that one could steal a little more site ... and go right on under Waterloo Bridge and look at St. Paul's [Cathedral], which seemed to be the one decent view'.[188] Gerald Barry agreed to the use of additional land although the Port of London Authority insisted that the building, situated on the water's edge, must be designed to collapse immediately if struck by a ship. Built on timber piles, the curving structure cantilevered out over the River Thames (Figure 3.24). For speed of construction, Drew designed an undulating, double-skin roof fabricated in compressed cork and aluminium (Figure 3.25). Manufactured off-site by the same Bristol aircraft company that had also manufactured her wartime kitchens, the lightweight roof was easily assembled on site.[189] Drew later explained:

> *We had been having quite a lot to do with aircraft factories at that time because of the involvements we had had with prefabrication in kitchen design. It seemed to us that by constructing the roof of a double skin of aluminium with a cork sandwich we could make one that could be quickly assembled on site by aircraft rivet technique.*[190]

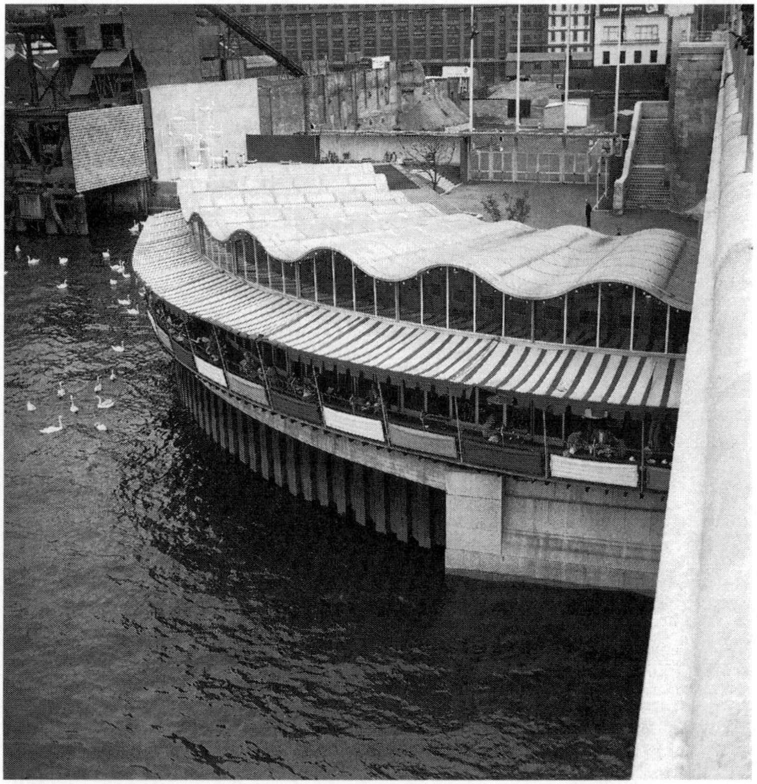

3.24 Riverside Restaurant from Waterloo Bridge with Paolozzi's *Fountain* in the background, Festival of Britain, South Bank, London, 1951

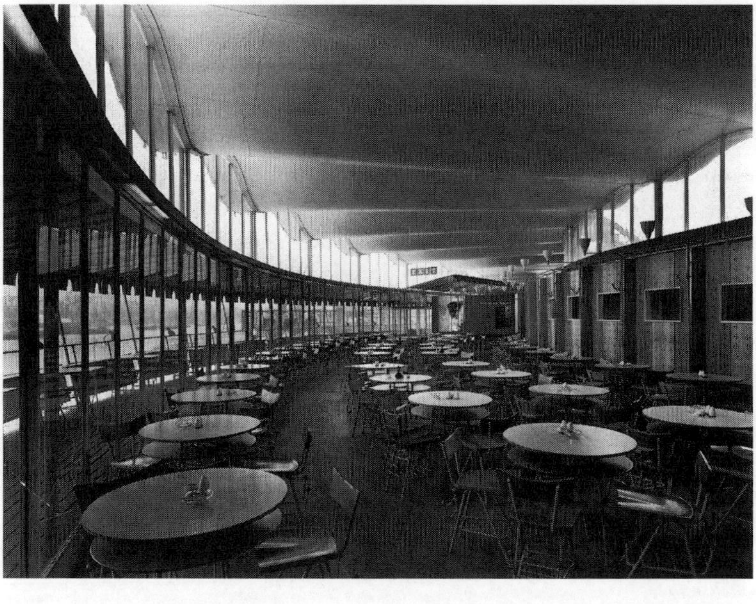

3.25 Riverside Restaurant Interior, Festival of Britain, South Bank, London, 1951

Making further use of her artistic network, Drew commissioned Ben Nicholson to design a *Festival of Britain* mural for the entrance of the Riverside Restaurant.[191] At Drew's suggestion, he created a curved piece to echo the curved entrance to the restaurant, composed of three hardboard panels painted with geometric forms in a colour scheme of greys and yellow to complement the restaurant.[192] Eduardo Paolozzi also created an artwork to add interest to the blank boundary wall adjacent to the restaurant (Figure 3.24). The *Fountain* comprised a free-standing, lightweight steel frame and a series of concrete cups carrying water down the framework. It was Paolozzi's first public sculpture and one the first in a series of collaborations between Paolozzi and Drew. For Drew, these associations were crucial to her work and she later observed that Paolozzi, Victor Pasmore and others accompanied her to CIAM meetings; 'There was not a division in their minds or ours … Art had to be part of the same spirit of the building'.[193] This collaborative spirit continued throughout Drew's later work, as Chapter 7 will investigate.

CONCLUSION

The influence of Drew's father ensured her work was committed to the social cause of architecture from the outset and she held a particular interest in the design of hospitals throughout her career. Drew's interest in innovative materials may also be traced back to her father and this too remained a constant in her work. Her wartime domestic work combined this social agenda with material innovation to limit the drudgery of British housewives and set the tone for her post-war work. Indeed, limited information remains of Fry and Drew's wartime experience, yet the period appears central in shaping their ideas for the future. Their artistic milieu in London, work with the CIAM network and town planning projects formed the basis of their subsequent careers.

Drew's work is characterised by a common-sense approach, a theme that emerges irrespective of design brief or budget. She recognised this as a fundamental aspect of her architecture, as she later observed, 'I am basically inclined to do what is sensible'.[194] This sentiment sets the tone for her work overseas with Fry, where her pragmatism was invaluable when faced with difficulties building in a post-colonial context.

NOTES

1 Elizabeth Darling, '"Into the world of conscious expression": Modernist revolutionaries at the Architectural association, 1933–39', in *Man-made Future: Planning, Education and Design in Mid-twentieth Century Britain*, (ed.) Iain Boyd Whyte (London and New York, 2007): pp. 157–73, p. 159.

2 RIBA Archive, F&D/13/2. E. Maxwell Fry, 'Old Man's Epilogue', undated.

3 Frank Knight cit. in Sile Flower, Jean Macfarlane and Ruth Plant (eds), *Jane B. Drew Architect: A Tribute from Colleagues and Friends for her 75th Birthday 24th March 1986* (Bristol, 1986): p. 42.

4 Catherine Burke, *A Life in Education and Architecture: Mary Beaumont Medd* (Farnham, 2013), p. 33.

5 Hugh O'Shaughnessy, Jane Drew's Eulogy, 3 August 1996.

6 Trevor Dannatt interview with Jackson, October 2011.

7 See for example: Catherine Burke, *A Life in Education and Architecture: Mary Beaumont Medd* (Farnham, 2013); Jill Seddon and Suzette Worden (eds), *Women Designing: Redefining Design in Britain between the Wars*, (Brighton, 1994); Lynne Walker, 'Golden Age or False Dawn? Women Architects in the Early Twentieth Century', http://www.english-heritage.org.uk/content/imported-docs/f-j/women-architects-early-20th-century.pdf. Accessed: 13 December 2012.

8 Frank Knight cit. in Flower et al, *Jane B. Drew*, p. 41.

9 John Morrison Archive. *Jane Drew Biography*, unpublished manuscript, p. 1. The couple married in 1907.

10 John Morrison Archive. *Jane Drew Biography*, unpublished manuscript, p. 1.

11 Shusha Guppy, 'Obituary: Dame Jane Drew', *The Independent*, 1 August 1996.

12 Maurice Down cit. in Flower et al, *Jane B. Drew*, p. 13.

13 Guppy, 'Obituary: Dame Jane Drew', 1 August 1996.

14 John Morrison Archive. *Jane Drew Biography*, unpublished manuscript, p. 19.

15 John Morrison Archive. *Jane Drew Biography*, unpublished manuscript, pp. 19–20.

16 UL Archive, Peter Dale Papers [uncatalogued]. Jane Drew interview with Peter Dale, 1992.

17 John Morrison Archive. *Jane Drew Biography*, unpublished manuscript, p. 18.

18 RIBA Archive, F&D/10/5. Jane Drew's certificate for matriculation as a student at the University of London, 30 September 1929.

19 John Morrison Archive. *Jane Drew Biography*, unpublished manuscript, p. 20.

20 Mary Crowley cit. in Burke, *A Life in Education and Architecture*, p. 41.

21 UL Archive, Peter Dale Papers [uncatalogued]. Jane Drew interview with Peter Dale, 1992.

22 Geoffrey Jellicoe cit. in Karen Parker, 'Jane Drew, Architect and Practical Idealist', in *Women Designing*: pp. 136–40, p. 137.

23 John Morrison Archive. *Jane Drew Biography*, unpublished manuscript, p. 18.

24 AA Archive, Student Register 1929–34.

25 John Morrison Archive. *Jane Drew Biography*, unpublished manuscript, p. 33.

26 AA Archive. Pantomime Programme, 1931.

27 AA Archive. Student Register, 1929–34.

28 See Darling, 'Into the world of conscious expression', pp. 157–73.

29 AA Archive. Student Register, 1929–34.

30 Dusa McDuff correspondence with author, 5 July 2012.

31 John Morrison Archive. *Jane Drew Biography*, unpublished manuscript, p. 21.

32 John Morrison Archive. *Jane Drew Biography*, unpublished manuscript, p. 37.

33 John Morrison Archive. *Jane Drew Biography*, unpublished manuscript, p. 26.

34 UL Archive, Peter Dale Papers [uncatalogued]. Jane Drew interview with Peter Dale, 1992.

35 Jane Drew interview with Barrie Fernandez, 8 April 1983. See Barrie Fernandez, 'Women in British Architecture: Jane Drew – Architect', 1983, unpublished BArch dissertation, Liverpool School of Architecture.

36 See Darling, *Wells Coates*, pp. 36–9.

37 Elsa Lanchester cit. in Elizabeth Darling, '"The Scene in which the daily drama of personal life takes place": Towards the modern interior in early 1930s Britain', in *Designing the Modern Interior: From the Victorians to Today*, (eds) Penny Sparke, Anne Massey, Trevor Keeble and Brenda Martin (Oxford and New York, 2006): pp. 95–105, p. 102.

38 Tait was employed by Alliston and Drew as an assistant for a few months in 1935 and 1936, see Dictionary of Scottish Architects, 'Gordon Thomas Tait', http://www.scottisharchitects.org.uk/architect_full.php?id=204366. Accessed: 18 March 2013.

39 Thornley left after a few months to work at his family's architectural firm in Liverpool. Thornley cit. in Flower et al, *Jane B. Drew*, p. 17.

40 'Competition Results', *RIBA Journal*, 44 (5 June 1937): p. 799.

41 RIBA Archive, F&D/15/5. Drew to Fry, 26 March 1945.

42 John Morrison Archive. *Jane Drew Biography*, unpublished manuscript, p. 26.

43 Gould, *Modern Houses in Britain*, p. 45.

44 Myles H. Wright (ed.), *Small Houses, £500–£2,500* (London, 1937), p. 52.

45 'St. Giles' Mount, Winchester. Architect Jane B. Drew (of Alliston & Drew), A.R.I.B.A.', *Architect & Building News*, 156 (25 November 1938): pp. 228–9.

46 Jane Drew interview by Karen Parker, 19 June 1988.

47 John Morrison Archive. *Jane Drew Biography*, unpublished manuscript, p. 33.

48 HL, Walter Gropius Papers in the Bauhaus Archiv, MS Ger 208 (745). Letter Fry to Gropius, 9 December 1941.

49 Krishna R. Dronamraju, *If I am to be Remembered: The Life and Work of Julian Huxley with Selected Correspondence* (London, 1993), p. 95.

50 HL, Walter Gropius Papers in the Bauhaus Archiv, MS Ger 208 (745). Letter Fry to Gropius, 9 December 1941.

51 Walker, 'Golden Age or False Dawn?', p. 24. This group also entered a competition for industrial housing, see *Industrial Housing in Wartime: Results of the Competition organised by the Royal Institute of British Architects* (London, 1940), p. 12.

52 UEA Archive, Jack Pritchard Papers, PP/7/1/7/8.

53 HL, Walter Gropius Papers in the Bauhaus Archiv, MS Ger 208 (745). Letter Fry to Gropius, 14 September 1940.

54 RIBA Archive, F&D/23/2. Letter Fry to Emma Spering Drew, 25 May 1941.

55 RIBA Archive, F&D/10/13. Letter Maud, 'Ceilwart Farm, Llanber Rd, Barmouth, NW Wales' to Drew, 19 August 1940.

56 Fry later reconsidered as his father (now living in New Jersey), who had been willing to act as Ann's guarantor while she was in America, wanted to arrange her schooling and accommodation. Beyond providing financial assistance, Fry did not wish Ambrose Fry to be involved and therefore decided that Ann should remain in England. See HL, Walter Gropius Papers in the Bauhaus Archiv, MS Ger 208 (745).

57 RIBA Archive, F&D/10/14. Letter Fry to Emma Spering Drew, 5 November 1941.

58 Jane Drew interview by Karen Parker, 19 June 1988.

59 Fry, *Autobiographical Sketches*, p. 165.

60 See for example RIBA Archive, F&D/18/2. Letter Fry to Drew, 31 March 1942; Letter Fry to Drew, 2 April 1942.

61 HL, Walter Gropius Papers in the Bauhaus Archiv, MS Ger 208 (745). Letter Fry to Gropius, 9 December 1941.

62 Dannatt stayed at the practice 'off and on' until September 1948. Flower et al, *Jane B. Drew*, pp. 24–9.

63 UL Archive, Peter Dale Papers [uncatalogued]. Trevor Dannatt interview with Peter Dale, 1992.

64 UL Archive, Peter Dale Papers [uncatalogued]. Trevor Dannatt interview with Peter Dale, 1992.

65 Dannatt cit. in Flower et al, *Jane B. Drew*, p. 27.

66 UL Archive, Peter Dale Papers [uncatalogued]. Trevor Dannatt interview with Peter Dale, 1992.

67 HL, Walter Gropius Papers in the Bauhaus Archiv, MS Ger 208 (745). Letter Fry to Gropius, 9 December 1941; *Rebuilding Britain* (London, 1943), p. 3.

68 *Rebuilding Britain*, p. 4.

69 *Rebuilding Britain*, p. 5. Elizabeth Darling suggests that the exhibition catalogue was written by Anthony Cox, see *Re-forming Britain*, p. 209.

70 *Rebuilding Britain*, pp. 8–9.

71 *Rebuilding Britain*, p. 24.

72 Jane Drew interview with Karen Parker, 19 June 1988.

73 RIBA Archive. F&D/12/1. Letter John [Gloag] to Fry, Gold Coast, 13 March 1943.

74 RIBA Archive. F&D/12/1. Letter Anthony Hudson Davies to Fry, Gold Coast, 2 April 1943.

75 The MARS Plan of London has been discussed in some detail and need not be repeated here. See for example: E. Maxwell Fry, *Fine Building* (London, 1944), pp. 83–114; Arthur Korn, Maxwell Fry and Dennis Sharp, 'The M.A.R.S. Plan for London', *Perspecta*, 13/14 (1979): pp. 163–73; John Gold, 'The MARS plans for London, 1933–1942: Plurality and experimentation in the city plans of the early British Modern Movement', *Town Planning Review*, 66 (July 1995): pp. 243–67.

76 Fry, *Fine Building*, p. 86.

77 Adams states, based on the work by Clarence Perry, that residential units should provide housing for a 'population for which one elementary school is ordinarily required, its actual area depending upon population density', see Adams et al, *Recent Advances in Town Planning*, p. 299.

78 Fry, *Fine Building*, p. 97.

79 Korn, Fry and Sharp, 'The M.A.R.S. Plan for London', p. 163.

80 RIBA Archive, F&D/15/2. Letter Drew to Fry, Africa, 16 April 1943.

81 RIBA Archive, F&D/15/2. Letter Drew to Fry, Africa, 24 October 1943.

82 Burke, *A Life in Education and Architecture*, p. 2.

83 Jane Drew cit. in Julian Holder, 'The Nation State or the United States? The Irresistible Kitchen of the British Ministry of Works, 1944 to 1951', in *Cold-War Kitchen: Americanization, Technology, and European Users*, (eds) Ruth Oldenziel and Karin Zachmann (Cambridge, MA, 2009): pp. 235–58, p. 237.

84 RIBA Archive, F&D/12/1. Letter Fry to K.E. Robinson, Colonial Office, 19 November 1943.

85 HL, Walter Gropius Papers, MS Ger 208 (746). Letter Drew to Ise and Walter Gropius, 6 February 1944.

86 HL, Harvard University, Walter Gropius Papers in the Bauhaus Archiv, MS Ger 208 (746). Letter Fry to Gropius, 16 March 1944.

87 Dannatt cit. in Flower et al, *Jane B. Drew*, p. 27.

88 Jane Drew, *Kitchen Planning: A Brochure of New Plans and Suggestions for Labour-Saving Kitchens* (London, 1945), pp. 6–9.

89 Drew, *Kitchen Planning*, p. 52.

90 RIBA Archive, F&D/15/2. Letter Drew, London, to Fry, 'c/o The Governor, Nairobi, East Africa', 21 November 1945.

91 This may have been designed by Lund Humphries, as the publishers had designed Drew's previous 'stylish letter heading', as Trevor Dannatt noted. UL Archive, Peter Dale Papers [uncatalogued]. Trevor Dannatt interview with Peter Dale, 1992. For the letterhead, see for example RIBA Archive, F&D/18/10. Letter Fry to Drew, 12 February 1949.

92 Jane Drew, 'Domestic Appliances (Powered and Non-Powered)', in *Design '46: A Survey of British Industrial Design as Displayed in the 'Britain Can Make It' Exhibition Organised by the British Council of Industrial Design* (London, 1946): pp. 15–21, p. 15.

93 National Archive, British Council Papers, BW 83/2. 'Jane Drew: 1946 Tour to Switzerland'.

94 Fry and Denby were reconciled over lunch at the Athenaeum Club and Fry wrote, 'You could hardly see us for olive branches'. RIBA Archive, F&D/18/5. Letter Fry to Drew, 28 March 1945.

95 Nicholas Bullock, *Building the Post-War World: Modern Architecture and Reconstruction in Britain* (London, 2002), p. 247.

96 E. Maxwell Fry, 'The New Britain Must be Planned', *Picture Post*, 10 (4 January 1941): pp. 16–20, p. 19.

97 Fry, 'The New Britain Must be Planned', p. 19.

98 Fry, 'The New Britain Must be Planned', p. 19.

99 RIBA Archive, F&D/18/4. Letter Fry to Drew, 30 October 1944.

100 Jane Drew, 'Editor's Foreword', *Architects' Year Book*, 1 (1945): p. 5.

101 For the third edition published in 1949, Ove Arup replaced Charles Reilly on the editorial board; for issue 5 Trevor Dannatt assumed the role of sole editor (he had been Assistant Editor since the first issue) and Drew moved to the editorial board.

102 Trevor Dannatt interview with Judi Loach, Docomomo-UK Meeting, 10 December 2012.

103 Desmond Bernal contributed an article to *Circle* on 'Art and the Scientist', pp. 119–23; UL Archive, Peter Dale Papers [uncatalogued]. Jane Drew interview with Peter Dale, 1992.

104 Trevor Dannatt interview with Judi Loach, Docomomo-UK Meeting, 10 December 2012.

105 Drew, 'Editor's Foreword', *Architects' Year Book*, 1 (1945), p. 5.

106 M. Christine Boyer, 'An Encounter with History: the Postwar Debate Between the English Journals of *Architectural Review* and *Architectural Design* (1945–1960)': pp. 135–63, p. 140, http://www.team10online.org/research/papers/delft2/boyer.pdf. Accessed: 20 February 2013.

107 Jane Drew, 'Housing', *Architects' Year Book*, 1 (1945): pp. 43–54, p. 43.

108 Percy Ford, 'British Housing Policy', *The Listener*, 545 (22 June 1939): p. 1304.

109 John Charles Walsham Reith, *Report of the New Towns Committee* (London, 1946), p. 68.

110 *The Listener*, for example, ran a series of weekly articles about homes and the task of rebuilding.

111 Earl of Dudley, *Design of Dwellings: Report of the Design of Dwellings Sub-committee of the Central Housing Advisory Committee*, (London, 1944), p. 9.

112 Plan Your Greater Britain – curiously withdrawn from print and not for review by 1948.

113 Julian Huxley, 'Did you hear that?', *The Listener*, 698 (28 May 1942): p. 682.

114 This resulted in the passing of National Parks and Access to the Countryside Act, 1949.

115 Percy Ford, *A Breviate of Parliamentary Papers, 1940–1954: War and Reconstruction* (Oxford, 1961).

116 Earl of Dudley, *Design of Dwellings*, p. 69.

117 Ministry of Health, *Housing Manual* (London, 1949), p. 23.

118 Ralph Tubbs, *Living in Cities* (Harmondsworth, 1942), p. 37.

119 See The Housing (Temporary Provisions) Act, 1944.

120 H. Ashworth, *Housing in Great Britain* (London, 1957), p. 23, p. 40. This figure includes private dwellings; Local Authorities built only 1,934 permanent dwellings in 1945.

121 The *Housing Manual* decreed that 'there is no really satisfactory substitute for a house with a garden for families with young children', see Ministry of Health, *Housing Manual* (London, 1949), p. 82. Meanwhile, the *Design of Dwellings* found them to have a 'depressing appearance', see Earl of Dudley, *Design of Dwellings*, p. 20.

122 Ministry of Health, *Housing Manual*, p. 82.

123 Postwar studies include: H. Kamenka, *Flats: Modern Developments in apartment house construction* (London, 1947); Walter Segal, *London Housing: A Comprehensive Survey by "Building" of the Post-war Work of the London County Council. Housing Flats, Equipment* (London, 1949).

124 Earl of Dudley, *Design of Dwellings*, p. 10.

125 Earl of Dudley, *Design of Dwellings*, p. 10. Percy Thomas also made this point in the BBC publication, *Homes for All* (Worcester, 1945), p. 21.

126 'Freda White', 'New Towns for Old', *Architects' Year Book*, 4 (1952): pp. 33–9, p. 34. This article sounds remarkably like Fry, who later remarked that the work of an architect could be likened to that of a 'neutral (albeit compassionate) doctor'. See Allan, *Berthold Lubetkin*, p. 322.

127 John Henry Forshaw and Patrick Abercrombie, *County of London Plan* (London, 1943), p. 77.

128 This was imposed because of the surrounding two-storey houses that were not to be overshadowed by high-rise blocks. See 'Flats at Lewisham. Architects: Fry, Drew and partners', *Architects' Year Book*, 4 (1952): pp. 156–9, p. 156.

129 Earl of Dudley, *Design of Dwellings*, p. 23.

130 Denys Lasdun, 'Second Thoughts on Housing in London', *Architects' Year Book*, 4 (1952): pp. 137–43, p. 138.

131 For further discussion, see for example: John Gold, *The Practice of Modernism: Modern Architects and Urban Transformation, 1954–1972* (Abingdon, 2007); Bullock, *Building the Post-War World*; Darling, *Rebuilding Britain*.

132 FLL, CIAM Collection, C006. Letter Sert to Giedion, 21 December 1949.

133 Theo Crosby cit. in Flower et al, *Jane B. Drew*, p. 54.

134 FLL, Ferrari Hardoy Papers, A017. Letter P.L. Cocke, Assistant Honorary Secretary of MARS Group, to Hardoy, 20 May 1948.

135 FLL, CIAM Collection, C006. Letter Fry to Sert, 20 May 1949.

136 The School was held from 8 August 1949. FLL, Ferrari Hardoy Papers, A017. Letter P.L. Cocke, Assistant Honorary Secretary of MARS Group, to Hardoy, 20 May 1948.

137 FLL, Ferrari Hardoy Papers, A069. 'First CIAM Summer School – Names & Addresses of Members', August 1949.

138 FLL, Ferrari Hardoy Papers, A069. 'First CIAM Summer School – London 1949'.

139 Reith, *Report of the New Towns Committee*, pp. 7–8.

140 Ashworth, *Housing in Great Britain*, p. 99.

141 Reith, *Report of the New Towns Committee*, p. 5.

142 Reith, *Report of the New Towns Committee*, p. 5.

143 The population level was quickly raised to an 80,000 maximum population, and further expansions were planned throughout the 1960s and 1970s. See Frederic James Osborn, *New Towns: Their Origins, Achievements and Progress* (London, 1977).

144 Frederick Gibberd, 'Harlow', *Builder*, 173 (19 December 1947): pp. 700–3.

145 Osborn, *New Towns*, p. 166.

146 Reith, *Report of the New Towns Committee*, p. 31.

147 Reith, *Report of the New Towns Committee*, p. 30.

148 W. Eric Adams, *The New Town of Harlow* (Harlow, 1954), p. 11.

149 Frederick Gibberd, 'Harlow New Town', *Architect and Building News*, 192 (19 December 1947): pp. 245–58.

150 S. Phillips Dales, 'Harlow', *Builder*, 173 (19 December 1947): pp. 700–3, p. 701.

151 Gibberd, 'Harlow New Town', p. 248.

152 FLL, Josep Lluis Sert Collection, E003. Letter Drew to Sert, 31 July 1947.

153 Paul Lester Wiener and José Luis Sert, 'A New City in Brazil', *Architects' Year Book*, 3 (1949): pp. 75–86, p. 86.

154 Valerie Fraser, *Building The New World: Studies in the Modern Architecture of Latin America 1930–1960* (London, 2000), pp. 206–11.

155 RIBA Archive, F&D/13/4. E. Maxwell Fry, 'Maxwell Fry and the Modern Movement', undated.

156 Jane Drew, 'Housing', *Architects' Year Book*, 1 (1945): pp. 43–54, p. 46.

157 Drew, 'Housing', p. 44.

158 Despite her concerns for efficiency there are some anomalies in the planning, such as the kitchen and bathroom being at opposite corners of the house, rather than stacked on top of each other to simplify drainage service runs.

159 Drew, 'Housing', p. 49.

160 Drew, 'Housing', p. 46.

161 Photographs of Jellicoe's housing were included in the first *AYB* and are similar to those proposed by Fry and Drew at Harlow.

162 Mark Llewellyn, 'Producing and Experiencing Harlow: Neighbourhood Units and Narratives of New Town Life 1947–53', *Planning Perspectives,* 19 (2004): pp. 155–74.

163 Tyrwhitt ensured that Fry was involved in proceedings, despite his post at Chandigarh, and he contributed a paper on the New Towns, 'The Idea and its Realisation' to the conference proceedings published the following year, see *The Heart of the City: Towards the Humanization of Urban Life*, (eds) J. Tyrwhitt, J.L. Sert and E.N. Rogers (London, 1952), pp. 87–9. For Tyrwhitt, see for example Ellen Shoshkes, 'Jaqueline Tyrwhitt: a founding mother of modern urban design', *Planning Perspectives*, 21 (April 2006): pp. 179–97.

164 'Discussion on Italian Piazzas', in *The Heart of the City*: pp. 74–80, pp. 78–9.

165 Sigfried Giedion, 'Historical Background to the Core', in *The Heart of the City*: pp. 17–25, p. 17.

166 J.M. Richards, 'The Failure of the New Towns', *Architectural Review*, 114 (July 1953): pp. 29–32; J.M. Richards, 'Failure of the New Densities', *Architectural Review*, 114 (December 1953): pp. 355–61.

167 Blum, *Oral History of John Donald Cordwell*, p. 69.

168 Christine Hui Lan Manley, 'Modern City versus Garden City: Housing at Harlow New Town', *The Influence of Fry and Drew*, unpublished conference paper, 2013.

169 Maxwell Fry, 'The Idea and its Realisation', in *The Heart of the City*: pp. 87–89, p. 88.

170 'Rayon Industry Design Centre', *Architect and Building News*, 195 (28 January 1949): pp. 76–9, p. 77.

171 'Rayon Industry Design Centre', *Architect and Building News*, p. 77.

172 Waters, 'In Search of Sir Gerald Barry', p. 42.

173 UL Archive, Peter Dale Papers [uncatalogued]. Jane Drew interview with Peter Dale, 1992.

174 Waters, 'In Search of Sir Gerald Barry', pp. 42–3.

175 UL Archive, Peter Dale Papers [uncatalogued]. Jane Drew interview with Peter Dale, 1992.

176 UL Archive, Peter Dale Papers [uncatalogued]. Jane Drew interview with Peter Dale, 1992.

177 'The Festival of Britain: Architectural Plans for South Bank Site', *Builder*, 177 (25 November 1949): pp. 686–90, p. 686.

178 Mary Banham, 'Foreword', in *The Festival of Britain*, Atkinson, xvii.

179 UL Archive, Peter Dale Papers [uncatalogued]. Jane Drew interview with Peter Dale, 1992.

180 Starrett and Greenwell met at Fry, Drew and Patners, and married in 1951.

181 Starrett worked on the Riverside Restaurant, Todd on the entrance tower, and Christopher Knight on the stairs and bridge.

182 UL Archive, Peter Dale Papers [uncatalogued]. Jane Drew interview with Peter Dale, 1992.

183 UL Archive, Peter Dale Papers [uncatalogued]. Jane Drew interview with Peter Dale, 1992.

184 Michael Bussell, 'The Use of Concrete in the Post-War Era', in *Preserving Post-War Heritage: The Care and Conservation of Mid-Twentieth Century Architecture*, (ed.) Susan Macdonald (Donhead St Mary, 2001): pp. 81–103, p. 94; UL Archive, Peter Dale Papers [uncatalogued]. Jane Drew interview with Peter Dale, 1992.

185 UL Archive, Peter Dale Papers [uncatalogued]. Jane Drew interview with Peter Dale, 1992.

186 The British Library, National Life Stories, C466/28. *Lynn Chadwick*, Interviewed by Cathy Courtney, 1995, p. 159.

187 The British Library, National Life Stories, C466/28. *Lynn Chadwick*, p. 159.

188 Jane Drew, 'The Riverside Restaurant', in *A Tonic to the Nation: The Festival of Britain 1951*, (ed.) Mary Banham (London, 1976): pp. 103–4, p. 103.

189 Some ten years after construction the restaurant remained watertight and in sound condition.The same riveted joints were later used in school buildings, lasting for 40 years before the rivets started to pop and could not be repaired. Peter Dale communication with author, December 2011.

190 Drew, 'The Riverside Restaurant', p. 103.

191 The mural was later moved to the VIP lounge at Heathrow Airport and is now owned by the Tate Gallery.

192 The Tate Gallery, 'Festival of Britain Mural, 1951', https://www.tate.org.uk/art/artworks/nicholson-festival-of-britain-mural-t07027/text-summary. Accessed: 29 March 2013.

193 UL Archive, Peter Dale Papers [uncatalogued]. Jane Drew interview with Peter Dale, 1992.

194 Jane Drew interview with Barrie Fernandez, 8 April 1983.

4

West Africa: Planning, Village Housing and New Schools for Ghana

THE PASSAGE TO BRITISH WEST AFRICA

Fry's career was diverted, repeatedly, by the major events of the twentieth century, not least, the First World War had delayed his studies,[1] and it was the declaration of war in 1939 that was to interrupt his practice, just at a point when he was obtaining considerable work and recognition.

Fry rightly declared that, 'War and architecture never go together',[2] although it was the effect of the war that would go on to shape and define the rest of his career. Whilst reluctantly winding down his practice, he made one last attempt to preserve it, by humbly asking the War Office for work. He secured a meeting with General Williams of Fortifications and Works who declined his architectural services, but offered Fry the post of Staff Captain in the Corps of Royal Engineers. Fry's home life was also rapidly changing. Fry and Drew first met at an RIBA committee meeting.[3] Drew was already divorced and establishing her vision of a 'women-only' practice by this point, and she ended up taking Fry in, amongst others, to her house in Woburn Square, which had become a kind of hostel for numerous artistic waifs and strays.[4]

In the War Office, Fry was effectively a civil servant working on supplies, logistics and the administration of war, organising the setting up of camps for soldiers and ensuring supplies were available.[5] He was posted to Derby and filled his time writing and whilst away from the distractions of London completed his first book, *Fine Building*. Fry loathed being separated from Drew, but also wanted to work on meaningful projects and volunteered for 'over seas' work at the earliest opportunity. He was not the only architect offered foreign posts whilst serving in the Royal Engineers, and some of his fellow students obtained similar positions, although it seems none were aware of this until after they had returned to the UK.[6] On the 25th April 1942 Fry and Drew married at Caxton Hall, Westminster followed by a lunch-time reception at their favourite venue, *The Cafe Royal*, attended by his best man Julian Huxley and Charles Reilly.

Shortly after, Fry sailed from Liverpool, shaken at the devastation enemy raids had wreaked upon his hometown,[7] and unaware of his final destination, which had not been divulged. Whilst on board he wrote, *Architecture for Children* and claimed that he completed the final chapter just as they arrived at their mystery destination 25 days later, five degrees north of the equator.[8]

The ship docked at Accra in the Gold Coast (later Ghana) and although Fry had wanted to leave the UK, he again found himself 'desperately unhappy. Jettisoned. Marooned in a tropical backwater'.[9]

He described that despite being 'the Royal Engineer factotum for the country with a large Humber car to take me about', that he 'was not overburdened with work'.[10] His friends kept in touch, expressing their occidental and romanticised imaginings of Africa, including Hans Feibusch (the mural artist at Sun House) who narrated the sojourn,

> *Since the outbreak of war you have been gradually disappearing into an official cloud, being whisked away into the war office first, then into the country, and now God knows where into the Orient; and although you are always very much present in my mind it is you as you were formally, that I see moving about, in offices full of gold braid, or building barracks, or now sweating in Africa*[11]

Reilly kept in touch too, lyrically picturing the African women, 'gaily clad … walking like queens and the men like Roman Emperors'.[12] There is little information about what Fry actually designed during this war period, other than some 'useless airstrips',[13] extensions to the European Club and a Boy Scout HQ.[14] We know he became proficient at snooker (all very colonial so far), and he also spent time in the 'black towns' (he was also a progressive liberal), but very little survives.[15] Fry found it 'increasingly boring' and perhaps having too much time to ponder his life he mused, 'I had sunk low in my own estimation and wondered what would be the end of me'.[16]

Drew, on the other hand, was very busy in London. She curated exhibitions, such as 'Rebuilding Britain'[17] and was enjoying socialising with a new group of friends including Henry Moore, Peter Gregory, Elizabeth Denby, and Kenneth Clark who was her 'best friend nowadays without doubt'.[18] She also lunched with Fry's first wife (Ethel Speakman), and relayed to Fry, 'it is curious as a relationship'.[19]

In addition to sitting on RIBA committees (which gave a 'sense of self-importance'),[20] she designed kitchens and stoves for the British Commercial Gas Association[21] as well as fake factories designed optimistically, to confuse enemy bombing missions. Intriguingly, she also claimed to have used her practice as cover for secretive MI6 work for which she was paid in cash and may have involved translating documents written in French into English, although the details are sparse. Ever the storyteller, she later commented: 'I wasn't allowed to talk about it then and I still don't let myself'.[22] She also obtained commissions from the Kuwait Oil Company for a new town in Ahmadi. The maserplan was prepared by Wilson Mason with Drew's commissions connected to health care such as a hospital, dispensary, clinics as well as schools and housing.[23] All European staff housing was to be air-conditioned as well as 'compact and designed to be run with a minimum of domestic help'.[24] The housing for the local workers was arranged in brick terraces with lightweight pitched roofs and walled rear compounds, all built in a separate 'Arab Village'. Fry wrote to her from Africa in 1944 stating that a fee of 4½ per cent of the construction costs and 'site visits dealt with separately' would be appropriate.[25] Fry describes these schemes as 'nearly completed' in the selection documents for Chandigarh in 1950, but there was no publicity or reports in the British architectural journals on these numerous commissions.[26]

TOWN PLANNING ADVISOR

In Ghana Fry was unexpectedly summoned to meet with Lord Swinton, the Resident Minister for West Africa. Swinton needed a Town Planning Advisor for the entire region and despite Fry's lack of qualifications in this area he agreed to do the job, providing three conditions were met:

1. That Jane Drew become Chief of Staff
2. He was granted three months leave prior to starting the post
3. That a research trip to Roosevelt's Tennessee Project was funded during the three-month leave.

It transpired that Drew had written to the Minister for Town and Country Planning, Henry Strauss, and met him for lunch at *The Ivy* where she, 'put in a word for her absent husband with the desired result'.[27] This is demonstrative of Drew's approach, she was a 'people person' with a likeable character, and able to turn situations to her advantage – or as Trevor Dannatt put it, 'she was a rainmaker'.[28]

Drew wanted the adventure but felt torn; she described it as 'a terrible decision' not wanting to leave her children behind who were living as evacuees in Wales with their long-term nanny and house keeper, Maud Hatmil.[29] She had also developed a successful practice in London, despite being 'bombed out totally once', where she 'lost the Picassos, Henry Moores and Bonnards' she had been given.[30] Designing for the African context was not new to Drew. In 1937, Alliston and Drew received a commission for a model village and school in Kenya. Drew discussed the requirements in detail with Peter Koinange, the son of the Kikuyu tribe's chief. This was her first experience of designing for a tropical climate and, although the scheme was unbuilt, it provided useful experience for her subsequent work. Fry tried to entice her to join him and sent a cable saying, 'he would do it if I would join him and I could have Nigeria[31] (that is, to do the plans for Nigeria, with Fry presumably planning the remaining three smaller colonies). Swinton wrote to the Colonial Office in Downing Street to arrange Fry's visit to America, an important research trip that would 'bring me livelier ideas of planning than I found among my own contempories here … I would like to fulfil the long wish to see the Tennessee Valley administration work and discuss with them the growth of administration from its small beginnings'.[32] Fry could combine the visit with Drew's, who was conducting research into coloured enamels for kitchen appliances in the States. Together, they considered how timber prefab, civil engineering, and solar energy for hot water could be deployed in Africa. They also visited Walter Gropius and Drew agreed that she would join the Africa project, flying direct to Gold Coast from America.

It was an 18-month appointment with generous tax benefits, furnished accommodation and first class return passage from the UK,[33] but forbade Fry from engaging in any private practice putting on hold any dreams of re-establishing a London office.[34] Fry was awarded a married person's salary whereas Drew had to make-do with a single person's wage that was further reduced, '£800 would be about right for a man but it would be too high for a woman and taking into account this sex differentiation we think £700 per annum would be the right figure'.[35] Eager to supplement the basic accommodation with some home comforts, they arranged for china from *Fortnum and Mason*, a radiogram and a typewriter to be shipped out, as well as a second-hand Ford V8 which was to be sent to the Port of Takoradi.[36] In addition a whole array of office equipment was requested, including six double elephant size drawing boards, pastel crayons ('these rather important'), reams of paper, paints, surveyors levels, stainless steel drawing instruments, scales, inks, filing cabinets and a sunprint frame with chemicals, all to be charged to the Colonial Development Fund.[37]

Fry's official title was Town Planning Advisor to the Resident Minister, and together with Drew as Chief of Staff, their role was to, 'provide draft or sketch plans for a selected number of the principle towns ... to advise the Government on the setting up of town planning legislation, and some kind of machinery for carrying it out, and for dealing with nearly everything that came our way'.[38] They were permitted to recruit their own team of four assistants and looked to former colleagues to join them.[39] William Holford was also making enquiries on Fry's behalf[40] and they eventually hired, amongst others, 'Mr. [John Dawe] Tetlow, the first fully qualified West African architect'[41] and later to become the first president of the Ghana Society of Architects.[42]

TOWN PLANNING FROM COLONIAL POLICY

The implementation of town planning is a complex and expensive task requiring considerable investment and reconciliation of often contradictory requirements and agendas. Whilst the early colonial planning was limited to European settlements designed to 'protect human health and welfare', the 'native segments of urban areas were generally ignored'.[43] However, by the late 1930s funding of large 'capital projects' was proposed, marking a shift in the late colonial agenda of the UK. Whilst, as Njoh notes, the impoverished colonies may have become a source of embarrassment to the UK government, it was development to stimulate economic growth that motivated the Colonial office.[44] 'Welfare' and 'Development' were politically nutritious and altruistic terms, but the underlying assignment was to 'to promote commerce with, or industry in, the United Kingdom'.[45] It was hoped that by increasing commerce with the colonies the significant trade deficit that had steadily increased after WW1 would be reduced. Construction projects were ideal in this regard as they stimulated manufacturing and shipping as well as professional consultative activity. It was proposed that 'backward areas must be helped to develop a prosperous economic life', with the means of purchasing British manufactured goods.[46] Furthermore this forced investment was presented as an altruistic gift and as token of appreciation for the 'spontaneous and wholehearted support' shown by the colonies during WW2.[47] The allocation of these grants did not, however, sway from the old principle that a 'colony should only have those services which it can afford to maintain out of its own resources',[48] in other words grants were only made to those colonies that had the economic means of reciprocating through increased trade. By 1940 these remits were considered too narrow, 'limiting and hampering' the ability to act in ways that would be transformative, not only in commercial terms, but through other means such as education and projects requiring recurrent funding, for example.

This prompted the change in the title of the Act to now include Welfare, and loans were replaced with grants resulting in many building projects in the aftermath of WW2. Furthermore, various reports were commissioned during WW2 specifically concerned with education, such as *Mass Education in African Society*[49] and a report from the *Commission on Higher Education in West Africa*[50] prompting a review of the existing provision which was woefully inadequate. There were only 43 recognised secondary schools in 1942 for the entire region of British West Africa – and a population that exceeded 27,000,000. It was in these two burgeoning sectors of Town Planning and Education, both of which were underpinned with substantial government funding, that Fry and Drew were able to offer their services and attempt to develop an appropriate architecture for the political independence that was to follow.

TOWN PLANNING THE MAJOR TOWNS OF BRITISH WEST AFRICA

Initially, most of Fry and Drew's work was pragmatic and based on organisational planning in the form of 'sketch plans' spanning across the entire region.[51] At Bathurst, Gambia, for example, the practice provided new layouts for drains, which previously flooded and even flowed the wrong way.[52] Although Fry claimed that there was no African architecture from which to draw any precedent he was invariably seduced by what he saw, describing Bathurst to Drew,

> *how charming it is. White walls enclosing gardens, wide grass grown streets, white robed men and gorgeously dressed Jollof women, all moving as if were in a dream. And a waterside road lined with colour washed old stone buildings with arcaded fronts on the one side and all sorts of delightful foreshore messes on the other*[53]

The report for Bathurst, co-authored with assistant, Betty C. Benson, echoed these views and was in favour of preserving most of the old town,

'We do not share in the common and now traditional view of Bathurst as being an "open sore", "blot on the Empire", etc. It appears to us to be an old and faded port of some charm retaining many of its original trading houses, not without architectural character, by comparison with Lagos, over-crowded only in part; and by comparison with Freetown, well preserved. This impression remains after several complete perambulations of all parts of the town'.[54]

They objected to the proposed removal of the administration out of the town, segregating the Europeans and amounting to what Fry and Benson described as, 'a clear desertion of the Africans, and would be so regarded. Its effect would take the heart of Bathurst ... '[55]

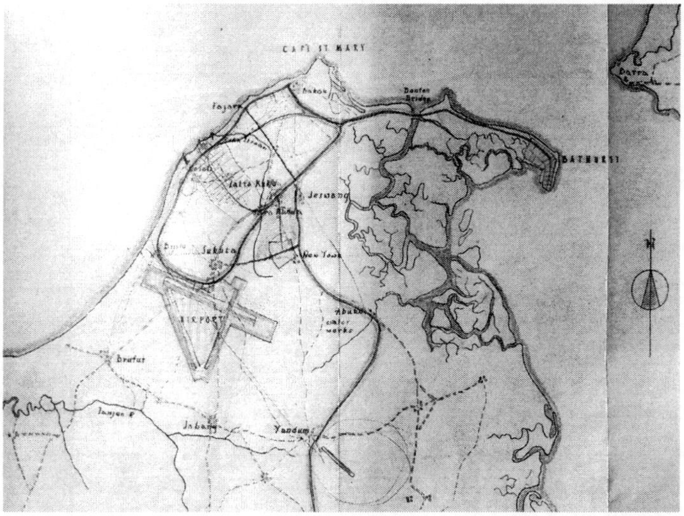

4.1 Planning the Region around Bathurst, Gambia, 1944

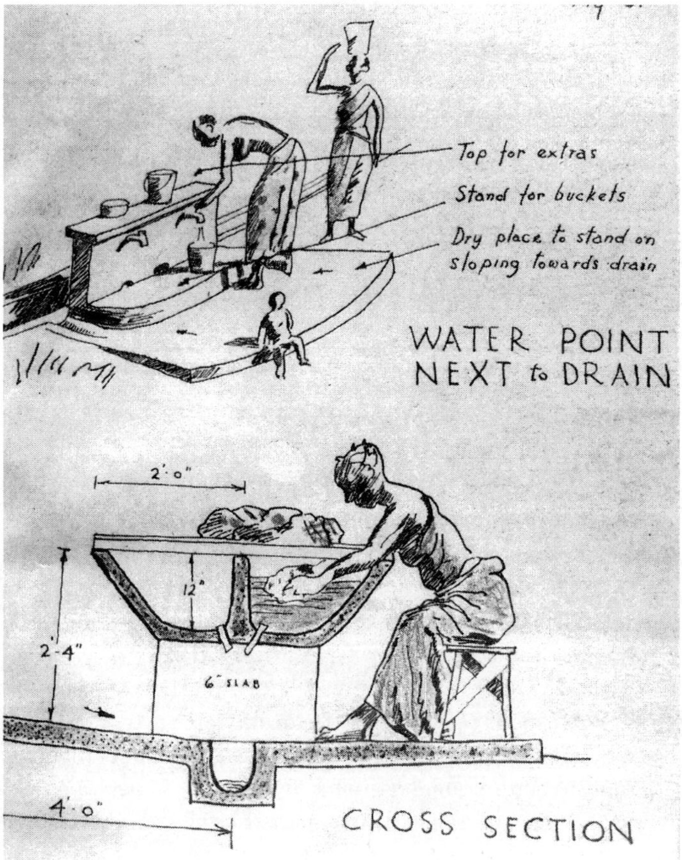

4.2 Design for Laundry and Water Point, Bathurst, Gambia, 1944

Although employed by the Colonial regime they did not shy from challenging its decisions and were generally sympathetic to the local populations. The European bungalows were not party to much discussion, once upgraded with WC's 'in place of bucket latrines they will in our opinion make very comfortable dwellings', a shrift overview that made way for their recommendations for the African housing. They found that 'the worst feature of native housing, apart from its lack of sanitation and drainage, is the absence of cleanable surfaces such as concrete floors provide ... improvement will come with new drainage and sewerage, but it might be considered whether, where complete rehousing is not possible, the laying of concrete floors might not mark a stage of progress ... '[56] Coloured drawings were also included of washing stations and basic dwellings, a forerunner to those in *Village Housing in the Tropics*.

The final aspect of the report considered a new airport and the implications this would have on the neighbouring villages resulting an unplanned expansion 'lacking adequate streets, open spaces, sanitation, hygiene and morals'.[57] They also added that a series of bungalows for the Europeans should be included close to the airport as well as hotel and golf course that could form an 'important stopping point in world air travel', perhaps even becoming a 'health resort in the dry season'.[58]

Their planning report on Freetown contains no plans or drawings merely descriptions of what Fry (and K. W. Farms, planning assistant) thought was required, such as a deep quay, improved sanitation, wider roads and slum clearance – it is barely more than stating the obvious. Fry felt that a town plan was a, 'complete luxury, until the port has a new deep water quay and a proper system of water storage'.[59] A plan had been prepared by the PWD as early as 1929 to improve the water supply, but remained unexecuted – until the arrival of the large development grant and introduction of people like Fry and Drew, there seemed to be a lethargy in completing tasks such as these in the colonies.[60] The matter was discussed in Parliament in 1944 with the Secretary of State for the Colonies questioned over the 'failure to develop the harbour of Freetown to a standard worthy of the British Empire'.[61] Freetown had the highest rainfall on the coast but still suffered water shortages in the dry season. Fry was reticent to propose dramatic changes, despite his desire to improve the slums and overall sanitation, he wanted little to change that would lessen the exotic experience of these towns. One of the most provocative suggestions Fry made for Sierra Leone was that a Town Planning Officer be appointed and that 'he should not be attached to so executive a department as the PWD but to the Development Branch of the political administration'.[62]

The Gold Coast served as a test bed as it was the location of the Resident Minister's and Fry and Drew's offices and it was here that they developed more detailed proposals for Takoradi, Sekondi and Kumasi, as well as outlining the planning legislation for a Ghanaian Ordinance, 1945.[63] They were not designing specific buildings but broad, strategic town planning proposals, such as linking the two ports of Takoradi, draining a marsh in Kumasi and proposing a bridge to connect the two parts of the town divided by railway tracks.[64]

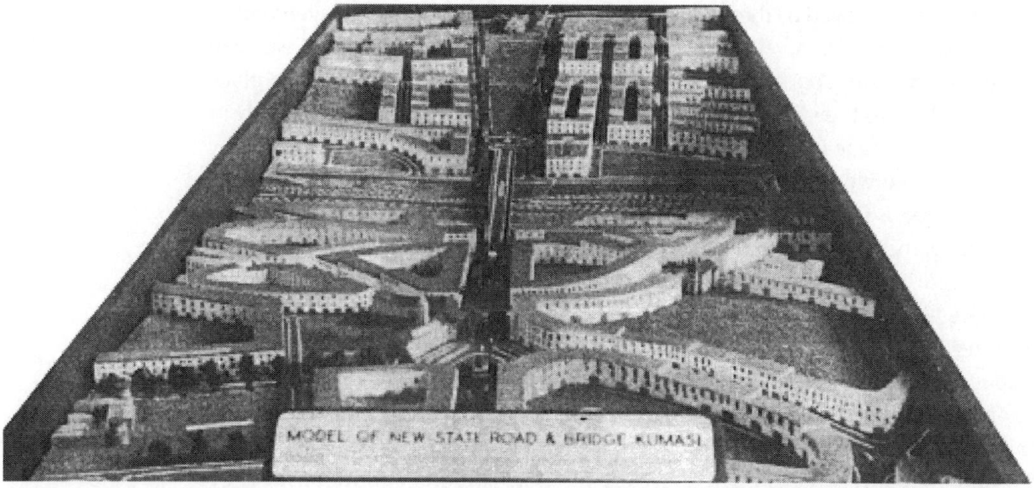

4.3 Plans for Modifying Kumasi's Railway, Ghana, 1946

Many of their proposals had been previously mooted but remained as memos and correspondence.[65] Fry and Drew also set about publicising their proposals through exhibitions, such as the one held in Kumasi in October 1945 following the submission of their plans to the Resident Minister in August of that year.[66] Their plans were to accommodate a population increase of some 30,000 people in just three years, but displayed a complete naivety in terms of addressing local politics and land ownership. They also got sidetracked into designing street furniture, an irrelevant luxury where most people needed fresh water.[67] They presented their findings in an embossed green leather-bound presentation copy to the *Asantehene* Osei Agyeman Prempeh II who prophetically noted how the proposals would prompt 'litigious struggles over rights in land ownership' that would follow.[68] When the exhibition opened extra police were drafted in to control the crowds as hundreds flocked to see how it would affect their neighbourhoods and ancient land rights, and as expected quarrels erupted over ownership and speculation at potential values.[69] Fry and Drew remained curiously silent over these events. In Accra, Fry proposed the demand for high quality housing should be met by dwellings in parkland leading into the centre of the town. It was not a formal geometric 'garden city' plan but adopted the same strategy of providing open gardens between low-density bands of housing, a standard colonial approach.

In the colonies this served as a means of segregating the colonisers from the colonized, satisfying the 'obsessive concern with "health" … the driving force behind planning in all colonial territories',[70] and as Fry explained his plan was approved by the health board, 'My plan for a Govt. Centre for Accra was approved by the Central Health Board this morning. That is quite important to us as it fixes this part of the town good and proper and stops all kind of nonsense over the siting of future government buildings'.[71] Although Fry may have been thinking about beautification, town planning was never undertaken for its own sake or for aesthetic betterment, and the British intended to use the city as a base for allied troops and subsequently needed a well-planned and 'orderly' layout for expedient deployment.[72]

The British had already introduced some of their planning legislation in Nigeria[73] but Fry described the problems there as, 'extremely difficult'; they made plans for small districts throughout Lagos but these were deemed inadequate in a matter of months due to the 'rush' of new residents and Government Departments wanting to develop certain large areas of land.[74] Fry's ambition for the planning of Lagos and surrounding towns was stunted by the rapidly changing conditions, the inability to enforce planning, the rate at which land was developed illegally and the lack of staff. Fry suggested that 'Mr. Waide, the town-planning officer of the Lagos Executive Development Board, should devote his time entirely to town-planning and that two other officers should be appointed as his assistants', so great was the workload.[75] Recognising the limits, they set about trying to organise transport routes and simple measures such as proposing sufficient open space to flank each road. Acknowledging the arduous task, and the limits of what could be achieved Fry conceded 'all we concerned ourselves with, was to provide a plan which would make a road structure likely to work, and which would safeguard the remaining open spaces, of which there are very few'.[76]

Their most significant contribution was on a much smaller scale and the numerous interventions they proposed to 'village design' were well received by many settlements, as well as the Colonial regime.[77] Fry and Drew's recommendations were concerned with the placing of roads around a village, the visibility of road junctions, social and public amenities as well as gardens and sports provision. They also designed latrines, washing stations and water wells. Drew described how the extended 'ribbon development along main roads' was particularly, 'rampant', and they wanted to avoid this by encouraging more development with village clusters.[78] They developed a broad overview of the entire region and gathered substantial data from their experiments and observations, but the idea that a team of six architects and planners could in any meaningful way resolve the planning issues of four countries spread over such vast areas was foolishly optimistic, but equally shows how the colonies were governed and the arrogant confidence that was displayed in the British Colonial system. Fry was eager to complete the work in Africa and to return home. As mentioned in Chapter 2, he wrote to Drew expressing his sentimental musings for an imagined England and for vacation in Wales;[79] Africa was only ever a temporary mooring and it was thoughts of returning home to the metropolis that spurred Fry's efforts.

THE MODERN ORIENTALISTS

There have been extended discussions about the notion of Modern architecture being proposed in colonial settings and the desire of architects to present white geometric forms with 'scientific' climatic modifications as somehow politically neutral, or at least not colonial.[80] Fry and Drew, in particular, have been frequently singled out as purveyors of this approach, and through their plans and designs they have become emblematic of the late colonial policy of Britain. Fry quickly dismissed any African attempts at 'architecture', and although he claimed they searched for architectural precedents,

> *there was none. Not in our own colonial buildings which were without character or the sort of response to natural conditions that we were seeking; nor in African building which taught us the value of shade but was of a passing order the beauty of which we could admire as it fell and decayed ...* [81]

Yet both he and Drew were seduced by what they saw as a raw, ruinous primitivism, and rather like Said's 'modern Orientalist', they were there to 'rescue' Africa, and for it to be 'restored to the present'.[82] This was not to take place through the 'scientifically advanced techniques of philology and of anthropological generalization' outlined by Said, but through the tangible medium of architecture with its own claims of technological advancement and building physics. Occupying an almost paradoxical position of being both seduced by their surroundings and at the same time wanting to 'improve' it Fry declared that,

> *We were fated to make a new architecture out of our love for the place and our obedience to nature, and to make it with cement and steel, asbestos sheets, wood above the termite line, glass, paint and some stone later, and not much else*[83]

Specialist construction techniques developed by Europeans for use in the hot climates of the colonies was not a new phenomenon,[84] but it was the mid-twentieth century that saw a significant interest in the 'scientific' development of construction to suit tropical climes. Fry and Drew gave the impression that they were trail-blazers, inventing a new architecture for the continent; however, they derived a considerable amount of their technical methods from others, not least in West Africa from the Public Works Department building guides and bulletins.[85] It is important to stress therefore, that although modern architecture was purporting to be politically neutral and distinct from the colonial endeavour, it remained an integral component. While the pomp of the classical facades had been displaced with rubble stone and rendered surfaces alluding to the neutral, the procurement, technology and finances remained in the hands of the colonisers. Despite this, the direction and focus of the building programmes had begun to shift. There had been some prior attempts to improve the general housing of Africans by the Colonial regime, but these were mainly reactionary attempts rather than predetermined efforts to raise standards and welfare.[86]

By the 1950s and the rapid approach to political independence, there seemed to be a British concern to improve basic dwelling and sanitation. Despite the Colonial system's notoriety for its bureaucracy and hierarchical structure certain individuals were able to conduct significant, almost maverick experiments in planning, and were given scope to test their ideas 'live in the field'. One such individual was the planner and architect Alfred Edward Savige ('Bunny') Alcock who was developing his own 'experimental housing estate' at Asawasi, Kumasi.[87] The estate was self-constructed by villagers adopting Alcock's methods of laterite block production and roof truss jigs, and his proposal also included shared sanitation facilities such as latrines and washing stations.

Alcock's designs exceeded the minimum building regulation requirements and 'the experiment has proved', he proudly noted, 'that a considerable saving in cost of

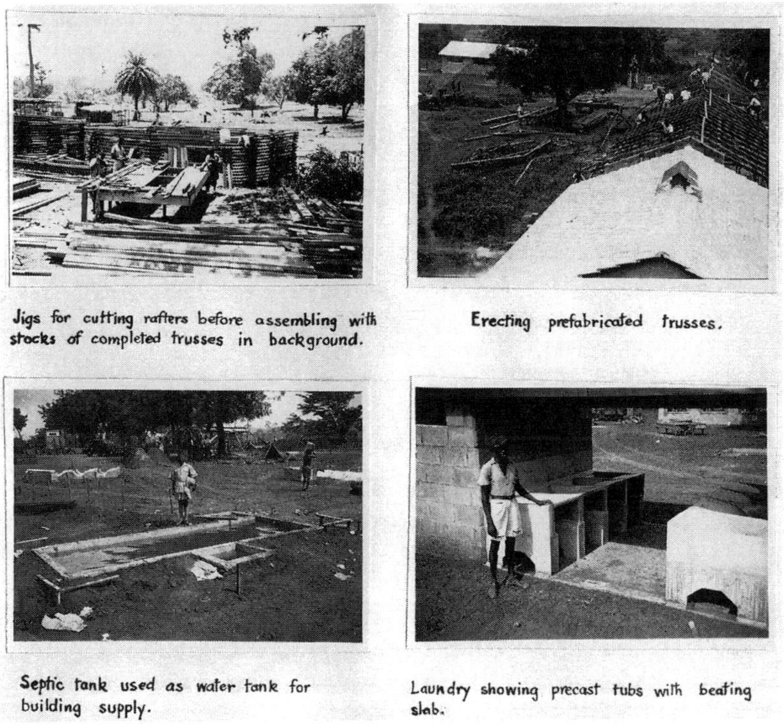

4.4 Alcock's experimental housing in Asawasi, 1945–6

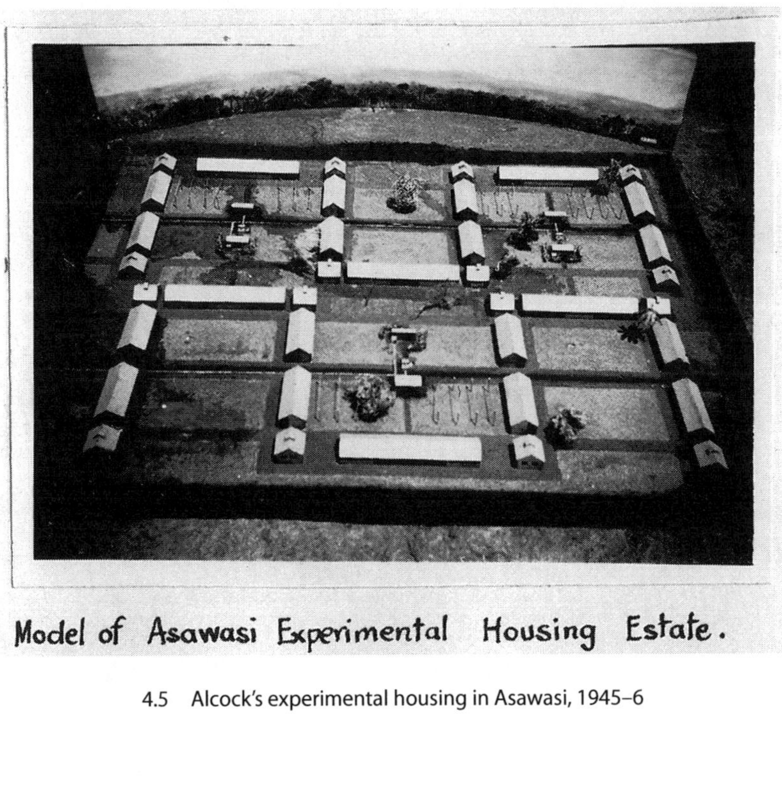

4.5 Alcock's experimental housing in Asawasi, 1945–6

construction can be made by the use of the unconventional materials and methods employed. Soil stabilization has proved a structurally sound method of building.'[88]

There was clearly some exchange and sharing of ideas, as Alcock acknowledged the help of Fry and Drew in his report,[89] and Drew also recalled to Fry how she had been 'thinking about Alcocks [sic] housing and the improvements we should like to see made … '[90] Indeed, many of Alcock's innovations were later adopted by Fry and Drew,[91] such as Fry's 1945 plan for Agbani Neighbourhood in Enugu which bears a strong resemblance.[92] Fry and Drew also developed the second phase of Asawasi, where a variety of housing types were proposed arranged in small clusters around sports grounds and community buildings.[93]

Fry described how he missed Drew's 'enjoyment of Asawasi which from being a little experiment has become a big scheme spawning all over the hillside.'[94] It is a clear forerunner of the Chandigarh sector.

Drew was especially focused on improving welfare and amenity, and she outlined some of the pragmatic objectives of their dwelling designs, many of which inform today's sustainability agenda, 'we had to collect water from the roofs of buildings to re-use when purified. We had to design buildings that could be used without air-conditioning. We had to work out shading devices. We would have to build in concrete for the structure, as no other material was available.'[95] In addition she set up extensive consultations with the future inhabitants and, having gained a basic grasp of the local languages, was able to communicate effectively with the Africans.[96] She explained that,

> *for any job it is worth consulting, where possible, all those who work in and use such buildings and get direct reaction, and to regard the building as something which will help to produce the required work and the required atmosphere*[97]

In order to explain their town planning proposals further, they organised further exhibitions in their office and invited the local dignitaries, 'who arrived in full regalia with their umbrella bearers before them … we had a certain success with the chiefs who were quite quick at seeing what improvements town planning could bring.'[98]

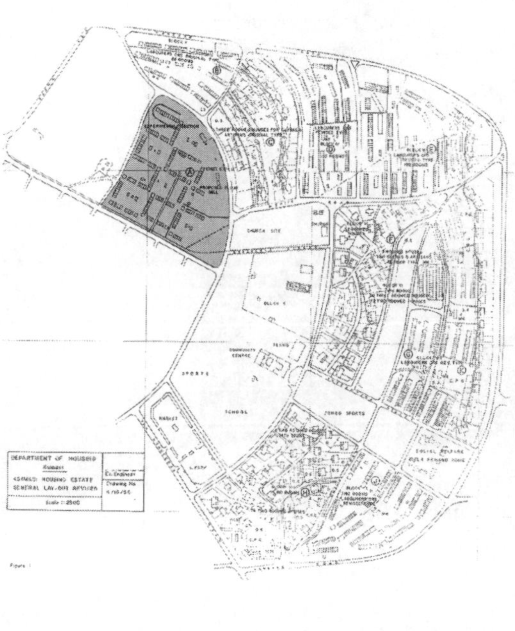

4.6 Fry and Drew's housing at Asawasi, 1946

They developed simple methods of providing, 'latrines, easy ways of digging wells. We found out what trees to grow.'[99] They looked to address the problems of water supply, sanitation and laundering because if these three elements could be improved, better health would follow, as well as reducing physical effort and time expended on such mundane chores. They also attempted to use local materials and building techniques, including Alcock's 'stabilized earth' which formed a durable, cheap building material that could be built using widespread skills.[100] However, Drew described how this method was rejected when, 'the Africans realised they [the houses] were in fact made largely of mud. Then we got a back-lash. It was political, if concrete was right for the Europeans it must be right for them'.[101]

A NEW MANUAL FOR A NEW DOXOLOGY

Their collection of model building solutions and experiments was compiled into a small guide for building in the tropics. The intention was not to write for the specialist engineer or hard-bitten *Sapper* – their desire was to provide advice and information to the 'District Commissioners and District Officers, the Chiefs and Native Authorities',[102] as well as what they described as 'the growing number of Africans alive to the future of the place they live in'.[103] It was to be a new publication that broke the ties from the old PWD ordinances that were primarily concerned with buildings for Europeans and set out some ideas for improving the African dwelling and village. By 1947 they had published this collection of notes as *Village Housing in the Tropics*.[104] It was to be an easy read, carefully laid out as a handy guide that could easily be carried around the colonies. It brought together the conclusions and practical outworkings of a number of specialist guides, whilst maintaining the feel of a personal notebook, far removed from the dry PWD building manuals. The results are wholeheartedly empirical, and it has a charming pioneer tinkering approach, full of 'low-tech', but always practical solutions.

It goes without saying that this publication was part of the Colonial enterprise, a small component in the machine of Empire that ensured British architects, contractors and consultants retained not only the key commissions and opportunities but, more importantly, were strategically positioned as the producers of knowledge and custodians of expertise in a new age of Development and Welfare rather than Empire and Military might. Again, Said's 'modern Orientalist' is particularly useful here in noting that once a method is developed by the Orientalist it is set to be perpetuated as, 'a common discourse, a praxis, a library, a set of received ideas, in short a doxology, common to everyone who entered the ranks,[105] and so it was with Fry and Drew's book. A system of building in the tropics, based on consolidating previous knowledge was neatly and succinctly presented in this trim tome which very quickly became orthodoxy for building in hot climates. Its universal applicability is encapsulated in both the front and back endpapers that proudly display a plan of British West Africa, alongside a map of the world, coloured to highlight the tropics as well as the territories of the British Empire, all sat below the centrally placed Britain.

The map demonstrates how the tropics were viewed, as a homogenous band, defined as the orderly tract between Cancer and Capricorn – nonchalantly and neatly spanning continents, oceans, climatic variances, altitudes, not to mention cultures, traditions and natural resources.[106]

The manual served another important function, namely that of raising Fry and Drew's profile and being deployed as an office advertisement, with its pragmatic approach reassuring potential clients and government departments, it also

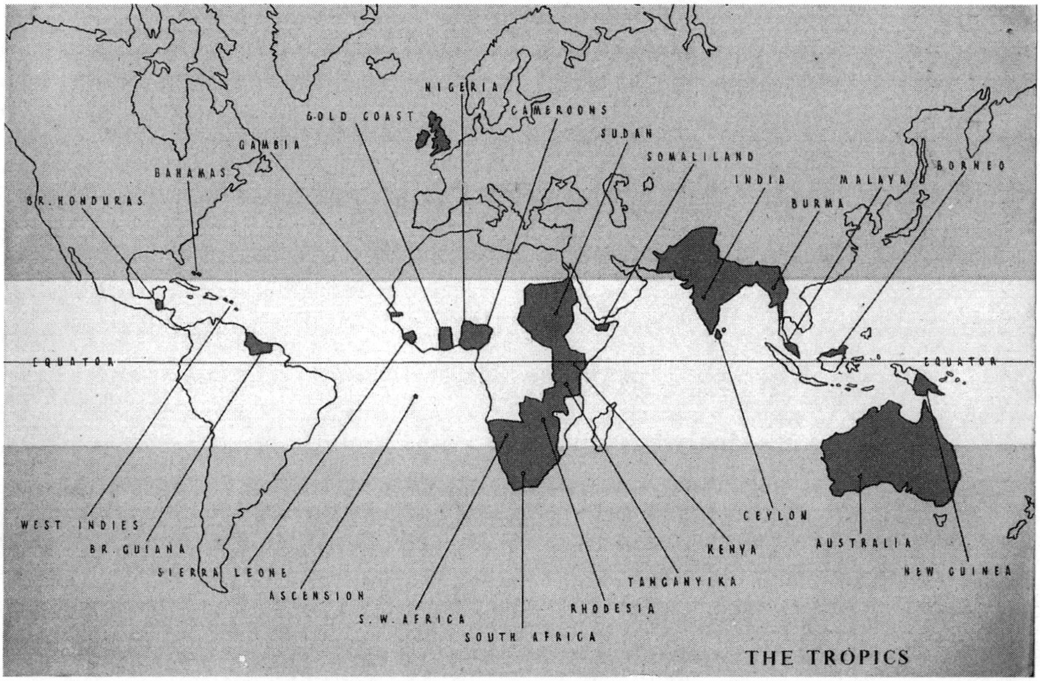

4.7 The Tropics as displayed in Village Housing in the Tropics, 1947

helped to congeal a set of ideas that were being developed by an otherwise disparate collection of individuals. Although Fry and Drew's book was published in 1947, it was a lecture delivered in 1948 by Sir Frank Stockdale, Comptroller for Development and Welfare in the West Indies at the RIBA, that brought together a number of key protagonists, experiences were shared and the canon of Tropical Modernism more formally established in the metropolis. The metropolis was not a refined and efficient machine where knowledge was perfectly collated and transmitted, nor was the notion of tropical architecture a coherent entity. For example, there seemed to be genuine astonishment that other architects were working on similar construction problems throughout the world. Fry exclaimed that, 'I did not realize, at the time when we were working in West Africa, that Mr. Gardner-Medwin and his party were doing similar work on the other side of the world'.[107] This seems extraordinary considering that the two of them were friends and former colleagues[108] and further highlights how the Empire was a disparate, disjointed entity not a polished or single conglomerate.[109]

According to Dannatt it was only when Drew saw the monopitch roofs and building types developed by Gardner-Medwin and De Syllas in the West Indies that she and Fry immediately adopted them for use in West Africa.[110] Prior to this moment their designs had shown a conventional pitched roof with a central ridge, as found on some of the early schools in Ghana discussed below.

FURTHER WORK IN THE GOLD COAST: HOSPITALS FOR KUMASI

The contract as Town Planning Advisors was rapidly coming to an end, but according to Fry, the Governor Henry Gurney stated that, 'the British Government has promised us £200m to develop the colonies after the war and we propose to have a good bite of it. We could send to London for one of those big stuck-up architects, however what about you two?'[111] It was an enticing prospect as the only other commissions on offer were approved post-war work consisting of a 'starchy English diet of schools and housing'.[112] By the end of 1945 Fry had managed to obtain commissions for seven schools in West Africa and he returned to London shortly afterwards with the 'surveys and schedules' to enable the design work to start.[113]

In London he assisted with the RIBA *Building Now* exhibition of 1946, but it only offered a conservative, weary selection of schemes that seemed so far removed from the pre-war efforts.[114] This convinced Fry to pursue the more profitable and interesting body of work in hotter climes, particularly as these endeavours were so generously underwritten with government grants. One of the first large-scale public schemes to be designed was a hospital in Kumasi, with Fry and Drew acting as Consultant architects working alongside the PWD. Fry managed to negotiate a significant fee, and the build cost was considered, 'on the high side by colonial standards'.[115] The working relationships were also strained, with Fry attempting to manage the project from London with limited staff on site. It resulted in an antagonistic and strained dialogue with the Colonial Office, who characteristically understated the relationship as being 'anything but easy'.[116] Fry's agenda at this time was to manage a number of projects in West Africa from a 'central' office in London. From the Colonial Office's perspective, this resulted in a 'lack of frequent consultations, misunderstandings and delays' worsened further by Fry preventing the site architect from making any real decisions.[117] The Colonial Office immediately began to look for an alternative within the hierarchy of the PWD.[118]

In an attempt to save the commission Drew flew out to Accra meet the new Governor, Gerald Creasy, but the decision was already made, and Creasy noted plainly, 'the contract should be terminated and the hospital should be designed by architectural branch of the Public Works Department, suitably reinforced.[119] The PWD had extensive experience in designing 'field hospitals' and wards, but this was a significant building standing at five stories, and the typical utilitarian PWD approach would not be appropriate. The completed building retains certain mannerisms of Fry and Drew's work, not least the concrete columns expressed proud of the main building-line, the precast concrete screens, wavy roof and the expansive glazing on the north-facing façade.

The most significant aspect of this project was its inclusion on the practice CV where, along with another hospital that they designed in Kumasi, it was prominently deployed to win further work in the region and as a means of convincing new clients that they had some competence of working on large projects in the tropics, in additional to their theoretical planning proposals.[120]

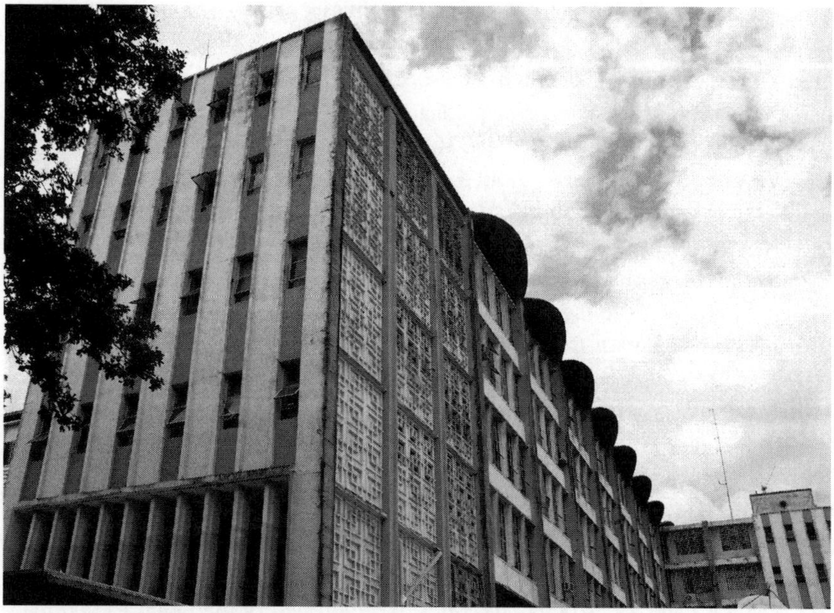

4.8 Kumasi Hospital, Ghana, 2012

MISSIONARIES AND MODERN SCHOOLS IN GHANA

The role and significance of a sympathetic client is often overlooked in the production of architecture; at Gold Coast Thomas Barton, Director of Education, fully supported Fry and Drew in their desire to develop a progressive architecture for the tropics. The new schools were to be more than sheds for learning, and were to serve a didactic purpose and Barton 'considered that architecture formed an integral part of the education and that this education should be as forward looking as possible'.[121] The school building programme was not an innocent enterprise and as Uduku notes they 'remained a powerful symbol or logical citadel, reinforcing local beliefs associated with the prestige of Western knowledge … '[122] A tripartite approach of state, church and education was deployed, entwined within the colonial enterprise of a foreign power, which undermined the validity of the existing 'bush school' and forced a particular agenda.[123] This was codified in the Colonial Office's *Education for Citizenship in Africa*, in 1948, which set out a case of 'training for self-government'.[124] It was through the schools that 'socialisation' was to take place, cultures, social classes and tribes were to be mixed and, it was proposed, the future leaders would emerge.[125] The method of developing the school system in Ghana was to expand the existing prestigious missions' schools and to develop teacher-training colleges that would satisfy the reciprocal demand for new teachers in the expanding education system (without any financial burden on the British Government). Although the schools may have looked towards a political independence and western educational models, for Fry the 'educationalists' were not the only clients; he claimed that they were also designing for 'the country in which we worked … not only the pupils, but the earth, soil, the rocks and the climate, playing equally upon the earth and in people … '[126] A design principle was developed for all the schools and colleges that they hoped would also 'give some feeling of security and social cohesion'.[127] In general, a planning pattern was proposed which used the administrative functions to form a gateway into the grounds, beyond which low-rise linear residential blocks were positioned to form an enclosed playground (or parade ground) as well as directing a visual focus towards the assembly hall or chapel and teaching spaces beyond.

The designs are not identical and each school is arranged to respond to its setting, but there was a desire to create a 'family resemblance', as Fry described to Drew, 'I think that standardising plan forms and details is going to be necessary from our as well as their point of view ... '[128] Initially Fry and Drew designed extensions to some existing schools, such as at Mfantsipim School, Cape Coast, where in 1947 behind the rather grand colonial style buildings they added a science block.

It was a rudimentary solution indebted to the PWD method of construction with overhanging eaves and a veranda walkway providing shade to a row of single rooms beyond.[129] The block includes the now famous pre-cast concrete balustrade, yet despite attempting to introduce an architecture that was distinct from the colonial regime, it was at this school, as well as in neighbouring Adisadel, that rioting broke out (by both students and African staff) in 1948, over the imprisonment of Kwame Nkrumah and the rest of the 'Big Six'.[130] Although it was proposed that a British-style education was to lead to self-governance, when changes began to happen rather quickly and in modes not envisaged by the British, the schools were carefully

4.9 Site Plan of Ho-Hoe, Teacher Training College School, 1948–49

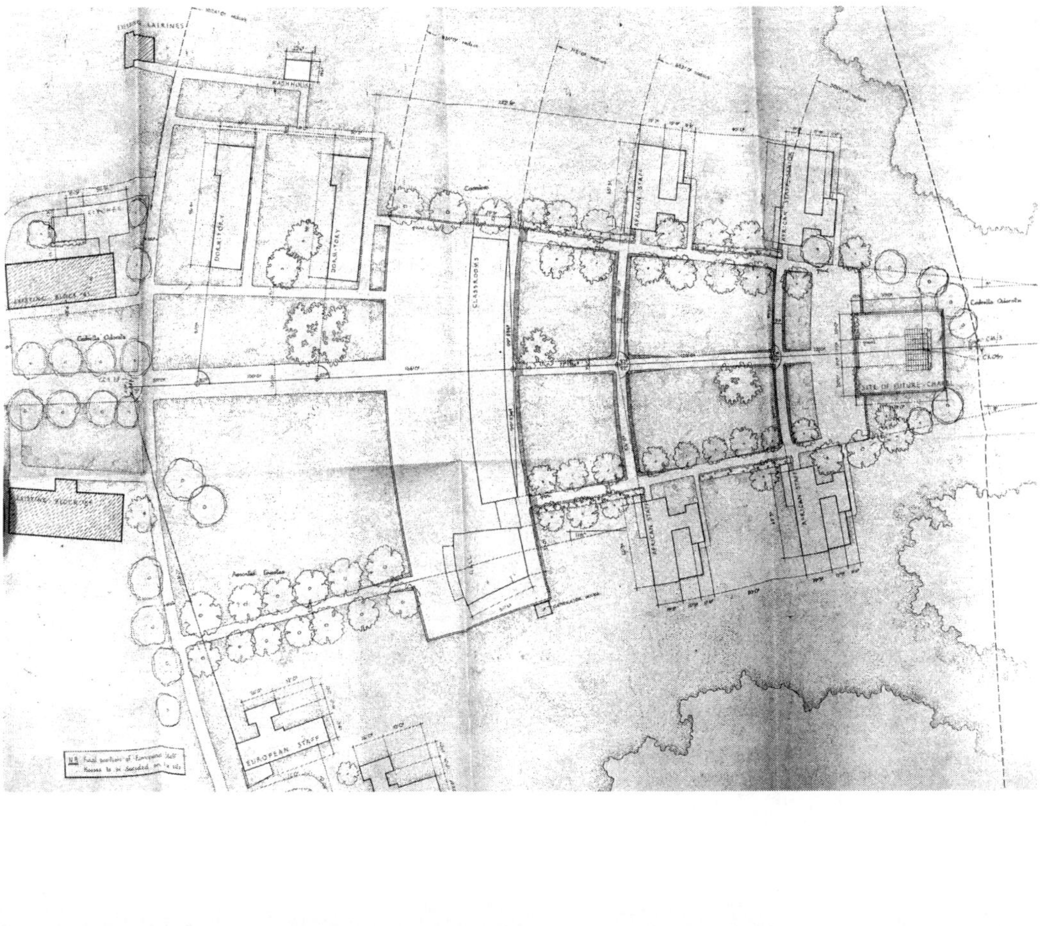

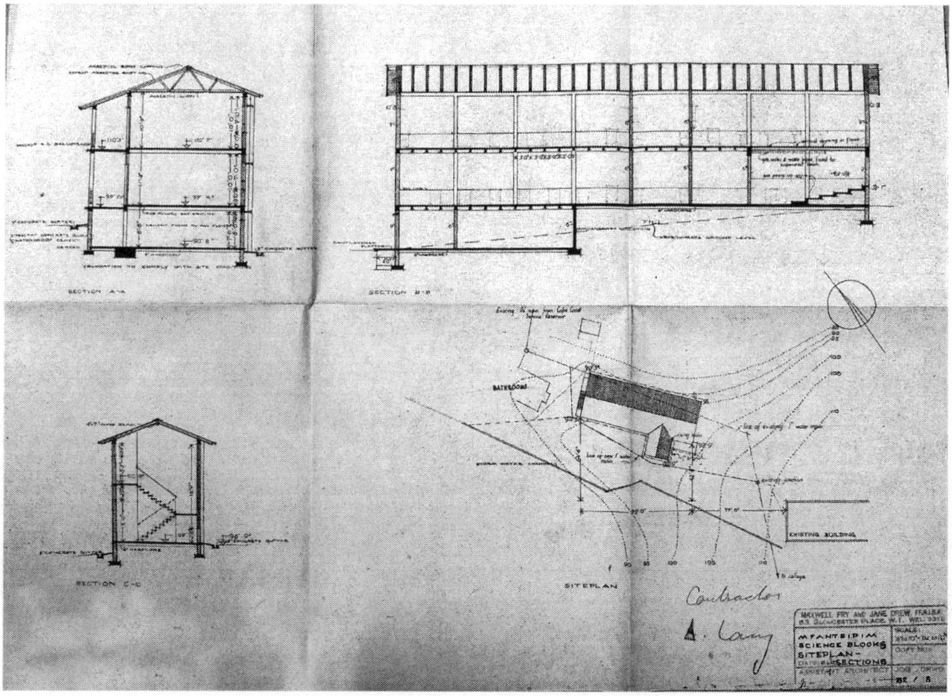

4.10 Drawing of Science Block at Mfantsipim, Cape Coast, 1947

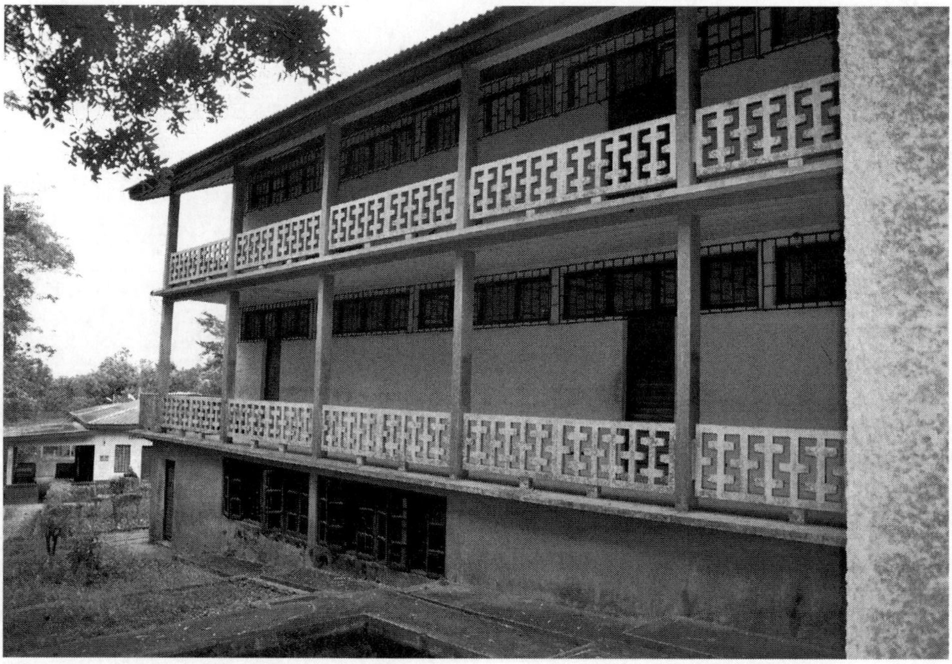

4.11 Photograph of Science Block at Mfantsipim, Cape Coast, 2012

monitored and watched.[131] Recognising their political significance, Nkrumah founded his own schools, employing and teaching the rioters who were expelled from the prestigious Cape Coast government schools.[132]

It was within this fraught arena that Fry and Drew were to develop an appropriate architecture for education, attempting of course to respond to climate, but more crucially, delivering an architecture that was not overtly 'colonial', yet responded to the colonial notion of a school. In light of these constraints and concerns over budget it is not surprising that the first schools designed by Fry and Drew were of an austere and basic design. They were generally built with rubble-stone and reinforced concrete walls topped with lean-to asbestos roofs. Shade was provided by external walkways and large projecting eaves along with window hoods and to a lesser extent the patterned concrete balustrade designs. North of Kumasi in Mampong, St. Andrew's College was constructed from 1948; a chapel and administration form the entrance gateway leading onto a central walkway that cuts through a lawned area with residences running perpendicular to the path which terminates at the dining hall.

Like all the other schools a bell tower (sometimes also used as a water tower) dominates the composition, and as Uduku suggests it was used to 'signify presence, authority and power'[133] and perhaps more than any other building type served as the ultimate symbol of colonial occupation.[134] The old colonial approach is also manifest in the staff houses that line the road leading up to the school. 'African Masters Houses' adopt the 'traditional' compound housing type whilst the considerably larger European Master's house (complete with 'boys quarters') is a bungalow with garden arrangement.[135]

4.12 St. Andrew's College, Mampong, 2012

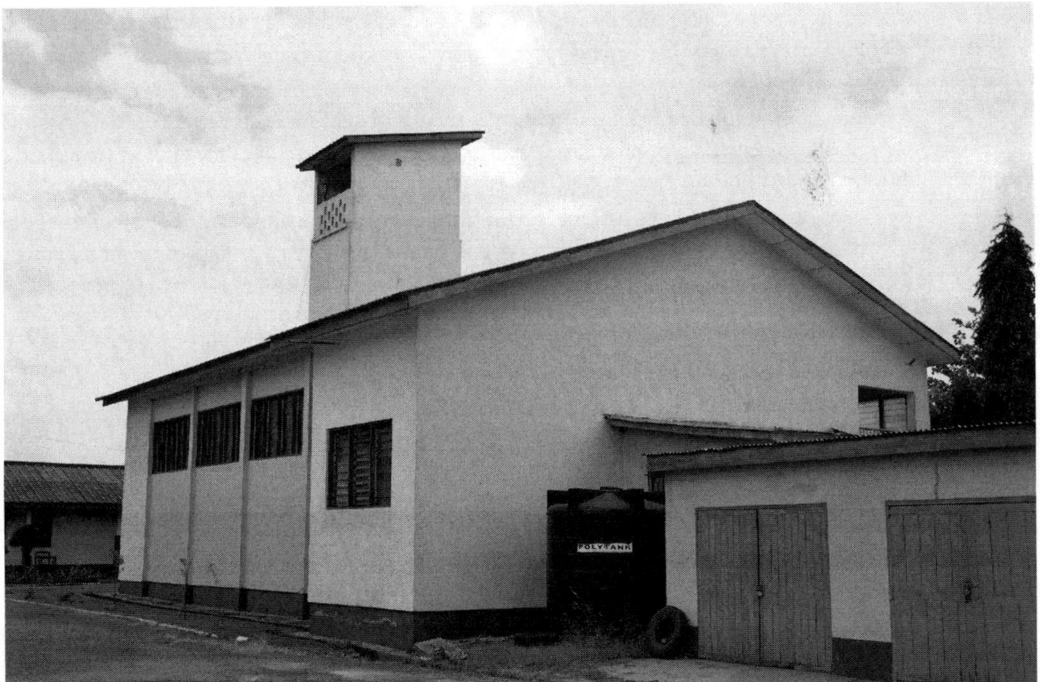

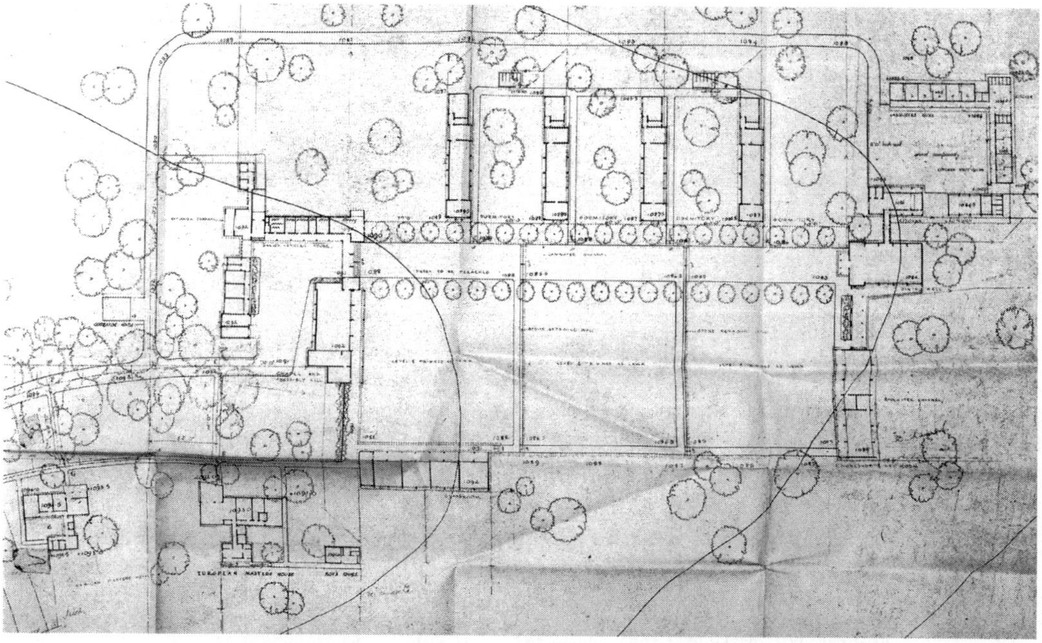

4.13 St. Andrew's College, Mampong, 1948

At Ho-Hoe college (1948) a similar housing solution can be found, although the residences for African and European are of a similar size, the Compound and Bungalow types are again deployed respectively, arranged around the informal but focused campus plan.[136] Fry and Drew did not discuss the housing distinctions; it was an accepted arrangement that despite the approaching independence and attempts to provide a modern and progressive architecture, remained unchallenged.

Although Drew claimed it was the remoteness of the Mampong site that dictated the simple design and detail (and the school is a little more secluded than many others) perhaps the ascetic nature of the architecture was due to it being a test case that relied on current building practices and limited contractor experience. A remote siting at high altitude was not just a means of avoiding mosquitoes, but was all part of the educational agenda.[137] Set within sites of natural beauty and with the bell tower serving as a visual and audible beacon, these institutions were ever present in the landscape overlooking the 'wilderness' and 'savage' from their raised citadels of knowledge. Lagae explains how there was ' … a deliberate attempt to create a world of one's own, a heterotopia as it were, where students could be formed in a milieu that protected them from all dangers and seductions that … pervaded the African urban environment'.[138]

The schools at Aburi and Somanya are further examples of this approach, set on plateaus amongst forests and ravines they are within easy reach of Accra yet in the remote mountain mist feel monastic and almost mythical; they are places of transformation, new allegiances, and initiation into a new mode of thinking. Huxley elaborated on this further stating that the main task of St. Monica's, Mampong was to 'make English school girls, on the way to becoming Christian mothers, out of Ashanti girls'.[139]

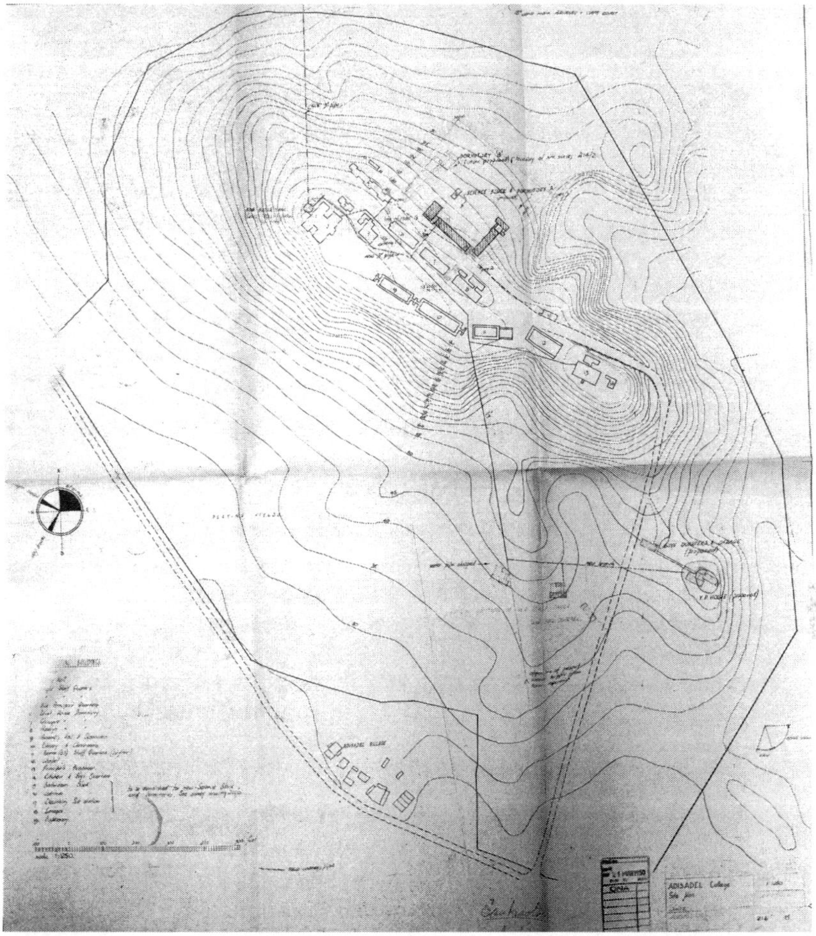

4.14 Site Plan of Adisadel College, 1950

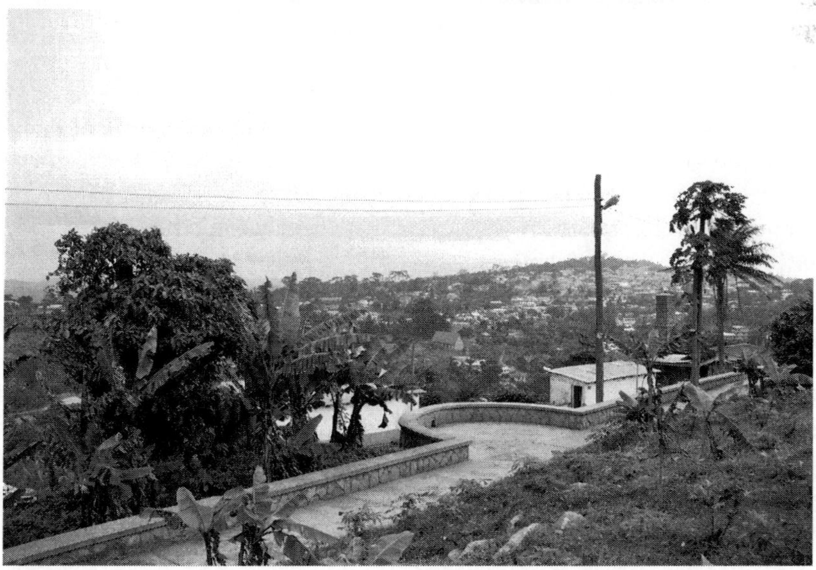

4.15 A view from Aburi showing the surrounding hills, 2012

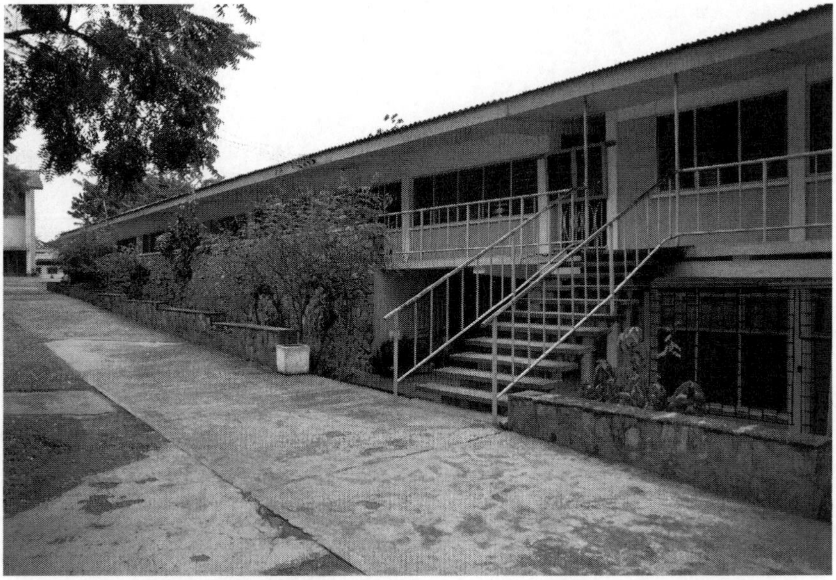

4.16 Somanya, set on an elevated position remote from village life, 2012

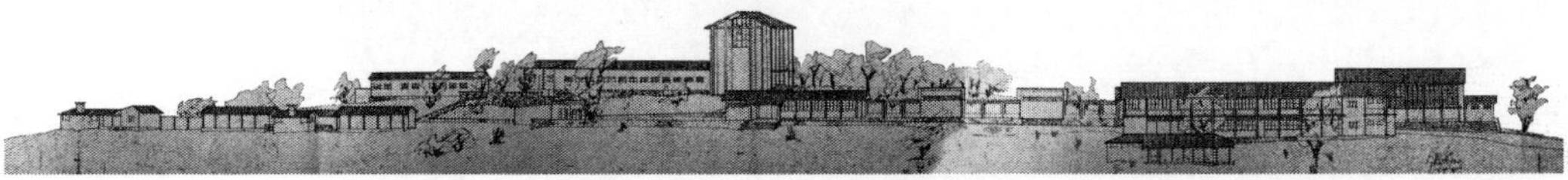

4.17 Elevational drawing of Somanya, c. 1950

In stark contrast to the basic design of St. Andrew's is the refined campus of Prempeh College also located on high ground in near-by Kumasi. Here the teaching rooms are connected by a dramatic curved walkway that traces the contours of the site, and looks into an enclosed garden towards the dormitory blocks. The bell tower is placed on a central axis running from the administration area, and the assembly hall is placed outside of the main school enclosure for shared community use. The facades are composed of projecting concrete box frames, brise soleil and delicate precast concrete infill screens that allow the movement of air whilst providing shade and enclosure. The classrooms have walls that are shaded by the external walkways and permit cross ventilation through a series of timber and glazed louvers. Fry described one of the site visits, 'I went up to Kumasi with Geoffrey Knight … Prempeh looks good and especially the coloured glass in the hall, garish but most satisfactory'.[140] On a further site visit to Aburi Fry and assistants celebrated their accomplishments,

'We went, yesterday afternoon late, Lang, Geoff and Myself, up to Aburi and it is wonderful. There are mistakes, but in the main it is magnificent – strong, dramatic and human. As the light failed in a dramatic sunset it became romantic to a degree, with the hall porch lit with concealed lights, and all sorts of other surprises to be found. The three of us wandered around for a long time. Little Lang filled with the sense of what he had done and how well he had done it, and when he got back in the house we drank Lang's health in Aste Spumante, and well he deserved it because there never was a better finished job'.[141]

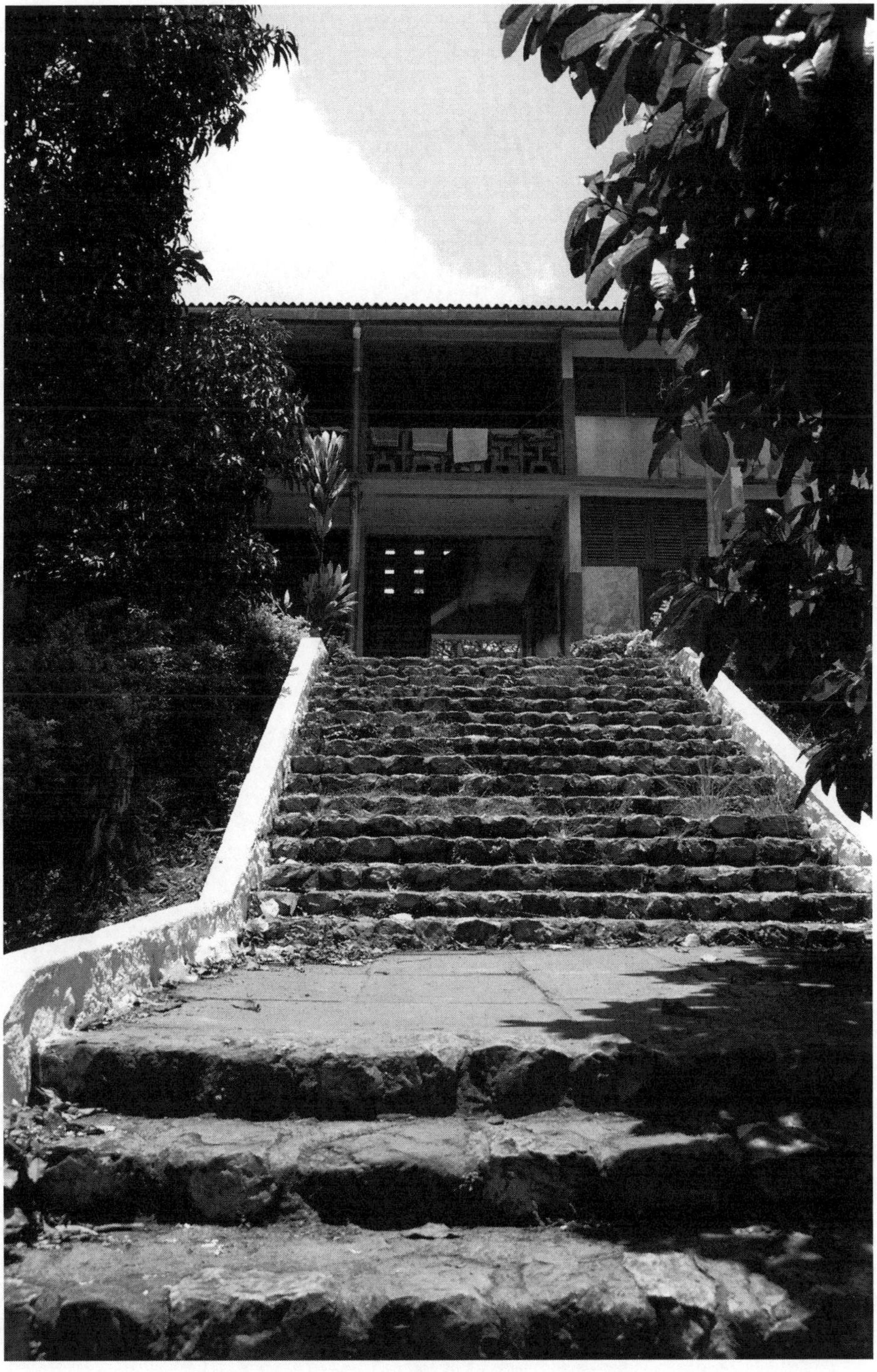

4.18 Akropong Presbyterian Teacher Training College, 2012

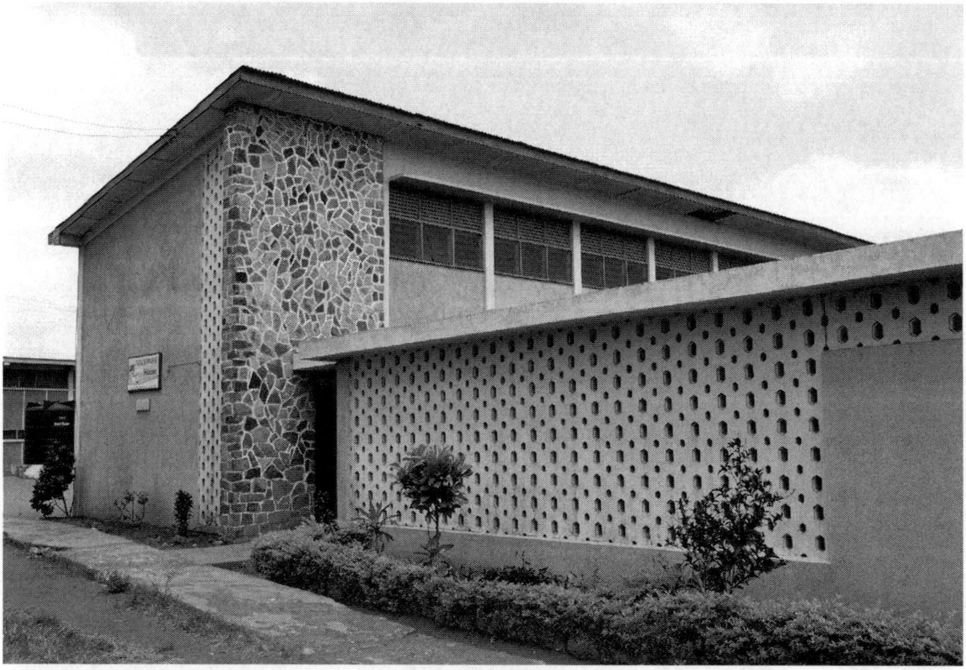

4.19 St. Monica's Dormitory Block, 2012

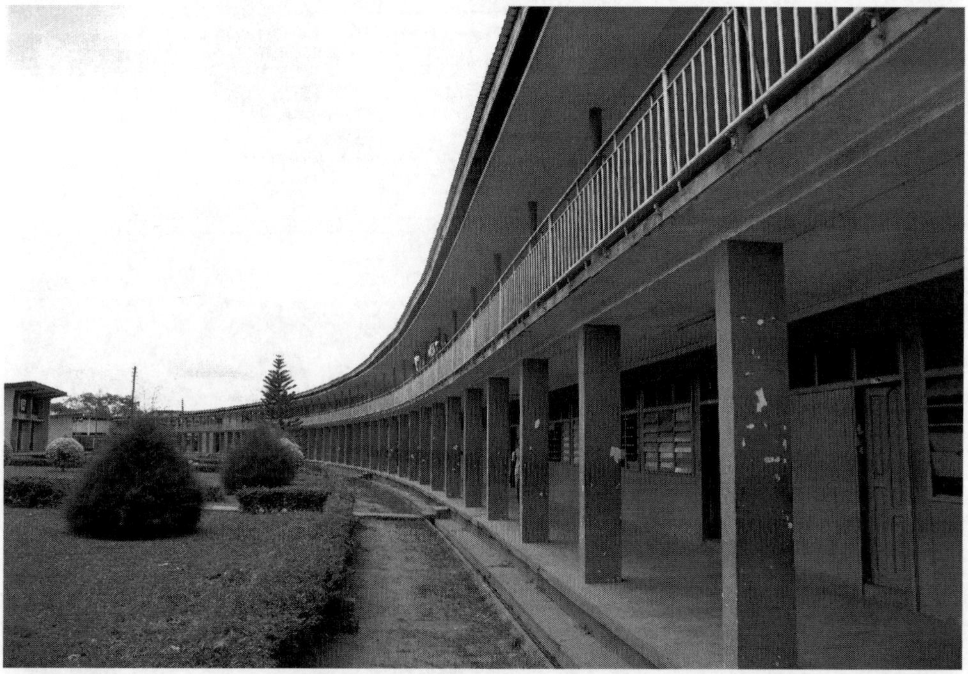

4.20 Prempeh College, Mampong, tracing the site contours, 2012

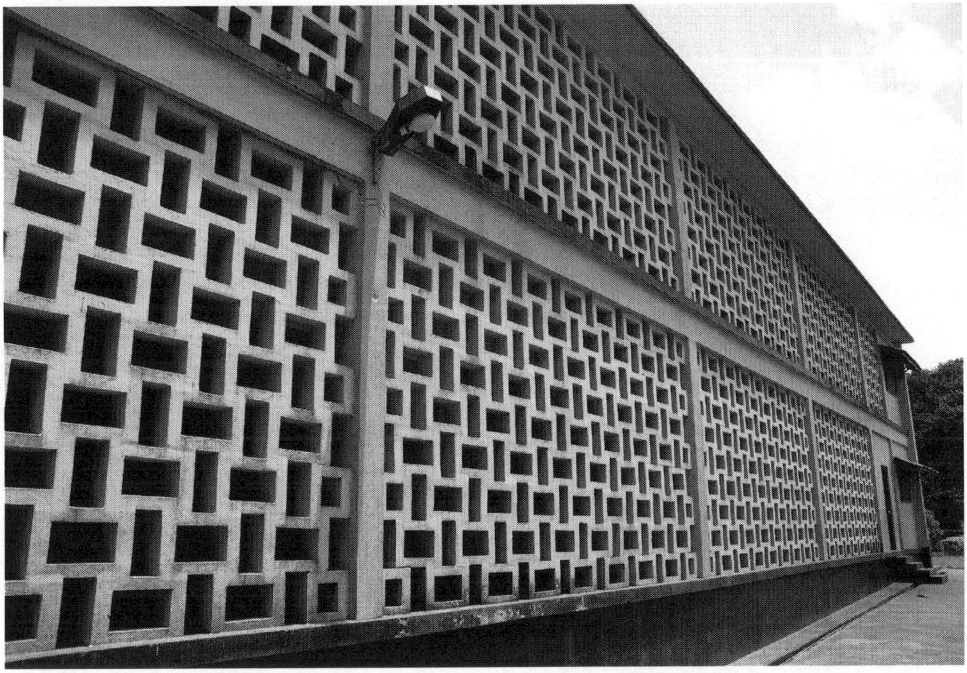

4.21 Prempeh College, Mampong, concrete scree, 2012

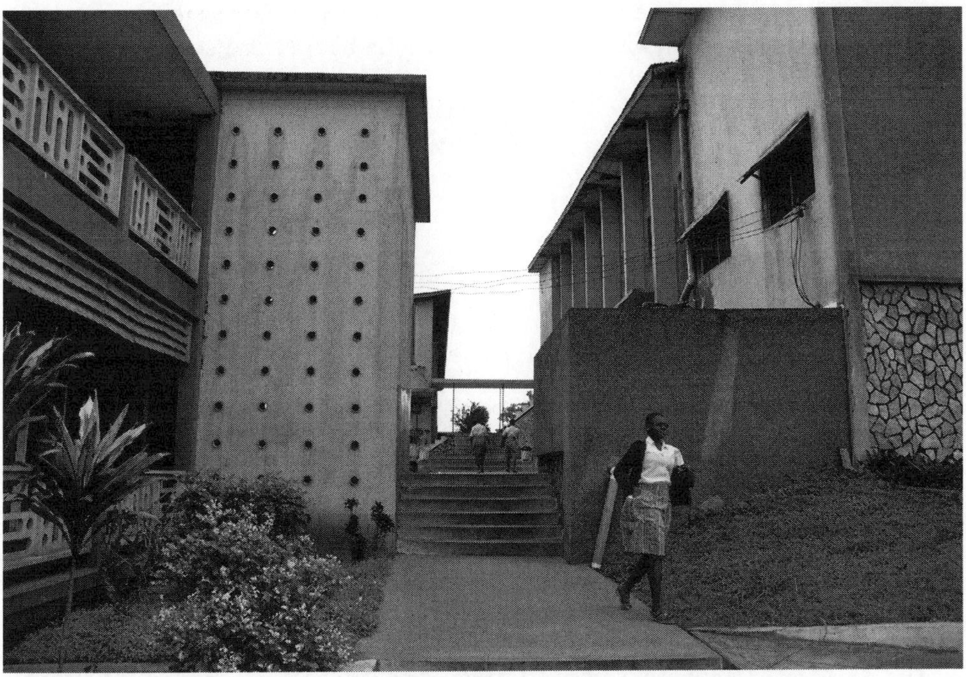

4.22 Aburi Girls' School, residences, 2012

The school is entered by passing underneath the administration which is arranged in a linear bridge like structure, counterpoised with a large watertower clad in brown mosaic that is set with the open play area. The chapel is placed centrally at the end of the yard with various teaching and dining blocks enclosing the compound.

Two of the residential blocks are called, 'Aberdeen' and 'Kilsyth' after the Scottish Presbyterians who supported the build. In Apowa, Takoradi they designed a Catholic Boys School, (although there is no discernible shift in the architectural agenda to reflect the denominational creeds) with the usual courtyards, gardens, meeting hall and pitched roofs extending over loggias and walkways. Fry described it to Drew, having recently returned from India, ' … We saw Apowa in a downpour. It is nearly finished. … It is small scale – very horizontal – but complete, and I think it has come off and will age extremely well'.[142]

Another notable early example for yet another denomination is the Wesley Girls' School in Cape Coast. The plan adopts a bold linear axis that runs through a garden courtyard and centred on the campanile/water tower with chapel below framed on either side by dormitories and classrooms. The open gallery access to the bedrooms provides shade and the balustrade is formed from precast units with additional timber louvers at high-level to cut out glare. Although the materials and precise detailing is indebted to European modernism, the general arrangement and building types are beholden to the old colonial barrack design as well as to the pioneering work undertaken by Leo de Syllas in the West Indies during WW2.[143] Whereas de Syllas used timber and rather clunky details Fry and Drew were able to generate a more expressive finish through the use of concrete. Liscombe notes how Fry and Drew attempted to 'modernize both African and colonial custom'

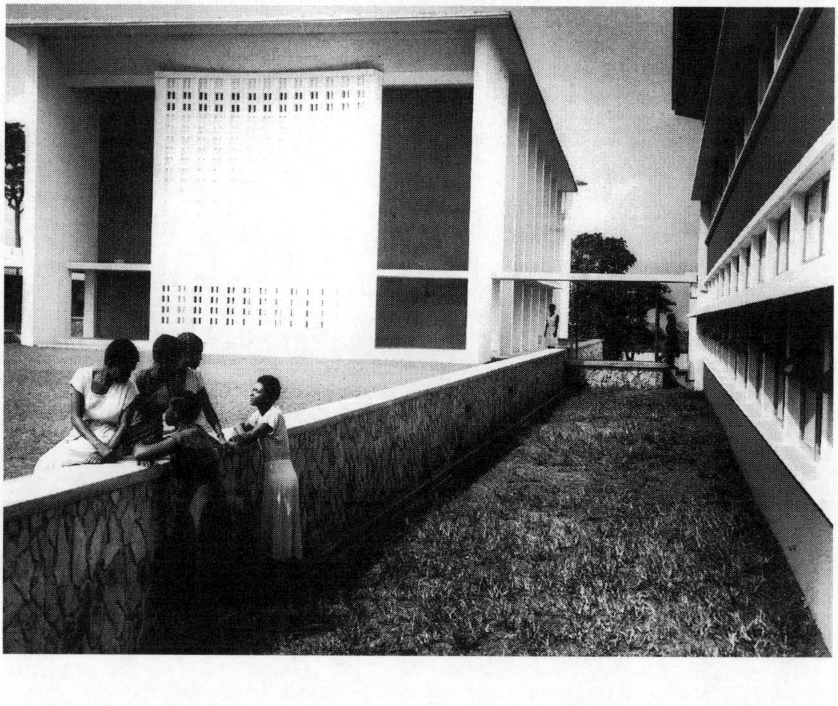

4.23 Aburi Girls' School parade ground and chapel, c. 1950s

and that their European abstract functionalist forms could somehow fabricate 'spaces for the political and socio-political interchange necessary to a legitimate postcolonial accord'.[144] Equally they were responding to pragmatic concerns and Drew stated that the architectural character of their designs was generated by 'the sunbreakers, grilles and other shading but breeze-permitting devices', as well as a desire to 'design in a way which, without in any sense copying African detail, gives a

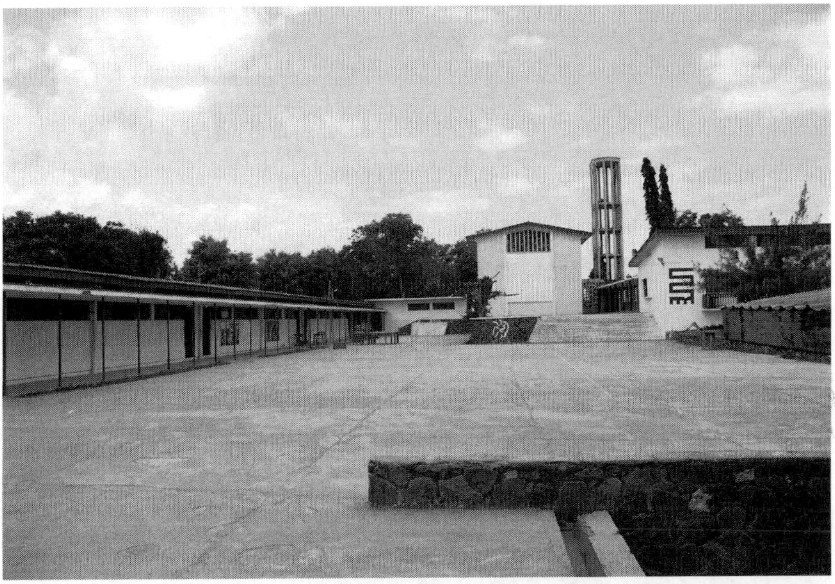

4.24 Apowa Boys' School, Takoradi, parade ground, church and campanile, 2012

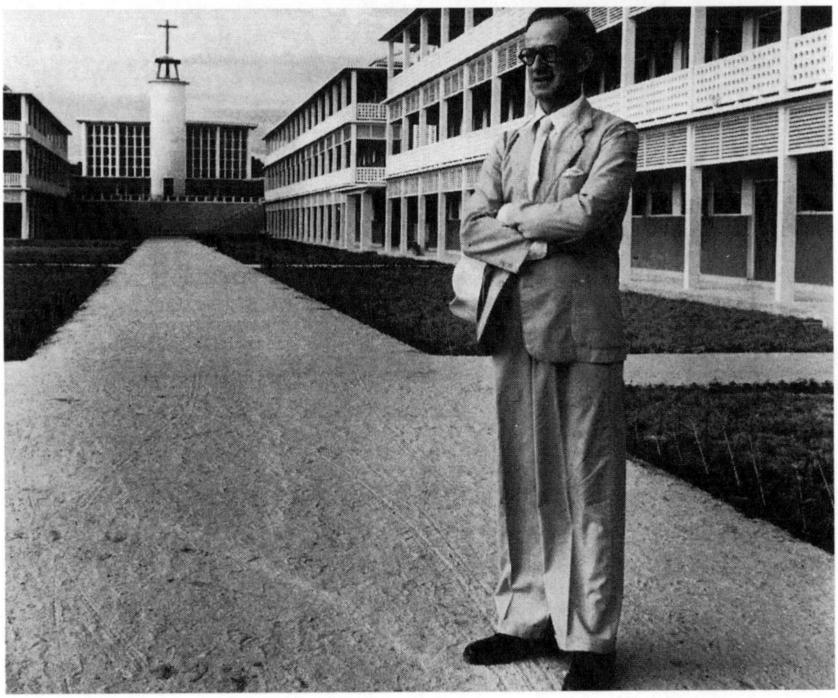

4.25 Wesley Girls' School, Cape Coast: Maxwell Fry in front of dormitory blocks and chapel

response which is African'.[145] For example, some of the grilles are simple tessellating shapes which were derived from the patterns found on 'pots, mats, baskets and cloth' that delighted Fry,[146] whereas others depict local figurative symbols, such as the Ashanti Stool at Prempeh and Opuku Ware.

War and fate had first taken Fry into Africa, but he developed an affinity for the continent, and with Drew's appetite for adventure and desire to work directly with her clients it quickly matured into a fertile arena that they quickly dominated. Their early town planning work and surveys exposed them to large tracts of West Africa where they were able to not only observe the varying conditions and become

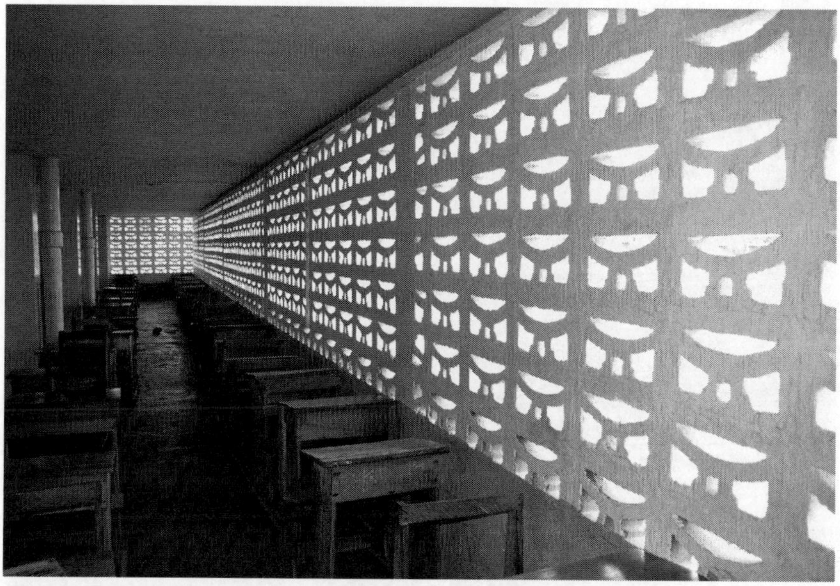

4.26 Concrete grill screen making reference to 'African patterns', 2012

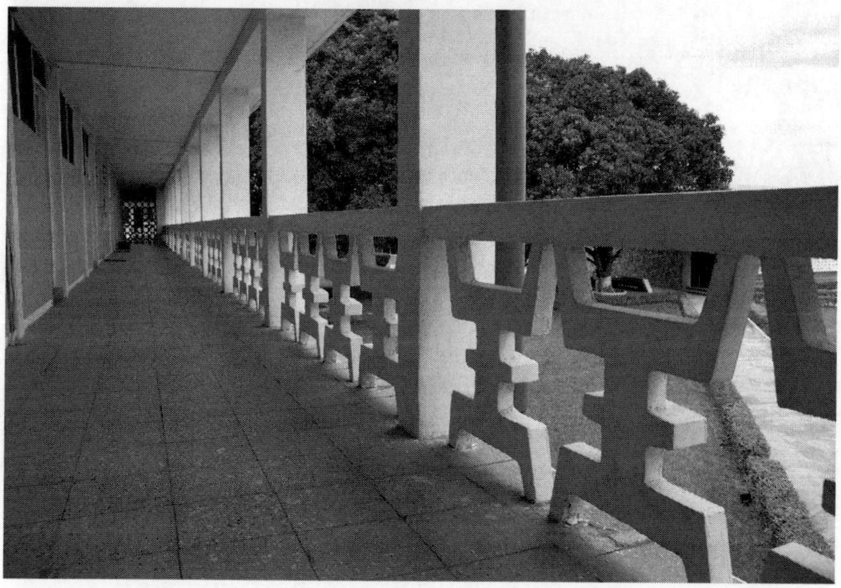

4.27 Ashanti Stool depicted in concrete screen at Opuku Ware, 2012

familiar with the towns and villages, but also to develop a large number of contacts and friendships. Fry and Drew were concerned with the African living conditions, as well as providing overall masterplans and sanitation improvements for the Colonisers. Some of their work could be considered naïve and rather far-fetched, not least their lack of understanding of local landownership rights in places like Kumasi. Yet, despite this, they also showed some sensitivity to the places they were working in, and did not simply dismiss all African towns and dwellings outright. The most notable example is in Bathurst where a careful preservationist approach was taken. They put their field notes and ideas into a small booklet that would help with the planning of the African village. Although the book was indebted to previous studies and knowledge, its success came from its accessibility and cartoon-like diagrams. With the sudden implementation of the Development and Welfare grants (and Fry's prior notoriety from Impington College) they were perfectly placed to be awarded the new school building projects in Ghana. They developed an architecture that was rather timid at first resembling PWD huts. After seeing the work others were developing in the West Indies and with the support of their client Thomas Barton a more expressive manner was introduced to the school designs and they inflected their work with 'local' motifs and references. They tended to follow the old tropical design principles with one-room thick buildings shaded by an external veranda walkway. It was the large projecting eaves (to shed water as well as provide shade), careful detailing and the harvesting of rainwater that has enabled these schools to endure the harsh climatic conditions and to remain in good condition. It was this work that established their reputation in the emerging and fertile canon of 'tropical modernism' and they would go on to work in Africa until the 1960s.

NOTES

1 RIBA Archive, F&D 13/5, 'A collection of texts by Maxwell Fry, 1958–1974', p. 1.

2 Maxwell Fry *Autobiographical Sketches* (London, 1975), p. 163.

3 RIBA Archive, F&D/20/1 'Maxwell Fry Memoirs, "England" Chapter 1', p. 4.

4 Other residents included Richard Llewellyn-Davies and Peter Moro.

5 Architects over the age of twenty-five were exempt from military service and were to be only employed in ways that utilized their professional training. They were used to complete the large works required by the armed forces. See Anthony Jackson, *The Politics of Architecture: a history of modern architecture in Britain*, (London, 1970), p. 78.

6 For example, Robert Gardner-Medwin was posted to the West Indies, Percy Johnson-Marshall to Burma, William Henderson to Turkey.

7 Maxwell Fry, *Fine Building* (London, 1944), 61

8 RIBA Archive, F&D/20/2, 'Maxwell Fry, Full Autobiography, 1985', p. 48.

9 RIBA Archive, F&D/20/2, 'Maxwell Fry, Full Autobiography, 1985', p. 54.

10 RIBA Archive, F&D/20/2, 'Maxwell Fry, Full Autobiography, 1985', p. 55.

11 RIBA Archive, F&D/12/1, Letter from Hans Feibusch to Fry, 30 November 1942.

12 RIBA Archive, F&D/12/1, Letter from Charles Reilly to Fry, 4 September 1942.

13 RIBA Archive, F&D/20/2, 'Maxwell Fry, Full Autobiography, 1985', p. 56.

14 RIBA Archives, F&D/18/3, Letter from Fry to Drew, 4th April 1943.

15 RIBA Archive, F&D 13/5, 'A collection of texts by Maxwell Fry, 1958–1974', p. 1.

16 RIBA Archive, F&D 13/5, 'A collection of texts by Maxwell Fry, 1958–1974', p. 1.

17 The exhibition was opened by Reilly, who in a letter to Fry described it as a 'grand show'; RIBA Archive, F&D/12/1. Letter from Charles Reilly to Fry, 17 April 1943.

18 See letters from Drew to Fry, 5 July 1942 and 1 December 1942, Letter from Drew to Fry, 1942, RIBA Archive, F&D/15/1.

19 Letter from Drew to Fry, RIBA Archive, F&D/15/1, Letters Drew to Fry, 28 August 1942.

20 John Morrison Archive. *Jane Drew Biography*, unpublished manuscript, p. 30.

21 John Morrison Archive. *Jane Drew Biography*, unpublished manuscript, p. 37.

22 RIBA Archive, F&D/25/1, 'Fragments of Jane Drew's Autobiography and relating documents, 1983', p. 13.

23 BP Archive, 68422, Kuwait Oil Company (KOC) – building programme 1947–1951.

24 BP Archive, 68422, Kuwait Oil Company (KOC) – building programme 1947–1951, booklet, para 7.

25 RIBA Archive, F&D/18/4 – Letter from Fry to Drew, 1st February 1944.

26 See *Selection of Architects and Town Planners for Chandigarh, Visit of P. N. Thapar and P. L Varma to Europe* file at Chandigarh City Art Museum, letter from Fry undated, but written prior to his visit to India in January 1951.

27 RIBA Archive, F&D/20/2, 'Maxwell Fry, Full Autobiography, 1985', p. 89.

28 Interview with author, 26th October 2011.

29 John Morrison Archive. *Jane Drew Biography*, unpublished manuscript, p. 38.

30 John Morrison Archive. *Jane Drew Biography*, unpublished manuscript, p. 38. Letter from Trevor Dannatt to Charles Reilly, UL Archive, D207/4/5, 'Regarding the bombing of Miss Drew's Office', 21 September 1944.

31 John Morrison Archive. *Jane Drew Biography*, unpublished manuscript, p. 38.

32 RIBA Archive F&D/12/1, Letter from Fry to K.E. Robinson, Colonial Office, 19 November 1943 'Correspondence to and from Maxwell Fry, 1942–1947'.

33 RIBA Archive F&D/12/1, Letter to Fry from Secretary of State, Colonial Office, 14 December 1943 'Correspondence to and from Maxwell Fry, 1942–1947'.

34 Drew continued her practice and was also permitted to continue with the Architects' Year Book.

35 The National Archives, CO 554/133/6, Appointment of Town Planning Advisor, Letter from C. H. M Wilson to A. B. Cohen, Colonial Office, 30th November 1943.

36 RIBA Archive F&D/12/1, Letter from Fry to Mssrs Griffiths McAllister ltd., 'Correspondence to and from Maxwell Fry, 1942–1947', 11 December 1943.

37 RIBA Archive F&D/12/1, Letter from Fry to Mr. Bratreet, Crown Agents to the Colonies, '"Correspondence to and from Maxwell Fry, 1942–1947"', 1 December 1943. In 1945 a Colonial and Development and Welfare Act was passed providing a total of £120m to be spent on various projects throughout the British Colonies, over a ten year period – this funded Fry and Drew's projects as well as later ones such as the schools and universities. See Colonial Office, 'Colonial Development and Welfare: Despatch Dated 12th November 1945 from the Secretary of State for the Colonies to Colonial Governments', (London, 1945).

38 Maxwell Fry, 'Town Planning in West Africa', *African Affairs*, 45 (1946): pp. 197–204, p. 197. Drew spent some time making alternations to the Government House in Freetown, for example. RIBA Archive, F&D/10/14 – To and from Jane's Mother, c. 1940s, JD, Accra, to mother, 'The Green, Bossington near Porlock, Somerset' [stayed there during war], n.d.

39 Peter Shepheard (1913–2002) wrote to Fry just before Christmas in 1943, 'I have thought very hard about your invitation to West Africa and have decided that, as things stand at the moment, Mary [his wife] and I feel we cannot join you'. RIBA Archive F&D/12/1, Letter from Peter Shepheard to Fry dated 8 December 1943, '"Correspondence to and from Maxwell Fry, 1942–1947"'. One of the main reasons for turning down the offer was the anticipation of the heat, with Shepheard stating that, 'I am no good myself at sticking even English heatwaves'. RIBA Archive F&D/12/1, Letter from Peter Shepheard to Fry, 8 December 1943, '"Correspondence to and from Maxwell Fry, 1942–1947"'. Perhaps Shepheard overcame his aversion to the heat, as he later put forward his CV to be selected for the Chandigarh project in 1950; see Chandigarh Art Gallery and Museum, 'Selection of Architects and Town Planners for Chandigarh', LT-0006.

40 They discussed someone called, 'Spreull', [probably David Wilson Spreull (1911–95), who trained at Liverpool School of Architecture and worked as a town planner in Jamaica from the 1940s until 1965] and with whom Fry had apparently once danced the Can-Can, and 'the other is a girl, – is that a bar?'.

41 Fry, '*Town Planning in West Africa*', p. 197. 'Freddie' Charles also worked for them at this time. John Dawe Tetlow (1913-?) studied at Liverpool School of Architecture between 1935–38.

42 A fact proudly recalled by Fry in J. M Richards, ed., *New Buildings in the Commonwealth* (London, 1961).

43 Ambe Njoh, *Planning Power; Town Planning and Social Control in Colonial Africa* (London, 2007), p. 59. See also Ambe Njoh, *Urban Planning and Public Health in Africa; Historical, Theoretical and Practical Dimensions of a Continent's Water and Sanitation Problematic* (Farnham, 2012), p202, and Bigon Liora, *A History of Urban Planning in Two West Africa Colonial Capitals; Residential Segregation in British Lagos and French Dakar (1850–1930)* (Lewiston, 2009).

44 Njoh, *Planning Power*, p. 62.

45 This quote was part of the 1929 Development Policy, quoted in Colonial Office, 'Statement of Policy on Colonial Development and Welfare', ed. by Colonial Office (London, 1940), p. 5.

46 F M Bourret, *The Gold Coast: A Survey of the Gold Coast and British Togoland 1919–1946* (London, 1949), p. 165.

47 Colonial Office, 1940, p. 4.

48 Colonial Office, 1940, p. 5.

49 Christopher William Machell Cox, *Mass Education in African Society* (London, 1944).

50 H.M.S.O., 'Report of the Commission on Higher Education in West Africa', (London, 1945).

51 Fry, 'Town Planning in West Africa', 197.

52 The planning problems of Bathhurst were discussed in Parliament in 1949, and described as 'frustrating'; it appears the plans were not being executed, see http://hansard.millbanksystems.com/commons/1949/feb/24/gambia-development.

53 RIBA Archive, F&D/18/19, 'Correspondence from Maxwell Fry to Jane Drew', 17 August 1944.

54 Maxwell Fry and Betty Benson, '*Draft Townplanning Scheme for Bathurst and the Kombo Area*' Office of the Townplanning Advisor to the Resident Minister, West Africa, 1946 (written in 1944), 3.

55 Fry and Benson, '*Draft Townplanning*', p. 3.

56 Fry and Benson, '*Draft Townplanning*', pp. 8–9.

57 Fry and Benson, '*Draft Townplanning*', p. 21.

58 Fry and Benson, '*Draft Townplanning*', p. 23.

59 Fry, 'Town Planning in West Africa', p. 198. Ships had to 'dock' at sea and rely on smaller boats to bring in/out the goods resulting in an inefficient and time consuming process.

60 See Colonial Office, 'Freetown: Canalisation of Streams and Surface Drainage (Typescript Correspondence, Annotated, and Memoranda Relating to Public Works in Freetown, between Acting Governor Cookson in Sierra Leone and the Colonial Office)', (London, 1929).

61 See http://hansard.millbanksystems.com/commons/1944/jul/12/sierra-leone-freetown-harbour.

62 E. Maxwell Fry 'Town Planning Scheme for Freetown' Accra, Government Press Gold Coast.

63 Njoh, *Planning Power*.

64 Maxwell Fry, 'Developing "the Most Beautiful Town in West Africa"', *West African Review*, June (1946): p. 625.

65 National Archives, CO96/753/14, Letter 30 March 1939, Governor of Gold Coast from Government House, Accra, to Malcolm MacDonald Development of Kumasi 1939.

66 See T. C. McMaskie, *Asante Identities: History and Modernity in an African Village, 1850–1950* (London, 2000).

67 McMaskie, *Asante Identities*, p. 209.

68 McMaskie, *Asante Identities*, p. 212.

69 McMaskie, *Asante Identities*, p. 201.

70 Anthony D King, 'Exporting Planning: The Colonial and Neo-Colonial Experience', in *Shaping an Urban World*, ed. by Gordon Cherry (London, 1980), p. 210. For further contemporaneous views on health and planning see Alasdair C. Sutherland, 'Regional Planning Problems in Ashanti', *Planning Outlook Series 1*, 3 (1955): pp. 16–28, and Raymond E. Dumett, 'The Campaign against Malaria and the Expansion of Scientific

Medical and Sanitary Services in British West Africa, 1898–1910', *African Historical Studies*, 1 (1968): pp. 153–97.

71 RIBA Archive, F&D/18/4, Letter EMF to JD, 25th Oct 1944.

72 See Antoni Folkers, *Modern Architecture in Africa* (Amsterdam, 2010), p. 51

73 Home, 'Town Planning, Segregation and Indirect Rule in Colonial Nigeria', Robert K. Home, 'Town Planning and Garden Cities in the British Colonial Empire 1910–1940' *Planning Perspectives*, 5:1, 23–37 and J. A. Omotola, 'Planning Law in Nigeria', *Third World Planning Review*, (1991), vol. 13.

74 Fry, 'Town Planning in West Africa', p. 201.

75 National Archives, CO583/274/5 'Town Planning and Housing 1944', Saving Telegram from the Governor, Nigeria to the Secretary of State for the Colonies, London, 13 January 1944.

76 Fry, 'Town Planning in West Africa', p. 201.

77 Robert Home, 'Town Planning, Segregation and Indirect Rule in Colonial Nigeria', *Third World Planning Review*, 5 (1983), pp. 165–175, p. 172.

78 John Morrison Archive. *Jane Drew Biography*, unpublished manuscript, p. 38.

79 RIBA Archive, F&D/18/4 – Fry to Drew, 1944, Letter EMF to JD, 30th October 1944.

80 Mark Crinson, *Modern Architecture and the End of Empire* (Aldershot, 2003), pp. 132–137.

81 RIBA Archive, F&D/14/4, 'Fry's Memoires', p. 16.

82 Edward Said, *Orientalism: Western Conceptions of the Orient* (London, 1978), p. 121.

83 RIBA Archive, F&D/14/4, 'Fry's Memoires', p. 16.

84 For example see, James Johnson, *The Influence of Tropical Climates on European Constitutions* (London, 1813). W. J Simpson, *The Maintenance of Health in the Tropics* (London, 1916). J Balfour Kirk, *Public Health Practice in the Tropics* (London, 1931). D. B Blacklock, *An Empire Problem: The House and Village in the Tropics* (Liverpool & London, 1932) and for a general overview, Jiat-Hwee Chang, and Anthony D King, 'Towards a Genealogy of Tropical Architecture: Historical Fragments of Power-Knowledge, Built Environment and Climate in the British Colonial Territories', *Singapore Journal of Tropical Geography*, 32 (2011): pp. 283–300.

85 Many of these guides were produced under the direction of the Engineer, Sir Hubert Edmund Walker.

86 See section in Chapter 5 on Tema Manhean.

87 See National Archives, CO96/781/1, 'Housing Schemes Kumasi, 1945–6'.

88 National Archives, CO96/781/1, 'Housing Schemes Kumasi, 1945–6', photographs of the construction process are also held in the Royal Commonwealth Society archives, Cambridge University.

89 See National Archives, CO96/781/1, 'Housing Schemes Kumasi, 1945–6'.

90 RIBA Archive, F&D/15/2, Letter from Drew to Fry, 6 June 1945.

91 Alcock also published a number of building guides, including A E S Alcock, and Helga M. Richards, *How to Plan Your Village* (London, 1953); A E S Alcock, and Helga M. Richards, *How to Build for Climate* (London, 1960).

92 See Figure 3 in Rhodi Windsor Liscome, 'Modernism in Late Imperial British West Africa: The Work of Maxwell Fry and Jane Drew, 1946–56', *Journal of the Society of Architectural Historians,* 65 (2006): pp. 118–215.

93 See Graham Tipple, *Extending Themselves: User-Initiated Transformations of Government-Built Housing in Developing Countries* (Liverpool, 2000), p. 246.

94 RIBA Archive, F&D/18/6, EMF, Kumasi, to JD, London, 31st July 1946.

95 John Morrison Archive. *Jane Drew Biography*, unpublished manuscript, p. 38.

96 RIBA archive, F&D/25/3, Drew, J. 'Reflections on My Life and Work', p. 3.

97 John Morrison Archive, *Jane Drew Biography*, unpublished manuscript, p. 43.

98 RIBA Archive, F&D/30/1, 'Jane Drew, Full autobiography', p. 74.

99 Maxwell Fry, 'Max Fry: Inspirations, Friendships and Achievements of a Lifetime in the Modern Movement', *Building,* 229 (1975): pp. 52–58.

100 These houses were also cooler than the houses in the city that, as a result of the bye-laws had to be constructed from concrete blocks.

101 RIBA Archive, F&D/30/1, "Jane Drew, Full autobiography", p. 74.

102 Jane Drew, Edwin Maxwell Fry, and H. L Ford, *Village Housing in the Tropics: With Special Reference to West Africa* (London, 1947): p. 3.

103 Maxwell Fry, 'Town Planning in West Africa', *The Architects' Year Book no.1* (1947): p. 72.

104 Harry L. Ford, remains a largely unknown figure. He worked with Fry on the plan for Enugu, see RIBA Archive, F&D/18/4, 'Fry to Drew, 1944', letter from Fry to Drew, 25th October 1944. He went on to develop a hydro-electric scheme for Kariba Township in Rhodesia during the early 1960s, and also planned a new town in Botswana in 1975, see Harry L. Ford, 'A New Town in Botswana', *Planner* 61 (1975): pp. 145–49.

105 Said, *Orientalism.*

106 Also entwined within this book are glimpses of their home-life – the book is dedicated to E. C. (Peter) Gregory the owner of the book's original publisher Lund Humphries. Drew was very close to Gregory and the friendship, it would seem, put some strain on Fry and Drew's relationship. See "Mr. E. C. Gregory". Letter from Jane Drew to *The Times* [London, England] 14 Feb. 1959: 10. *The Times Digital Archive*. Web. 13 June 2012. See also John Morrison Archive. *Jane Drew Biography*, unpublished manuscript, pp. 34–35. Fry described it as a 'marital triangle with Peter (E. C. Gregory) in the third, and yet commanding position', RIBA Archive, F&D 14/4, Maxwell Fry's Memoirs, 'Building', undated, p. 61.

107 Maxwell Fry commenting on the lecture by F Stockdale, Robert Gardner-Medwin, and Leo De Syllas, 'Recent Planning Developments in the Colonies', *Royal Institute of British Architects Journal,* 55 (1948): pp. 140–48, p. 148.

108 Gardner-Medwin worked for Fry immediately when he left University. See Iain Jackson, 'Tropical Architecture and West Indies: From Military Advances and Tropical Medicine, to Robert Gardner-Medwin and the Networks of Tropical Modernism', *The Journal of Architecture,* 18 (2013): pp. 167–195.

109 See Robert Home 'Transferring British Planning Law to the Colonies', p. 403 for further discussion on the transfer of planning models from West Indies to West Africa.

110 Interview with Trevor Dannatt, London, 26 October 2011.

111 Quoted by Fry in *Building*, 'Max Fry – inspirations, friendships and achievements of a lifetime in the modern movement', 229 (31 Oct 1975): pp. 52–58, p. 56.

112 Atkinson, G. Anthony. "British Architects in the Tropics". *Architectural Association Journal*, 69 (1953): pp. 7–21.

113 RIBA Archive, F&D/18, Letter from Fry to Drew, undated, 1945.

114 See 'Building Now Exhibition', *Architects Journal*, 103 (2 May 1946): pp. 345–347.

115 National Archives, CO 96/805, Proposed erection of hospital at Kumasi, 1946–1949.

116 National Archives, CO 96/805, Letter from The Secretariat Accra, to J. K. Thompson, Colonial Office, 21 Sept 1948, London, 'Proposed erection of hospital at Kumasi, 1946–1949'.

117 Ibid.

118 Ibid.

119 National Archives, CO 96/805, Telegram to Secretary of state for the Colonies from G. Creasy, 3 November 1948, London, Proposed erection of hospital at Kumasi, 1946–1949.

120 See 'Kumasi Central Hospital and nurses' training centre, Ashanti, Gold Coast', *Builder*, 181(21 September 1951): pp. 376–381 and 'Kumasi Central Mental Hospital', *Architect & Building News*, 200 (5 July 1951): pp. 14–20. Neither article mentions Fry & Drew, only the PWD architect A. G. Paton.

121 Jane Drew, 'West Africa', *Architectural Design*, (1955): pp. 137–149, p. 137. *Tropical Architecture in the Humid Zone* (1956) was also dedicated to Barton.

122 Ola Uduku, 'The Colonial Face of Educational Space', in *White Papers, Black Marks: Architecture, Race, Culture*, ed. by Lesley Naa Norle Lokko (The Athlone Press, 2000), p. 46.

123 See Mark Hanna Watkins, 'The West African "Bush" School', *American Journal of Sociology*, 48 (May 1943): pp. 666–75.

124 Advisory committee on Education in the colonies, *Education for Citizenship in Africa* (London, 1948).

125 A. R. Thompson, *Education and Development in Africa* (New York, 1981).

126 RIBA Archive, F&D 14/4, Maxwell Fry's Memoirs, p. 13.

127 Drew, 'West Africa', p. 139.

128 RIBA Archive, F&D/18/19, – Fry to Drew, undated, EMF, 'Office of the Town Planning Adviser to the Resident Minister, West Africa', to JD, 30 December 1945.

129 See Churchill College Archive, ARUP 8/21.

130 See National Archives, CO 964/13 Gold Coast, Commission of enquiry School Riots.

131 See National Archives, FCO 141/5079, Schools in Ghana.

132 See National Archives, FCO 141/5079, Schools in Ghana.

133 Uduku, 'The Colonial Face of Educational Space', p. 53.

134 The clocktower is used as the title of the Preface in Antoni Folkers, *Modern Architecture in Africa* (Amsterdam, 2010), so dominant is its presence throughout Africa.

135 See Churchill Archives Centre, ARUP 8/24.

136 See Churchill Archives Centre, ARUP 8/26.

137 The colonial policy was to build at an altitude believed to be beyond the reach of the mosquito, see Philip D. Curtin, 'Medical Knowledge and Urban Planning in Tropical Africa', *American Historical Review*, 90 (1985): pp. 594–613.

138 Johan Lagae, '"Montcassin, Montserrat Or … An Alcazar?" Architecture, Propaganda and Everyday School Practices in the College Du Saint-Espirit in Bujumbura (Burundi)', in *Colonial Architecture and Urbanism in Africa*, ed. by Fassil Demissie (Farnham, 2012), p. 287.

139 Elspeth Huxley, *Four Guineas: A Journey through West Africa* (London, 1954), p. 135.

140 RIBA Archive, F&D/18/13, Fry to Drew, 1953, EMF, Accra, to JD, Chandigarh, 9th August 1953.

141 RIBA Archive, F&D/18/14, Fry & Drew, 1954, EMF, Accra, to JD, Ch. 30th May 1954.

142 RIBA Archive, F&D/18/14 – Fry & Drew, 1954, EMF, Accra, to JD, Ch. 28th May 1954.

143 See Jackson, 'Tropical Architecture and the West Indies: from military advances and tropical medicine, to Robert Gardner-Medwin and the networks of tropical modernism', pp. 167–195.

144 Liscome, 'Modernism in Late Imperial British West Africa: The Work of Maxwell Fry and Jane Drew, 1946–56', pp. 118–215.

145 Drew, 'West Africa' pp. 137–149, p139

146 Fry, *Art in a Machine Age*, p. 12.

5

The Development and Reassessment of Tropical Architecture in West Africa

UNIVERSITIES FOR THE COLONIES

Fry and Drew's involvement with West Africa did not cease with their school work in Ghana. They continued to work throughout the region on numerous projects that were commissioned in the wake of the *Report of the Commission on the Higher Education in the Colonies*, (referred to as the Asquith Report)[1] published by the Colonial Office. The report recognised the growing demand for university education in the colonies but haughtily noted that 'education should be adapted to the environment and mentality of the people'.[2] There had been various attempts at forming educational establishments in West Africa, but by 1945 there were only four institutions of higher education that had the status of 'university' in the entire Empire[3] – i.e. recognised as such by British universities.[4] The British government sought to fulfil what it described as its 'moral obligations as trustees of the welfare of Colonial peoples' by implementing a 'programme of social and economic development', which included educational facilities. According to Asquith, the outcome of these enterprises would 'lead to the exercise of self-government' by the Colonies.[5] Universities were seen as being an important part of nation building, a *civilising* prerequisite to self-governance. The report noted, 'in the stage preparatory to self-government universities have an important part to play; indeed they are indispensable'.[6] As a result a number of new colleges were proposed for the West Indies, West Africa, East Africa and the Sudan, and Malaya, which would eventually become self-awarding universities once deemed to have reached the required standard.[7]

A dedicated Commission was established for West Africa and having taken account of population levels, land area, languages, cultures and traditions, stated that 'the British West African colonies are themselves an empire' and that, contrary to the argument of selecting a single site for all West Africa, 'the general needs of these populations cannot be met wholly by one centre of higher education even for a short time'.[8] They recommended the city of Ibadan should form the site of a new college, and that the existing buildings at Achimota College in Ghana, and Fourah Bay College in Sierra Leone (which would also serve The Gambia), could be extended and improved. Ibadan was selected as it was less congested than

Lagos, and with a burgeoning population of 400,000 a new hospital was required in the city which might be planned in conjunction with the medical school.[9] It was also an 'African city'; it had existed independently of the Colonial regime and as such suited the nation building and nationalist agenda of the political landscape. That said, the existing site conditions and occupants were not considered important and Elizabeth Huxley described it as 'untouched bush, forest and farmland'[10] before adding that, 'Nigeria has no style or tradition either to inspire or constrain. The architects had a true *carte blance*'.[11] The Committee also gave scant recommendations with regards to the architectural requirements, stating that 'the best expert advice will be sought and that the buildings will be of fine architectural standard, fitting for the first University of West Africa, and providing inspiration to its staff and students'.[12] The buildings were to be 'of simple construction, capable of modification and extension as conditions require … residential buildings should provide individual rooms for each student as well as common rooms'.[13] There wasn't a unanimous decision however, and a minority report, co-authored by Fry's friend, Julian Huxley, made strong claims for a centralised university in West Africa (located at Ibadan), with the other three colonies providing only 'territorial colleges'.[14] The minority report was accepted, no doubt because it was less expensive and because West Africa was generally viewed as a coherent unit by the British, although the decision provoked revolt, especially in Ghana where plans were immediately made to self-fund a university college in Cape Coast.

IBADAN UNIVERSITY: 'THIS BUILDING PALAVER WASTES TOO MUCH TIME AND ENERGY'[15]

A number of high-profile British architects applied for the West African College commission, including Herbert Baker and Grey Wornum, but their reluctance to employ 'native' staff, and reticence to offer training to help formulate a construction industry in the region precluded their consideration.[16] The Inter-University Council eventually short-listed three architects, Edward Payne of Sir Aston Webb & Son, Hugh Casson, and Fry & Drew, following an interview they noted,

> *Mr. Payne was unsuitable and that Casson had not anything like the experience or organization to justify the entrusting to him of such an immense project. Maxwell Fry had the great advantage of wide local experience and he has a very big organization at his disposal and seemed quite confident that he could in addition to his other commitments undertake the Ibadan project.*[17]

In addition to sending copies of books, drawings and ideas for tropical architecture, Fry set out comprehensive notes on how they proposed to manage the project, and from the outset was adamant that the work could be largely supervised from London. He also proposed employing African staff in London 'where they would be exposed to the full mental rigours of an architect's office at work'.[18] An enthused Kenneth Mellanby (Principal-Designate) wrote to Fry in confidence immediately after their meeting,

> *we felt with your experience and standing you were the obvious choice ... We hope that when you have planned the whole layout it will be possible for some individual buildings to be carried out by such young men as Hugh Casson who we also saw so that they have a chance of making their name.*[19]

The only major reservation that concerned the committee was Drew. In the Inter-University Council notes of October 1947, they claimed that 'Jane Drew was not *persona grata* in certain quarters in West Africa but it was agreed that this did not constitute a serious obstacle to the selection of Maxwell Fry. The panel considered that it would not be proper or practical to ask Maxwell Fry not to employ his partner on this project'.[20] It is not entirely clear why Drew was unwelcome – but this comment is indicative of how in this small colonial community the rumours and perceptions of Drew quickly spread without evidence or justification presented. Fry and Drew accepted the commission but from the outset it was a difficult process. There were multiple changes, lack of sufficient data and the channels of communication from client to architect lacked the required hierarchy and structure for such a large undertaking. Some of the correspondence relating to the design process is relayed below demonstrating how the project was contested and debated. Fry and Drew attempted to steer the project but the campus that exists today was far from their initial vision. Mellanby played a key role, and by June 1949 was already highly critical of their proposals,

> *They [the architects] have just sent us some perfectly absurd plans for buildings, which would be dark, hot stuffy, and not water-tight, and also it is very difficult for us to get them to realise the implications of the difficult things they do, when it comes to planning for African labour and material.*[21]

The relationship between client and architect quickly deteriorated and Mellanby confessed 'how much I dislike our Architects',[22] the main problem seemed to be Fry and Drew's desire to create a low density campus with the buildings spread out over considerable distances. Mellanby rejected the proposed plan because it was considered inconvenient, expensive to maintain and expensive to construct.[23] Fry and Drew wanted adequate ventilation around their buildings, and cited the design of University College of the West Indies campus as inspiration for this scheme but after considerable pressure from Mellanby developed a more concentrated solution. So concerned was Mellanby that he (perhaps informed by the events at Kumasi Hospital) recruited 'regional architects' from the PWD, such as J.E Evens, who challenged Fry and Drew's approach in favour of conservative solutions, more in-line with the colonial PWD agenda than setting a modern architectural outlook.[24] The relationship waned to the point that the building committee 'set up a guarantee fund, and contributed in all a sum of several thousand pounds from their own salaries'[25] so that alternative architects could be employed without delay. When Fry attended site and worked directly with the clients there seemed to be rapid and congenial development with both parties happily collaborating, Fry described one of his site visits,

I have passed my first day here and the atmosphere improves. I have just had a long talk alone with Kenneth [Mellanby] starting with much nervousness on both sides but getting freer as it proceeded [sic] and ending with an invitation to swim in the pool with him ...[26]

Members of the committee felt however that when Fry returned to London their requests were ignored. In one letter to Fry they reveal that the plans were not even shown to the committee, 'for fear that they would damage the reputation of your firm still further'.[27] Furthermore, Fry seems to ignore his own climatic design advice resulting in 'half of the students' rooms are facing south-east contrary to our [the clients] desire'. They demanded that Fry visit Ibadan immediately to resolve the problems 'on the spot' to 'avoid endless wrangles and unsatisfactory discussions by post'.[28] Again, with reference to the Kumasi Hospitals, the clients' confidence was further dinted by the 'reports which we have received of your firm's work in the Gold Coast have been most unfavourable, and we hope therefore that a similar situation will not arise in Nigeria'.[29] Mellanby even tried to convince the Inter-University Council to employ additional architects as a means of circumventing Fry and Drew, 'Think of every college in Oxford or Cambridge designed by Sir Giles Gilbert Scott! I feel we should accept the Architects' layout, which seems to me satisfactory, and let them build one hall of residence, and then have an open competition for further designs ... I have not suggested this to the Architects, who will no doubt become hysterical.[30] Despite these early wrangles by 1950 the scheme was progressing reasonably well, but according to Mellanby,

The main snag Fry is up against is that every time we go out to the site, we see the ghastly village buildings which Jane designed, and which stimulated from an eminent French visitor the expression: "But the English used to be very good at designing tropical houses: these contain more mistakes than I thought possible in any one building".[31]

Due to extended delays and concerns over design quality, Mellanby decided, without informing the architects, to substitute Fry and Drew's staff housing for PWD housing types T62 and T63.

The Registrar later justified this decision to an infuriated Drew stating that they were, 'much more economical and which experience had showed were more satisfactory in their design' than those designed by Fry and Drew.[32] The committee's strong dislike and mistrust of Drew continued, Mellanby claimed that 'Fry in London is quite out of our, and it seems under Jane's control'[33] he was also highly critical of Fry, describing his management as, 'undoubtedly extraordinarily incompetent. They seem to be such complete amateurs

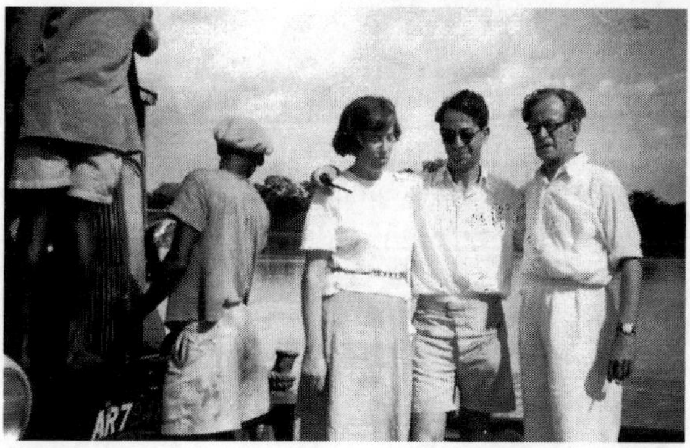

5.1 Mellanby, Fry and Drew; with Drew appearing not to like Mellanby very much either, c. 1940s

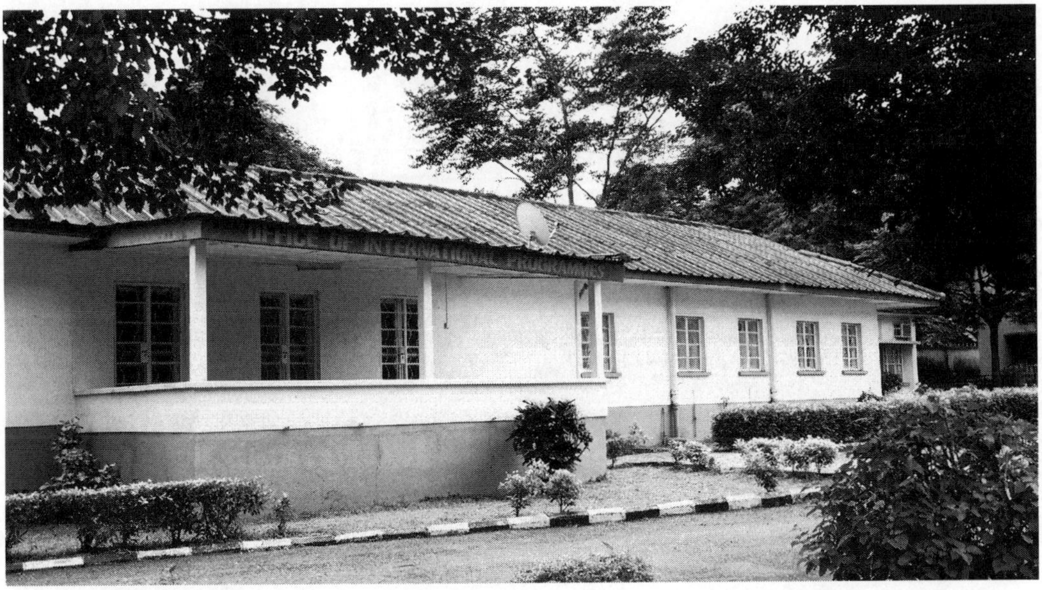

5.2 PWD Staff Housing at Ibadan, 2012

when it comes to the discussion of any technical problem, and Fry is quite capable of suggesting some experimental technique which he does not understand'.[34]

After threats of arbitration a meeting was held in London.[35] Fry was to appoint a 'resident architect' to be based on site and capable of driving the project forward in consultation with Mellanby. The role was filled by A. Halliday, but upon his arrival the Building Committee was very quick report its concern, 'that a man of such junior status has been sent', noting that, 'Mr. Halliday has no tropical experience and is not authorized to make any other than very minor decisions regarding the work'.[36] Mellanby was more forthright in his letter to Walter Adams, Secretary of the Inter-University Council for Higher Education in the Colonies, describing Halliday as, 'practically useless, and I understand from the Government Architect here there would have been no difficulty in getting an experienced person for the job if the firm of Fry, Drew and Partners had been prepared to pay a reasonable salary … incidentally as we have to pay the fares of the Architects, it seems a waste that we should have to pay for a useless young man, and even worse, that we should have to pay for a partner whom we do not wish to have here … ', again referring to Drew.[37]

Halliday was not a complete loss, as he developed the overall masterplan involving the relocation of the Library building away from the main campus area and was on site to monitor the construction which was awarded to an Italian contractor, Cappa and D'Alberto (who were already working in the region).

The campus is located on the north-east outskirts of the city covering some five square miles. Although Fry wanted the campus layout to be based on the UCWI Mellanby forced a redesign that would concentrate the plan, 'the disperal of the present plans would be intolerable … in hot or wet weather the maximum concentration is desirable … '[38] The site is approached along a ceremonial drive which leads to the administrative offices, complete with tower (and clock) as well

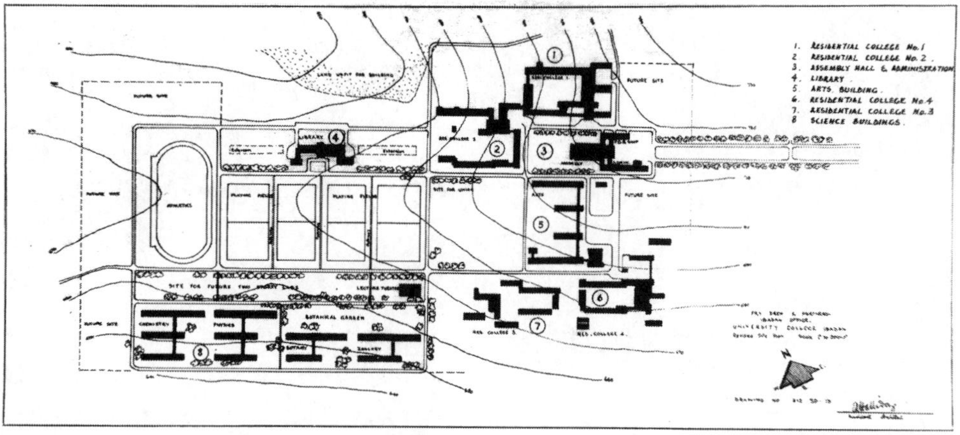

5.3 University College of Ibadan, plan

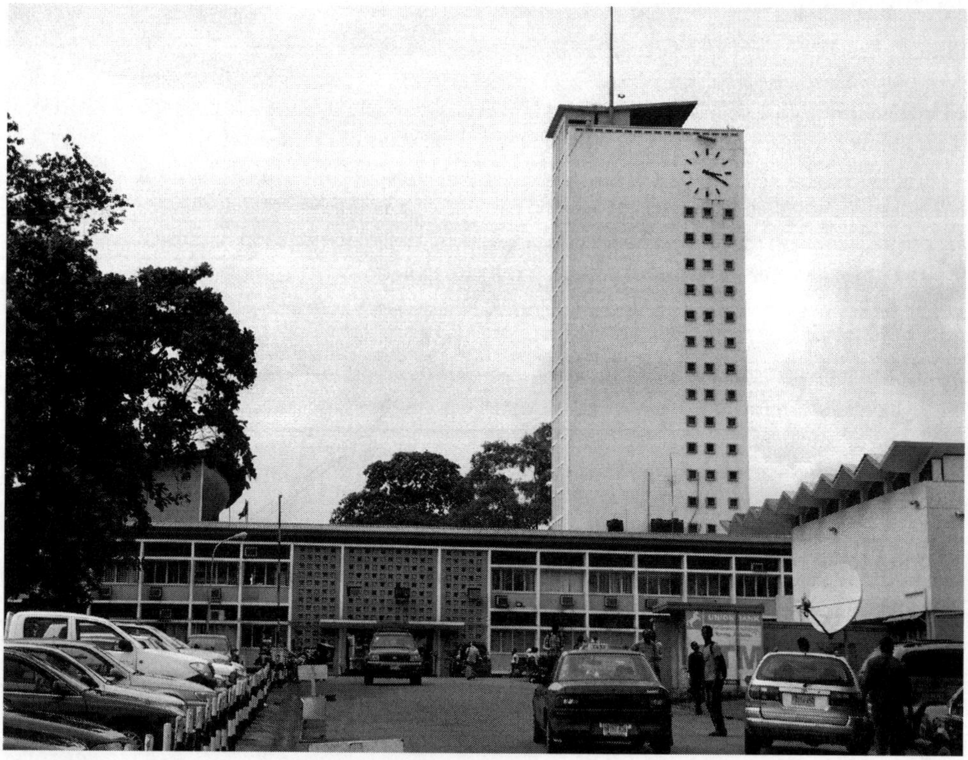

5.4 Entrance, Administration and Tower at Ibadan University, 2012

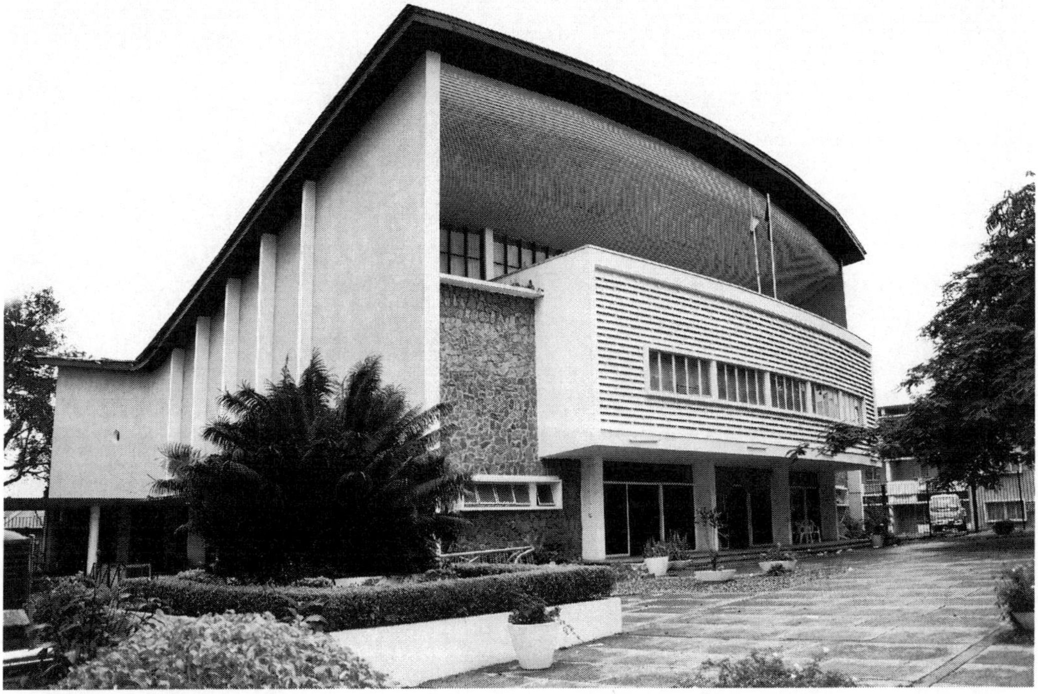

5.5 Trenchard Hall, funded by the United Africa Company, 2012

as an assembly hall paid for by the United Africa Company form the entrance gateway. The hall (known as Trenchard Hall) is a bold symmetrical design with a curved projecting roof that floats above a concrete rectilinear balcony set within a random rubble wall. Located amongst the administration buildings and tower it helps to complete the arrangement and serves as a gathering place for the university being used for ceremonies and performances. As Huxley explained, 'the UAC is known for hard headed, even ruthless efficiency and would not spend its money on speculations or frills',[39] but they wanted to be associated with the new university and donated the sum of £60,000 for the construction of the building. The donation was part of their public relations and they were keen to make the offering 'before the result of the mineral royalties enquiry was made known. It would be an advantage to allocate the Company's contribution to something concrete; the buildings perhaps to be named after the Company'.[40] The UAC was a powerful firm operating throughout West Africa, and whilst this philanthropic gift was generous, it was also offset against tax, and served as a permanent reminder of the firm's power and presence in that region as well as trying to curry favour with its workforce and future politicians.[41]

The rest of the campus is arranged in a series of courtyards made up of residences, teaching spaces and lecture halls. Each residential hall houses around 200 students and the familiar precast concrete balustrades of the Ghanaian schools are reused both as handrails and as shading devices at high-level in front of the balconies and access walkways. A highlight is the Sultan Bello Hall; after passing through the entrance

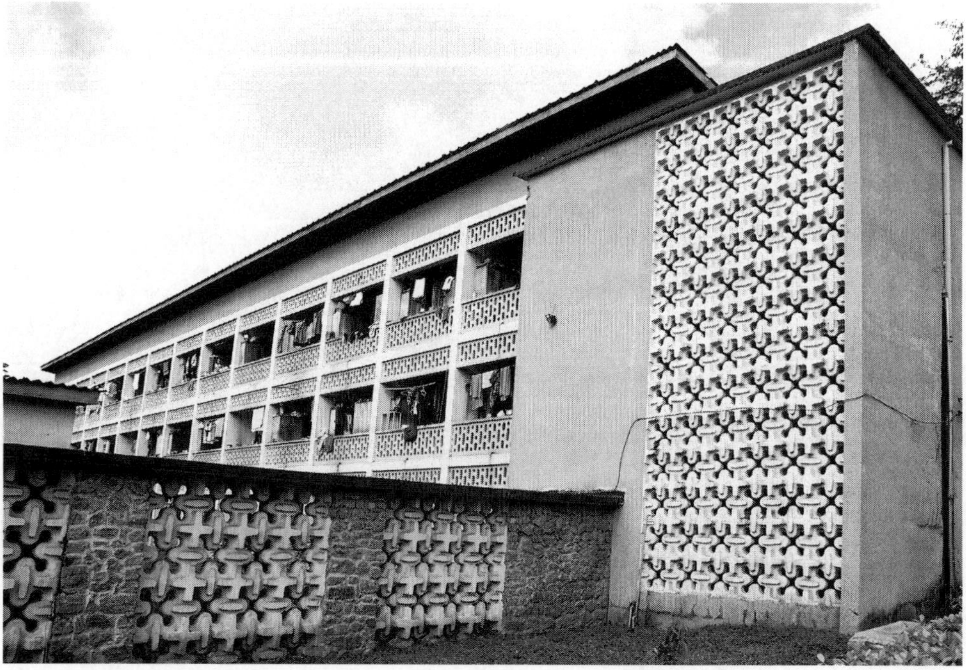

5.6 Student Residences at Ibadan University, 2012

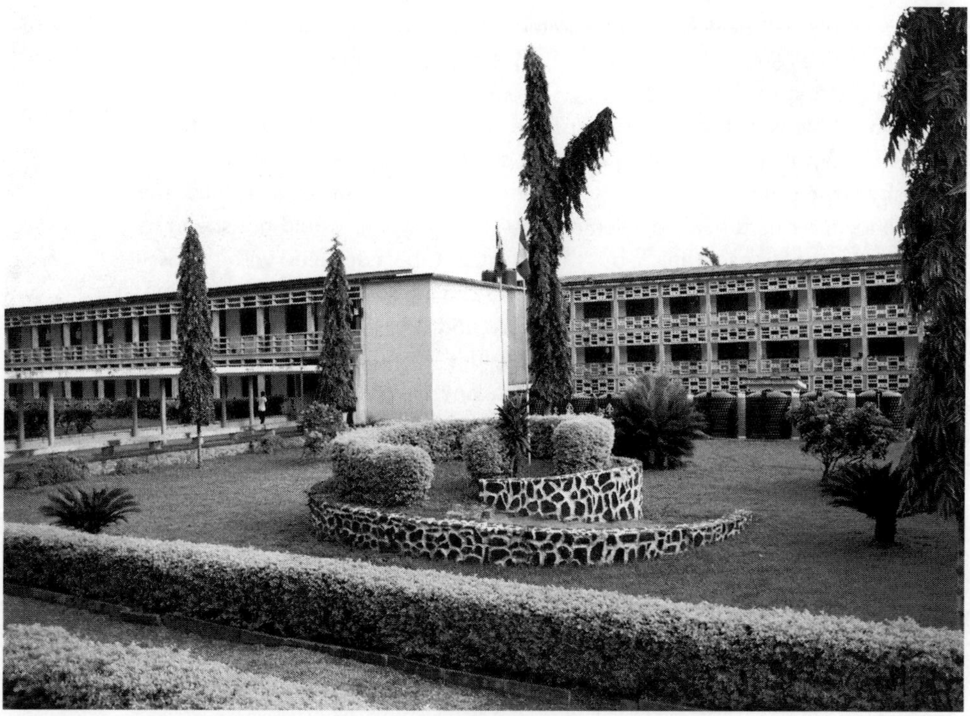

5.7 Interior Courtyard of Student Residences at Ibadan, 2012

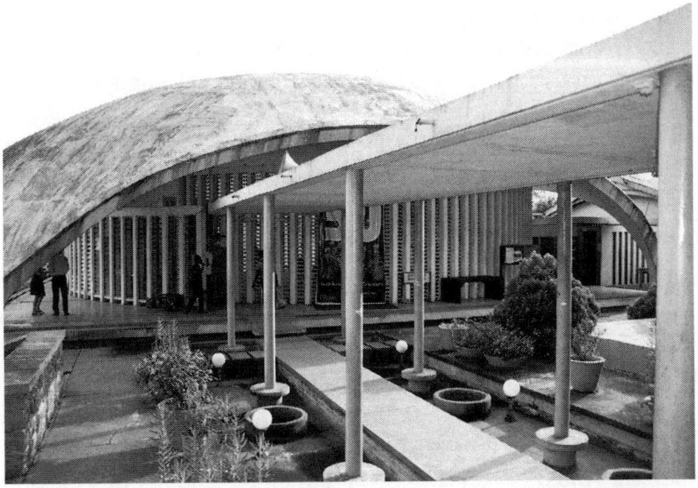

5.8 Sultan Bello Hall, 2012

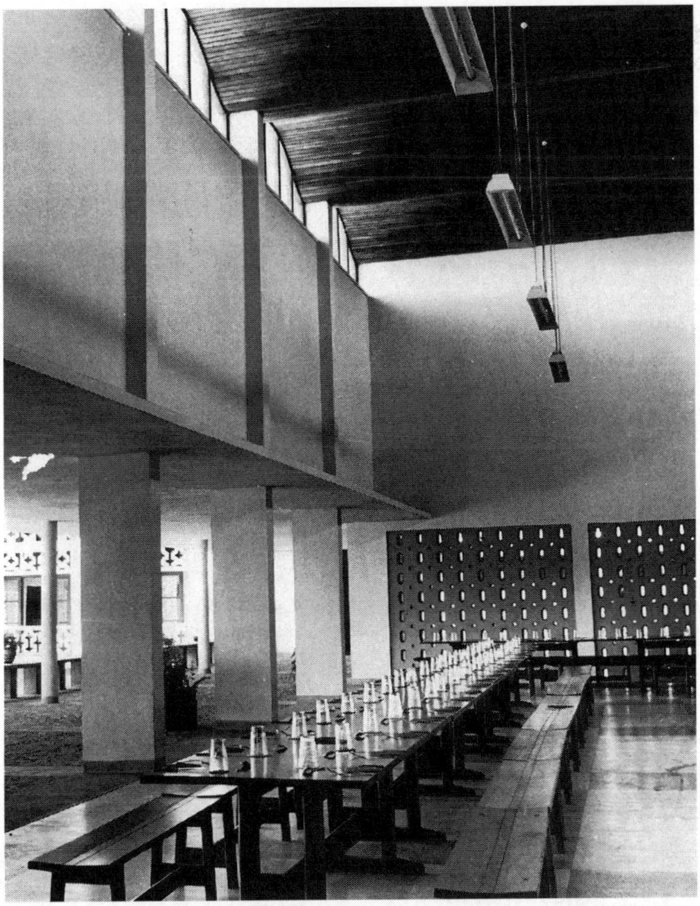

5.9 Mellanby Dining Hall, 1956

portal a raised narrow pathway bridges the ponds and gardens leading towards a perforated concrete drum structure topped with a large concrete domed roof.

The courtyard gardens of the Hall create a secluded and peaceful enclave for the residents, heavily indebted to the model of the Oxbridge colleges that the large communal dining halls also allude to. The precast concrete motif is liberally applied to any vertical surface including stairwells and even on the garden walls. The panels had become more than shading devices and were being deployed as symbolic motifs. This method reached its zenith at the library building (curiously positioned in front of the sports pitches) where the entire outer façade is composed of the precast concrete screens. It forms a striking effect, and coupled with the insect screen forms an effective semi-permeable skin with comfortable interior temperatures. Access balconies run behind the concrete screens serving as further shading devices to an inner glazed façade, which permits light into the book stacks and offices located across five floors.[42] When Fry visited in August 1953 he thought the library was 'big and strong' and he was pleased with Trenchard Hall but pondered their profuse application of the pre-cast screens, 'too much decoration at Ibadan, too much lace. I should have liked it soberer. Perhaps it will be its character'.[43] The following year he again wrote to Drew concerning the library and that, 'the grille work is really too light in section'.[44] Fry wanted an architecture that was defined through form, horizontal line and geometric pattern,[45] but the balustrade motif had modulated from being a playful component into a cladding material that enveloped the entire campus. The later buildings at Mfanstipim School were also heavily decorated with ornate balustrades and facades.

The Building Committee did acknowledge that 'material progress' had been made on the scheme since Fry took personal responsibility for it and increased his number of site visits, however, they were a little taken aback when Fry informed

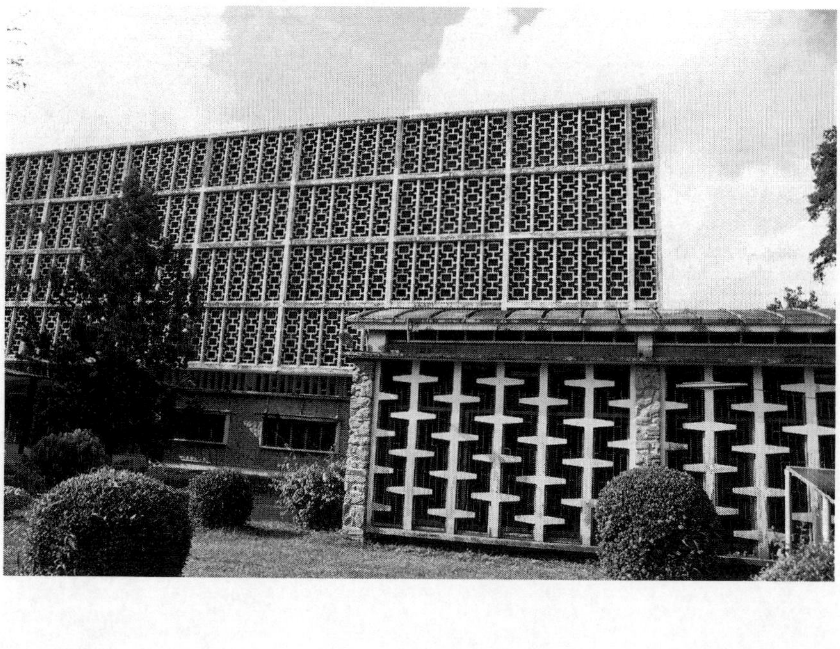

5.10 Library at Ibadan University, 2012

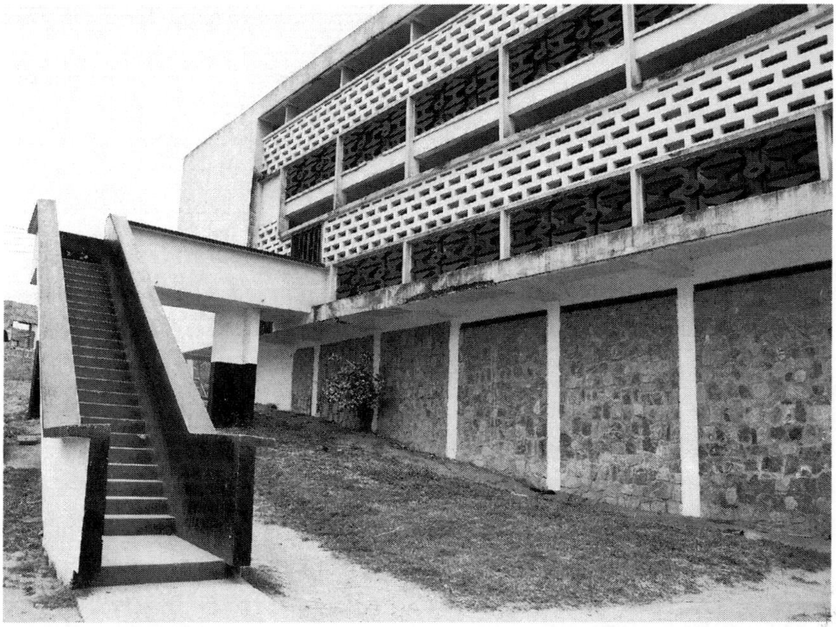

5.11 Mfantsipim School, Cape Coast, 2012

them of his intention to take on a new project, which would take three years of his time, and where he would be expected to live on site – in India. They wrote to Fry purporting to be, 'seriously disturbed by the news' and that 'the phase of the programme which is now about to begin is the most important of all, and that therefore all possible provision must be made to ensure its smooth and expeditious working'.[46] Fry was almost blasé in his response to the clients concerns, claiming that the majority of the work was complete and that 'the burden of what now needs to be done falls now on the London Office', adding that 'we will both visit Ibadan as the work dictates … '[47] For Fry, the excitement seemed to be in the initial design and 'problem solving'. Once this stage was completed he was content to let others manage the construction whilst he moved on to new schemes. Fry and Drew did return to the Ibadan site numerous times during its construction and received frequent progress reports from the London office. In a rare upbeat moment Fry decreed the completion of the campus as 'the crown of our careers, of mine at least'.[48]

COMMUNITY CENTRES

An emerging building type that Fry and Drew developed in the 1950s was the Community Centre. The idea was to produce a shared building without any particular fixed function to galvanise and foster a sense of community belonging. This concept quickly developed following the 1937 Physical Training and Recreation Act which extended powers and grants to UK local authorities to develop social and physical amenities, particularly on new residential estates.[49] During the inter-war years there was concern over how the working classes were to use their increased leisure time, as the Government report summarised,

> *... the future is to see an extension of mechanisation, a further reduction of working hours is likely to follow ... .but it is one thing to have spare time and another to know how to use it wisely. Already it is clear that there is little tradition of leisure amongst large classes of those to whom it has come, and the increase of leisure confronts us with a new social problem.*[50]

Coupled with this 'problem' of increased leisure were the effects of the significant rehousing programmes that were taking place as a result of 'de-slumming'. Flora Stephenson noted that, 'the residents will have severed connections with organisations which formerly gave them their social, cultural, or recreational outlets. The complete break with former surroundings, neighbourhood meeting places, and old friends will leave them at loose ends during their leisure hours. It is natural therefore, for the Community Centre idea, is itself a new approach, to develop most quickly where living conditions are non-traditional'.[51] A similar tactic was deemed necessary for the colonies, where new mechanised production, industrial agriculture and mining resulted in migrant labour and rural workers moving into rapidly urbanising centres. The Colonial Social Welfare Advisory Committee met in 1944 to discuss the establishment of 'social centres' in various colonies and conducted a survey on the existing provision of 'reading rooms' and 'social meeting places' throughout Africa.[52] They discussed the Government's proposal to 'establish 10 Social Centres to provide social amenities for Government employees in the districts [of Gold Coast]',[53] but for employees of private companies they would have to rely on their employers to fund any such facilities. They reported that, 'Cadbury Bros., Ltd., have undertaken to erect and equip two village halls at Sunum and Berekum as a contribution towards the scheme of mass education'. Whereas in the UK the onus was on providing appropriate leisure, the centres in the colonies prioritised 'education and health [rather] than recreation'.[54] The United Africa Company was prepared to entirely fund the Accra Community Centre, subject to the condition that, 'the right type of European leadership – a professional community centre leader or social worker – was guaranteed, the Board might consider bearing the whole cost of such a centre. It would be a condition that the venture should bear the Company's name in some form'.[55]

Fry and Drew were awarded this commission (with Theo Crosby as architectural assistant) and developed a design dominated by a large hall set within an entrance courtyard and decorated with a mosaic by Kofi Antobam. Antobam was a British trained Ghanaian artist, who depicted a group of the *Ga* people in traditional dress with the message, 'it is good we live together as friends and one people'. The large mural is clearly visible at some distance and as Crinson notes is used to, 'represent reassuring images of pre-colonial rural life and a unified nation; they imply continuity even if the location, appearance and function of the building they ornament is far from the life depicted in these images'.[56] The carved wooden door delivered a similar message. At Tarkwa the African Manganese Corporation funded

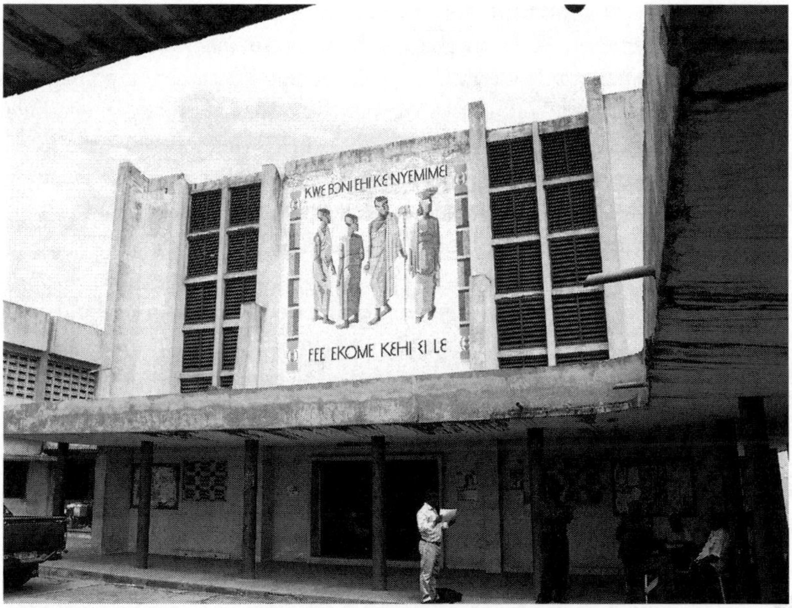

5.12 Accra Community Centre, 2012

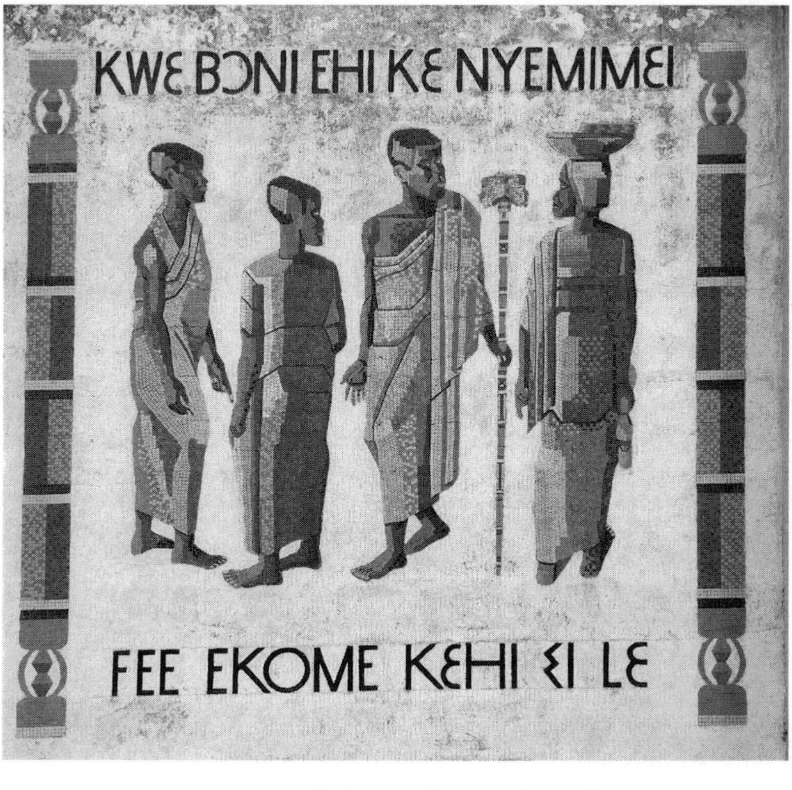

5.13 Accra Community Centre, Mosaic by Kofi Antobam, 2012

a community centre for its mineworkers and their families. The developments were not entirely philanthropic ventures, there was a paternalistic overtone to all these buildings. Not only did the corporation dominate the financial structure and transportation within the region, they also determined what activities their staff engaged in outside of work. This married perfectly with the Government's desire for Community Centres to be a place where 'moral' recreation could be organised along with adult educational classes, health advice and 'highlife' music performances to 'prevent young men and women from falling into nightclubs' social evils'.[57]

The Tarkwa Community Centre, like the Accra building, is a bold composition, set in the prime location of the settlement, in front of a playground and overlooking the rusting metal rooftops of ramshackle houses. It contains the usual facilities, including a hall, bar, committee meeting rooms and smaller classrooms opening into an enclosed courtyard. The main building is dominated by the brise soleil on the front façade and along with the piloti, exaggerated cornice-gutter and elevated entrance, the building has a distinct feel of Le Corbusier's Mill Owners' Association building in Ahmedabad. The interior is broken up with small courtyards, lightwells and louvered walls that encourage cross-ventilation as well as visual links and views through the building out to the

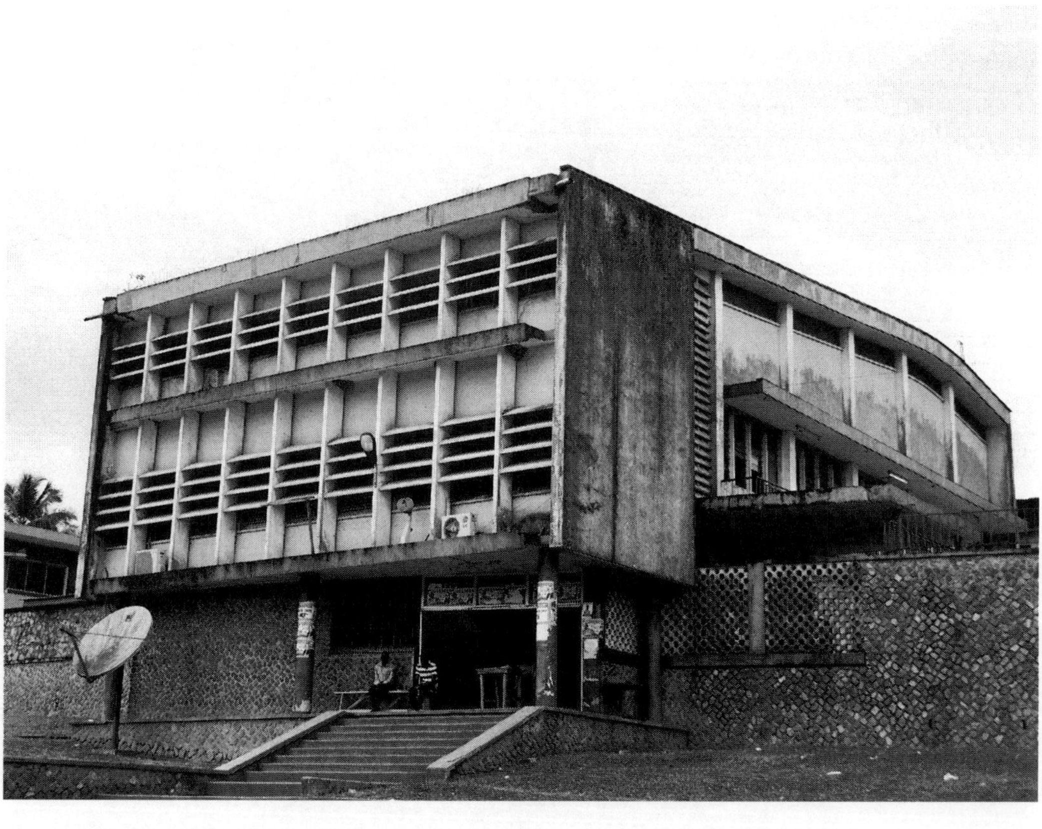

5.14 Tarkwa Community Centre, 2012

village beyond. The building is in such stark contrast to the surrounding streets and houses, serving as a symbol of the progressive and forward looking image that the mining company wished to portray.

TEMA MANHEAN

The University College of Ibadan and most of the Ghanaian schools were constructed whilst Fry and Drew were employed in India, with the projects managed by their new partners and their more experienced architects based in the London office. Fry and Drew both returned to the continent on numerous occasions and continued to seek work there whilst in India.

One of the major projects of the time was the development of a new port close to Accra. It was deemed an urgent requirement to replace the 'time-honoured, primitive, but nevertheless effective' surf boat harbour system that was not suited to the large-scale shipping of cargo and was considered perilous by passengers.[58] In addition to a new port the plan was to exploit the substantial bauxite deposits by establishing an aluminium production plant powered by a hydro-electric scheme and to develop a new town.[59] The favoured site was only 18 miles from Accra and as a result, the inclusion of a new town into the wider-scheme could offer some much needed expansion room for the capital, following the twin-city model that already had precedent at Sekondi -Takoradi.

Ghana was well placed to fund so-called 'national projects', having a substantial savings pot amassed through limited spending during the war and the levying of income tax, however, it did not have all the finance it needed to complete the Volta River Project, and the British Government agreed to pay up to £57m, providing 'at least 75 per cent of the [aluminium] output' was offered to the UK market first.[60] It was a proposal that both managed to appeal to the nationalist and imperial cause. The Convention Peoples Party (CPP) political success in 1951 and the election of Kwame Nkrumah, whose desire for rapid industrialisation and 'modernisation' gave the scheme fresh impetus.[61] Nkrumah proposed that a dam and hydroelectric power station with a capacity of over 800,000 kW be built in Akosombo on the Volta River to serve the Tema industrial settlement project.[62] The Tennessee Valley Authority project served as an important precedent, as did other developing nations and their five year plans, such as India and its Bhakra Nangal dam project, however as Huxley notes the main problem was to 'reconcile the investment of so large a sum of foreign capital with the coming political independence ... '[63]

In addition to generating electricity and producing aluminium the plan was to fund and set the standards for dwelling in West Africa too;

> *for the first time in West Africa, it was decided to create a community which could enjoy all the advantages of modern civilisation – well-designed houses, a well equipped hospital and comprehensive health, social and cultural services, pipe-borne water and underground sewage system, well-laid-out and lighted streets, up-to-date markets and stores ...*[64]

A new town of 84–90,000 people was proposed[65] and it was envisaged that this population would be employed in the docklands and harbour. The new population would come from all over Ghana, resulting in a 'gigantic piece of social engineering for uniting people of different social and cultural backgrounds into an integrated social, commercial and industrial community'.[66] There had been very little government involvement in housing throughout all of British West Africa, with most of the housing efforts directed towards providing housing for the European population.[67] The Volta Project presented the opportunity to buck this trend and to develop quality housing and facilities, perhaps in an attempt to settle the civil unrest and strikes that erupted in Accra during 1948 (as discussed in Chapter 4).

RAPID MODERNISATION AND EMBALMING THE PAST

Tema Village, an existing fishing village located on the proposed new town centre and harbour area was treated differently than the other villages due to be re-sited in the region. Rather than providing a basic core house and cash compensation, each house was to be replaced, funded exclusively by the government, with no cash compensation being offered.[68] This was an attempt to induce the villagers to move quickly to the designated new village and not to remain as squatters on the old site. In addition, Tema was also to be seen as a model development built to high standards which the self-build approach could not deliver. The proposal, designed for around 5,000 residents (a figure which would eventually rise to 12,000[69]), was to move the village three miles east down the coast to a new settlement renamed, *Tema Manhean*.

The existing village was located between two lagoons, which provided the natural boundaries of the settlement; to live outside of that zone was unacceptable for a Tema-born person and the government's decision to 'move' the town was met with considerable indignation. Indeed, this situation lasted for seven years. Although the village was less than a day's travel from the capital it remained a rural community heavily dependent on fishing and small scale agriculture. The existing housing was thought of as 'little more than shacks built of timber and corrugated iron', along with 'adobe thatched with palm fronds',[70] with 'few, if any permanent buildings'.[71] Rather than simply providing new homes for the residents within the larger masterplan, this community was to be kept distinct, living in a separate part of the New Town in an attempt to 'preserve the tradition and custom of the village'.[72] Despite Nkrumah's desire for 'modernisation', this community was to be conserved and protected, almost as if they were a fixed entity to be embalmed as a symbol of a bygone time.[73]

The residents of Tema, the *Ga* and *Adangbe*, adopted separate dwellings for male and female family members, with complex definitions of dwelling, household and family units.[74] Onokerhoraye provides an acute description of these households as 'a group of people who live together and eat from the same pot', although they may not all sleep under the same roof, indeed one 'house' may contain numerous family groups living in separate rooms, but nevertheless still in the same dwelling-unit.[75] This mode of living would present a challenge to the planners and surveyors so used to associating a particular definition of the family with a single dwelling. Ascertaining the number of family members per building was not clear-cut, resulting in guess work and inaccurate counting – the results of which were to inform the requirements of the new village.[76] It was decided that the housing was to be replaced on a like-for-like basis, with the number of rooms in the old village determining how many rooms one would receive in the new.

THE HOUSING PLAN FOR TEMA MANHEAN

The initial plans for the Tema new town were prepared shortly after the Engineers' proposals for the dam and harbour. The main contributors were the Town Planning Advisor, Alcock (who had already worked with Fry and Drew on village planning),[77] Town Planning Officer for Gold Coast, Helga Richards and planner Denis C. Robinson.[78] Together they developed a pragmatic arrangement based around the requirements of the port area and utilisation of the existing road routes. The planning was likened by Alcock to UK town plans developed following the New Towns Act, 1946,[79] and the layouts certainly adopt some of the same desires, namely the social amenities provision.

'Communities' and 'Neighbourhoods' as the building blocks for planning, were also proposed by Ernst May (1886–1970) for Kampala, Uganda, circa 1947. 'Such demarcation', claimed May, 'making it easy even for the more primitive African

5.15 Initial Plan for Tema, 1950s

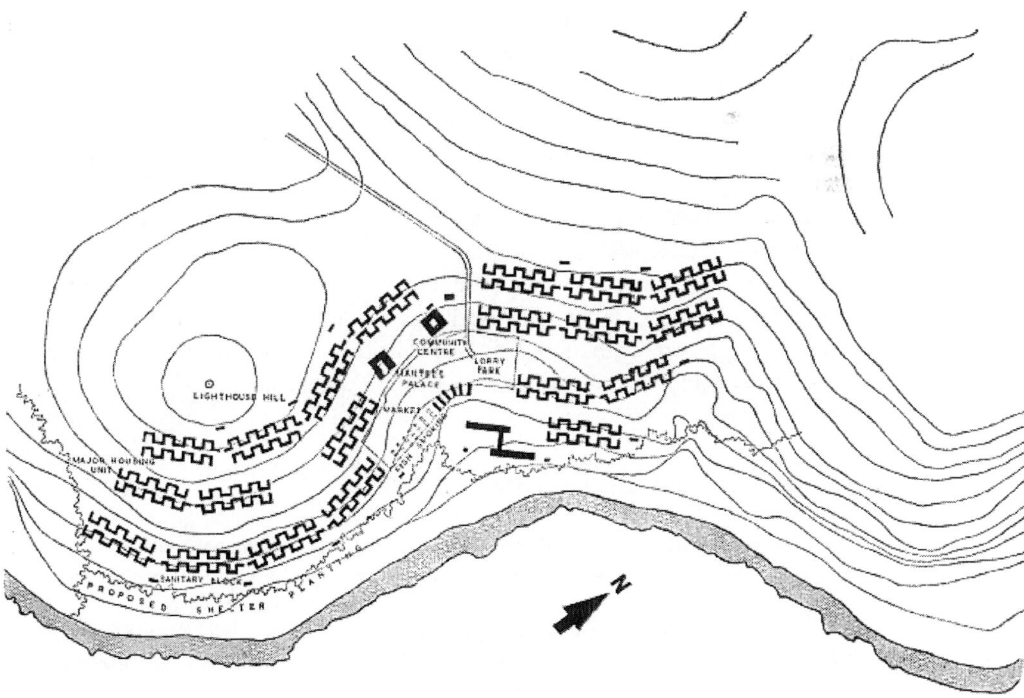

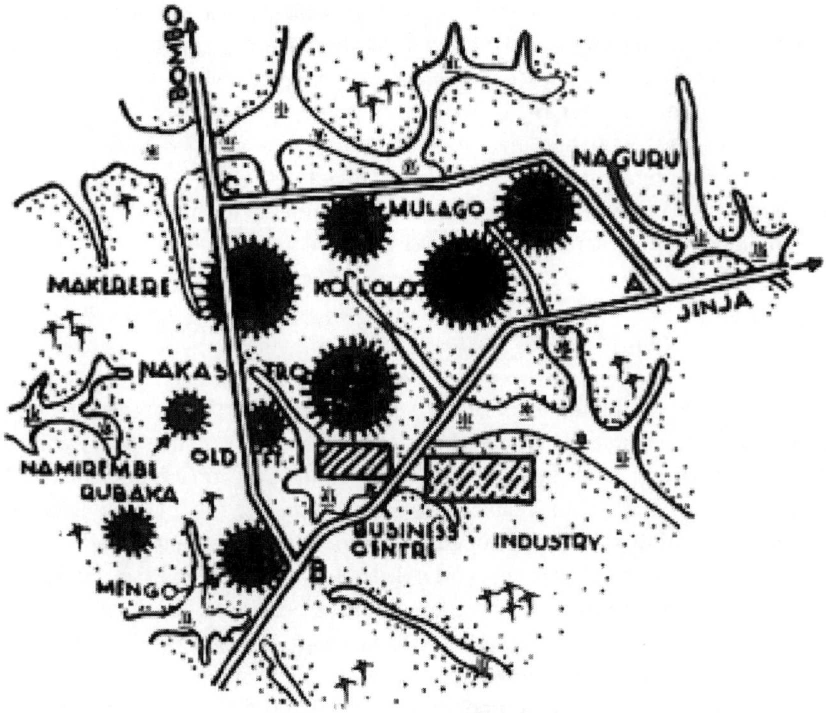

5.16 Ernst May's plan for Kampala, 1947

to conceive the boundaries of his home district, and to take an interest in its development in competition with other Neighbourhood units'.[80]

The Neighbourhood unit was seen as an organisational device, a means of ordering and defining the local population into neatly defined distinct groups. From the planners perspective this mode of operation was also deemed as the most appropriate, attempting to design for 'social structure', and borrowing from Max Lock's agenda where, 'the "raw materials" of a plan are the citizens of the place'.[81] Transposed into a colonial context it became a means of cataloguing and organising the local population. Furthermore, in Africa this method of planning was extrapolated into forming a suitable bridge between the rural small-scale settlement and emerging urbanised living patterns of much larger estates. It is of note that Fry, May and Lock all published articles in the same edition of the Architects' Year Book, 1947, further suggesting a body of thought, and an exchange of ideas was actively taking place and shared within the colonial and emerging post-colonial territories.

BACKGROUND TO THE FRY & DREW PLAN

> *we would be willing to offer you services which may be what you require. Chandigarh has been an example of how well our proposed arrangements can work. We suggest that we be consultants so far as the overall plan is concerned and responsible for seeing that the detail town planning is done by resident staff in the Gold Coast. We could also undertake to design such of the more important buildings as you may think proper …*[82]

Fry and Drew's reputation for building in hot climates was now firmly established and coupled with their relationship with Alcock they were the prime candidates to design Tema Manhean. This relatively small part of the overall project could have been easily designed by Alcock, it could have also simply formed one of the Neighbourhoods within the newly proposed Communities, but from the outset it was physically isolated and treated as distinct. Moreover, Alcock was particularly concerned about the social impact of migration and pending industrialisation; he acknowledged that despite the provision of funding there was 'no basic data [about the socio-economic status of those to be rehoused] and our background is all guesswork. Great social change is going on and we have not got the necessary knowledge and experience to forecast its future trends'.[83] It was Fry and Drew's prior experience in the region managing to secure substantial commissions from the Development and Welfare Initiatives including the numerous schools and community centres that set them apart as being most able to satisfy Alcock's concerns.[84] In addition to their 'expertise', their rapport with villagers elsewhere was considered a useful attribute, and it was hoped they could help to pacify the hostility of the Ga as well as providing much needed insight to discern future housing requirements. An exhibition complete with scale model of the overall Volta River proposal was constructed and toured the various villages and provinces to raise African goodwill.

The Tema Manhean project has been somewhat overlooked,[85] not least because Fry and Drew were working in Chandigarh between 1951–1954, and perhaps as a result of their complex business arrangements, resulting in some uncertainty over design authorship.[86] As Fry and Drew were busy in Chandigarh, it is unlikely that in the first instance they were the lead designers for Tema Manhean and that the work

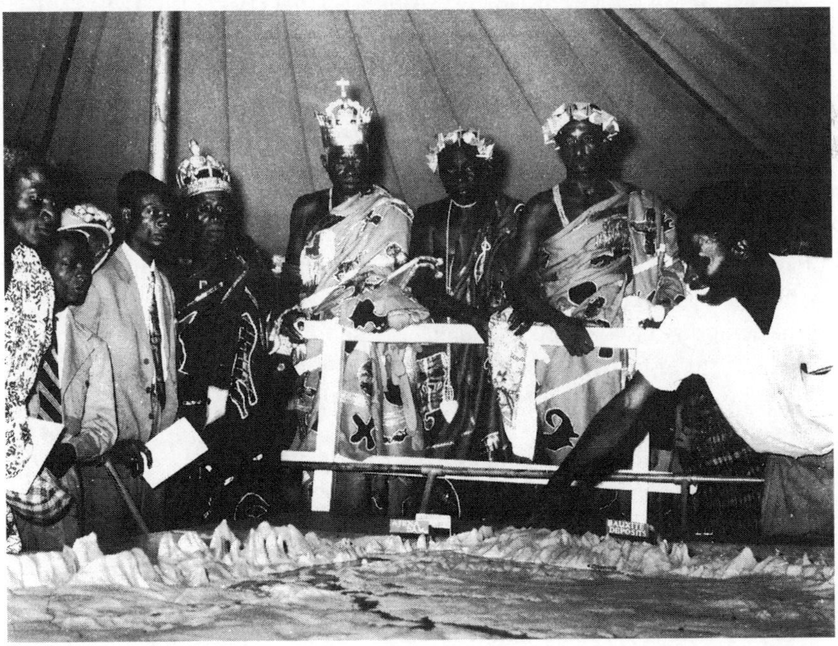

5.17 Volta River Project Touring Exhibition with Model, c. 1950s

was overseen by their senior staff members and partners, Drake and Lasdun, until Fry and Drew returned to the UK in 1954.[87]

Fry's notions of planning at this time involved what he called a 'correction of a *gridiron* plan'.[88] Grid planning was adopted to suit the 'compound' house layouts, that is, courtyard-like arrangements with a central enclosed yard. Fry's concern was that these were laid out indiscriminate of the topography, 'often with disastrous results'.[89] His solution was simple, and already tested in the school designs, 'keeping to contours where possible and substituting paths for unnecessary roads'.[90] He also rejected the 'unrealistic' garden city planning and made the case that planning is a human, rather than a mathematical problem.[91] The solution was to be found in the grouping and site planning of architecture whereby, 'it is the arrangement of the house units into social groups of real significance that is important; and it is this that is so often neglected. The minimum house is *not* enough in itself ... '[92] In addition to the grouping of houses, Fry sought to offer housing solutions supplemented with other facilities, but equally he recognised that the act of planning a house, especially for a society undergoing such flux (what he called, 'backward areas'[93]) faced the real risk of not meeting future requirements. Fry's solutions included, 'concentrating on the shell of the housing, perhaps just the roof and the supports'.[94] They had learned from their experiences in Chandigarh that it was almost impossible to house the poorest residents with adequate architect designed housing built with limited government funds. It was also an acknowledgement that the architect did not need to 'provide everything' and that with some infrastructure the resident was more than capable of developing their own home.

PLANNING TEMA MANHEAN: RECONSIDERING THE COMPOUND HOUSING TYPE

Fry and Drew initially proposed what they called an 'Open compound' housing solution for the Tema Manhean plan. This was a variation on the 'traditional' compound house, modified to encourage more cross ventilation, and to serve as what Fry called an, 'intermediate type that will fit the broader requirements of the differently organised life of the future ... '[95] In other words he saw the open compound as a means of enabling a mode of living that was somewhere between the African and European arrangements.

The compound was a type frequently used in Ghana, taking the form of mud-walled structures of the remote villages and later concrete or swish buildings designed by the PWD. It consists of a defensive walled enclosure with a series of rooms arranged around part of the periphery and an open external space used for cooking, laundry and safe keeping of animals at night. It was a flexible type easily modified to suit the changing requirements of the occupiers and providing secure and secluded exterior space within its confines. Initially, it was condemned as being climatically poor, but the problems were more to do with the lack of services, rather than the house design. Drew and Fry initially described them as, 'unsatisfactory from health and hygiene viewpoints ... the houses are often too dark, damp, and

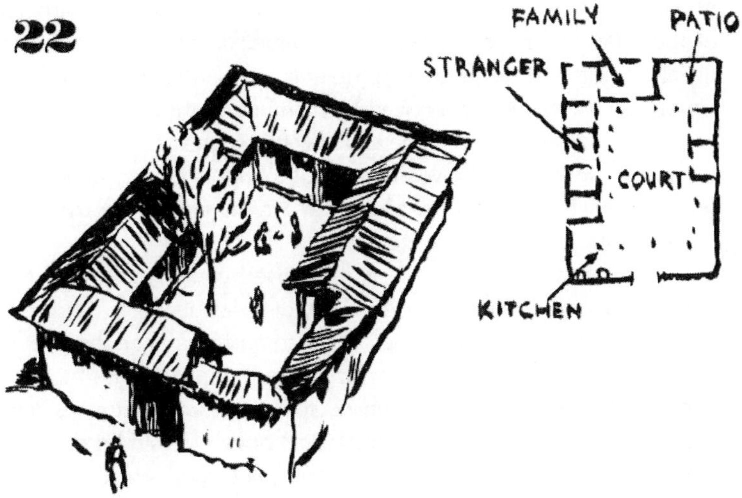

5.18 Fry's sketch condemning the Compound Housing Type, 1947

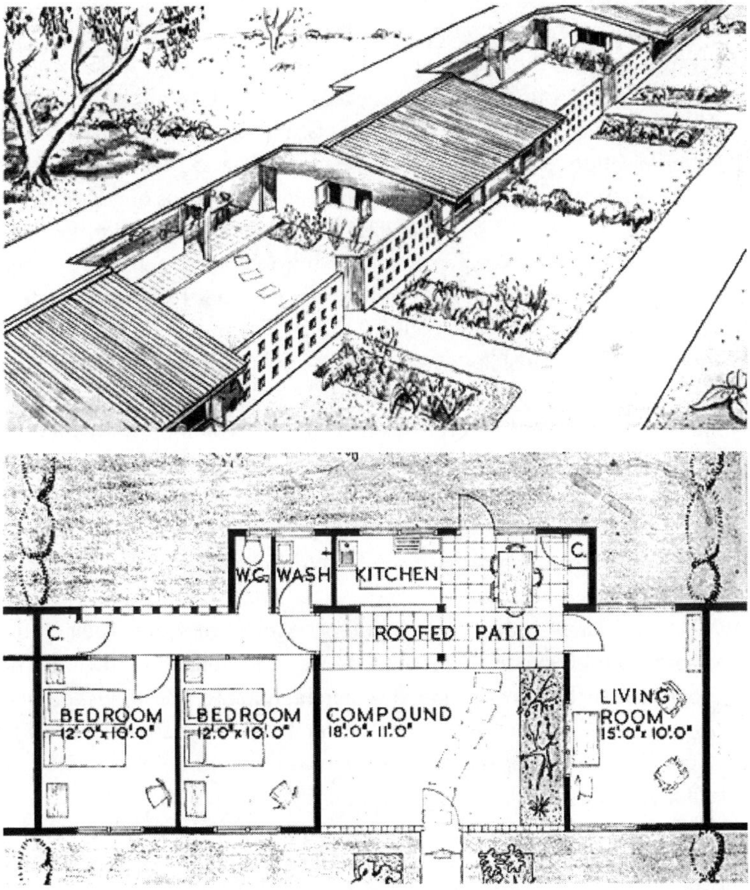

5.19 Initial 'open compound' design for Tema Manhean by Fry and Drew, 1956

under-windowed … the problem is to provide suitable houses within the incomes of the villagers, answering present-day needs in the transitional stage of their development'.[96] None of these problems were insurmountable, and although Fry and Drew initially denounced the compound they recognised the symbolic importance of this type of home and sought to generate an alternative derivative. They had also designed several compound houses for African staff at numerous schools. Their first plan for Tema Manhean was based on what they called an 'open compound' in a terraced arrangement.

This resulted in a series of semi-enclosed courtyards linked together with the neighbouring property for economy of construction. It is not clear how this solution would cater for the diverse and complex living arrangements of the Ga, but the African members of the committee thought this housing type coupled with the proximity of the new town would prompt some of the residents to, 'better themselves',[97] or in other words, individual houses would be occupied by family members of both sexes, as per the European model. This was the preferred method at Tema New Town where it was decreed that, 'housing accommodation shall be non traditional. The tribal compound has no place in Tema and is replaced by the private family dwelling'.[98] However, the problems over occupancy numbers and a general rejection of the proposals by the village prompted a new design approach to be developed. Although many European architects and officials working in Ghana found the compound to be climatically problematic and physical representations of the unfathomable living arrangements, as early as 1947, Fry was rethinking their initial rejection of the compound; 'first, people and their needs; second, climate and its attendant ills; and third, materials and the means of building'.[99]
The compound house became what Fry called an 'especial object of study', and suggested that 'if we approach it as anthropologists we stress its semi-tribal or "extended family" occupation, its communal hearth, its arrangement of small rooms round a courtyard, its self-sufficient, wall-enclosed unity. This, we say, is the expression of a way of life that must be respected'.[100] It was a flexible type that was easily modified to suit the changing requirements of the occupiers and it provided secure and secluded exterior space within its confines. Fry would even go on, rather tenuously, to try to associate the compound with the modernist objective of the 'free plan', claiming that, 'within the compound-wall the Indian or African has already designed open plans as free as those of his modern brother; exterior and interior being sometimes hardly definable. Needs of security, not climate, led him to build protecting walls'.[101]

Fry and Drew's ideas on the tropics were not fixed. Their writing reveals, as one would expect with experience and reflection on the outcome of experiments, shifts in agenda and architectural approach. *Village Housing in the Tropics* was written relatively soon after their initial sojourn into Africa. It was an exploratory publication, and was by their own admission not intended for building professionals.[102] By the time they had published *Tropical Architecture in the Humid Zone* in 1956 considerable advances in the field had taken place. Not least, the conferences on Tropical Architecture held in Venezuela (1947), Lisbon (1952), London (1953), Durban (1957) and UNESCO had organised a symposia in

Delhi (1952) and another one in Uganda (1955), the foundation of the Tropical Architecture course at the Architectural Association had been established (with Fry undertaking some of the teaching, more in Chapter 6) and extensive publications featured in the architectural journals. The general canon of Tropical Architecture had been formalised and clearly defined, with European architects setting the agenda, leading the technological advances and delivering nearly all of the large building projects. In addition, further design data and solutions were being produced by the Building Research Station who appointed a Colonial Liaison Officer, G. A. Atkinson in 1948. This post, partially funded by the Colonial Office was to advise architects working in the tropics and to help oversee the *West Africa Building Research Institute*.[103] Atkinson was to investigate not only climatic design, but to also develop more cost effective modes of building in the tropics through scientific experimentation of building techniques and materials. His findings published from 1950 as the *Colonial Building Notes*, with the intention of disseminating these results and techniques around the globe. In addition to the house designs, a more general revision of the Tropics was taking place acutely demonstrated in Fry and Drew's publications. As previously mentioned the front and endpapers of the *Village Housing* book presented the tropics as a homogenous band but by 1958 a modified series of plans replaced the overly simplistic efforts of the previous decade. The nonchalant and linear tract of Cancer and Capricorn were now replaced with wavy lines and climatic variances were carefully plotted onto the map. Population was considered, as was rainfall and wind – all key factors in determining architectural solutions besides temperature. Alongside these climatic studies is a map attempting to show religious beliefs, perhaps in deference to Fry and Drew's desire to firstly observe people and different ways of living as well as the technical knowhow of construction.

For Tema Manhean, a revised plan was produced that substituted the terraced housing for small groups of individual closed compound houses weaved around the principle routes, in small clusters of six and ten residences. Fry and Drew changed their approach from the overly simplistic solution of one housing type 'fits all', proposing three standard types that were capable of further modification to suit the residents' requirements and easily modifiable over time.

Like at Chandigarh (see following Chapter), a number of 'prototype' houses were constructed and consultation with the future residents sought.[104] As a result of this exchange, the housing designs were rejected after being deemed 'undignified', by the villagers. The hostility towards the scheme resulted in 'violent acts' and the 'destruction of the prototype houses by a section of the youth ... '.[105] It was not the architecture *per se* that they were reacting against, rather it was seen as a symbol of the forced migration, although the single pitched roofs were singled out as the main problem.[106] The roofs were easy to construct and did not require an expensive ridge detail, however, this type of roof was associated with cheap self-builds, whereby corrugated iron sheets were simply placed on the walls and 'kept in place by loading with stones'.[107] As a result of the villagers' intervention they were replaced in the final scheme with double pitch roofs.[108] The villagers were not expecting replacement housing like-for-like, they wanted an architecture that was

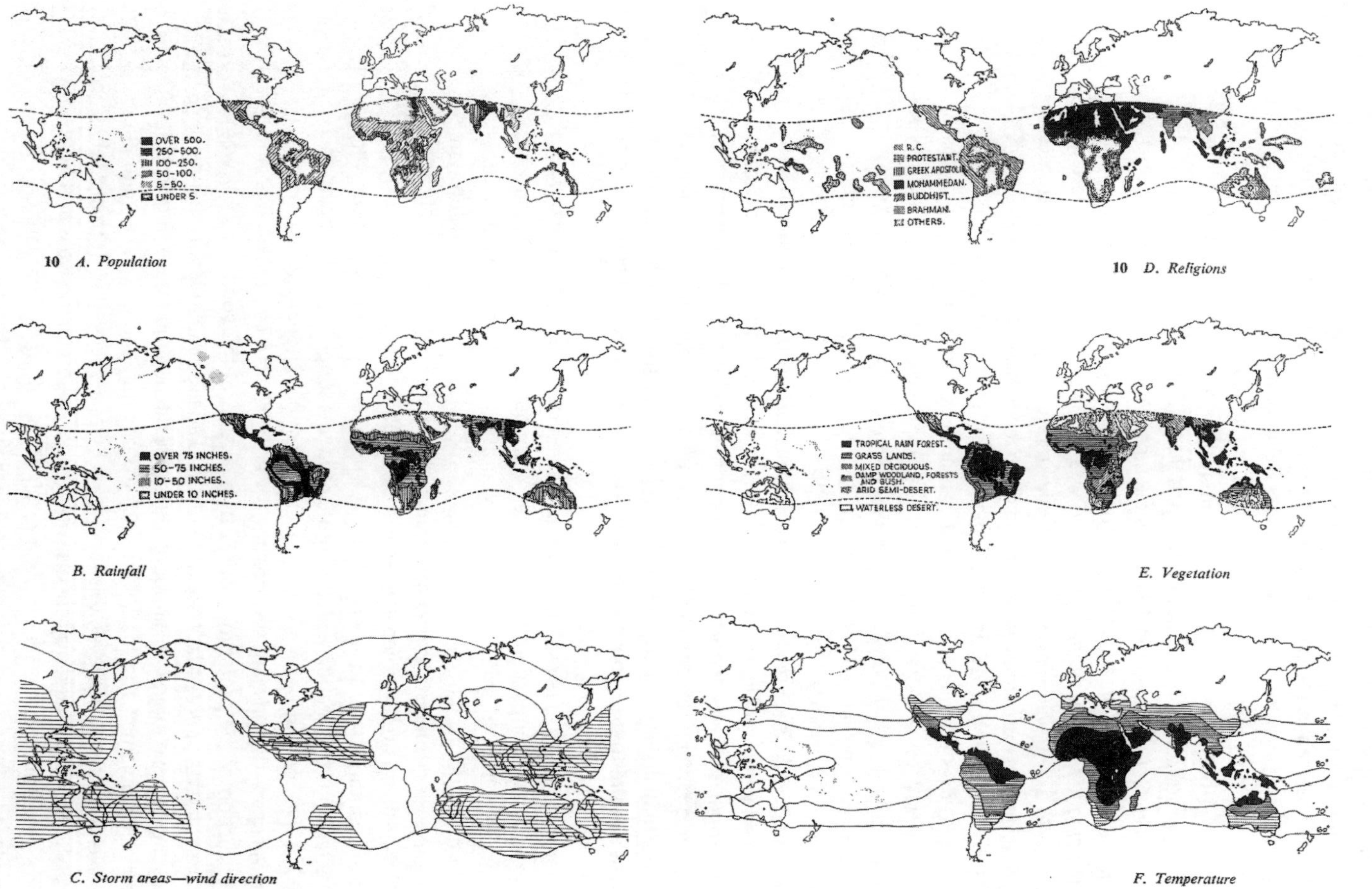

5.20 Reviewing the effects on climatic design, new map from Fry and Drew Tropical Architecture in the Humid Zone, 1956

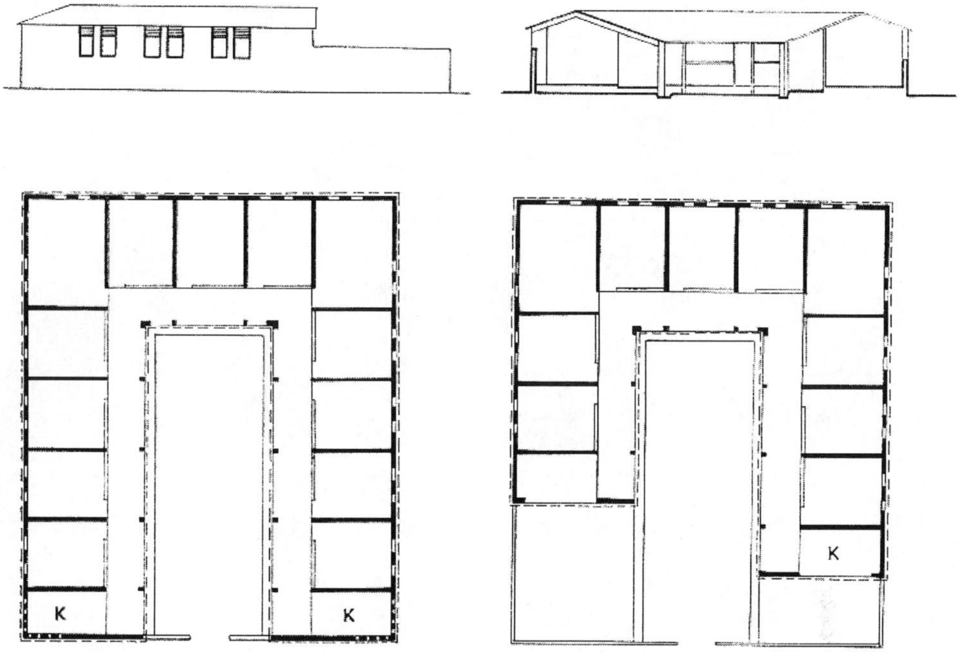

5.21 Revised housing plans for Tema Manhean, 1950s

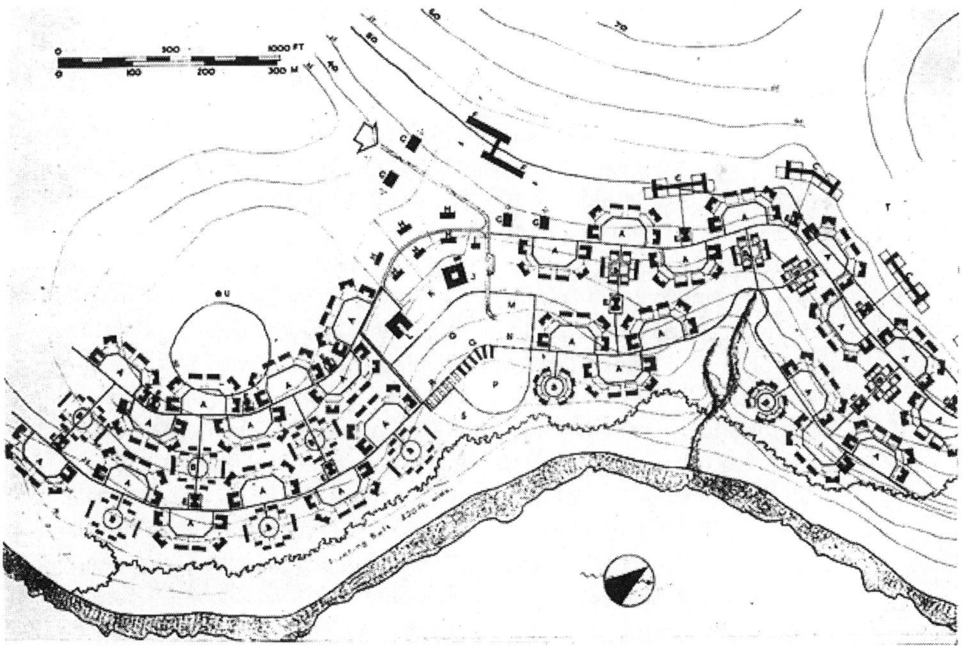

5.22 Revised Plan of Tema Manhean, incorporating the closed compound housing, c. 1950s

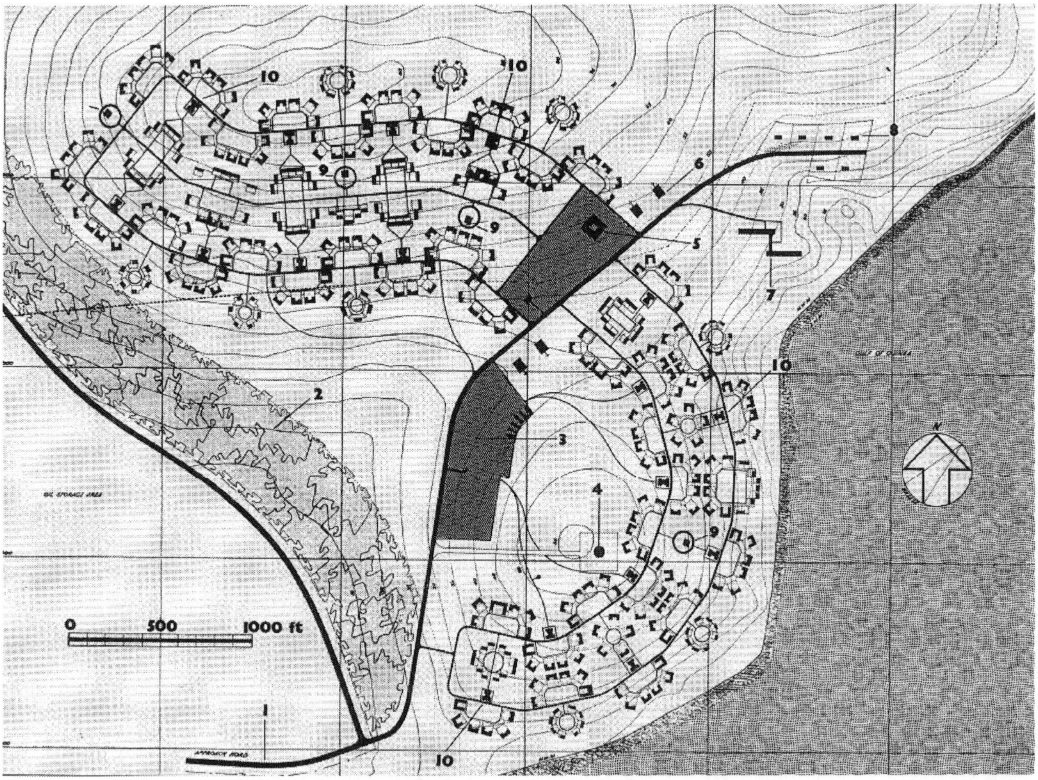

5.23 The final planning arrangement for Tema Manhean, 1958

associated with their perceptions of dignity and prestige, that is, in the manner of the European PWD bungalow.

In addition to the basic compound type, laundry and sanitation blocks would be provided for groups of houses to share. The plan was further revised following concerns from the residents that the houses were too close to the sea. These demands were promptly met to keep the villagers on side and to expedite the construction process. A. E. S. Alcock resigned from his post of Town Planning Advisor, and it was anticipated that 'with him will go Helga Richards', leaving the future of the project in a precarious position.[109] 'I sniff that', mused Atkinson, 'while Nkrumah and Botsio are interested in town planning there may well be less enthusiasm elsewhere … '[110] Furthermore with the loss of dynamic individuals like Alcock and Richards, coupled with the large number of vacancies that had yet to be filled, they were faced with appointing individuals who were not deemed suitable.[111] They proposed forming a system whereby British senior officers would remain head of the planning team, 'while promoting a Gold Coaster as head of the planning service and advisor to the minister (Given good relations, the senior non-African would guide, advise and befriend his Gold Coaster colleague)'.[112] Furthermore, Atkinson wanted the planner Charles Abrams (1902–1970) to 'influence directly' the situation by contacting Otto Koenigsberger or Robert Gardner-Medwin,[113] presumably selecting these two individuals because of their involvement with the UN housing reports that were produced throughout the 1950s.[114] This 'intervention' resulted in Koenigsberger (who was working for the London School of Hygiene and Tropical Medicine at the time) contributing to the Volta Project reports and providing some planning solutions, along with the American and first planner of Chandigarh, Albert Mayer (1897–1981) who designed a smelter settlement at Kpong,[115] and Architects' Co-Partnership who were tasked with reducing costs of the proposed housing solutions.[116]

By 1959 the villagers had, reluctantly, begun to evacuate their old homes, selling their corrugated iron roofs and moving into their new houses in Tema Manhean. It was an experimental development that was not to be repeated, being subsequently deemed, 'impracticable for the Government to finance the building of one house per family but schemes are in operation to assist the private individual'.[117] These proposals, which were developed by Koenigsberger, included a 'roof loan scheme by which a man who has provided himself with the walls of a house can be assisted to purchase the necessary roofing materials' along with other proposals that would help with site clearance, provision of services and the establishment of building societies.[118] This was a significant step for 'welfare and development' and would encourage the community to work on their own settlements rather than distant 'experts' prescribing a scheme.

Fry and Drew's overseas work dramatically altered from this point, largely as a result of changing political objectives and the inadequacies of large planned residential quarters. A new building type was emerging however, namely the large office block and company headquarters. Fry was less confident in their overseas work continuing, 'it is a very different thing being given a job and trying to get one […] and I am not good at sticking my toe in the door'. With political independence already won by Ghana, and Nigeria's just months away he pondered, 'the big spending era is nearly certainly coming to an end'.[119]

NOTES

1 The 1945 report was based on a series of previous government recommendations. See Nduka Okafor, *The Development of Universities in Nigeria* (London, 1971). Parts of this chapter have been previously published in *Planning Perspectives*, 'The planning of late colonial village housing in the tropics: Tema Manhean, Ghana', online 5th February 2014, DOI: 10.1080/02665433.2013.829753

2 See Nduka Okafor, *The Development of Universities in Nigeria* (London: Longman Group, 1971), p. 83.

3 These were the Royal University of Malta, with a history dating back to 1592, The Hebrew University of Jerusalem, The University of Hong Kong, established in 1911 and the University of Ceylon which incorporated two early colleges founded in 1870 and 1921.

4 C. Asquith, *Report of the Commission on Higher Education in the Colonies*. (London, 1945), p. 8.

5 Asquith, *Report of the Commission on Higher Education in the Colonies*, p. 10.

6 Asquith, *Report of the Commission on Higher Education in the Colonies*, p. 10.

7 See L. J Lewis, 'Higher Education in the Overseas Territories 1948–58', *British Journal of Educational Studies*, 8 (1959): pp. 3–21.

8 H.M.S.O., 'Report of the Commission on Higher Education in West Africa', (London, 1945), p. 60.

9 H.M.S.O., 'Report of the Commission on Higher Education in West Africa', p. 65.

10 Elspeth Huxley, *Four Guineas: A Journey through West Africa* (London, 1954), p. 186.

11 Huxley, *Four Guineas*, p. 186.

12 H.M.S.O., 'Report of the Commission on Higher Education in West Africa', p. 171.

13 H.M.S.O., 'Report of the Commission on Higher Education in West Africa', p. 128.

14 H.M.S.O., 'Report of the Commission on Higher Education in West Africa', p. 139.

15 National Archives, BW90/314, Letter from K Mellanby to Sir Hector Hetherington, 24 August 1950, University College, Ibadan, Nigeria, June-Aug 1950.

16 National Archives, BW 90/309, Copy of Minutes, 33599/9/1/47.

17 National Archives, BW 90/309, Letter from Walter Adams to Dr. Trueman (from inter-university council) 28.10.1947.

18 National Archives, BW90/309, Fry's pitch letter to the Inter-University Council, 9 Oct 1947.

19 RIBA Archive, F&D/12/1, Letter from Dr. Kenneth Mellanby dated, 10 November 1947. For a detailed account of Mellanby's version of events see, Kenneth Mellanby, *The Birth of Nigeria's University* (London, 1958).

20 National Archives, CO 554/150, Higher Education in West Africa, Nigeria, Ibadan College Architects. Notes from October 1947.

21 National Archives, BW90/311, Letter from Mellanby to Adams, 25 June 1949.

22 National Archives, BW90/311, Letter from Mellanby to Adams, 1 July 1949.

23 Cambridge University was offered as a comparison, but its campus covers less than half the area of the proposed Ibadan plan, and had ten times as many undergraduates.

24 National Archives, BW90/311, Report of The Committee, September 1949.

25 National Archives, BW90/311, Letter from Mellanby to Walter Adams 8 October 1949.

26 RIBA Archive, F&D/18/10, letter from Fry to Drew, 2nd October 1949.

27 National Archives, BW90/311, Letter from F.P.G. Hunter (Secretary and Registrar) to Fry and Drew, 20 December 1949, 'Building: detailed correspondence with architects and contractors'.

28 Ibid.

29 Ibid.

30 National Archives, BW90/311, Letter Mellanby to Adams, 1 July 1949, I.U.C. Building & Site, 1949.

31 National Archives, BW90/312, Letter Mellanby to Adams, 30 January 1950. I.U.C. Building & Site, Jan–Jun 1950.

32 National Archives, BW90/312, Letter from F.P.G. Hunter (Registrar) to Drew, 13 April 1950.

33 National Archives, BW 90/311, Letter from Mellanby to Adams, no date.

34 National Archives, BW90 / 312, Letter Mellanby to Adams, 30 January 1950. I.U.C. Building & Site, Jan-Jun 1950.

35 National Archives, BW90/314, Notes from College Building Committee Meeting, 24 June 1950, chaired by Sir Alexander Carr Saunders,, London, University College, Ibadan, Nigeria, June-Aug 1950.

36 National Archives, BW90/316, Letter from A. I. Bowman, acting Registrar to Fry and Drew, 6th December 1950, London, University College, Ibadan, Nigeria, Part 5 1950.

37 National Archives, BW90/316, Letter from Mellanby to W. Adams, 7th December 1950, London, University College, Ibadan, Nigeria, Part 5 1950.

38 National Archives, BW91/311Mellanby, Memorandum to members of provisional council, 6 August 1949.

39 Huxley, *Four Guineas*, p. 140.

40 Unilever Archives, UAC/1/1/2/10 fo. 523, Port Sunlight, Board Meeting of 6 February 1951. Minute No.6147.

41 Ibid.

42 J. M Richards, ed., *New Buildings in the Commonwealth* (London, 1961).

43 RIBA Archive, F&D/18/13, EMF, Ibadan, to JD, Chandigarh, 14th August 1953.

44 RIBA Archive, F&D/18/14. EMF, Ibadan, to JD, Ch. 25th May 1954.

45 Fry explained the almost arbitrary approach of the straight line, 'I may explain my preference for horizontal lines by pointing to the obvious structural reasons for them, but for my particular enjoyment in playing about with lines as I do, there is no logical explanation'. RIBA Archive, F&D/13/1, Texts by Edwin Maxwell Fry 1920–51 'Article – E. Maxwell Fry September 1951'.

46 National Archives, BW90/317, Letter from A. I Bowman, acting Registrar to Fry and Drew 16th January 1951, London, University College, Ibadan, Nigeria.

47 National Archives, BW90/317, Fry to the Registrar, 16th January 1951.

48 RIBA Archive, F&D 14/4, Maxwell Fry's Memoirs, p. 61.

49 National Advisory Council for Physical Training and Recreation, *Memorandum on the powers of local authorities on the Physical Training and Recreation Act 1937*, (1937, London).

50 Ministry of Education (Great Britain), 'Community Centres', (London, 1945), p. 3.

51 Ibid., p. 3.

52 See National Archives, CO 859/113/6, Social Welfare Community Centres.

53 National Archives, CO 859/113/6, Social Welfare Community Centres, *Colonial social welfare advisory committee: Community Centres* (1944).

54 Ibid.

55 Unilever Archives, UAC/1/1/1/2/8, United Africa Company, Board Meeting of 8 July 1947. Minute No. 3569 p.159. The final cost was around £32,000 and the centre was opened on 4th March 1951by the Governor Charles Arden-Clarke, See Unilever Archives ref. UAC/2/20/3/6/1/2, 'Gold Coast UAC News' 15 March 1951.

56 Mark Crinson, *Modern Architecture and the End of Empire* (Aldershot, 2003), p. 148.

57 Nathan Plageman, 'Everybody Likes Saturday Night: A Social History of Popular Music and Masculinities in Urban Gold Coast/Ghana, c. 1900–1970', (Indiana, 2008), p. 211.

58 United Kingdom Trade and Industrial Mission to Ghana, 'Report of the United Kingdom Trade and Industrial Mission to Ghana', (London, 1959), p. 21.

59 Other places were considered for aluminium production, including a site in North Borneo. See Notes initialled A.H.P, National Archives, CO 96/828/5, 18th January 1950, Volta River Survey Part 1.

60 National Archives, CAB 129/57/2, Confidential Cabinet Paper, 12th November 1952, 'Volta River Aluminum Scheme'.

61 Kofi Diaw and Einhard Schmidt-Kallert, *Effects of Volta Lake Resettlement in Ghana-A Reappraisal after 15 Years*, 9.

62 Yuri Smerti, *Kwame Nkrumah*, p. 202.

63 Huxley, *Four Guineas*, p. 157.

64 Amarteifio, D. A. P. Butcher, and David Whitham, *Tema Manhean, a Study of Resettlement*, 5.

65 Robinson, 'Planning in the Commonwealth: Tema, the New Port of Ghana', p. 90.

66 Amarteifio, *Tema Manhean, a Study of Resettlement*, p. 5.

67 Any forays into African housing territory were as a result of emergency rather than general improvement. For example, the government built 684 dwellings in Kumasi when the *Zongo* (strangers) town was demolished following an outbreak of bubonic plague in 1924.

68 Amarteifio, *Tema Manhean*, p. 7.

69 Another village called Sakumo was also located on the same coastal stretch. What became of that village and its residents is not known. Why was Tema picked out for preservation and this seemingly similar village erased?

70 United Kingdom Trade and Industrial Mission to Ghana, 'Report of the United Kingdom Trade and Industrial Mission to Ghana', p. 23.

71 Robinson, 'Tema', clause 6.2.2. no pagination.

72 Robinson, 'Tema', clause 6.2.2, no pagination.

73 A similar situation is presented in Chandigarh, India, and the desire to 'preserve' the adjacent 'pre-Chandigarh' village of Kensal. See Vikramaditya Prakash, *Chandigarh's Le Corbusier: The struggle for Modernity in Postcolonial India*.

74 Graham Tipple, Amole Bayo, David Korboe, and Helen Onyeacholem, 'House and Dwelling, Family and Household: Towards Defining Housing Untis in West African Cities'.

75 A G Onokerhoraye, *Public Services in Nigerian Urban Areas*.

76 See Amarteifio, *Tema Manhean* for details of the counting procedure.

77 Alcock had extensive experience of working in West Africa, being Town Engineer in Kumasi. 1936–45, and then Gold Coast Town Planning Advisor from 1945–56.

78 See Otto Koenigsberger, 'Obituary Denis C. Robinson (1921–1981)', p. 119.

79 H.M.S.O., 'New Towns Act, 1946'.

80 Ernst May, 'Kampala Town Planning', in *Architects' Year Book 2*, ed. by Jane Drew (London, 1947), pp. 59–63.

81 Ruth Glass, ed., *The Social Background of a Plan: A Study of Middlesborough* (London, 1948).

82 RIBA Archive, F&D/1/2, Letter from Jane Drew, 15 January 1954. She seemed to have overlooked the numerous problems of working at Chandigarh in this pitch letter …

83 National Archives, CO96/829/2, Alcock quoted in a letter from G. A. Atkinson to William Holford 19 May 1951,Volta River Survey Resettlement.

84 See Colonial Office, 'Statement of Policy on Colonial Development and Welfare'. Drew also describes a number of their commissions in Jane Drew, 'West Africa'. H.M.S.O., 'Report of the Commission on Higher Education in West Africa', (London, 1945), Advisory Committee on Education in the Colonies, 'Education for Citizenship in Africa', (London, 1948).

85 For other recent work on Tema, please see Michelle Provoost, *Exporting New Towns: The Contextual v's the Universal Approach*, unpublished paper, 2012 and Viviana d'Auria, and Bruno De Meulder, 'Unsettling Landscapes: The Volta River Project', *OASE 82 L'Afrique C'est Chic,* 82 (2010), 115–38.

86 Lindsey Drake had responsibility for some of the work in Ghana – he may have been the lead designer on Tema Manhean before Fry and Drew returned from Chandigarh.

87 See Norman Creamer, 'The Work of Drake and Lasdun of Fry, Drew, Drake and Lasdun; Work in the Tropics', *Architectural Design,* 28 (1958), pp. 55–74.

88 Maxwell Fry, 'Town Planning in West Africa', *The Architects' Year Book no.1* (1947): p. 68.

89 Ibid., p. 68.

90 Ibid., p. 68.

91 Maxwell Fry, and Jane Drew, *Tropical Architecture in the Humid Zone* (London, 1956), p. 107.

92 Ibid., p. 117.

93 Ibid., p. 107.

94 Ibid., p. 107.

95 Fry, 'Town Planning in West Africa', p. 68.

96 Drew, *Village housing*, p. 30.

97 Fry, *Tropical Architecture in the Humid Zone*, p. 109.

98 Robinson, 'Tema' clause 3.3.1, no pagination.

99 Fry, *Tropical Architecture in the Humid Zone*, p. 23.

100 Fry, 'Town Planning in West Africa', p. 68.

101 Fry, *Tropical Architecture in the Humid Zone*, p. 253.

102 Drew, *Village Housing*, p. 72.

103 See National Archives, BW90/1212, 9 June 1948 Circular Dispatch, Building in Tropical Countries, 1945–8. The post was created by the Department of Scientific and Industrial Research, working in collaboration with the Colonial Office.

104 See Iain Jackson, 'Maxwell Fry and Jane Drew's Early Housing and Neighbourhood Planning in Sector-22, Chandigarh', *Planning Perspectives,* 28 (2013): pp. 1–26.

105 See Amarteifio, *Tema Manhean,* p. 6.

106 See Norman Creamer, 'The Work of Drake and Lasdun of Fry, Drew, Drake and Lasdun; Work in the Tropics'.

107 A E S Alcock, 'Changes in the Use of Building Materials and Products for Low Cost Housing in the Tropics', *Planning Outlook,* 6 (1969): pp. 38–45, p. 39.

108 Creamer, 'The Work of Drake and Lasdun of Fry, Drew, Drake and Lasdun; Work in the Tropics', p. 72. According to Trevor Dannat, Fry and Drew originally proposed double pitch roofs in Africa and only started using the monopitch after observing Leo De Syllas' work in the West Indies. Interview with Jackson.

109 UL Archive, D147/M13, Letter from G. A. Atkinson to Charles Abrams, 2 October 1956.

110 Ibid.

111 Articles were published in British journals attempting to seduce and recruit potential planners to work in Ghana, such as, Alan G. Wood, 'Palm Trees and Iced Beer', *Journal of the Town Planning Institute*, 45, March (1959): pp. 117–119.

112 UL Archive, D147/M13, Letter from G. A. Atkinson to Charles Abrams, 2 October 1956.

113 Ibid.

114 See Robert Gardner-Medwin, 'United Nations & Resettlement in the Far East', *Town Planning Review,* 22 (1952): pp. 283–98. Gardner-Medwin also had extensive experience of planning and architecture in the West Indies as a result of his time in the Royal Engineers during WW2

115 His solution was a rehashed version of the aborted Chandigarh plan, complete with superblocks.

116 Volta Rover Project Preparatory Commission, 'The Volta River Project, Vol1', Published for the Governments of the United Kingdom and of the Gold Coast (London, 1956). Ola Uduku, 'Modernist Architecture and "the Tropical" in West Africa: The Tropical Architecture Movement in West Africa, 1948–1970', *Habitat International,* 30 (2006): pp. 396–411.

117 United Kingdom Trade and Industrial Mission to Ghana, 'Report of the United Kingdom Trade and Industrial Mission to Ghana', (London, 1959), p. 23.

118 United Kingdom Trade and Industrial Mission to Ghana, 'Report of the United Kingdom Trade and Industrial Mission to Ghana', p. 23. For a comprehensive survey of self-build techniques in the developing world, see Graham Tipple *Extending Themselves: User-Initiated Transformations of Government-Built Housing in Developing Countries*

119 RIBA archive, F&D/18/17, EMF, 'on the plane to Kano', to JD, 'c/o Shell Oil Co, Singapore', 18 February 1959.

6

Chandigarh and the Tropics Revisited

BACKGROUND TO THE EAST PUNJAB COMMISSION

Gandhi's closing days and the dawn of Indian Independence were clouded by the schism between Muslim and Hindoo that lead to the partition of the Punjab in a welter of blood and horror, with refugees innumerable upon each side, and dislocations everywhere.[1]

The arrival of two Indian gentlemen to 63 Gloucester Place in November 1950 would have gone largely unnoticed by the architects and assistants so accustomed to the international visitors, friends and journalists who frequented Fry and Drew's office. Working on behalf of the Indian government these two visitors were on a special mission to recruit a team of architects and planners to design a new city and to supervise its construction for the newly independent nation. Armed with a list of architects to visit which included Berthold Lubetkin, Peter Shepheard and F. R. S Yorke,[2] the two Indians, a planner-cum-administrator P.L Varma and engineer P.N. Thapar, were also to explore possibilities in France and possibly, Holland, Sweden, Belgium, Italy, Germany, and Switzerland, if necessary. Top of the list in the UK, however, were Fry and Drew, not least because of their tropical experience. After listening to the Indians' proposal, Fry was not enthused and was very reluctant to accept what was on offer. The practice was developing into a successful business with the potential to capitalise on the post-war UK building boom and they had a number of prestigious commissions 'on site' including the Festival of Britain and Ibadan University in Nigeria. A condition of the Indian project was to reside in India for at least three years which would be a major interruption and contradicted Fry's preferred model of setting up a remote office that could be managed from London. Drew, on the other hand was very enthusiastic and following a 'sit in' by Varma and Thapar suggested that they all visit Le Corbusier in Paris to persuade him to reconsider a previous offer made by the Indians. Drew was instrumental in swaying Le Corbusier and his cousin Pierre Jeanneret, and according to Kavinde, 'Le Corbusier would not have accepted the commission to design the capital of Punjab without the initiative and interest of Max and Jane'.[3] After returning to London with the news that their new collaborator was Le Corbusier, the thrilled London employees hoped the entire practice would decamp to India,[4] but this was prohibited by the clients who insisted Fry and Drew only employ local staff with the intention of training them.[5]

This prompted an office revolt, but undeterred, Fry travelled out to India just a few weeks later.[6] Furthermore, (as briefly mentioned in Chapter 5) Fry and Drew felt that the junior partners were not sufficiently experienced to run and manage the practice in their absence and a new partnership was hastily formed with Denys Lasdun and Lindsey Drake.[7] Drew described it as a 'firm' rather than a 'partnership'[8] and Lasdun and Drake acted as office managers with each side of the partnership retaining their own commissions. Fry claimed, in an attempt to pacify their clients in Africa, that 'our firm has been strengthened by their joining us'[9], and although permission to delegate their responsibilities was granted by Sir Alexander Carr Saunders,[10] Kenneth Mellanby regretted not intervening,

> *Fry wrote to me and mentioned this India job in a personal letter, and I realise now that I ought to have insisted that he did not take it at that stage. Unfortunately his letter gave the impression that he was only going to pay a short visit, and would spend a substantial time coming to Africa, and as he said that taking on the Indian work would necessitate Miss Jane Drew remaining in the London Office*[11]

Perhaps Fry did not initially envisage spending three years in India, or underplayed it as a minor commission. Rather optimistically Mellanby contemplated 'whether pressure might not be brought on the Indian Government to delay their side of the work'[12], so concerned was he that Drew would be left in charge. He went on to state, 'her recent visit has shown us that while agreements can be maintained by personal contact with Maxwell Fry, if she is there, they are made in spite of her rather than with her assistance.[13] Despite these concerns, Fry went to India in January 1951, with Drew following later in the Spring after she had finished her work for the Festival of Britain project. Accepting the commission would also mean leaving their children behind again in the care of their nanny.

NEW TOWNS IN INDIA

The Partition of India and the loss of Lahore to Pakistan prompted a new State Capital to be planned in east Punjab, but this requirement must be set within a broader context of new towns and planning within India. Throughout the twentieth century a series of planning and housing improvement ordinances were proposed and executed, along with various Improvement Trusts, PWD projects, prefabricated housing and model solutions.[14] By the 1940s several towns were being designed under Otto Koenigsberger as Director of Housing, who contributed to a number of significant schemes along with consultants from Europe and America, such as Werner M. Moser, Frederick Adams and Roland Greeley, as well as Indian planners such as S. N. Joglekar, Dharam Singh Kler and P. L. Varma (the future administrator of Chandigarh).[15] It was however the Partition and subsequent migration and refugee problems that made new housing all the more urgent and prompted the United Nations Housing Mission to South-East Asia[16] as well as the Low Cost Housing exhibition in Delhi, with model village designed by Jaqueline Tyrwhitt.[17] Planning and new housing was deployed both as a means of improving health and everyday life, and as a tangible expression of progress using the most 'modern' expertise from Europe and America. The new towns were to be tangible metaphors for Jawaharlal Nehru's modern India emerging from her colonial bondage and demonstrative of future ambition.[18] The name of the east Punjab town was to be Chandigarh – literally *Goddess of Power Town*, further demonstrating the nationalist agenda and incorporating a Hindu deity reference into the secular-socialist democracy. Although local architects were preferred (certainly by Nehru, and arguably more suited to delivering the agenda of such a project), the first planner to be appointed for Chandigarh was the American, Albert Mayer (1897–1981) along with Julian Whittlesey, Clarence Stein and Matthew Nowicki (1910–1950).[19] Mayer was already working as a planner in India following service in WW2 and developed a friendship with Nehru,[20] but following Nowicki's untimely death in a plane crash in August 1950 a replacement partner was urgently sought.[21]

The result however, would eventually cost Mayer the commission and, as Pererea explains gave Varma and Thapar the opportunity to conduct their European recruitment tour.[22] The intention was to find a new partner for Mayer, but when Varma and Thapar managed to convince Le Corbusier and the others, Mayer's role was rendered superfluous. Furthermore the Indians wanted a full-time resident design team (something Mayer could not commit to) and losing Mayer would alleviate the financial strain of paying fees in US dollars.[23]

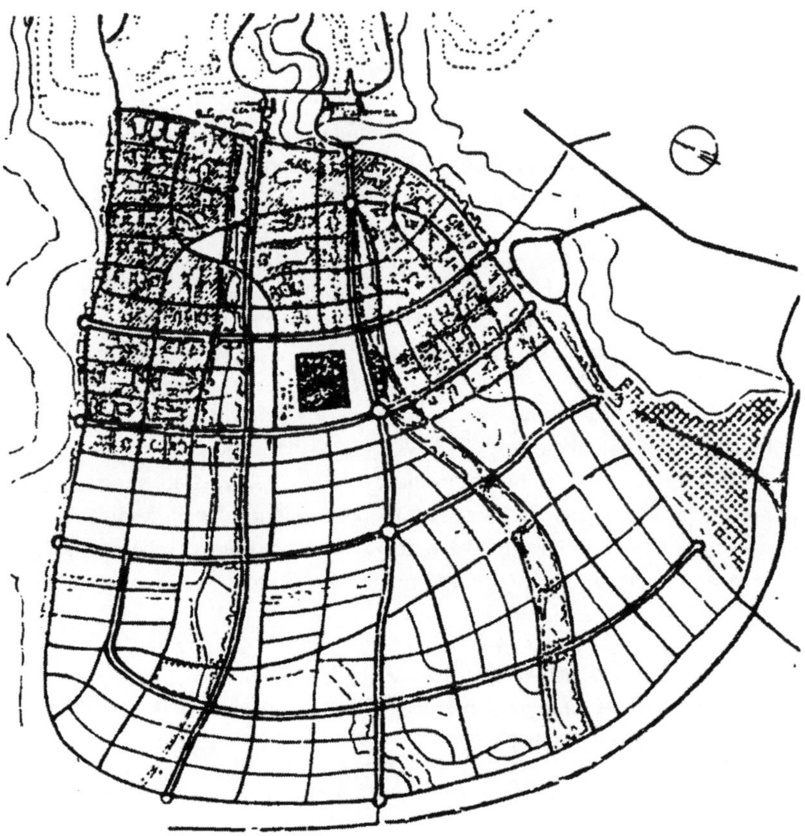

6.1 Albert Mayer's plan for Chandigarh, c. 1950

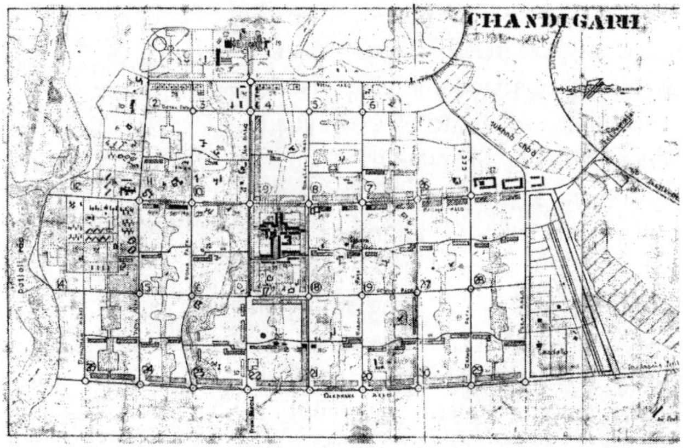

6.2 The revised Corbusian Chandigarh Masterplan, c. 1951

Although in the popular renditions of Chandigarh's story, Mayer's plan has been reduced to little more than a footnote,[24] it was the outcome of substantial research. Mayer had been working in India since 1945 and his contribution to the Chandigarh plan was the product of six months research into low-cost housing.[25] It was this foundation upon which Le Corbusier was able to build his own ideas and *plan* the city in little more than a couple of days.[26] The intention was for the new team to collaborate with Mayer, but as Stein noted, 'what is being built is a compromise with Corbusier, if such a thing as compromising with Corbusier is possible … '[27] Mayer's ideas were quickly eroded and although Fry was called on for support, he felt that the plan was, 'on the sentimental side of things, being an adaption of what is known as the Radburn plan',[28] whilst describing Le Corbusier's plan: 'the sheer audacity of which, took my breath away but one that, from every angle, seemed to answer the basic needs of city dweller, and I accepted it as such'.[29]

FROM MASTERPLAN TO BUILDING DESIGN

With the masterplan drawn up and Mayer effectively forced off the project, the remaining four architects had to decide how they were to work together.[30] Drew described how Le Corbusier was,

> *anxious to define his own role in the project. Namely, that he would be responsible for the design of the Government Centre which at that time comprised of the High Court, The Secretariat, The Assembly Chambers and The Governor's Palace*[31]

That is, the more prominent, civic and 'glamorous' works whilst the others were left with the, 'awful task of drafting the laws and doing the low-cost housing, health centres, hospital, schools, some shopping areas, etc.'.[32] It was an uncomfortable working relationship at first. Drew maintained her passion for the project but Fry was extremely reluctant. His archive reveals a rather strained relationship with Jeanneret, especially when they shared quarters at Shimla.[33] He found Jeanneret to be,

> *A decent man of his type but with fewer mental and cultural resources than ever I met with. He was Parisian as a man might be a Cockney, a man not only limited by his milieu, but unaware of it limitations, and though he had been Corbusier's help-mate for time out of mind up to the moment of his break with him, he reflected less of it than did Sancho Panza of Don Quixote, what though the cap fitted*[34]

Both men were awaiting the arrival of Le Corbusier and were reluctant to make a start on the scheme without him. To pass the time they engaged in Hindi lessons, but as Fry brought along a bottle of whiskey to each evening lesson they picked-up very little.[35] Whilst at the hotel Fry studied the levels and contours of the Mayer plan, and found that the Capitol buildings would not be visible on their ceremonial axis, suffering from the same fate as Lutyen's work in New Delhi. Jeanneret, meanwhile, was according to Fry,

> *found in a corner trying his hand at house plans ... I thought he should be studying the plan to prepare his Master for discussions to follow, but such was not the case*[36]

Fry was anxious to make a start, but the long days with very little to do, coupled with Jeanneret's reclusive company, lead to him regretting the commission,

> *I was beginning to positively hate Clark's, Pierre, Shabby Simla, and myself, and though Jane's arrival was imminent decided on retreat and wrote a letter to Thapur, in which I spoke of Pierre's isolation and lack of communication, of Verma holy man's evasion and of Thapur having not taken the trouble to see me even though he lived only two hundred feet higher up the cliff*[37]

When Thapur received the letter he immediately came to pacify Fry, and managed to convince him to stay. It wouldn't be the first time that Fry would threaten to leave, he found it difficult communicating with the two Indians, but he did eventually grow to like them and associated their qualities of 'toughness and sensibility' almost as being 'Yorkshire traits'.[38]

Le Corbusier was to visit India, 'for a few months of winter in good weather' and as such, had very little to do with the day-to-day running of the project on site, 'the entire thrust of dealing with the Indian situation was left to this English couple [i.e. Fry and Drew]'.[39] Whilst the notion of teamwork and collaboration was theoretically part of the CIAM agenda, in reality the Chandigarh project was highly individualistic. As previously discussed, Fry did not draw any distinction between Town Planning and Architecture and was reluctant to design buildings to fit within another architect's masterplan.[40] In part this principle was maintained as he and Drew were able to plan the 'interior' layouts of each sector, which were not party to the overriding city grid, nor to Le Corbusier's Modulor system, which they also refused to adopt.[41]

Regardless of the design team's bickering, the major requirement of the city (and India generally) was good quality and affordable housing, with rental charges, 'kept in tune with earnings'.[42] It was decided that a typical residential sector should be designed in the first instance rather than designing a smaller amount of houses spread across several sectors. Each housing sector would, generally, serve a specific socio-economic group and this approach is manifest through the size of the houses and the occupation density. Chandigarh was to be an administrative town and as such a large portion of the new population would be Civil Servants and government workers, most of whom were offered subsidised housing.[43] The client administrators apparently researched housing types at Oxford[44] and each rank was allocated one of 13 housing designs, 'to each of which a rather arbitrarily fixed cost had been allotted and was adhered to, making design further difficult'.[45] It is not clear exactly what kind of research was undertaken at Oxford, nor when this study took place.

Chandigarh was to house the Punjab administration with an initial population of some 150,000 people, which it was envisioned would eventually rise to 500,000. In addition to the pragmatic requirements of settlement (the practical, almost mundane requirements to even begin the project such as roads, water, building materials and eventually electricity should not be underestimated), the city was to also embody the spirit of India's Independence. The largest dwellings with the lowest densities (25 persons per acre) were built in sectors adjacent to the Government buildings in Sector-1.

The housing density is, generally, directly proportional to its distance from the Capitol Complex. As the sector number increases so does the density of the housing, resulting in the lower rank government workers located the greatest distance from the Capitol Complex.[46] Some have argued that, because of this, semantically the plan fails to connect with Nehru's vision for a democratic, socialist India, i.e. the poorest (and arguably the most exploited) are the furthest removed from the seat of power – and all based, apparently, on upon a method devised in Oxford![47] However, the greatest housing densities (around 75 persons per acre which is five times greater than the density at Harlow) are found in the Sectors

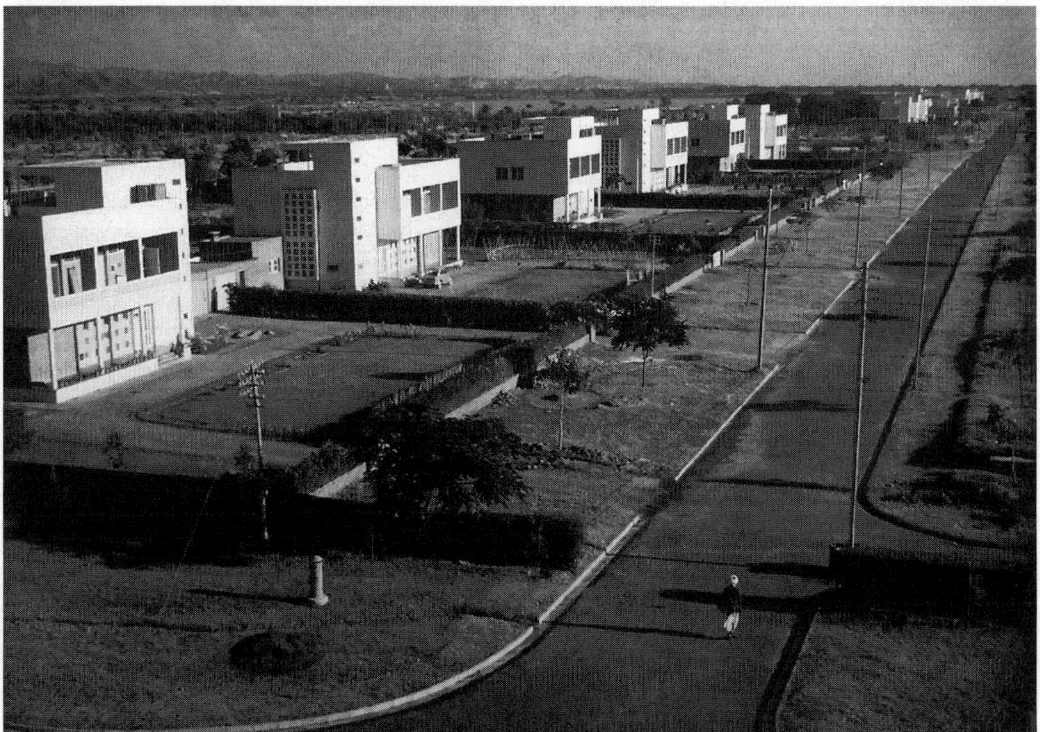

6.3 Large villas designed by Fry and Drew in sector 4, 1954

surrounding Sector-17 (the 'city centre'), providing opportunities for trade and employment, and a quantity of people to support the shops and markets. It would have made little sense to house tens of thousands of people around Le Corbusier's monuments regardless of the political significance of such work. In addition, there are small houses for junior civil service staff located in Sector-7, which is located close to Sector-1 so the plan is not as hierarchical nor as uniform, as it may first appear.[48]

SECTOR-22: CONSTRUCTING THE CITY

'The whole of the sector is developed internally. It is entered at four points only, and all houses are approached from inside the sector and back on to the fast motor roads. Yet no one is more than 400 yards from a bus stop, and within the sector every one may move from house to shopping, to school, to recreation – to all the day-to-day activities in life – in safety and with pleasure'[49]

The decision of where to begin such a large undertaking was governed by practical considerations and decisions made by the PWD before the architects were even appointed, and Varma in particular had a clear strategy for how the city was to develop. From his prior experience in the PWD he was familiar with the management of large projects and outlined a 'Programme of priorities for construction', which stated that, Government officials' and staff housing (including elementary schools and shopping centres) were to be undertaken as soon as possible, followed by

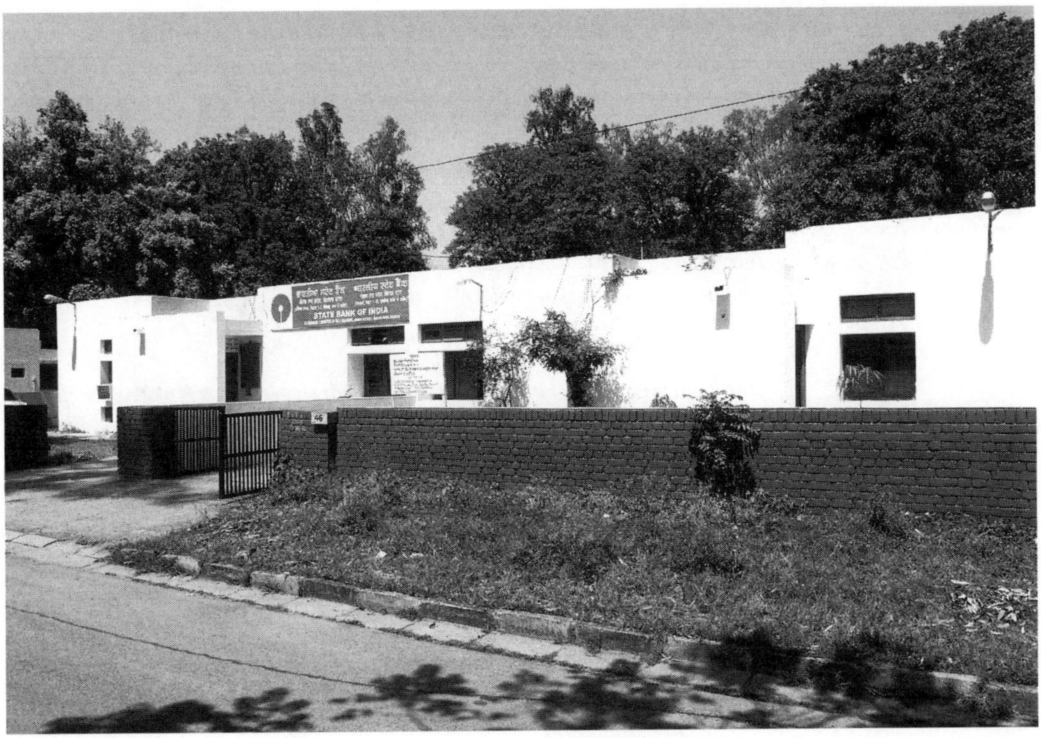

6.4 Sector-7 Housing designed by Fry and Drew, 2012

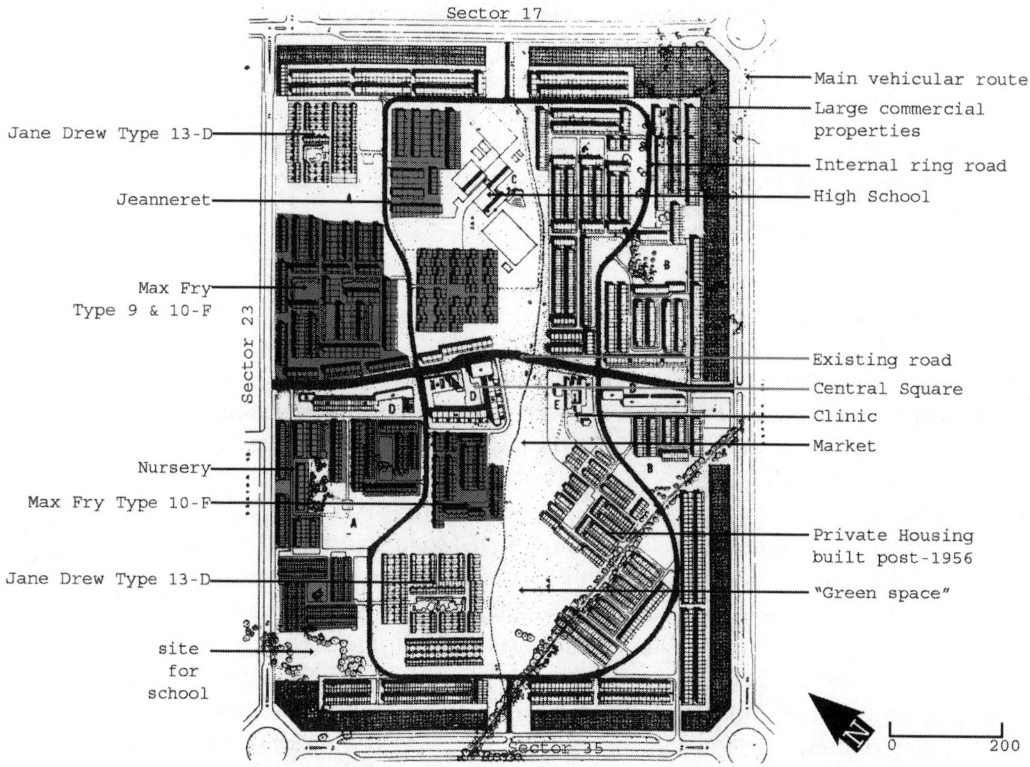

6.5 Plan of Sector-22 designed by Drew

temporary Government office accommodation and then two high schools and a 30 bed hospital.[50] His programme seems to have been closely followed, and set the agenda for the construction sequence. Although Varma was not overly concerned with the architectural proposals, he was part of the committee that agreed to the 'burning of 5 crores of bricks' (that is, 50,000,000) in advance of the project starting in November 1949, thereby largely forcing the architects' decision to use them.[51]

As Sector-22 was adjacent to the proposed bus station in Sector-17 and was, in addition to civil servants, to contain the large numbers of manual workers, clerks and shopkeepers required to establish the settlement, it seemed like the ideal place to start. Drew described it as a,

> *fairly low class sector which … has a large area of open space. It is planned as are all sectors in Chandigarh to look inwards and be fairly self-contained. The traffic roads are round the perimeter of the sector and are designed to take fast moving traffic which is not encouraged within the sector … the greens give a clear view of the Himalayas and contain the educational and recreational features of daily life, that is to say, the swimming baths, nursery schools, health centre, day school and so on, the idea being that, within a quarter of a mile of the dwelling, there should be green and school facilities …* [52]

Each sector has a series of planned open spaces that contain schools, clinics and other such public and community buildings. The edges of the sector take a defensive role with the larger commercial structures such as hotels and large shops 'protecting' the dwellings located within, from fast moving traffic and associated noise. The aim was to create a peaceful 'village-like' environment for the residents to 'move from house to shopping, to school, to recreation – to all the day-to-day activities in life – in safety and with pleasure'.[53] This has been largely achieved and there are adequate and pleasant spaces for recreation within each sector.

In effect, there were two Chandigarhs under construction. Le Corbusier's buildings forming one, and the mass housing schemes for clerks, office workers, and manual trades being the other. Drew was firmly camped in the latter vehemently proclaiming that, 'the first thing to know about Chandigarh is that it is no vainglorious national projection, but a sober necessity for a shattered state gathering its remnants together to consider the future'.[54] Drew was being naively optimistic if she genuinely believed their work at Chandigarh was anything other than a symbolic gesture to the housing requirements of India following Partition. Chandigarh was not a social housing project – but was fundamentally concerned with setting up a branch of government, courts and administration for the Punjab. However, pragmatically, the project did set a precedent and ambition for the minimum standards in Indian housing. The living conditions for many were greatly improved, and Chandigarh was the first city in India where every *legal* house, 'had water borne sewage and a supply of pure drinking water and electricity' as well as toilets and bathing facilities.[55] The Sector layouts for the lower ranks can be considered an extension of the ideas pursued in the social housing schemes Fry and Denby undertook in the 1930s, where they attempted to provide not only the basic housing but other wider community facilities, such as nurseries, clinics, social clubs and vast tracts of 'open space'.

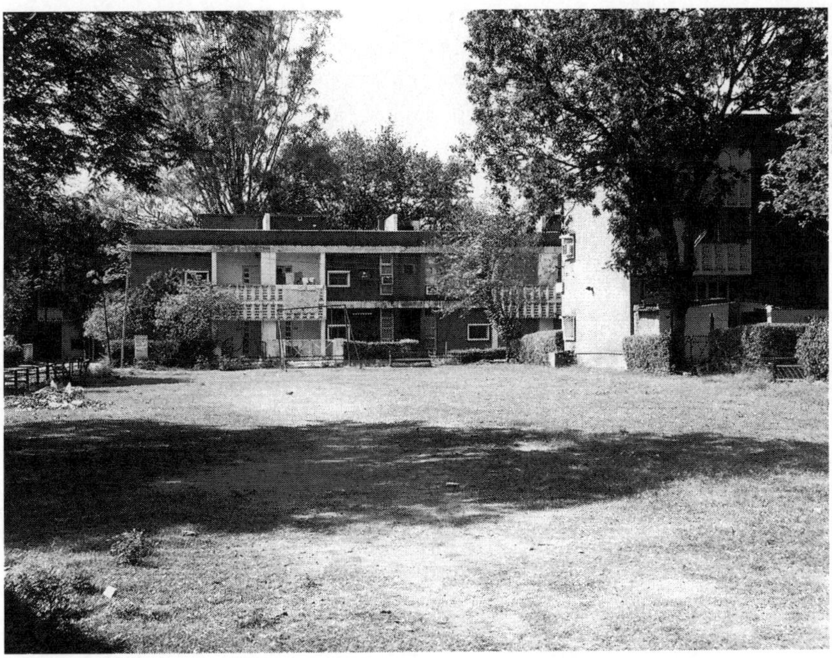

6.6 Open spaces within Sector-22, 2012.

Sector-22 was to be a heavily populated sector and would effectively function as a town in its own right, eventually housing almost 20,000 people. Newly arriving residents and visitors would simply have to cross the road from the bus station and enter into the new thriving settlement, without having to traipse through the building sites located throughout the rest of the city. It was both a practical solution and a public relations exercise. As such, the planning, facilities and housing received more attention from the three European architects on site than many of the Sectors that followed. A hotel was one of the first buildings to be constructed to house the journalists, bureaucrats and architecture-tourists who were proudly shown around the budding city. Prior to the construction of Le Corbusier's sector-1 buildings, as far as the outside world was concerned, Sector-22 was Chandigarh.[56] The success of the city rested on this sector attracting future residents, business and positive press reports.[57] By building one sector as a kind of 'model settlement', the designs and costings could be tested, markets and an economy could be quickly established and a flavour of the city disseminated.

THE NEIGHBOURHOOD UNIT AND THE CHANDIGARH SECTOR

The sector interiors adopt a picturesque approach to planning and are in stark contrast to the formality of the city grid. The masterplan of the city simply left the interior layouts of the sectors blank, to be 'filled in' by Fry and Drew as the city developed. Speed and efficiency of transportation was not the goal here, rather the relaxed, 'pack-horse' meandering of the seemingly un-planned. The influence of these designs can be, in part, traced back to Fry's work from the late 1920s when he was working with Adams and Thompson. Again, *Recent Advances in Town Planning* was to influence Fry's work in Chandigarh. One chapter sets out the criteria for forming a *Neighbourhood Unit*, the quality of which, it was suggested is, 'even more important than the quality of the house, in connection with securing wholesome and pleasant housing conditions'.[58] The outline criteria for forming a Neighbourhood Unit is listed below:

1. Provide housing for a population … for which one elemental school is required, its actual area depending on population density.
2. The Unit should be bounded on all sides by arterial roads sufficient for … 'through traffic'.
3. A system of small parks and recreation spaces … should be provided
4. Institution sites suitably grouped around central points or commons
5. One of more shopping districts, should be laid out in the circumference of the unit, preferably at traffic junctions and adjacent to similar districts of adjoining neighbourhoods
6. An internal street system: suitable for circulation about the unit and to discourage use by 'through traffic'[59]

There is a remarkable similarity between the 'Chandigarh Sector' and the description of the 'Neighbourhood Unit' – the list above perfectly aligns with Chandigarh's Sector-22. Although this has never been acknowledged by Fry; he was extremely reluctant to discuss the projects he did prior to his 'Modernist' work.

The Neighbourhood planning idea was persistent theme both in America and amongst the CIAM cohort in Europe, it was considered a universally applicable method of designing.[60] In the 1940s Drew used the *Architects' Year Book* as a forum for these topics and the first issue of 1945 included an essay by Drew on Housing and one on planning by Jaqueline Tyrwhitt.[61] The thrust of Drew's arguments targeted the 'semi-detached' housing boom of inter-war Britain, but her sketches showing tightly arranged terraced housing set within large open spaces containing schools and civic buildings, strongly resemble what she later proposed at Chandigarh, and also bear a strong resemblance to the garden-suburb-modernist blend.

Tyrwhitt's essay included a plan depicting a South African Neighbourhood Unit that also shares a very close similarity to the Chandigarh Sectors, including the central open space containing the civic amenities and faster roads around the periphery. She and Fry ran a MARS summer school at the AA in 1948 so they were likely to have engaged in debate and exchanged ideas on planning in the tropics, as Fry and Drew had recently published their seminal text on the subject and Tyrwhitt had just edited the popular book, *Patrick Geddes in India* in 1947.[62] Although Tyrwhitt's book on Geddes disseminated these notions more widely (and within the modernist circles that she associated with), his work was well known and had been recently published in various architectural journals.[63] We can also find applications of what Geddes called *constructive surgery* in Sector-22; in particular an existing crooked road that Fry and Drew integrated into their plan of

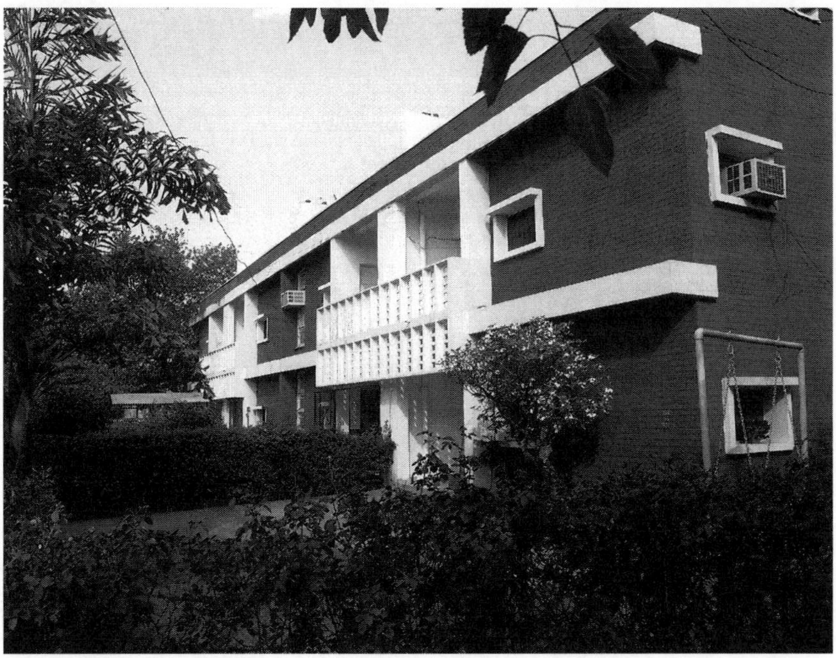

6.7 Housing in Sector-22, set amongst squares, parks and open landscape, 2012

the sector.[64] Whereas in 1950 Mayer viewed the site as a 'blank sheet of paper', Fry and Drew, as in West Africa, became seduced by the *primitive,*

> *beauty is everywhere; inherent; no more in the courtyards than in the swelling tree trunk; no less in the sweetly arching ironwork of the well-head than in the mild-eyed milk white bullocks that wait their turn. All is beauty; timeless*[65]

The retained road in Sector-22 informed the design for the rest of the plan, including the *Chowk* or central 'piazza off which come the tiny traditional bazaar streets' and nearby informal markets.[66] Running centrally north-south through the sector was an open green space forming a common, park and space for clinics, schools and community buildings. The distinct areas of housing were proposed in clusters sharing small access roads, 'with our band of open space in the sectors secured, we planned closely in urban formation, using terraces freely ... '[67] Sector-23 also incorporated the 'leisure valley', and the grid-iron plan was contorted as a result suggesting that they viewed the sector grid more as a guide rather than as a dogmatic and fixed prescription.

Drew had always modestly viewed her work as 'quiet';[68] it was not formalist, or sculpturally expressive. She is credited for the layout of Sector-22 and (as in

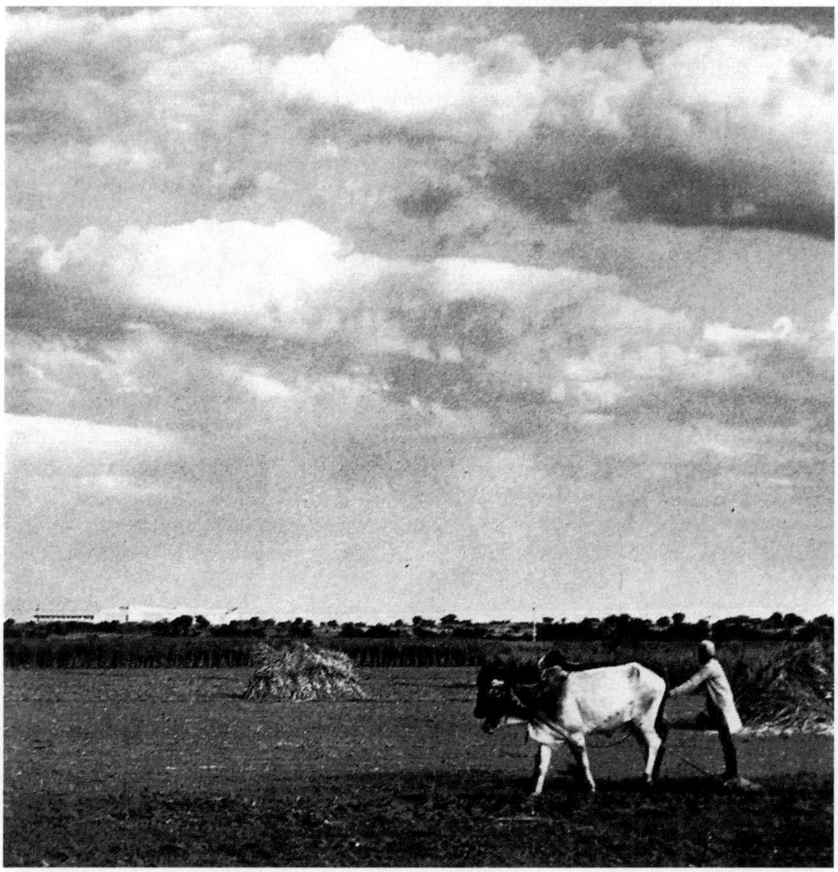

6.8 Timeless beauty or a blank sheet of paper? c. 1954

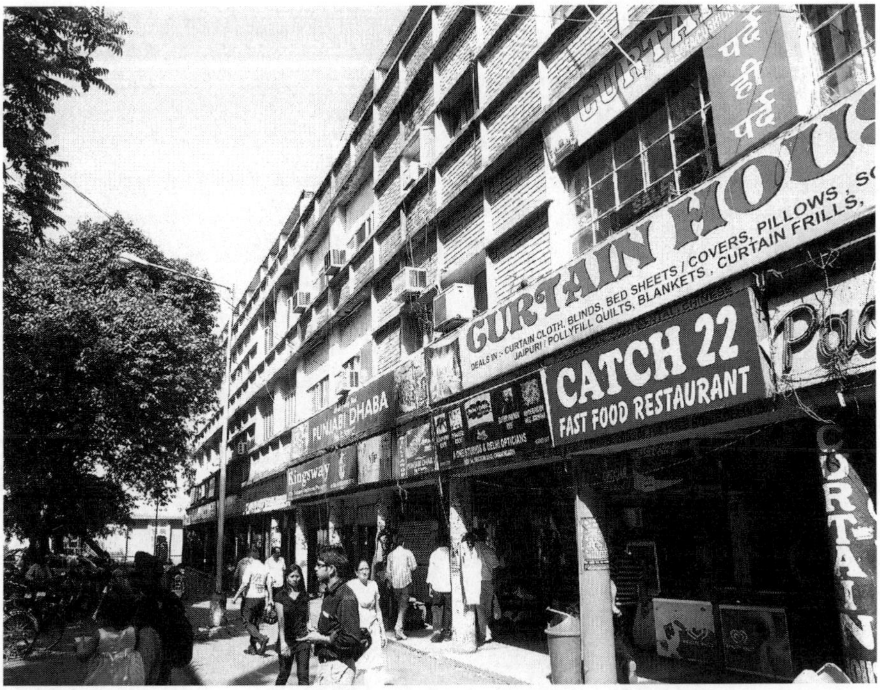

6.9 Sector-22 Shops – a Chowk, 2012

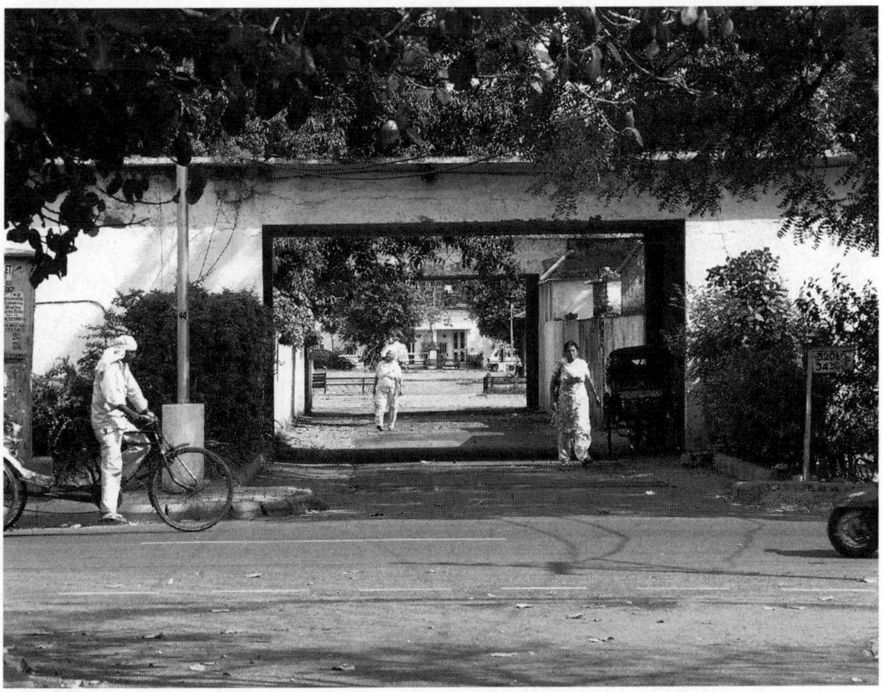

6.10 Housing Clusters, 2012

Africa) took more of an anthropological approach to design, attempting to distil what the clients required and, perhaps in contrast to the appearance of some of her buildings, sought an architecture that was homely and practical. As a result her work has been largely overshadowed by the machismo and more flamboyant displays at Chandigarh.

Minnette de Silva described Drew's humanity, and how she, 'was always deeply committed to the sociology of architecture … '[69] It was Drew who consulted the 'end users' of the city and tried to formulate some useful data from which the designers could derive their solutions. In her draft autobiography Drew recalls how she 'sat with medics for hours trying to figure out solutions'[70] for

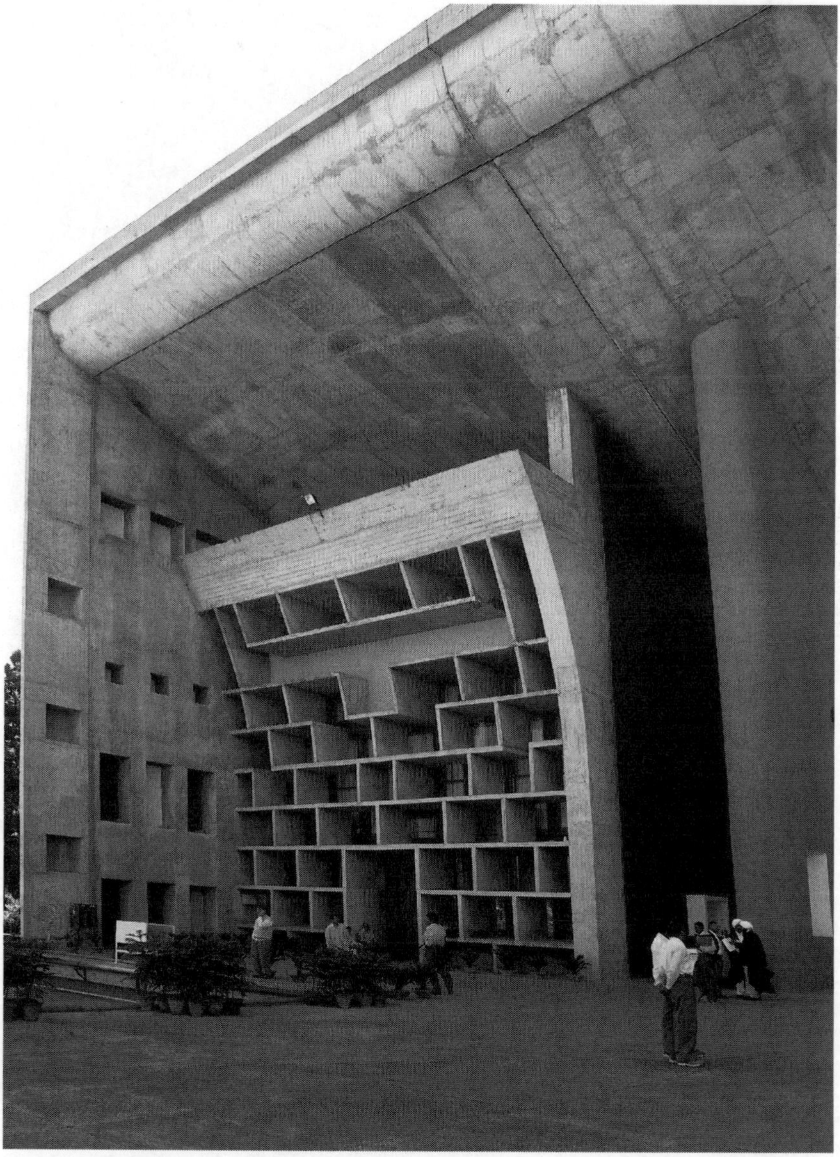

6.11 Le Corbusier's High Court Building, Sector-1, 2012

the Chandigarh hospitals and clinics, and how she consulted with the poorest workers over their needs. Coupled with their previous research into housing and small neighbourhoods, they made further, if limited attempts to respond to the Indian context. Mayer had previously warned Fry that there was very little statistical information available for Chandigarh and that the Indian circumstances 'demanded much more in the way of creative interpretation, or transfusion and synthesis of modern principles and thinking into the Indian scene, present and future'.[71] Without the data or means of procuring it, Fry and Drew set up more empirical studies involving the construction of mock-ups, informal interviews and discussions, as well as being mindful of the severe economic restraints that they faced. Fry expanded on the consultation process and claimed, 'we developed Sector-22, Jane and I, working as none of the others did, directly with the shopkeepers, the cinema owner and all the others concerned ... '.[72] Fry described how the shop designs were a,

> *sort of simple affair they could manage with their own means but conforming with our overall designs, and so successful was the outcome that they willingly built for us with their own money covered ways connecting their colonnade with the booths for the still poorer stallholders*[73]

The result was a combined design effort with the architects acting as 'facilitators' rather than form-makers. In light of this Fry and Drew's work cannot be simply viewed neo-colonialist architecture dressed up in modernist facades. The tropical architecture debate must acknowledge the reciprocal exchanges between the various clutches of client/engineer body, architect and end-user.

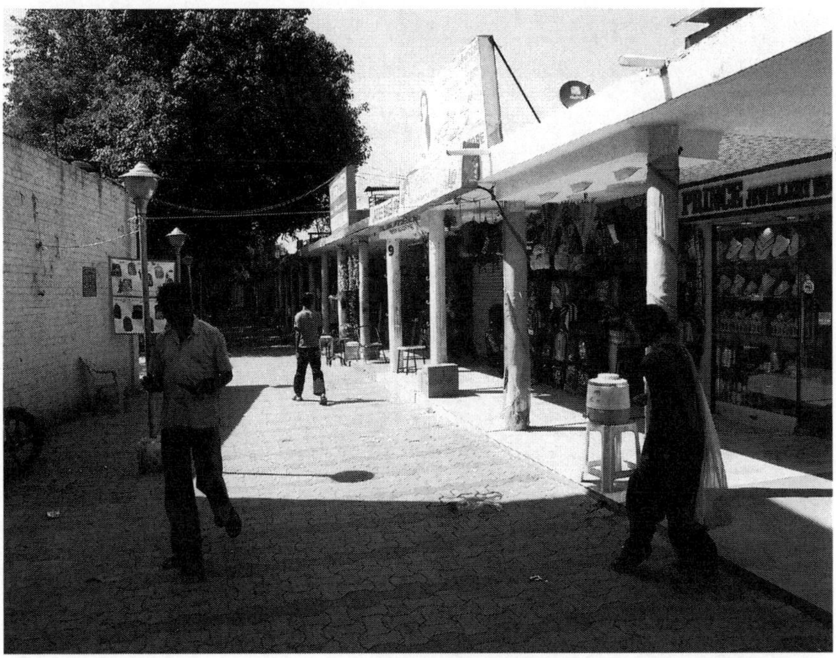

6.12 Shops in Sector-22, 2012

HOUSING AND CLIMATIC CONCERNS

Fry and Drew did not view climatic response and modern architecture to be mutually exclusive explaining that the, 'modern movement was translated and adapted for the sub-tropics with its base rooted in climate'.[74] Again at Chandigarh Fry observed that climate was,

> *the determining factor in Chandigarh architecture, and so it should be. There is no surer way to a suitable architecture, and one that is in accord with the deepest realities of the country: for it is climate that dictates agriculture, moulds customs and affects even religion. Climate is a great element in India*[75]

Chandigarh has six distinct seasons with significant diurnal temperature ranges and but it was to be, 'protection from the sun and from the dust-laden winds of the hot season' that was the 'architectural imperative, the rest was secondary'.[76] Whereas in West Africa their design approach attempted to induce cross-ventilation and reduce the thermal mass of the roof by designing a ventilated lightweight double-skin,[77] at Chandigarh they opted for larger spaces, greater volumes and shading. They avoided facing the dwellings south-west and opted for 'few and small windows on the exposed fronts and no openings of any size not protected by overhanging verandas'.[78]

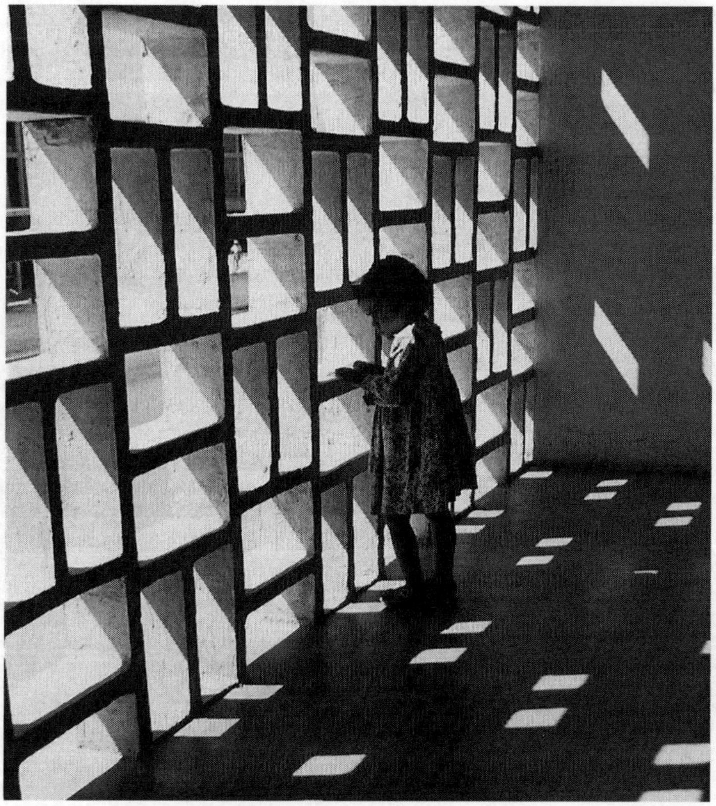

6.13 Shading, ventilation and inexpensive construction, housing in Sector-22, c. 1954

Extremely tight financial controls and a basic material palette also heavily influenced the designs. Availability of land and land-value was of less importance, so if cheaper materials were used, then in theory, larger houses could be supplied, thus helping with the climatic comfort. Brick was used almost exclusively as it was the cheapest available material (as previously mentioned largely due to the government purchasing large quantities in advance of the project). Concrete and glass were considerably more expensive and as a result the size of the window openings was reduced and 'wood shutters and louvers rather than glass windows' were specified.[79]

This palette suited Fry's approach to design. He had rejected the use of concrete as a facing material,

> *while my fellow architects took what I thought to be the easy course in buildings of mass concrete and violent even menacing proportions … I sought for what materials still bore the natural and human impress, using brick, slate, burnt tiles, timber, but of necessity brick in the laying of which the last building skill still flourished*[80]

The initial designs were humble and basic but before extensive construction of a house type could commence prototypes were built and then lived in, criticised and improved. The intention was for the new residents to 'experiment with new types of dwelling'.[81]

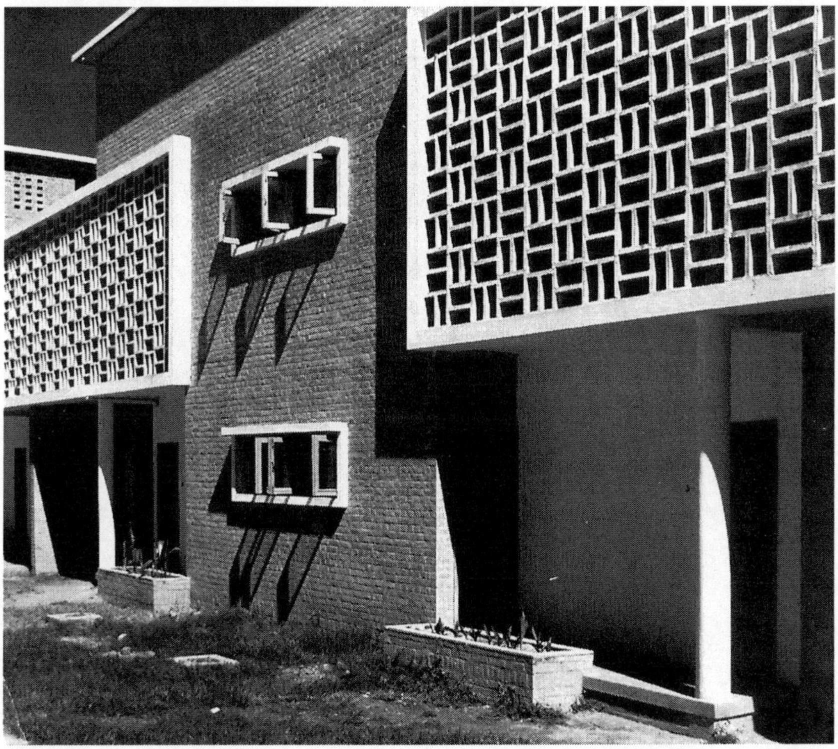

6.14 Prototype Housing in Chandigarh, type 9-F 'traditional living', c. 1954

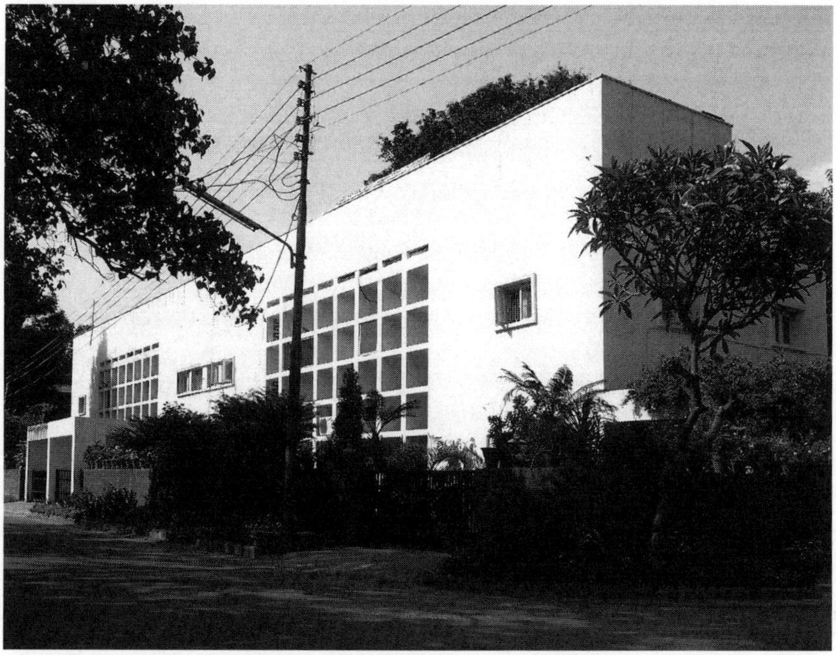

6.15 Prototype Housing in Chandigarh, type 9-F 'modern living', 2012

Fry and Drew were working directly with the future, albeit limited section of the population, empirically testing their designs. There were extensive discussions surrounding so-called 'traditional' aspects of the dwellings, especially regarding the provision of servants' entrances (the lower classes also employed staff), cleaning of WC's, *purdah* screens to balconies and kitchen worktop heights. The 'modern' alternative house layouts simply had fewer passageways for 'sweepers' and no *purdah* screens,

> *it became clear very early that tradition was not important except where it followed the climate and habits of living ... it should be understood that the character was produced from serious thinking about the facts of the situation, not from a wilful desire to be different*[82]

Fry viewed it more of a battle, 'a matter of money and space versus custom',[83] proudly announcing, 'custom lost to a new design without sweeper's doors and passages of the *purdah* screen. Fry and Drew maintained that climate was the main design driver, but *custom* was also seen as something to be eradicated, unless it neatly aligned with their aspirations for the architecture. The preference for sleeping outdoors on the roof during the summer, for example, was 'always' encouraged by the architects and the *barsati* justified their desire for flat concrete roofs. The debate was polarised so that only the 'modern' and the 'traditional' existed; regardless of the rhetoric about climate and sociological study, what they perceived as 'old' or 'custom' was always portrayed as degenerate unless it complied with the modernist agenda of their architecture. In addition, there was not really any experimentation in the housing, as only two options were presented, and neither would have a dramatic effect on how the houses were lived in. Despite this, the mere consultation of a future inhabitant was forward-looking and demonstrative of Fry and Drew's desire to collaborate, albeit in a flawed and narrow manner.

Fry and Drew did not discuss the *kind* of architecture they were producing, preferring to view it as a product of their functionalist response to climate, budget and sociological study. Fry took this very seriously and stated that the, 'integrity of intention is everything, and there is no place for what is meant merely to amuse or to be fashionable'.[84] They made no reference to the other schemes in India, such as those undertaken by Mayer, and the extensive, often innovative work by Koenigsberger[85], nor to the colonial studies of H. V. Lanchester.[86] There was a certain obstinacy to their approach, not wanting to acknowledge external contribution or precedent. Nehru on the other hand, despite his rhetoric of wanting a modern city, had a different approach when it came to housing for the poor, his view was that, 'our cheap housing schemes should be thought of chiefly in terms of providing sanitation, lighting and water supply', before adding, 'we can add to this as occasion offers and resources are available. Even good huts would be infinitely preferable with these amenities than solid construction'.[87]

Drew designed the Type 13 'peon' housing and was convinced that Nehru's low aspiration could be bettered.[88] The modest white rendered dwellings consisted of three main rooms, plus a shower room and WC set within the substantial rear courtyard.

Economy was achieved through adopting a terrace layout and by omitting the roof altogether from the outside WC. A covered external space adjacent to the kitchen could also be used for cooking and laundry during the dry

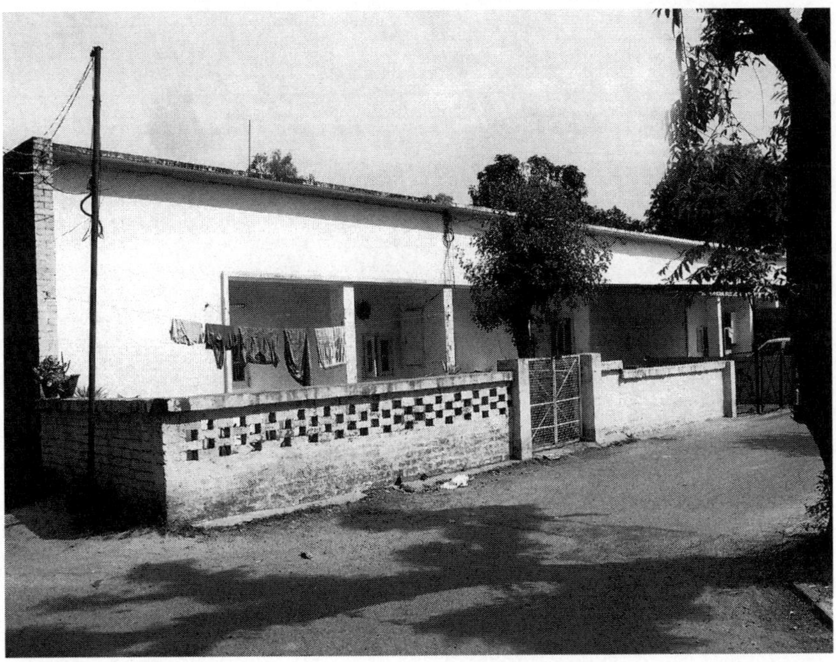

6.16 Type 13 Housing by Drew, 2012

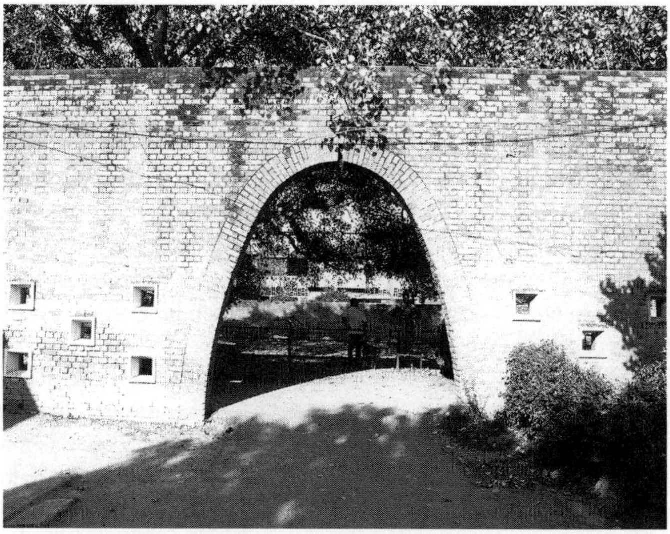

6.17 Type 13 Housing by Drew, 2012

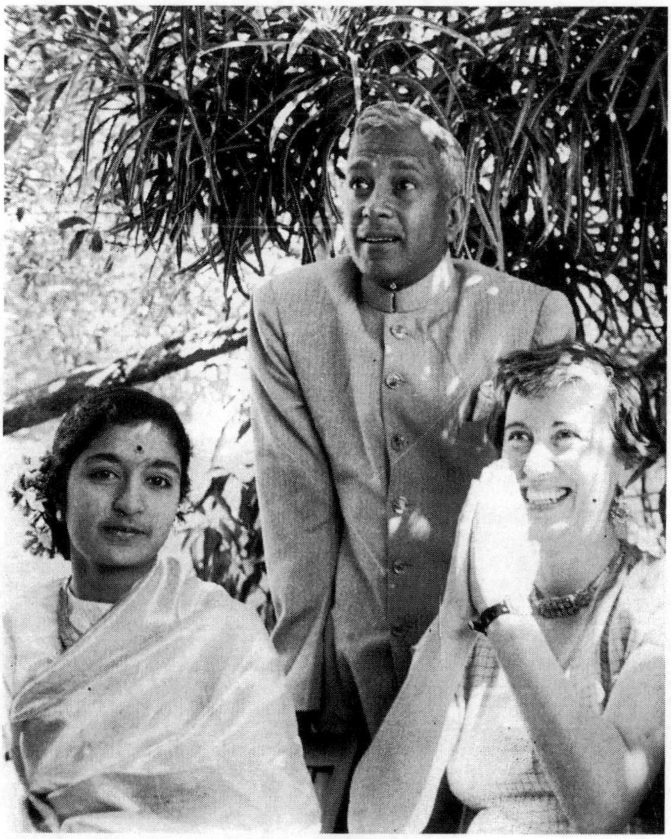

6.18 Jawaharlal Nehru, Indira Nehru and Drew

seasons. Nehru and Drew had a close friendship and he teased her with the half compliment proclaiming it as, 'the only cheap housing he had seen that did not look cheap'.[89]

The houses can be thought of as the *Existenzminimum* of Chandigarh – the most basic, yet efficient response to the problem of dwelling. It was not, however, a new solution to the problem. In 1950 the *Architect and Building News* ran a number of articles on housing in India, including a scheme built in Delhi that displayed a remarkably similar plan, albeit in semi-detached arrangement.[90] It is curious that Fry and Drew did not refer to or cite these solutions. It was however the overall arrangement and the 'village' setting, arranged around pedestrian scale streets, defined by monumental parabolic gateways, that set the work at Chandigarh apart. The intimacy of the 'streets' coupled with the entrance threshold creates a very private, even defensive development that is secluded and personal.

Some of the early prototype housing designed around 'tradition' still remain in Sector-22. The 'tradition' was supposedly expressed on the two-storey Type 9-F housing through extensive precast concrete screens (*jalis*). These were intended to prevent women located inside, being seen by people outside. Here, it was claimed, 'tradition' was informing the architecture of the front façade. It was a technique deployed extensively at Ibadan University with the intention of providing some shade, increased ventilation as well as pattern and relief to the expanses of masonry. Internally, Fry responded to the clients' demands for additional circulation space so that cleaners could access the WCs without entering the main rooms of the house.[91] The servants had a distinct zone within the house (shown hatched on the plans) giving them access to the kitchen, store and sanitation areas.

Cleaners were even given their own entrance to the staircase to enable them to access the upstairs WC without entering the house proper. The intention was to prevent 'contamination' of the house by the lower *caste* servants. Joshi unfairly perhaps, criticised these plans as being 'inefficient', but they were made in an attempt to cater for the clients' needs.[92] As well as providing a 'traditional' solution a subtly 'modern' variation of the type (9-FB) was also developed so that the inhabitants could trial both solutions. The exterior incorporated a dramatic white rendered brise soleil, replacing the verandah and *jalis* screen. A more 'efficient' (cheaper) plan was also possible as less circulation was required and the plot width could be reduced as a result of these changes enabling a greater density as well as economic savings. In effect there was very little difference between both variations internally – there was not a 'modern' or 'traditional' type – just Fry and Drew's narrow interpretations. The variety of housing types required in small numbers at this early stage caused problems for Fry,

> *there were not enough houses of any one type to enable us to design complete districts of our own type houses. Inevitably there was a mixing of interests and only an approximation to a comprehensive design was achieved ...*[93]

Drew also agreed and thought the montage of types and architects was 'one of the failings of Sector-22'.[94] She went on to conclude that it may have been better to

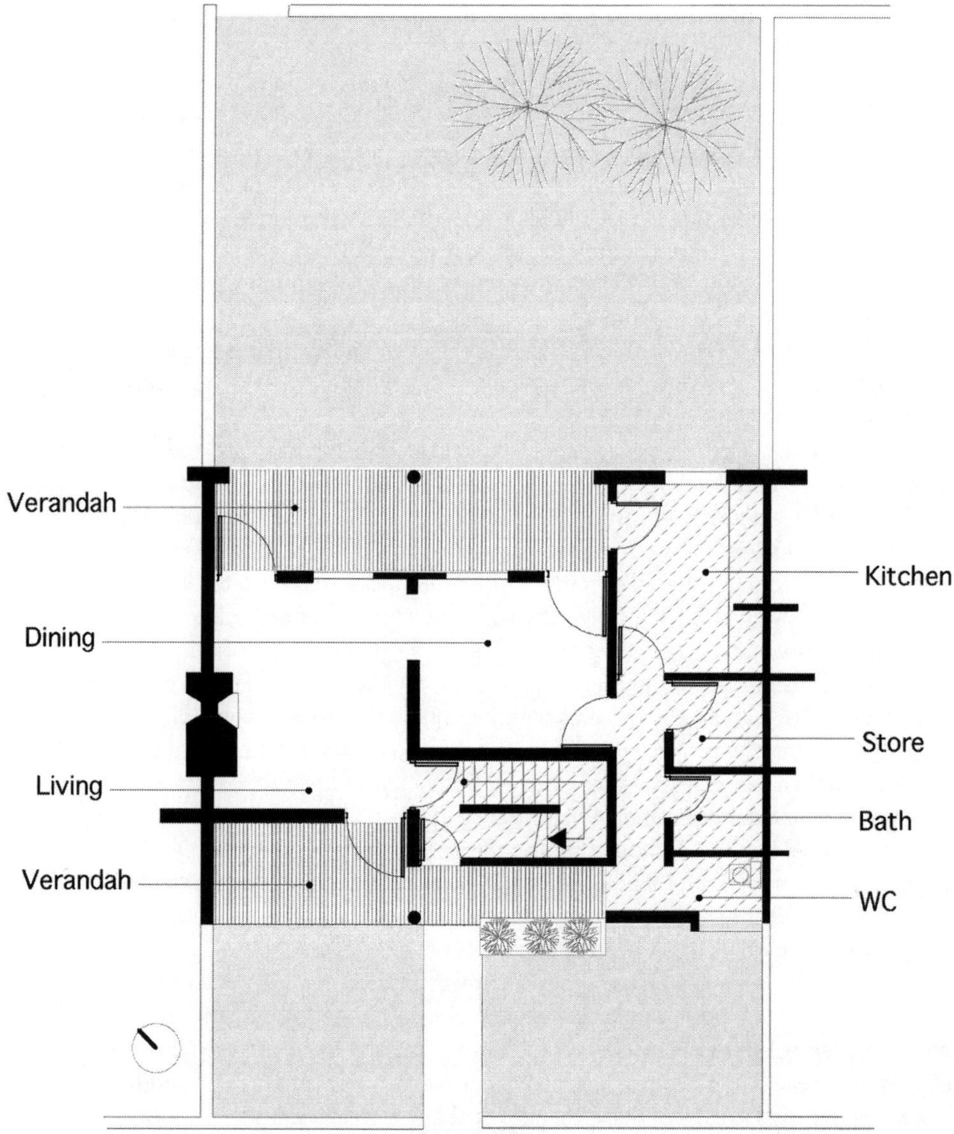

6.19 Type 9f Housing Plans

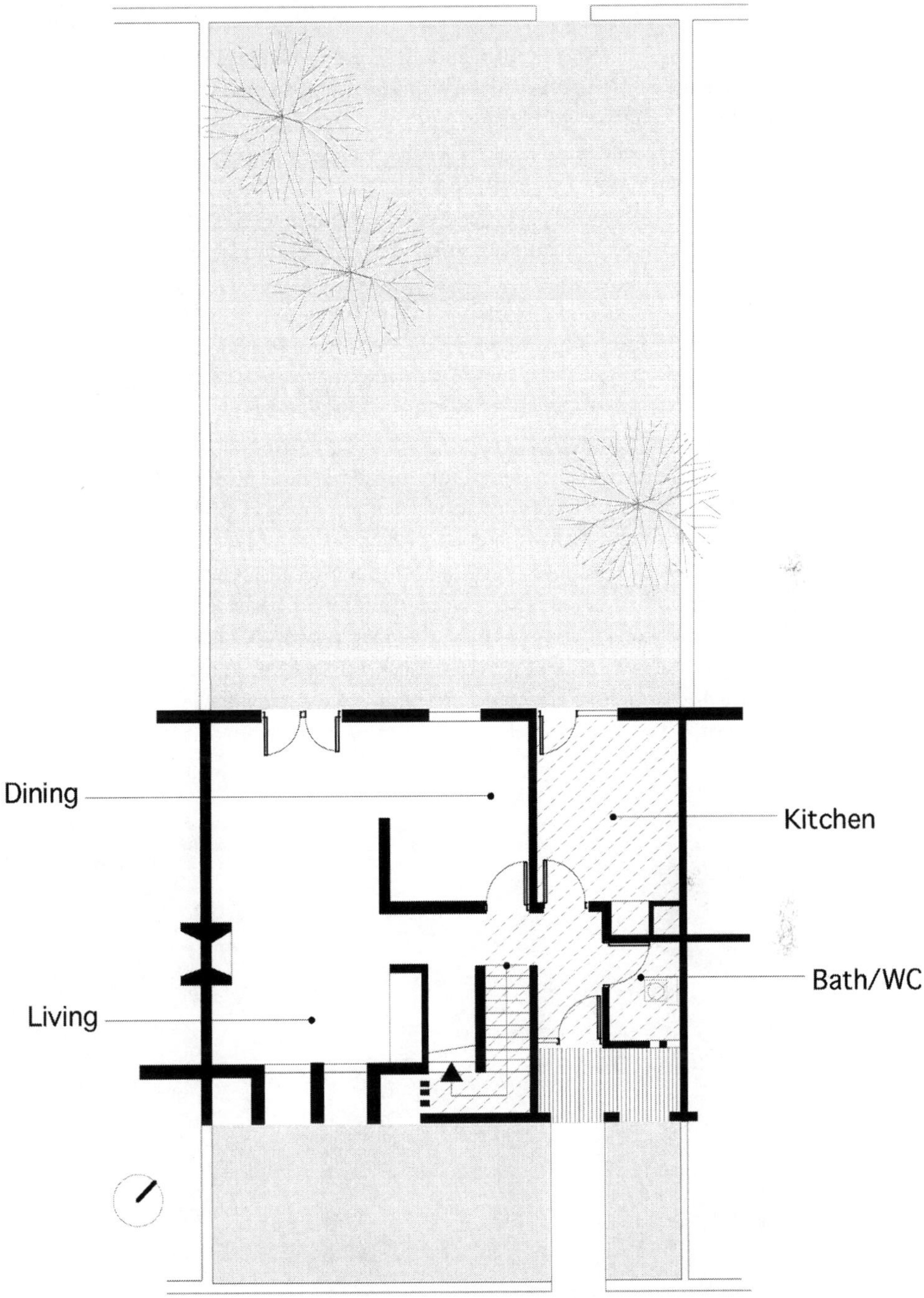

6.20 Type 9f Housing Plans

vary the architect for each sector but not within any one sector. They did not want Chandigarh to be likened to the UK 'New Towns', where many architects were used resulting in, as Fry wryly put it, 'an endless fidgety variety … to make the confusion beyond question'.[95] They viewed house design as being made up of individual units, rather than placing several housing types within one form – which would have been possible with Mayer's superblock. The diversity we now see in Sector-22, however, should be considered a strength and creates some variety as well as identity and 'place', and the experience of the sector is not a motley collection of housing types but a welcome and navigable ensemble of types, residents and styles.

Eventually attempts were made at an even cheaper 'Type 14' house, to cater for the cobblers, sweepers, laundrymen and so on, but even these failed to house the poorest workers, many of whom were artisans. The housing problem was an oversight with regards to budget, 'we tried to make provision for them', protested Fry, 'but in a certain sense we failed. There was no economy upon which we could do it, even with the smallest houses'.[96] This problem was not specific to Chandigarh and was certainly beyond the control of the architects. Although Madhu Sarin has produced convincing arguments about the housing crisis in Chandigarh,[97] it must be offset against the unmanageable influx of refugees at the time and the noble effort that went into providing temporary housing and 'go-downs' by the administration.[98] The UN report favoured self-build because it could remove the labour costs of construction enabling the government to concentrate on providing utilities and raising construction standards. This was taken up in Chandigarh and the Type-14D housing has proved to be highly suited to extension and modification. Other highly innovative solutions have also been developed as Prakash's research into mobile shops revealed.[99]

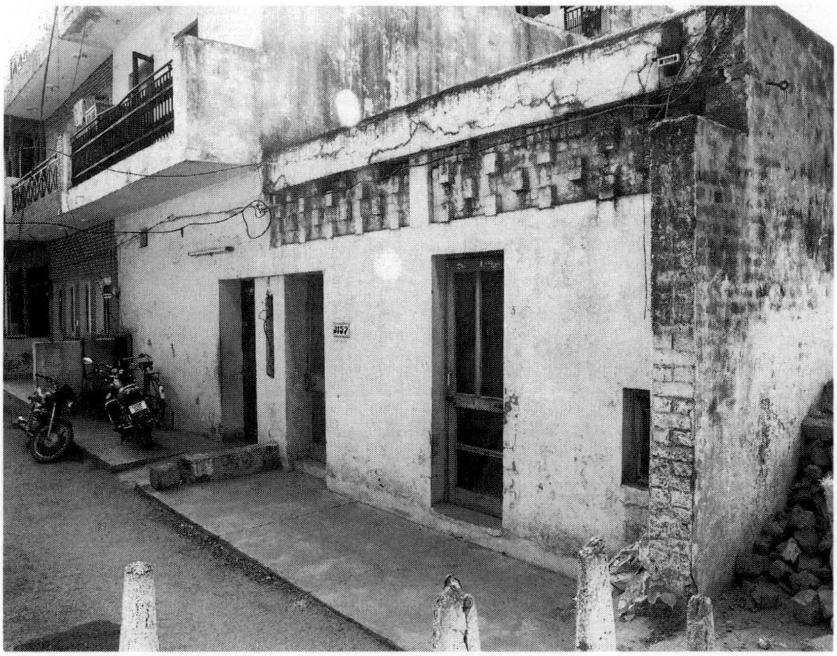

6.21 Housing Type 14D, 2012

GOVERNMENT PRESS BUILDING

A number of other commissions besides housing were completed by Fry and Drew in Chandigarh and many were significant buildings that have been somewhat overlooked. One of the first non-residential structures to be constructed was the Government Press Building.

The Printing Press was an important Indian institution, from the Goan presses of the sixteenth century through to the British Presses in the 20th, they were important places for the production of propaganda, political tracts and faith inspired publications. Anthony King viewed them as colonial building types, they were indeed an outworking of the government, making tangible what was discussed, as well as standardising and setting dates, holidays and festivals. It was a means of imposing a particular order and producing all the triplicate forms required by the excessive bureaucracy. Including a press at Chandigarh was demonstrative of the city's ambition – to be a centre for the production of statutes and political policy. The press also enabled the production of knowledge, the printing of text books – it was a means of controlling and distributing approved facts and histories. The type of architecture chosen for such an institution is therefore loaded with symbolic connotations –what was the printing press of a free and postcolonial India to look like, how would it mirror the political regime that it voiced? Rather than the cosmic and mystical forms preferred by Le Corbusier, Fry proposed a modern and efficient factory building – borrowing from US and European daylight factories and presenting the state as a modern, transparent, illuminated and efficient regime. The workplace was to be no longer subservient to the machine, and dirt and ill-health was to be expelled from the modern workplace.

The building offered an opportunity to test tropical building techniques on a larger scale than dwellings, and the extensive use of expensive glazing was further demonstrative of Fry's belief in his ability to modify climate through design, as well as to indicate the prestige that this building held. The main facade, facing the Madya Marg is composed of floor-ceiling glazing, arranged in three horizontal bands set between the floor plates.[100] Solar gain is minimised by the north-east orientation, whilst gaining maximum benefit from the shadowless north light. The plan is composed of two interlocking square courtyards forming a band of narrow rooms each benefiting from cross-ventilation and access to the open corridors located within the courtyard. A further highly innovative climatic device is the inclusion of adjustable louvers. The entire south-facing facade has horizontally orientated louvers to cut out the direct sunlight, whereas the north facing façade is double skinned with an inner leaf of vertical louvers that again can be manually modified to control ventilation and daylight.

The system enables the building to be completely opened up in the dry months maximising ventilation whilst controlling glare and sunlight penetration, and completely shut down during the monsoon. M N Sharma was given the responsibility for the detailing of the louver mechanism, which is a more elaborate aluminium version of the timber system detailed in *Village Housing in the Tropics*.[101] The building has a similar resemblance to Antonin Raymond's Golconde dormitory in Pondicherry, labelled 'the first modernist building in India'

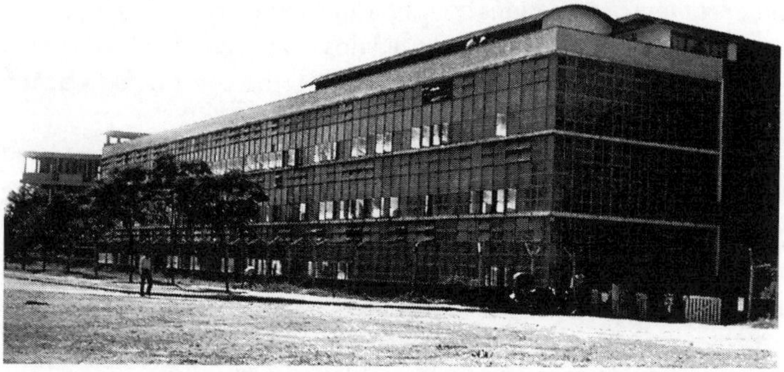

6.22 Government Press Building, c. 1954

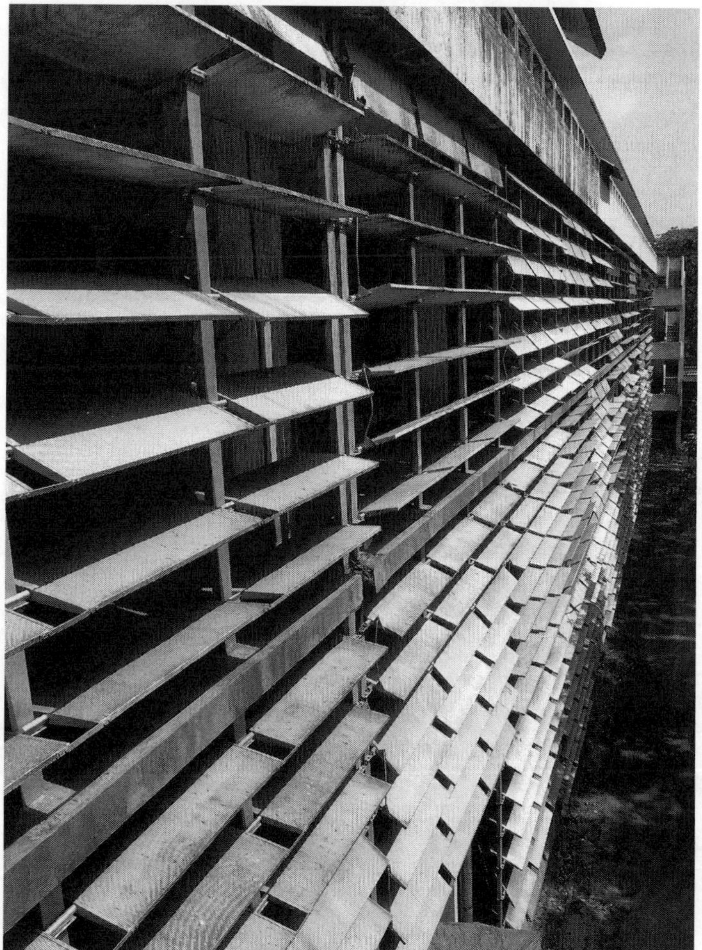

6.23 Government Press Building, Climatic controls, 2012

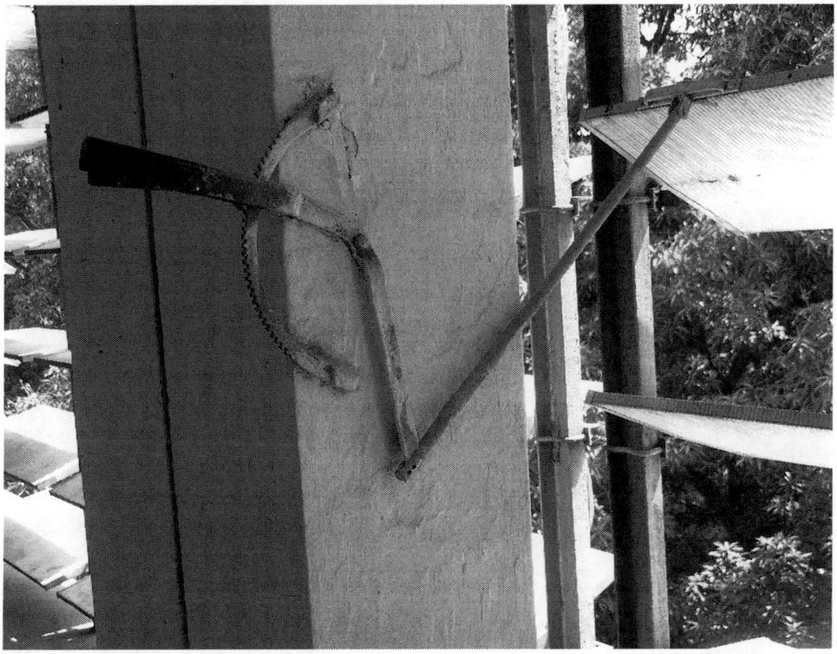

6.24 Government Press Building, Climatic control, 2012

that set the modern precedent of using the louver across the entire façade, and must have influenced Fry's proposal.[102] But whereas Raymond used concrete panels to form the louvers Fry achieved the same feat using the more climatically vulnerable, but daylight enhancing, glazing. The Government Press building is more akin to Fry's 1930s architectural outlook, as espoused in his writings of that time, with the internal linear production process perfectly suited to the sleek factory-like finish. This building shows Fry at his best, utilising new technology, generating crisp forms and details that respond to function, and allowing the proportions of the building components to generate the architectural intent. Fry was attempting to formulate a design based heavily on the Bauhaus model, yet partially adapted to suit the climate with the individual worker making the final adjustments to the architecture. The interior perfectly suited Fry's functionalist ambitions as reflected in the interlocking squares of the building's plan; the distinct processes of delivery, design, printing and binding efficiently flow around the building.

As one approaches the entrance, the glazing gives way to undressed stone (a cheap and freely available building material) and brick, albeit punctuated with Fry's trait square pre-cast concrete window surrounds, and the characteristic cantilevered canopy.[103] There is a grand double-banked staircase leading to offices stepped back from the main parapet line to preserve the crisp rectangular forms when viewed from below, and the ubiquitous modernist roof-top staff canteen with views out towards the Himalayan foothills.

This building could have been a non-descript factory shed located on cheap land at the edge of the city, but instead, it was given a prime location on the Madhya Marg, generously proportioned and expensively detailed. Along with the

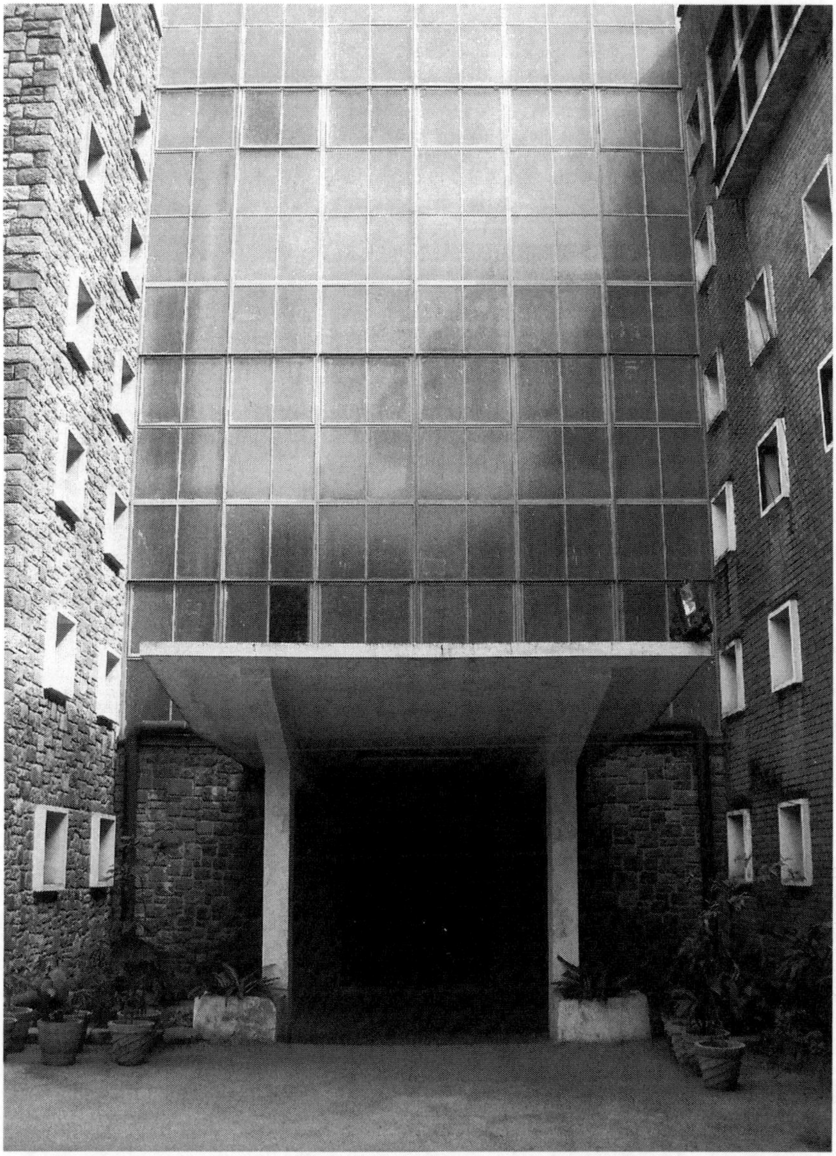

6.25 Government Press Building, entrance, 2012

model housing solutions in Sector-22, these set-pieces were utilised to help 'sell' the newly emerging city. The building is also an outworking of Nehru's socialist agenda, with the modern 'factory' becoming a symbol of worker empowerment and advancement – it is not the palace that is located in the centre of this city, but the proletariat factory. The printing press is also the voice of government, issuing its decrees as well as dispensing knowledge through the production of textbooks. Its clean, open and sleek appearance generates a romantic image of a modern, sanitary and efficient machine working on behalf of a transparent government, for the people.

EDUCATIONAL BUILDINGS

Providing educational buildings had not been a high priority during the British rule of India, but they formed central hubs in the Chandigarh Sectors. Most of these buildings did not feature on the Clients' Programme of Priorities document, so it is likely that Fry and Drew suggested their inclusion and wanted them to be integrated into the residential sectors rather treated as distinct from them.

Taking a cradle to the grave approach, facilities were provided for all ages, starting with kindergarten. Small nurseries were built, designed by Fry, the entire front façade designed to be opened, blending the inside with the outside. The built fabric was reduced to a column and roof and the children were to spend as much time as possible outside in the garden area with the nursery acting more as a pavilion or shelter during exceptionally hot or wet weather. They are also set within extensive gardens, now bordering on jungles, as places for exploration and discovery.

Drew's Model Secondary School in Sector-23 was initially designed as a Boys school. It is arranged around a large central courtyard which is shaded by the buildings, and to aid cross-ventilation the building is just one room deep. The most distinctive aspect of the school is the entrance, with its pre-cast concrete screen, by then a famous Fry and Drew motif used previously on most of their education buildings in Ghana and Nigeria. The screen creates a secure interior that benefits from the breeze and light, whilst being quick and cheap to construct. Unlike the schools in Ghana however, there is no ceremonial axis focusing on a congregational space. Instead a cloister with internal courtyard garden provides shade and enclosure as well as protection from the dust-laden winds.

Drew also designed the Girls school equivalent in Sector-18, constructed slightly later, which benefits from some additional features not found in the earlier rendition. Unlike the school in Sector-23, it is not arranged in a courtyard fashion, rather two long strips of classrooms are placed perpendicular to each other, giving each classroom a clear view out onto the playing fields and auditorium. Climatically, this is advantageous as it resulted in a very small south-facing facade, and all the rooms face north or northwest, using the semi-exterior corridor as a sunbreak. It also results in a successful series of spaces where students can gather outside of their classrooms protected from the sun and monsoon, whilst still having good views out into the school grounds. There is an open air theatre, and outside classrooms that benefit from being in the shade of the building, as well as being subtly enclosed with a curved concrete wall.

Both school designs, but the Girls' School in particular, attempt to remain 'open' and 'outside'. Visually, if not physically, the spaces are connected to the outdoors with an extended view wherever possible. This was part of the drive for health and fitness that accompanied the modernist social agenda, and schools were seen as key places for modelling healthy living exploiting all opportunity for light and exercise.[104] It was in the later model that a more radical approach was taken with greater emphasis on outdoor teaching. The open air school movement was in full swing following WW1 in the UK and by 1937 there were over 150 open-air

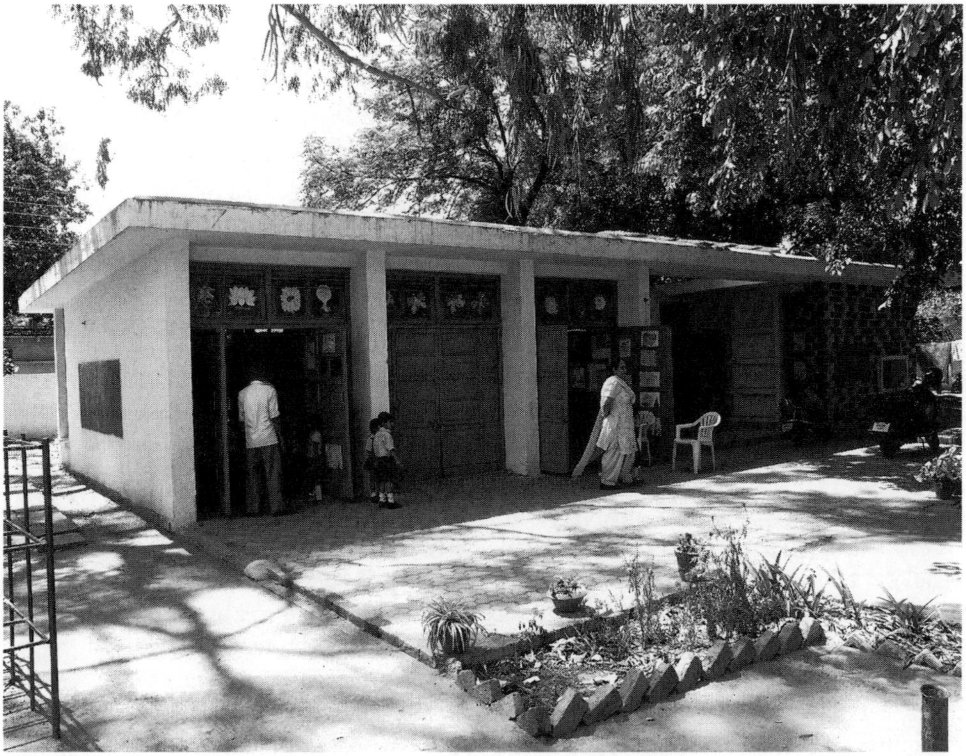

6.26 Government Nursery in Sector, 2012

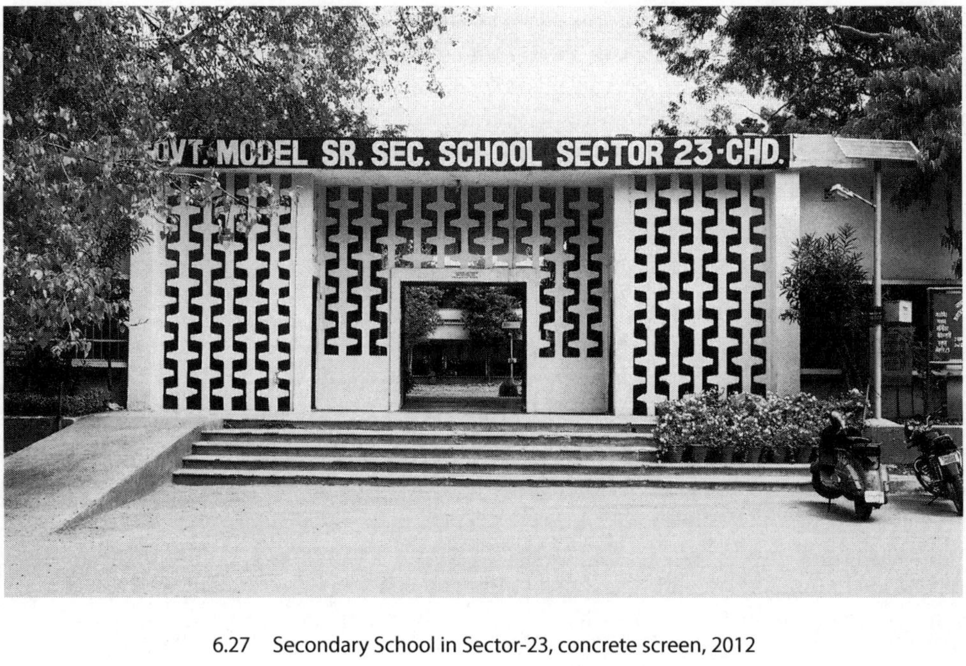

6.27 Secondary School in Sector-23, concrete screen, 2012

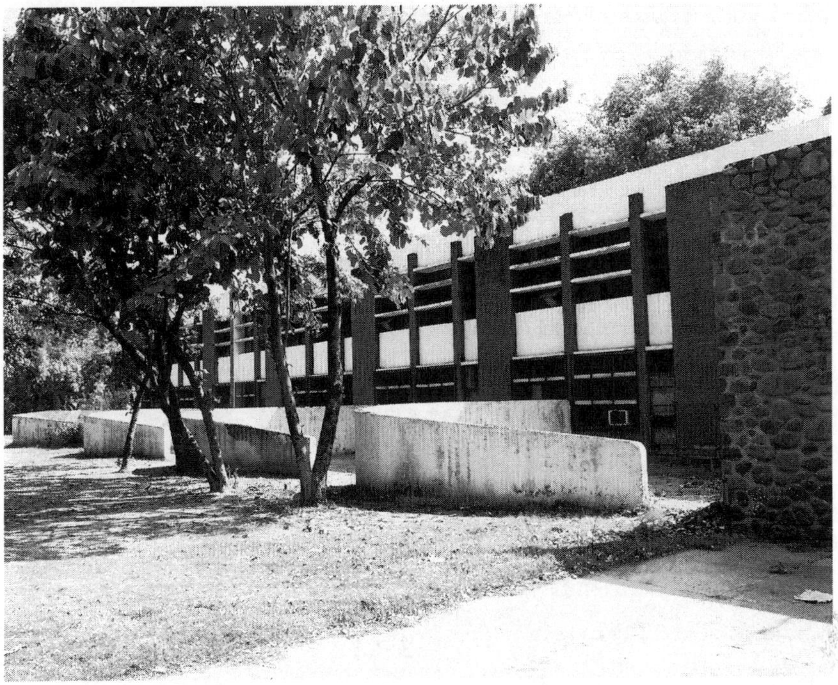

6.28 Secondary School in Sector-18, 2012

schools in the UK,[105] set up in an attempt to counter TB. Coming from a medical family and taking a keen interest in health, Drew would have been aware of these outdoor, sanatorium inspired schools. They were not just for sick children, but were becoming part of the everyday educational process and the school architecture was modified to facilitate this. It is likely that Drew with her concern for liberty, education and anti-institutional tendencies was driving these arguments and attempting to re-think how education could take place beyond the classroom. Unlike in West Africa where the schools were remote from the towns and villages, at Chandigarh the schools form central hubs within the sectors and were deployed to unite and instigate a community made up of newly arrived and diverse population groups.

The notion of the school under the tree was more generally accepted within India, whereas in Africa, the 'bush-school' was rapidly considered outmoded and unsatisfactory. Furthermore, both of Drew's schools are located next to other significant institutions that could supplement the formal education process, such as open–air theatres, yoga centres, gardens and other smaller museums and collections. Education was to be a blend of exercise and exposure to cultural institutions preferably all experienced in the open air.

There was significant investment in educational building at Chandigarh, responding to the lack of opportunity that had previously existed throughout Punjab. In addition to its role as a government administrative centre, there was also a desire for Chandigarh to excel in scholarly activity. The two colleges in Sector-11 (one for each gender) are indicative of this drive, and both were treated almost as independent university campuses with accommodation set within cloistered courtyards. Fry designed both colleges and the influence of West Africa is apparent in both cases, as is the English quadrangle model.

Two-storey brick structures are extensively deployed supplemented with a projecting concrete grill on the facades. Whereas in West Africa the precast concrete elements were small scale and usually formed part of the balustrade here

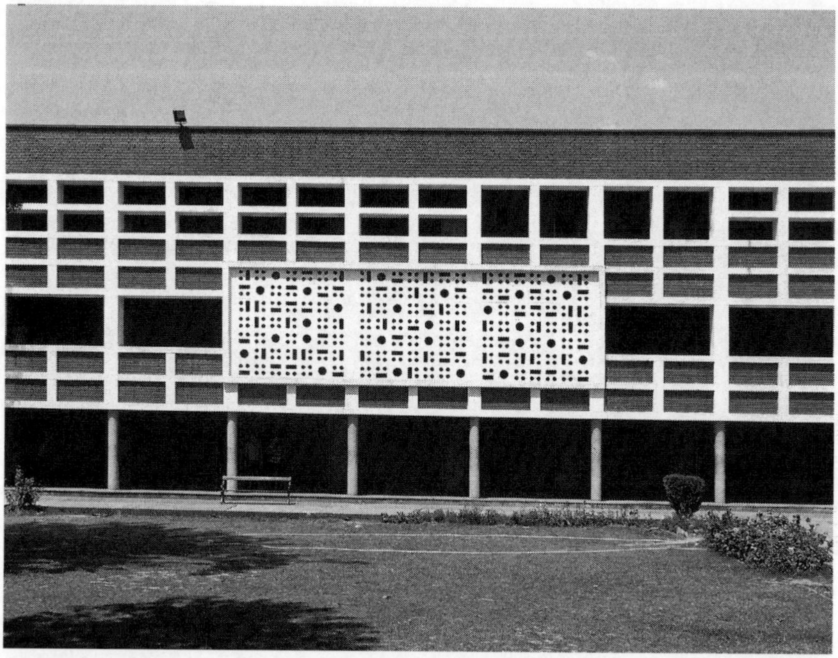

6.29 Government Colleges in Sector-11, 2012

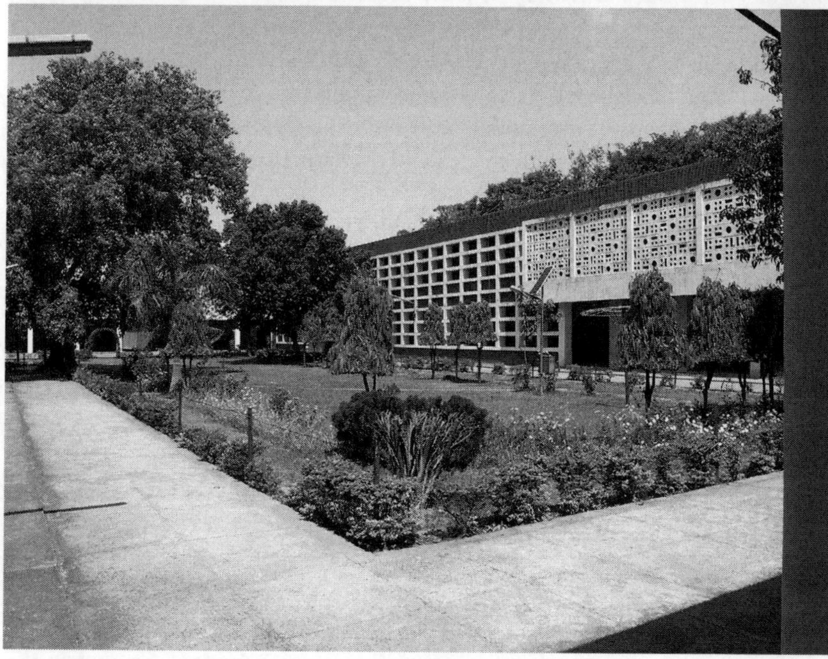

6.30 Government Colleges in Sector-11, 2012

it is exaggerated and transformed into a decorative motif, perhaps its restraint a response to Fry's own critique of his work at Ibadan which he thought had 'too much decoration … too much lace'.[106] Gables were built from random stone and other walls were fully rendered, again providing opportunity for sculpture and applied decoration.

Since his work in the 1930s, Fry consistently demonstrated his ability to take a particular motif, such as a balcony, sun-breaker or window surround and to replicate it across a facade, with a soothing, almost metronome-like quality. At a particular moment he breaks the monotony, and introduces some syncopation with an entrance, or by the use of a decorative counterpoint to the otherwise fiercely geometric arrangement. The Women's College, for example is dominated by the striking white rendered library building. Each level steps out above the one below and is architecturally given its own treatment. The sunbreakers are not purely utilitarian but are arranged as geometric components of a composition, setting up a pattern that moves from columns at ground floor, through punchy square window surrounds at first level, and then a further projecting grill at the top. These architectural devices each initially developed to provide shade, encourage cross ventilation or to raise the building off the ground, are here used, in part, as decoration and to create a bold and prominent architectural composition. Whereas the library at Ibadan appears to be draped in a fine lattice of concrete, at Chandigarh it is hewn from a solid block and the climatic components of tropical architecture have become something of an applied style.

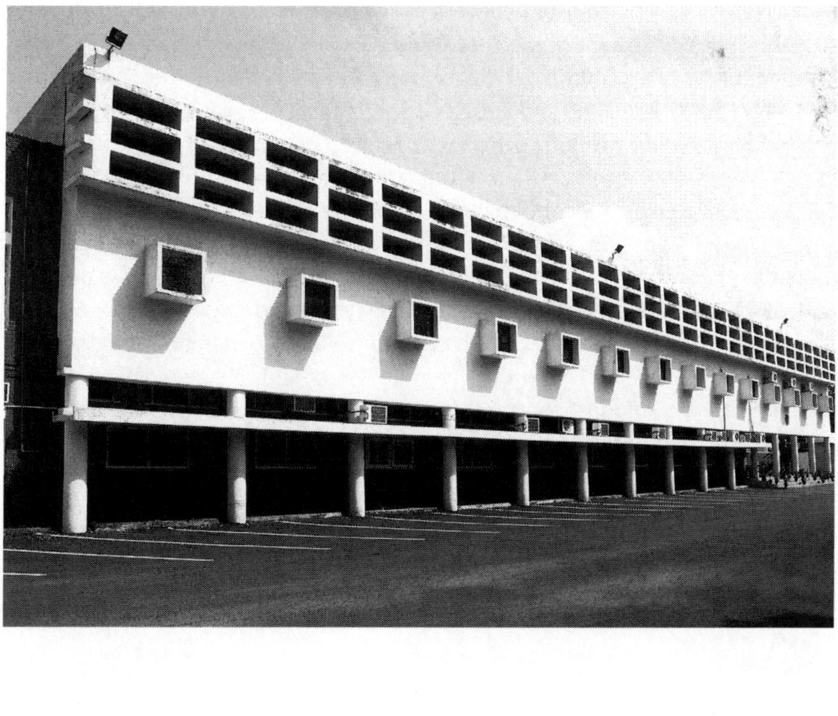

6.31 Library and the Women's' college in Sector-11, 2012

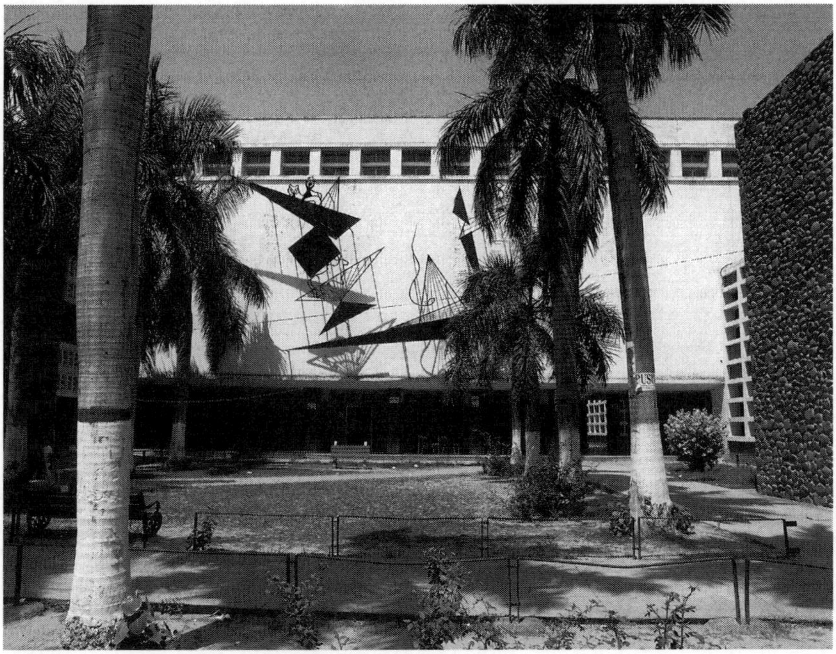

6.32 Sculpture on the Boys' College, 2012

HEALTH: HOSPITALS AND HEALTH CENTRES

Drew, with her prior experience and interest in health-care took primary responsibility for the hospitals and clinics of Chandigarh.[107] Health was to be improved not only through improved domestic sanitation but also through education and exhibitions. The health centre was offered as a new building type to facilitate this, replacing the converted dwellings previously used for doctor's surgeries – it was to be a place that could be visited for education, advice and, importantly, as a meeting place for women outside of the domestic setting. A dedicated maternity clinic and Tuberculosis centre were included along side an exhibition hall and lecture theatre. Each part of the centre had its own entrance to enable informal circulation through the building as well as for discrete access. The aim was to create a centre that was accessible, as well as ensuring maximum ventilation and sunlight – this was not the unregulated domain of the private practitioner but a dedicated and profession setting that would resemble the stark and clinical operating theatre. Drew particularly enjoyed these projects; working directly with users and clients, and through consultation generating solutions that respond to their needs. The health centre in Sector-22 is characteristic of Drew's restraint, the only flourish being the concrete butterfly canopy over the main entrance, supplemented with artworks in the interior.[108]

From this building she went on to design the General Hospital in Sector-16, a large project for such a small team of designers. A basic architectural vocabulary of rendered walls and undressed stone was deployed, as well as a precise and dramatic concrete screen set into the main block. The plan aims to maximise ventilation and views out and is arranged from a central spine corridor with the various wards and departments running perpendicular from it. A nurses' hostel is built adjacent to the hospital with exterior access walkways and shared living-kitchen facilities. The architecture is basic and austere, barely venturing beyond the rudimentary need of shelter, but governed by efficient circulation and arrangement of programme.

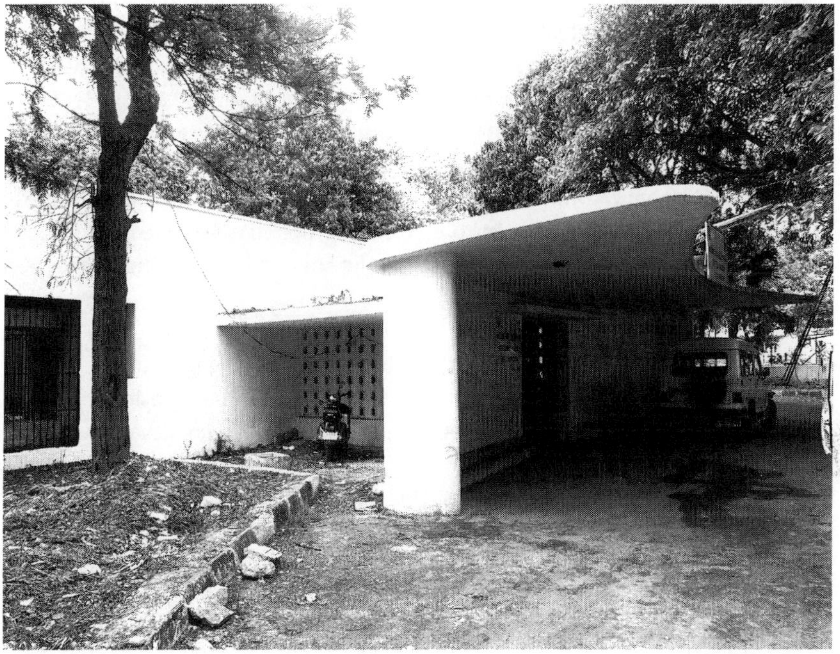

6.33 Health Centre in Sector-22, 2012

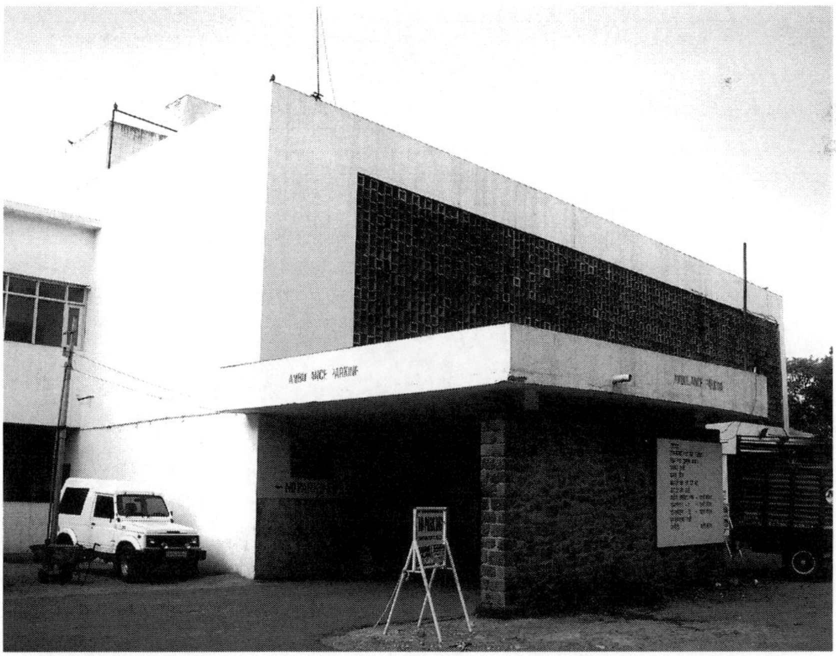

6.34 General Hospital in Sector-16, 2012

WORKING WITH LE CORBUSIER

Only Le Corbusier's High Court building had been completed in Sector-1 before Fry and Drew left Chandigarh, but they did see his other buildings when they later returned to India.[109] Fry described the High Court as a, 'bold conception embodied in a design of great nobility and fine proportions', before commenting on its supposed climatic abilities, 'a single high-arched roof, a parasol indeed, under which the building sits, protected further on the sun-side by finely adjusted vertical and horizontal sun-breakers'.[110] Writing about the project much later in life, Drew recalled how, 'the shuttering on the High Court despite Corbusier's pretty careful detailing had been badly done, supervised by Jeanneret'.[111] The casting was more akin to rough structural work rather than a final finish, justified by Le Corbusier as being like the, 'wrinkles and the birth marks, the crooked noses, the innumerable peculiarities' of people, further excusing the finishes as being, 'human; they are ourselves, our daily lives … '[112]

Jencks claimed that Le Corbusier's work at this time permitted, 'a certain amount of interpretation and execution by others',[113] implying that not everything was resolved, or complete. Indeed, the work in Sector-1 has the feel of being abandoned and ruin-like. Fry felt that the presence of the Himalayas, forming a backdrop to the Sector justified Le Corbusier's 'imperious dimensions',[114] and that whilst some of his closer friends thought 'he had exceeded the limits within which individual buildings communicate with one another. I know that he meant to go to the limits … '[115] The idea was that the Assembly Building and Secretariat (where the laws are made and administered) directly facing the High Court (where the laws and justice are dispensed), would form a kind of visual equilibrium, offset by

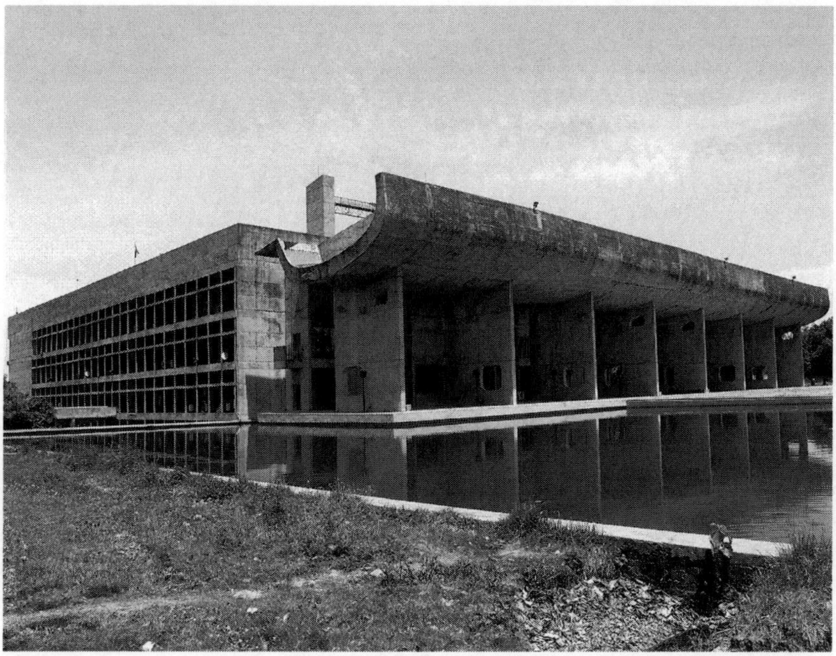

6.35 The Assembly Building in Sector-1, 2012

the (unbuilt) Governor's Palace, and a series of monuments and landscape imbued with political meaning and metaphor.

Fry acknowledged that in buildings of this type there is always a, 'justifiable element of exaggeration … and this fact could justify the use of brute concrete as being elemental and immemorial',[116] and felt that within the 'framework' of Sector-1 the, 'buildings are superb monuments, especially the High Court' and he accepted that the use of concrete was appropriate. However, he stressed that, 'nobody is going to make me admire an untreated face of inert concrete. What my eyes tell me I will respect. Nor can I agree that the interruption of official consequence in the form of the balconies and recesses in the length of the Secretariat Building is architecturally successful.[117] Although Sector-1 was considered a political landscape, with deep, cosmic philosophical intent, Le Corbusier described it to Drew in more profane, wanton terms, with the 'the little tufts of bushes set in hollows or hillsides he described as being inspired by the delight of hairs in women's armpits'.[118]

Fry and Drew had mixed, even contradictory views on Le Corbusier, and his work. Whilst Le Corbusier was highly suspicious of Jane Drew when they first met, they developed a close friendship at Chandigarh.[119] In a candid and revealing account of Le Corbusier's practice, Drew explained how Le Corbusier would start work whilst in Chandigarh, ' … at nine having breakfasted he went to the office … he used the expression "serrez les fesses" meaning squeezed his buttocks together in concentration', unorthodox, but perhaps an effective method for standing at the drawing board for long periods.[120] When in Chandigarh, Le Corbusier tended to work alone and the relationship with Jeanneret was still not sufficiently reconciled for concerted work, 'he did not collaborate with him [Jeanneret], though he told me he liked to collaborate preferably with a woman'.[121] Drew would go for long

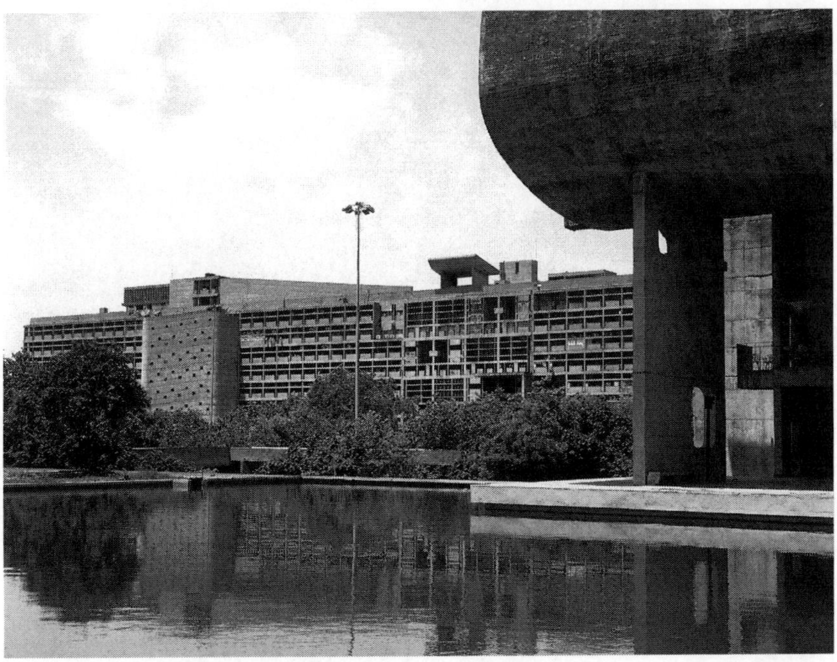

6.36 The Secretariat in Sector-1, 2012

walks with Le Corbusier around the Chandigarh camp and would admire the 'homogenous villages', animals and mountains.[122] The closeness of Drew and Le Corbusier did not go unnoticed, Eulie Chowdhury claimed, 'Romance Bloomed',[123] and Fry bemoaned, 'on the site I felt often estranged from Jane a man inadequate for her needs and weighing too little measured against Le Corbusier'.[124]

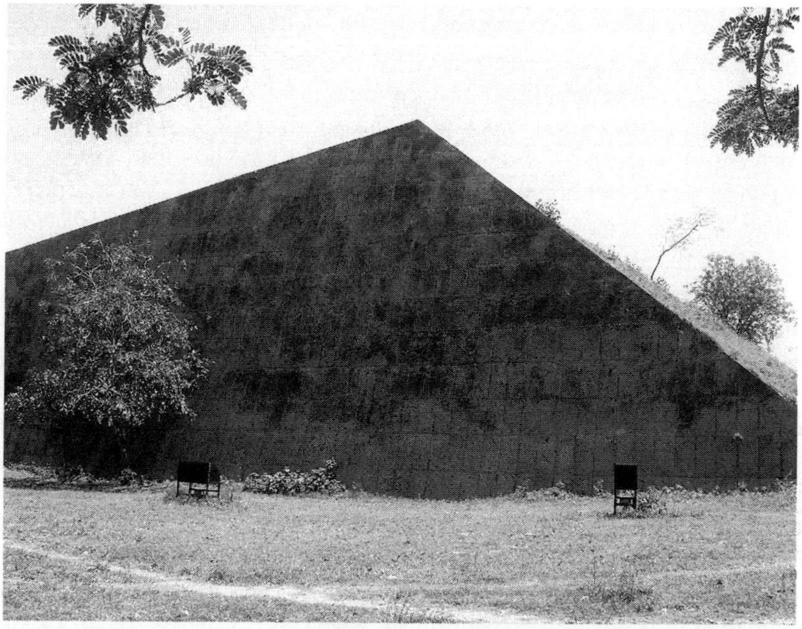

6.37 The landscaping and tufts, 2012

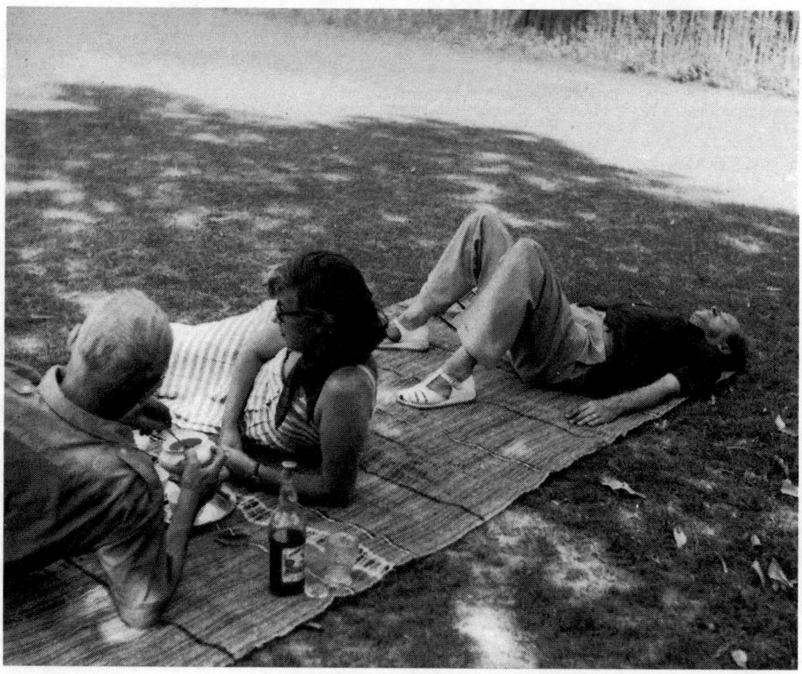

6.38 Le Corbusier, Jane Drew and Maxwell Fry reclining on a rug

Although Fry did enjoy Le Corbusier's companionship and reminisced that, 'back on site our best company was Le Corbusier. He lodged near us when he came, he was often with us drinking and talking into the night, of life and poetry and love'.[125] Or as Drew recalled it, ' … that was when he talked chiefly about himself. I once said he could not make five sentences without bringing himself in and he admitted it was difficult'.[126]

Although Drew felt that Le Corbusier was not particularly well-read[127], she did acknowledge 'the breadth of his knowledge, his experience in addressing the problems of housing in underdeveloped countries and the power of his personality'[128], before clarifying, ' … despite his greatness, he made many mistakes – as does anyone who tries anything new. Among these were the concrete brise soleil to his buildings which acted as a heat sink, radiating heat all night, without cooling, before reheating in the sun the following day'.[129] Drew in particular enjoyed his integration of art and colour, and described how Le Corbusier was able to use it to, 'heighten the importance of a room, doorway or position. I have already said that all Corbusier's [buildings] were sculpture, but he used tapestries to add richness of the courts of law and enamel work to make the doors of the assembly building more wonderful. He understood the Indian love of jewellery and colour. Not the tender colours of sweet peas, but jewel-like strong colours. Indians love colour and enjoyed its use'.[130]

She went on to explain that, 'to learn to love Corbusier's work and indeed to criticise it, we have to be receptive to quite a new set of symbols … his art works are worth a good deal of study and will give you considerable emotional pleasure. They are the works of an exceptionally talented man'.[131] Despite her admiration of the man and his work, even Drew felt that, 'towards the end of his life he himself

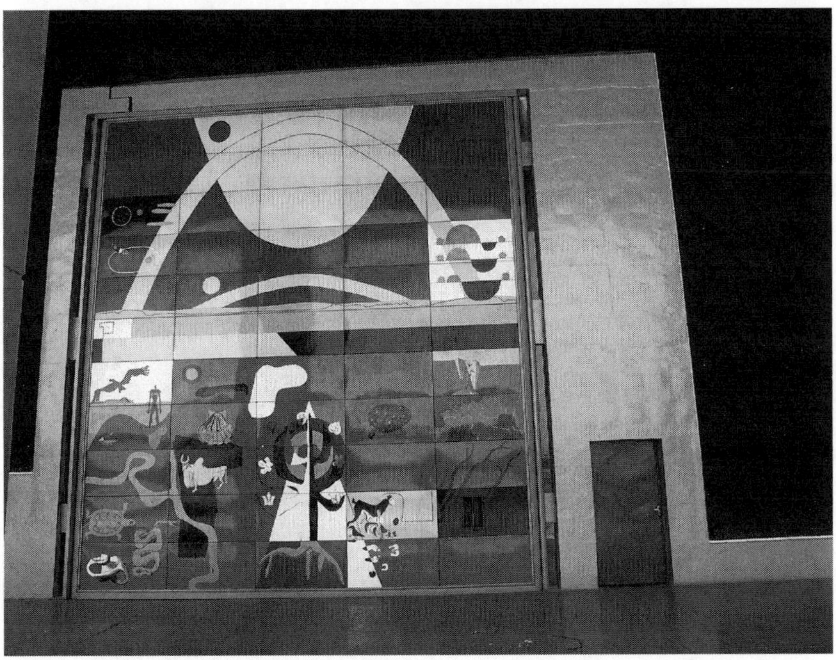

6.39 The Assembly Building Ceremonial Door, 2012

had got lost, believing too much in his own powers and his own symbols which are difficult to understand',[132] and giving few people any praise, 'he despised the rich people in Ahmadabad who had given him so much work'.[133]

Fry later remembered him as a loner, and almost as a mystic, 'His face, his whole bearing was that of a man set apart, hermetic, to be recalled to the world only by its needs of him, and by these tokens, didactic, authoritarian, a man, with more disciples than friends'.[134] Fry also challenged many of his design methods, recognising that for Le Corbusier the Modulor, for example, could be 'a chariot to heaven', but for others, 'a bus to a dusty terminus'[135] and frankly stated that he, 'talked a fair amount on nonsense in his time ... there were always those who were a little sceptical'.[136]

They were both grappling with the enigma that is Le Corbusier. His work, especially in India, manages to enthral and repel, appear both functional and whimsical, vastly overscaled and in other parts human and personal. Along with his work, he cannot be easily fathomed or readily understood. Perhaps these contradictions and ambiguities reached their peak during the design of Sector-1.[137] Despite the difficulties of working with Le Corbusier, Drew 'worked tirelessly' for the entire city and Sector-1 in particular to be recognised as a World Heritage Site, an honour that is yet to be bestowed.[138]

THE CHANDIGARH OFFICE: TRAINING INDIAN ARCHITECTS

Drew received regular updates and gossip about their London office from Theo Crosby who reassured Drew that, 'you will be pleased to hear that the office is running smoothly now – everybody having got used to the new arrangement. But, hell, it is dull without you'.[139] Without the attraction of Fry and Drew the office lacked its usual hum and stream of visitors. Crosby bemoaned Philip Johnson not visiting the office whilst in London forcing the office to 'follow his progress in Astragal ... '[140] Fry and Drew slowly built up a drawing office based on site employing young Indian architects, some of whom had worked in Europe.[141] The Indian architects lived and worked in tents initially until an office and rudimentary housing was constructed.[142] The team returned to Simla in April, 1952, presumably to escape the heat and to concentrate on the working drawings. Eulie Chowdhury's (the only other women on the team besides Drew) description of Shimla conjures images of a town clinging to the days of the Raj, 'back in civilization and though we continued to work hard there were dinners and cocktails in the evenings with an elegance undreamt of in Chandigarh'.[143] They all returned in November, to the Jeanneret designed 'temporary' studios (which are still being used as a museum dedicated to Le Corbusier). It was within these offices that a live architecture school was created for the team of young assistants with Sector-22 as the working-prototype. Once the trainees were deemed fully equipped in the nuances of tropical architecture, they could then design and oversee their own sectors using Sector-22 as the mould.[144]

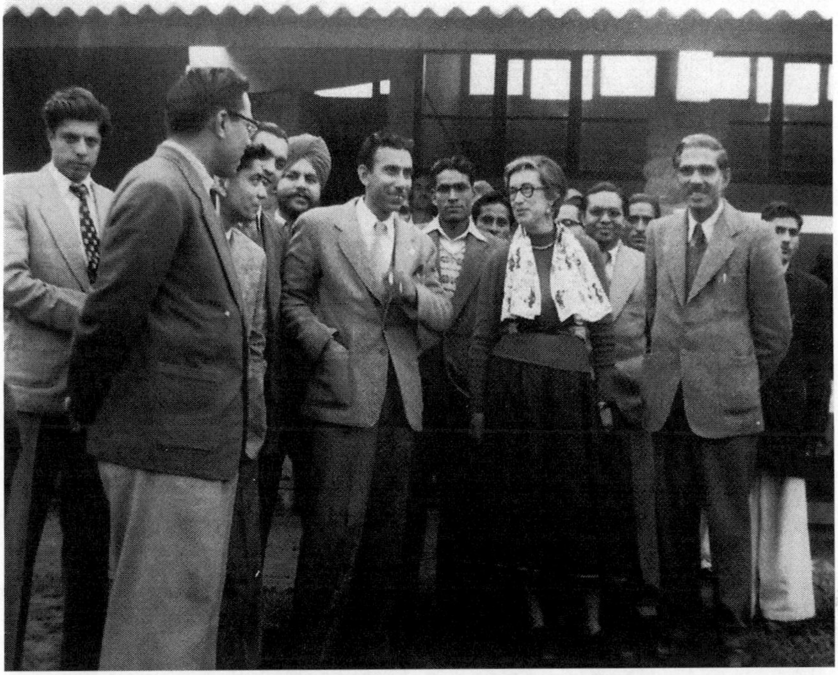

6.40 Drew with her Indian Assistants, c.1954

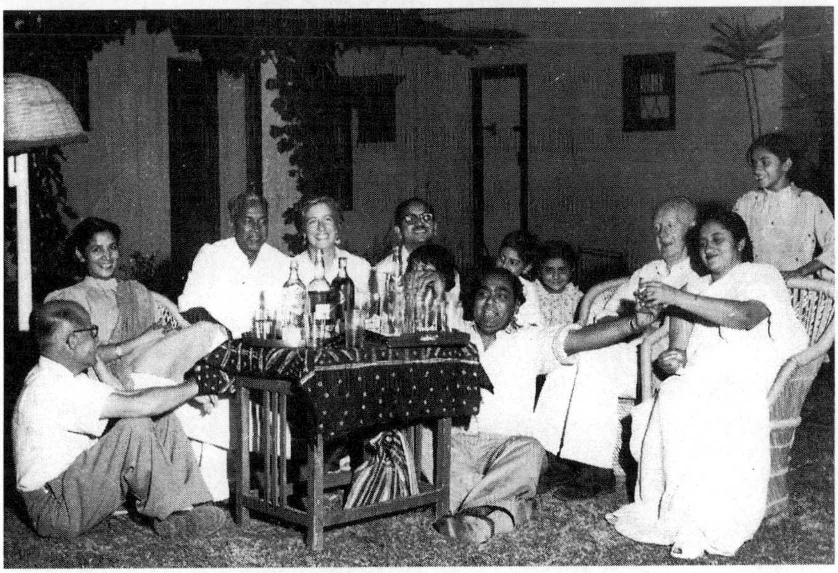

6.41 Drew with Nehru, Pierre Jeanneret, Peter Gregory and friends at the Chandigarh site, c. 1954

The Indian architects were given considerable responsibility and Fry and Drew made it a policy to 'give the various junior members of the staff work for which they are individually responsible'.[145] This was the only way it was possible for the small team to design such large numbers of buildings within such a short period. Drew ran 'a night school' for the Indian architects and it was down to her and Fry to manage the workloads, create the design teams and effectively take the role of practice managers.[146] They developed empirical building experiments, testing ideas such as brick arches and vaults, the remnants of which still survive outside the drawing office in Sector-18. Drew even dispensed what Fry called 'Raj Justice' to the villagers, acting as arbitrator and judge over disputes and quarrels.[147]

At one point the work demands became too great for the fledgling Chandigarh practice – Drew considered the office to be 'understaffed for the quantity of work it is doing'[148] and listed the work that each assistant was undertaking. M. N. Sharma (1923–) (who later became Chandigarh's first Indian Chief Architect in 1966) was in charge of the design and construction of a police station, housing, press building, offices, a hostel, nursery schools in Sectors 22 and 23, as well as supervising the construction of a cinema.

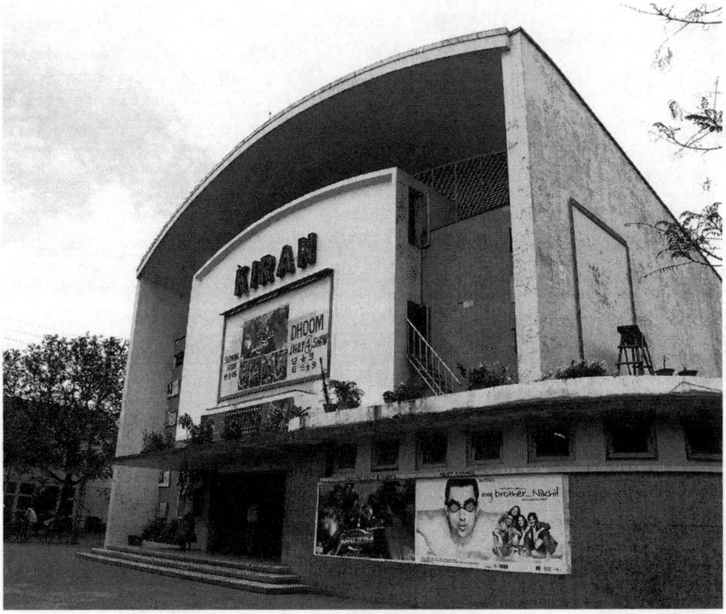

6.42 Kiran Cinema in Sector-22, 2012

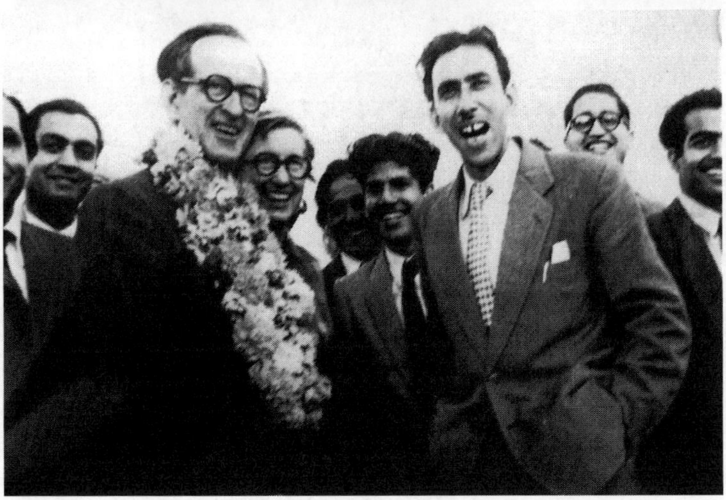

6.43 Fry and Drew with some of the Indian Design team including the first Indian Chief Architect, M N Sharma, c. 1954

Drew expressed her concern that 'we may lose Shama since he has applied for another post'.[149] Aditya Prakash (1923–2008) was also, incredibly, solely responsible for the design of a maternity hospital as well as housing and petrol stations. These were not minor, insignificant commissions even for experienced architects with a design team.[150]

In addition they were supposed to monitor the privately funded developments (only 50 per cent of the projects/houses were Government sponsored[151]), but Fry refused, as he lacked the resources, and fallaciously claimed, 'we are not great believers in external control of architecture'.[152] Within a few months however, when the private developments were under construction he quickly realised the error of his lenient ways. Many of the new houses developed what was considered a 'deliberate parody' of the government houses, and the more expensive ones tended to be, 'over complicated both in form and decoration, and the application of varicoloured external ornament becomes occasionally raucous'.[153] In other words, even though individuals were commissioning buildings that 'responded to climate' and with a uniform material palette it was ultimately the way they looked, and the fact they hadn't been designed with European approval that mattered. Inevitably, 'frame controls' were introduced and rules regarding windows, doors, balconies and terraced properties were enforced. Despite their professed lack of concern over architectural aesthetics, they quickly developed an aesthetic agenda when designs that did not conform to their notions of good taste (i.e. modernist) were proposed.[154]

As Fry's time in Chandigarh was coming to an end, he became increasingly concerned about the future of his Indian staff, namely architects working under the direction of engineers. Fry thought that the engineers of the PWD had, 'taken over the role of the client, in conjunction with the Chief Administrator'[155] and in a not too dissimilar fashion to Le Corbusier, (who used his friendship with Nehru as his means of getting his own way on numerous matters) wrote to Nehru to express his anxiety,

> *I have hesitated a good deal before deciding to write to you but I believe that this generation of architects is of such value to India, and to the oncoming stages of your national plans, that their continued suppression within the PWD system seems to me a sad waste of good creating spirit that should at all costs be prevented ... you will produce yes men eating their hearts out in private instead of lively responsible architects creating a new background for Indian life.*[156]

Nehru replied immediately to Fry, 'I entirely agree with you that our young architects should be encouraged in every way. I shall be glad to see your plan for the reorganisation of the building side of the PWD I am greatly interested in this matter ... I should myself like to meet you before you go back to England'[157]

Fry produced an organisation diagram which was sent to Nehru. This changed the hierarchy of the PWD and enabled the 'chief architect' role to be established in Chandigarh independent of the engineers, a role fulfilled by Jeannerret until 1966. Despite the difficulties they faced on the project; working with Le Corbusier, evasive clients, lack of realistic budgets and staffing levels, they still found the arrangement to be satisfactory as to tender for more work based on the same working practices, such as the Tema Manhean project in Ghana.

REFLECTING ON CHANDIGARH AND FUTURE DIRECTIONS IN THE TROPICS

Chandigarh was the culmination of 30 years work for Fry. The city manifests a number of ideas he and Drew pursued, stemming from the early works with Adams and the influence of Elizabeth Denby and her desire for integrating housing schemes within a broader context of social reform and the inclusion of amenity. An important characteristic of Sector-22 in particular is the open space, parks and play areas, containing clinics, schools and health centres as part of a comprehensive approach to community planning. Despite this, the individual houses may lack the radical edge of Fry's social housing schemes from the 1930s, not least because of the small scale of the dwellings and the lack of apartments. However India was a different setting with inexpensive labour and cheap land values not forcing a high density dwelling solution. In addition, the role played by the City Engineer and the PWD was very significant – their 'programme of priorities' and their foresight in the supply chain of building materials ensured that the building work was not delayed. It made best use of Fry and Drew's time on site, but also dictated their architectural designs and generated a vocabulary that they may not have picked given the choice.

The low-rise low-density option may have seemed prosaic and too conservative for Fry (which is possibly why he rarely spoke about or publicised his work at Chandigarh), it was too close to the Garden Suburb solution that he wanted to avoid; but for Drew it offered the chance to work closely with 'the users' and to offer genuine, if modest improvements in facilities, sanitation and services. Her housing designs, especially for the poorest residents provided many with quality housing that exceeded the expectations of Nehru and today is still highly regarded and in good condition. The 'internal' streets and generous gardens/outside spaces provide lots of flexible space whilst ensuring a suitable degree of privacy and security.

The internal layouts of all the houses adopt conventional arrangements, with the facade offering in some cases, features that mitigate against the bright sunlight and heat. However, despite Fry and Drew's claims about climate driving the design, the same houses were arranged to form 'village squares' and parks with the same facades facing in all directions. As a result, their climatic ability has been seriously compromised. It would seem that Fry and Drew accepted this in order to form the tight knit housing arrangements with each house overlooking or adjacent to an open space.

In a similar fashion, whilst they claimed to have made attempts at catering for 'traditional' living, in reality only a very limited number of experimental houses were constructed. It seems that aspects of 'tradition' were only incorporated or encouraged when it suited the aesthetics of the modern movement. The programme, material availability and funds were such that Fry and Drew were left with the task of generating as much floor space as possible for the funds available for that housing type. Indeed, their earliest solutions, resembling small-scale colonial bungalows, included separate houses for servants – even in the modest type-9 houses. This approach was quickly abandoned and it became apparent that the best solution was to offer as much volume as possible, with high ceilings, and to let the residents define how the space is to be used.

Their time at Chandigarh was not easy, the living conditions were demanding, especially in the early days, and they struggled with bureaucratic obstacles and testing working relationships. According to Kanvinde, they experienced, 'humiliation' from some of the 'high-ups in the P.W.D. which had caused some feeling of frustration towards the end of their term. I did realise then that they were not leaving on a happy note'.[158] By the end of his contract it would appear that Fry was about done, and bluntly stated,

> *'I was glad to go, despite the interest of it and the friendships it brought. It was not the happiest time of my life and would have been the happier without Corbusier whose presence hung over me as it attracted Jane'.*[159] *Drew was not always content either and received a letter from Crosby who noted that, 'I hear you have been ill and that you are unhappy'.*[160]

They seemed to be beset with health and work problems. It all became too much for Fry, and expressing regret over the entire episode pondered,

> *the three years I spent there seemed to me to have been wasted when set beside what I might have done had I stayed in England ... for if there is a second tide in the affairs of men then I lost it going there ... there are buildings done at a certain time that make the final reputation of an architect and these I might have had and lost, and losing them made me angry with myself.*[161]

When he returned to the UK, it seemed that everything had changed. He had to reintegrate into his practice, rekindle old client relationships and adjust to the changes that were taking place in the profession. Ove Arup wrote to Drew whilst she was still in Chandigarh and stated that 'I have not seen as much as Max as I would like, I think he is happy to be back, although slightly bewildered as to what to do next'.[162] Fry saw his future in Africa, but whilst in the UK his self-pity and disappointment reached its zenith, 'I lack some of the ingredients of a big impressive architect ... to dreamy, literary and basically unassertive, not anyone's idea of a big London architect, no, not even my own idea of one'.[163] Yet, opportunities in other areas in addition to building were quick to materialise. He became involved in the planning and later directed the Tropical Architecture course at the Architectural Association, which he believed would 'help the book and us and will not be too onerous'.[164] William Holford was one of the 'Panel of Advisors'[165] and devised the lecture programme and course content, and although Fry may have initially thought it an easy option the syllabus included an extensive list of thirty lecture topics including shade and ventilation, through to 'Tropical living habits'.[166]

By May 1954 Fry had returned to West Africa where he hoped to gain more commissions and having not set foot on the continent for some time had developed a giddy nostalgia for the place,

> *Dear sweet old Lagos – the whitewashed walls, the heavy trees, the heat, the pinky evening glow. How nice it is. I feel at home at once. … I gather that our firm will be welcomed here – the old firm with the solid rep. So I will come back and call about some work and we will open a proper office no doubt. …* [167]

But, within four days he had reverted to type, 'the first flush of arrival over I fall into some depression wondering what this life is all about, what future for these Africans with so much materialism and greedy politics.[168] Drew had developed several new project opportunities in India including large offices for Burmah-Shell in Calcutta. Fry, however, was reticent and in letters written to Drew almost every couple of days he outlined his desires and concerns, 'I don't really know … Jobs all over India means a permanent office and a good man in charge. Local surveys, local builders, etc. Do we want to set up in this way and enter into the details of an Indian practice?'[169] They had just got additional work in Nigeria and openings in Ghana for 'more schools at Tema and possibly a hotel at Accra'.[170] Fry outlined his vision for the practice, 'I am personally keen on working up the England practice again. How far can we, or dare we spread? How much Indian work could be done in London? How much time are we prepared to spend in the year in India. For me 3 months in the winter is the top, if not too much … '[171]

But Drew was eager to continue working in India and the exchange of letters continued throughout the summer. Fry again stressed his reticence, 'About Indian practices, this is what I feel. With taxation what it is we cannot afford to put much capital into making an office there … . You see W. Africa remains a very good field where we are known and respected and this we shouldn't neglect for something further afield. … I quite expect that when you leave India you will feel more like this too'.[172] Drew was to return to the UK in July 1954 and Fry wrote to her to remind her to 'bring back anything you think we should treasure and some Corb pictures … '.[173] The amount of incomplete work remaining was a major concern to Drew, particularly regarding the hospital.[174] Most of their projects in Chandigarh were not completed when they departed and they continued to receive questions about finishes and details from their Indian collaborators.

IRANIAN OIL WORKER HOUSING

Although Fry struggled to readjust to the architectural climate of Britain when he returned from Chandigarh in 1954, Drew was presented with numerous opportunities. Following the British and American sponsored *Coup* in 1953 oil production had recommenced in Iran, and the Anglo-Iranian Oil Company was now operating in a consortium with other American and European companies. It was this change that opened up opportunities for new architectural firms to 'tender' for work, whereas previously the AIOC projects in Iran had been awarded to Wilson Mason and Partners.[175] Drew explained, 'India led to my being invited to plan townships, really villages for the oil consortium in South Iran'.[176] In addition to her own commissions she was employed as a consultant and in 1955 visited numerous towns such as Agha Jari where she advised on housing design.[177]

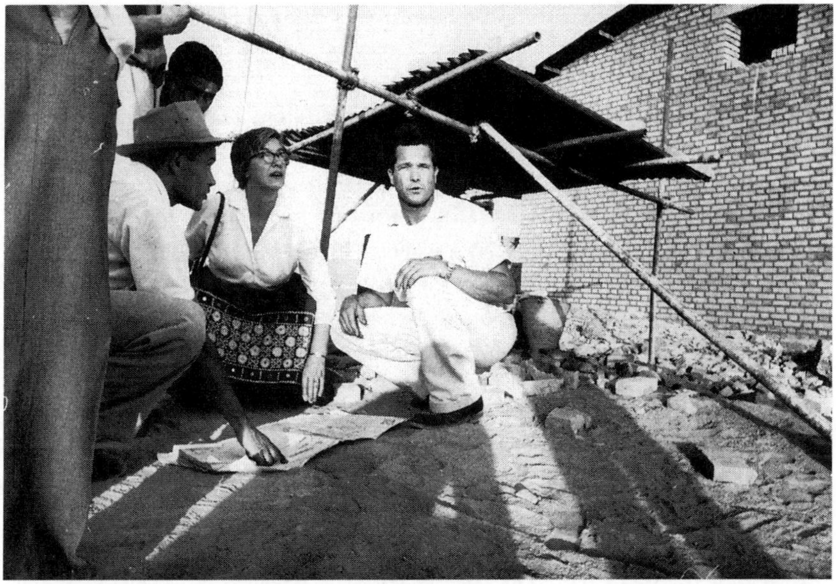

6.44 Drew and Andrew Young visit Agha Jari, 1955 Reproduced from the BP archive. © BP Plc

Drew adopted her usual method of data gathering and consultation, 'first I must study the areas, the medical statistics, climate, peoples' way of life and building materials. Extreme dry heat poses its own problems and oil seems to turn up in areas not usually habitable for man'.[178]

She was seduced by the landscape as well as the existing architecture,

> *The tribes' architecture in these areas was either of the tent variety or, where it was more permanent, it was in mud and the rather beautiful forms which mud architecture takes reflected an innate dignity and a power of imagination and dreaming which has no counterpart in Europe.*[179]

She sought to respond to this by rejecting the standardization that the existing oil company housing had deployed, but other than some experimental labourers' houses defaulted to a type developed in West Africa and India. At Masjid-i-Suleiman the houses are of rudimentary design, and staff at different positions are provided with commensurate accommodation. A similar approach was adopted for a further commission at Gachsaran, 'a township I had to design from nothing, it looked like a page from Dante's "Inferno". Flames leapt miles into the air, the rocks at extraordinary angles, and the place was barren'.[180] Despite Drew's desire to break the standardised approach to housing, the dwellings were arranged in terraces set back-to-back on a grid pattern, surrounding a small central park. A hospital and school were also constructed at the settlement.

As in Africa, Drew claimed that she 'invented an architecture and town design to meet the needs of the people with much dust screening and decking'.[181] The most basic housing type was arranged in a terraced formation with barrel-vault roofs paying homage to the existing mud dwellings in the region, but the more common arrangement was a 'compound' house with enclosed courtyard that could be used to keep animals, accommodate future property extensions and provide private enclosed space.

The labourers' housing was constructed in locally available rubble stone and adopted a similar approach to the 13D type housing Drew designed at Chandigarh with parabolic entrance gateways defining thresholds and carrying services across the roads.[182] Just like at Chandigarh a further 12 types were developed, one for each grade of staff,[183] 'my brief was clear: each house type had a budget, air-conditioning only affordable for the top brass'.[184] The result is a hybrid solution borrowing from the Tema Manhean compound house and the type 13D Chandigarh housing solution of dense construction and clearly defined small neighbourhood streets that provided shade and a sense of community in a previously uninhabited place. After consulting the 'excellent medical records', Drew noted that for the labourers 'eye disease was the chief enemy, mostly insect-borne' and subsequently devised a system of 'self-closing trash cans built into the dust screen before the entrance doors'.[185] Of course these were very small interventions but demonstrate a compassion for the people she was designing for. At community level she introduced small 'bazaar' centres into the design and made provision for a 'tree nursery with sweet earth and drip culture'.[186] The trees were planted in front of the houses where Drew claimed, 'even the poorest Iranians seemed to care more for the show they put up for visitors than convenience. So I knew the fronts would be well-kept'.[187] The diurnal temperature changes were so great and coupled with the intense rail storms 'most roofs had cracked' after a short period, whereas at Ibadan the same design was perfectly functional. Drew developed an alternative 'roof combination of high and low technology of Basra reeds and aluminium on tensile wires' attempting to create a roof that had very little thermal mass and capable of withstanding the rigours of expansion and contraction as well as intermittent but heavy rainfall.

Drew frequently described her career as something that happened to her, as if she was a passive agent, rather than a series of events that she in some way

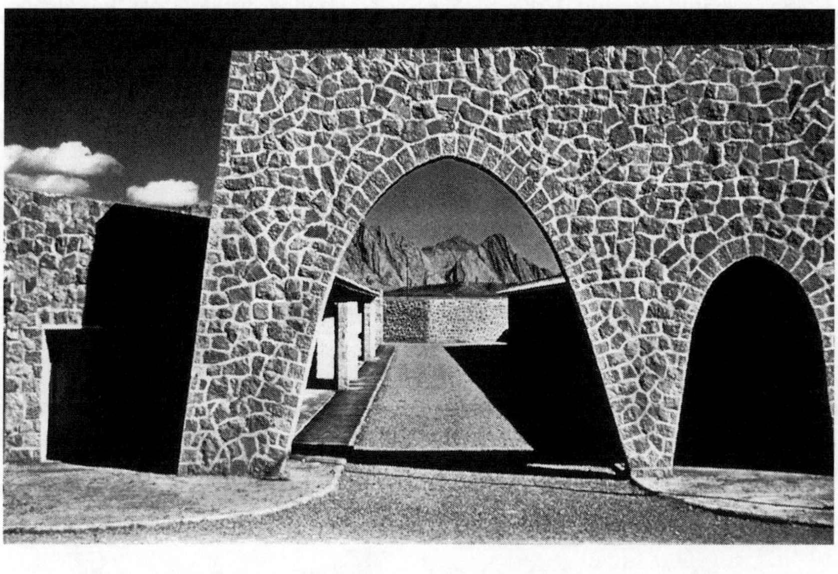

6.45 Housing in Iran based on the Chandigarh Sectors, c. 1960

brokered; 'again one thing led to another and I when I had completed my work I was approached by a firm in Tehran to design Shiraz University. They would manage the contract and supervise'.[188] The firm was the Architects Cooperative with whom Drew developed a close working relationship. Together they worked on various schemes, including a new town plan for Mashhad and a masterplan for the University of Mashhad (Ferdowsi). The Mashhad plan included Fry and Drew's name on the drawing but their involvement was limited. Drew seemed to enjoy the hospitality associated with her visits and almost colonial existence that persisted in the closed communities of the expat oil workers, 'how easily one slips into life where there are servants and food comes automatically. I spent last night staying in the V.I.P. suite in Gach Saran Guest house where a party was given in my honour with drinks on the roof terrace'.[189]

Her visits gave her the opportunity to conduct post-occupancy evaluations to the projects she had previously designed and she wrote to Fry that, 'I think that I should visit Gach Saran and get all the opinions about the houses (over 800 built) also I should do the British Council lecture at Shiraz (I have just had a cable saying they want me to)'.[190] Drew was intrigued by how her houses were inhabited and when she returned again later that year further comments were made to Fry, 'every one seems delighted with the houses which is nice. ... Yesterday I went to M.I.S. where I walked round for the last time. The garden of the nurses [sic] hostel has grown up well. In the evening I flew to Tehran over the snow clad mountains. Now I am sitting in the bright sunshine on Elisabeth Dalley's verandah waiting for a car to take me to Teheran'.[191] It was important to her that her designs were functional and liked by their inhabitants. Her trip to Teheran was to meet with the Architects Cooperative. This method of working suited Fry and Drew. They liked to remain based in London whilst working with a trusted partner on site. Along with Michel Ecochard and Ray Kingston, Fry and Drew were employed as 'consultants' and both attended a presentation given to the Shah on the Mashhad town and university plan.[192] They were used as establishment figures and a means of bringing prestige to the projects, acting almost as a brand for tropical architecture. The designs were not built at that time, as Drew explains, 'alas, this great scheme came to grief with the arrival of the Ayatollah, as did their Open University I was engaged for in Tehran'.[193]

OIL AND FINANCE IN THE TROPICS

Despite Fry's reservations and failed attempts to convince members of the London office to move to India they accepted two further commissions there in the wake of the Chandigarh scheme – the Oriental Bombay building in Mumbai and the Burma Shell in Kolkata. Fry returned to India in September 1954 to work on the Burma Shell project that was to occupy a prime plot overlooking the Maiden. Fry was concerned with the elevation and developing 'suitable wall construction and finish' to respond to the fully air-conditioned interior. A response to technology and materials drove the architectural solution with Fry describing his approach: 'dry construction, fibreglass lagging and non-active glass windows cutting out 40 per cent heat – outer skin of Kotah stone or marble – all very light'.[194] The result was a podium-tower arrangement with a fiercely applied fenestration grid. It is all perfectly resolved and finely tuned, with tile modules and window patterns seamlessly integrated, but this (one of the first multi-stories in Kolkata) is the architecture of anywhere and everywhere.

Although their previous work in the tropics was concerned with inexpensive housing, schools and village layouts with passive cooling and low-tech devices, these later projects were of a different type. These buildings were not at the 'development and welfare' end of the spectrum but were for international business; for although the tropics had been presented as 'rural' and 'primitive' (to appease the romantic imaginings of the Occident) they were also key nodes for the 'flow' and 'exchange' of people, capital and international business often driven by the profits and speculation of the oil industry. The two projects in India were funded by oil business and with the discovery of oil in Nigeria in 1956 and its looming impending independence a string of building projects followed there also with foreign investors keen to partake. A large number of bank buildings were commissioned with ex-pat practices awarded the projects throughout Ghana and Nigeria. A group of practices renowned for their 'progressive' tropical-modernist style designed the largest projects, such as Barclays DCO at Lagos by James Cubitt, Investment Bank Lagos by ACP, and The Co-operative bank at Accra by Kenneth Scott. Fry and Drew also designed a Co-operative Bank at Ibadan, which included '10 storey office block, large hall for 1000 and cold store and shop, all on a very long site with roads on 3 sides'.[195]

Vertical brise soleil provides shade to the windows and the south facing facades have no openings to reduce heat gain. Other flourishes include the concrete wave

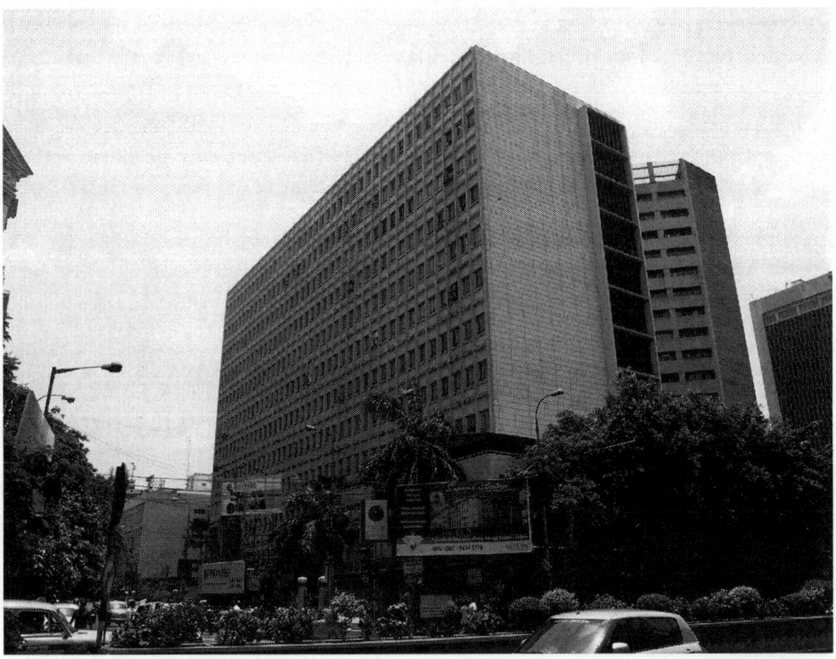

6.46 Burma Shell Building in Kolkata, 2012

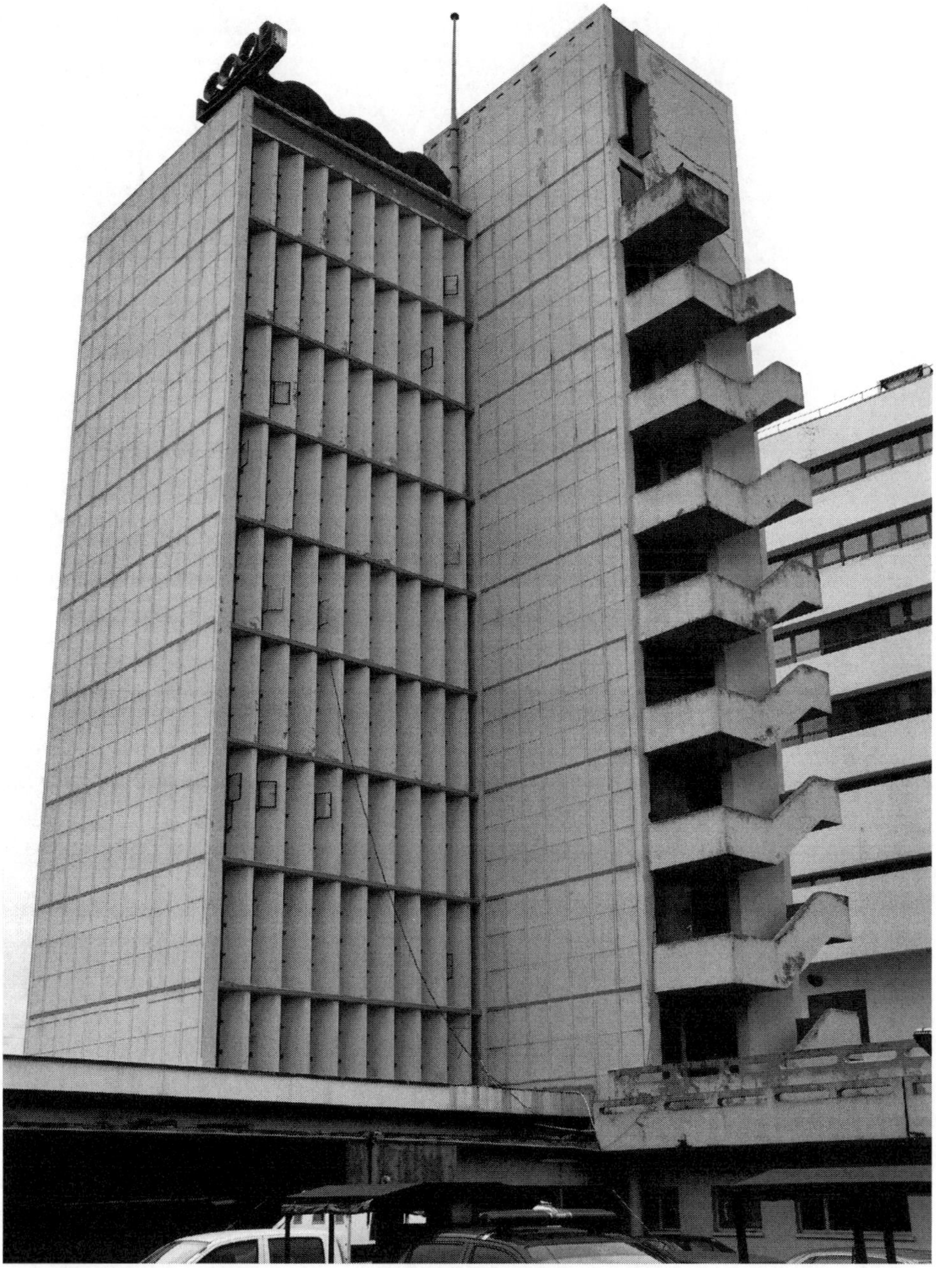

6.47 Co-operative Bank at Ibadan, 2012

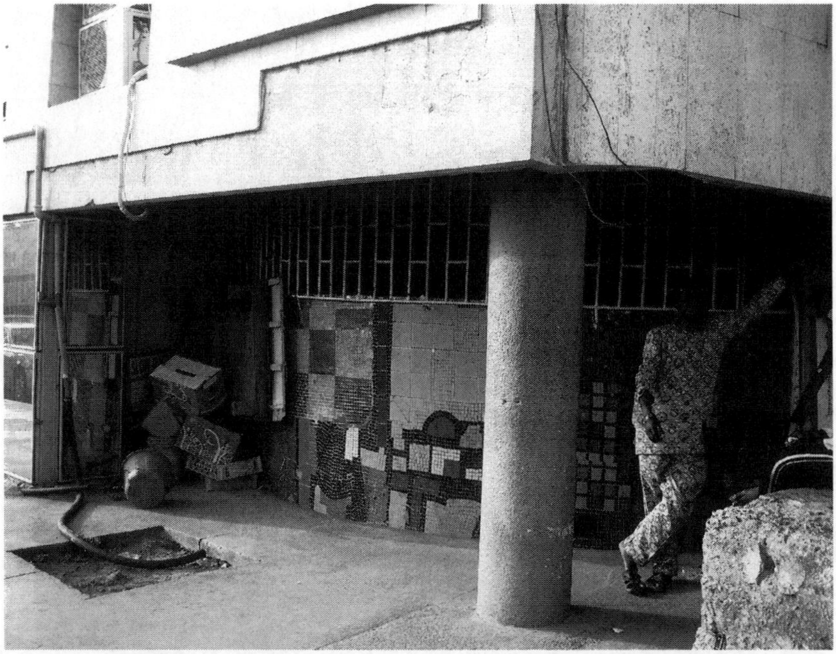

6.48 Co-operative Bank at Lagos, Mosaic, 2012

roof (similar to that at Kumasi hospital) and the Isokon-like cantilevered staircase, which helps to break-up the rectilinear form. They also designed a similar building for the Co-operative Bank on the Lagos Marina with adjustable louvers creating a permeable double- skinned façade. Most of these banks incorporated local art, and Fry and Drew commissioned a ceramic mural at ground floor wrapping round the corner of the building. Further work connected to the oil industry followed, with Drew working for the Shell Oil Company in Singapore and again in Nigeria where they designed an office building for BP. The BP office is a rather squat structure at five stories tall with projecting framed cornice. Again, a strong concrete grid dominates the main façade with the windows set back and further shaded by fixed louvers at the head of each window. Drew wrote to Fry that the BP office, '*looks very good*. The door details are not good and the photomural is very dissapointing [sic] but B.P. are delighted – I was introduced to the Prime Minister as the architects wife and he hardly noticed me ... Lagos is electric with excitement ... ' The letter was written just four days before Nigeria's Independence and Drew remained in Nigeria to join in the celebrations.

They continued to design a small number of schools (including three in Lagos), but the practice emphasis was now firmly fixed on the large multi-storey office buildings for international corporations, and the architecture of which, depending on the availability of air-conditioning, moved more towards a sealed mechanically cooled envelope with only mild leanings towards passive climatic design and solar shading. *Tropical Architecture in the Dry and Humid Zone* included a very short chapter on commercial architecture, where Fry and Drew acknowledged the challenge, 'if they [civic and commercial buildings] wish to avoid looking like any other modern building group anywhere in the world they will pay close and particular attention

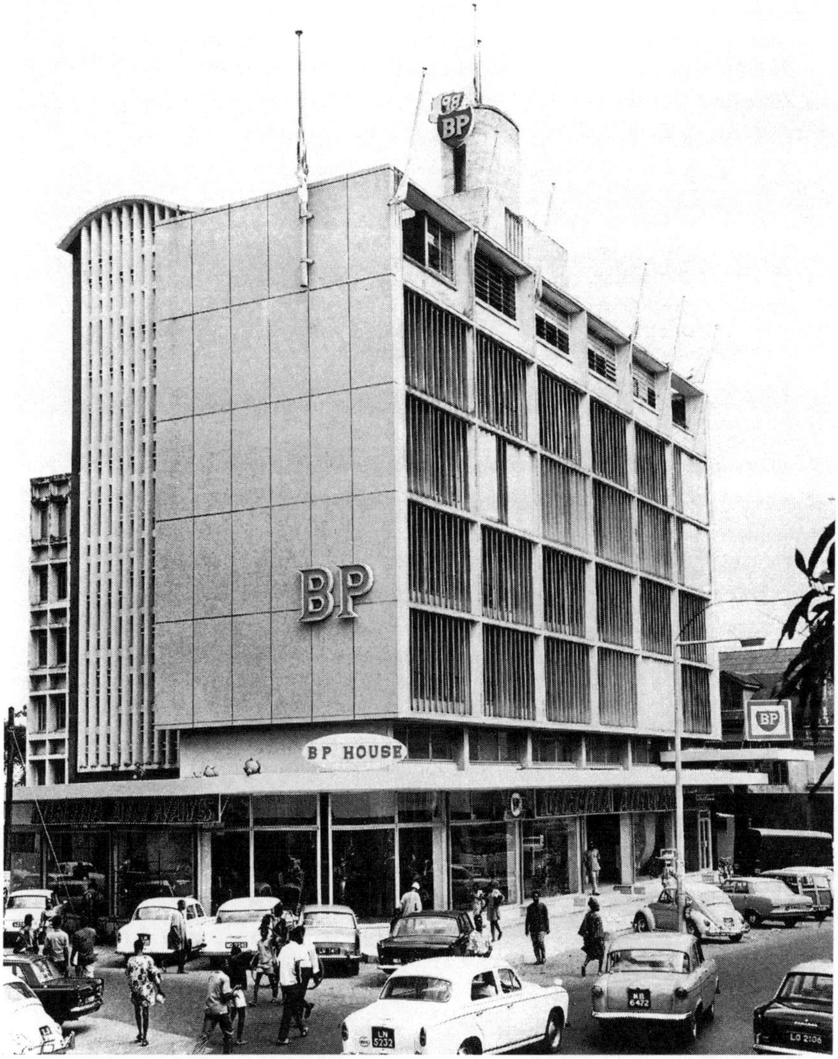

6.49 BP Offices, Lagos, 1962
Reproduced from the BP archive. © BP Plc

to the effects of their climate … '[196] but they didn't propose a detailed architectural response to the specific problem of commercial architecture in the tropics. The office building was a homogenous type regardless of its geographical placement. The development of air conditioning, which they discussed in some detail in *Tropical Architecture* was far too tempting for architects striving to maximise rental space and provide offices that were of 'international' standing. It seemed like an easy means of liberating the built fabric from its role as climatic modifier, as well as releasing the architect to pursue more elaborate forms and facades.

NOTES

1 Maxwell Fry, 'Chandigarh: The Capital of the Punjab', *Journal of the Royal Institute of British Architects*, 62 (4 January 1955): pp. 87–94, p. 87. Parts of this chapter were previously published in *Planning Perspectives*, vol 28, issue 1, 2013, pp 1-26.

2 Chandigarh Art Gallery and Museum, *Selection of Architects and Town Planners for Chandigarh, Visit of P. N. Thapar and P. L Verma to Europe* file.

3 See Achyut Kanvinde in *Jane B Drew Architect: A Tribute from Colleagues and Friends for Her 75th Birthday 24th March 1986*, Sile Flower, Jean Macfarlane, and Ruth Plant, eds., (Bristol, 1986), p. 68.

4 RIBA Archive, F&D/30/1, Jane Drew 'Full Autobiography', p.101.

5 Ravi, Kalia, *Chandigarh: The making of an Indian City*, (New Delhi, 1987), p. 39.

6 Many thanks to Ms. Ann Colin (Fry's daughter) for this information. Telephone conversation with author, 9th December 2011.

7 RIBA Archive, F&D/4/2, 'India', p.3. An agreement was made that Denys Lasdun and Lyndsey Drake would manage the practice in Fry and Drew's absence and 'take our place in the firm, to act for us, rather than our junior partners for whom the responsibilities were too great'. The profits were to be divided 37.5 Fry and Drew/62.5% Lasdun and Drake (whilst Drew was still in the UK), and then dropping to 10% Fry and Drew – 90% Lasun and Drake for the duration of the Chandigarh project.

8 RIBA Archive, F&D/30/1, Drew 'Full Autobiography', p. 101.

9 National Archives, BW90/317, Letter from Fry to the Registrar, 16th January 1951, Ibadan, Nigeria, 1951.

10 RIBA Archive, F&D/4/2, Fry, E. M., 'India', 13 February 1983, Car Saunders was the Chairman of the Inter-Universities Council., p. 13.

11 National Archives, BW90/317 I.U.C. Building & Site, 1951, Letter Mellanby to Adams, 19 January 1951.

12 Ibid.

13 Ibid.

14 See for example, H. V. Lanchester, 'Architecture and Housing in India', *Architectural Design and Construction*, (May 1942): pp. 94–95; H. V. Lanchester, 'Indian Cities and Their Improvement', *Architectural Design and Construction*, (June 1942): pp. 116–120; Partho Datta, 'How Modern Planning Came to Calcutta', *Planning Perspectives*, 28 (2013): pp. 139–47.

15 See Otto Koenigsberger, 'New Towns in India', *Town Planning Review*, 23 (1952): pp. 95–132, and Rachel Lee's, unpublished PhD Thesis on Koenigsberger in India.

16 Robert Gardner-Medwin, 'United Nations & Resettlement in the Far East', *Town Planning Review*, 22 (1952): pp. 283–98.

17 Tyrwhitt, Jacqueline. 'The Village Centre at the Exhibition on Low Cost Housing, Delhi, 1954', *Ekistics*, 52, no. 314/315 (1985): pp. 430–431. See also Ellen Shshkes, *Jaqueline Tyrwhitt: A Transnational Life in Urban Planning and Design* (Farnham, 2013), pp. 158–163.

18 See Kalia, *Chandigarh.*

19 Mayer was keen to develop a plan based on Clarence Stein's ideas of community planning and regularly consulted with him over the project. See Kermit Carlyle Parsons, ed., *The Writings of Clarence S. Stein* (Baltimore, 1998) and Tridib Banerjee, 'U.S. Planning Expeditions to Postcolonial India', *Journal of the American Planning Association*, 75 (2009): pp. 193–208.

20 Mayer was Planning and Development advisor to the government of Uttar Pradesh, 1948–1960. Whittlesey had also developed a masterplan for Calcutta and Mayer had prepared a plan for Bombay.

21 A number of other publications discuss the genesis of the city and the reader is invited to consult Norma Evenson, *Chandigah* (Los Angeles, 1966), Chris Gordon, and Kist Kilian, *Chandigarh: Forty Years after Le Corbusier* (Amsterdam, 1992) and Ernst Scheidegger, Maristella Casciato, Cerena Nievergelt, and Stanislaus Von Moos, *Chandigarh 1956: Le Corbusier, Pierren Jeanneret, Jane B. Drew, E. Maxwell Fry* (Zurich, 2010). For Matthew Nowicki's proposals see Lewis Mumford, 'Nowicki's work in India', *Architectural Record*, 116/3, (1954): pp. 153–9.

22 See Chandigarh Art Gallery and Museum, *Selection of Architects and Town Planners for Chandigarh, Visit of P. N. Thapar and P. L Varma to Europe* file, and Kalia, *Chandigarh* and Fry, 'Le Corbusier at Chandigarh', in *The Open Hand: Essays on Le Corbusier*, ed. by R. Walden (Cambridge, 1977).

23 Mayer's work in India requires further research. According to http://www.lib.uchicago.edu/e/su/southasia/mayer.html he continued to work in India until 1960 – in which case did he accept payment in Rupees? See also C. Rand, 'City on a tilting plan' *The New Yorker*, (30 April 1955), p. 36. According to Rand the foreign exchange issue also prevented American architects such as Frank Lloyd Wright being offered the post.

24 See Gordon, *Chandigarh* and *Chandigarh 1956*.

25 Mayer outlined his ideas and research in a series of briefing documents issued to the Punjab government. See various notes and appendices issued by Mayer throughout December 1949, Chandigarh Art Gallery and Museum *Selection of Architects and Town Planners for Chandigarh, Visit of P. N. Thapar and P. L Varma to Europe* file and UL Archive, D688/2/1/2, Robert Gardner-Medwin, *United Nations Mission to S. E. Asia*.

26 According to Fry in *'Le Corbusier at Chandigarh'*

27 C. Stein letter to Gordon Stephenson, 16 December 1951, Kermit Carlyle Parson, ed., *The Writings of Calernce S. Stein* (Baltimore and London, 1998), p. 533.

28 RIBA Archive, F&D/4/2, Fry, 'India', 13 February 1983, p. 12.

29 Ibid., p. 21.

30 See Kalia, *Chandigarh: The making of an Indian City*.

31 RIBA Archive, F&D/4/1, Drew, 'Le Corbusier and the relevance of his work in Chandigarh', pp. 3–4.

32 John Morrison Archive. *Jane Drew Biography*, unpublished manuscript, p. 6.

33 RIBA Archive, F&D/4/2, Fry, 'India', 13 February 1983, pp. 10–11.

34 Ibid.

35 Ibid.

36 Ibid., p. 13.

37 Ibid., p. 15.

38 Maxwell Fry, 'Chandigarh: The Capital of the Punjab', *Journal of the Royal Institute of British Architects*, 62 (4 January 1955): p. 87

39 See Achyut Kanvinde in *Jane B Drew Architect: A Tribute from Colleagues and Friends for Her 75th Birthday 24th March 1986*, Sile Flower, Jean Macfarlane, and Ruth Plant, eds., (Bristol, 1986), p. 68. Le Corbusier managed to secure a very lucrative contract at Chandigarh; he was to receive a 4% build-cost fee for each building he designed, plus £35/day whilst in India, plus an honorarium of £2000 per annum

40 RIBA Archive, F&D/4/2, Fry, 'India', 13 February 1983, p. 2. This is a view he had held since working with Thomas Adams, See Adams, *Recent Advances*, 1932.

41 Fry, 'Le Corbusier at Chandigarh'. Although Fry always remained sceptical of Le Corbusier, Drew held him in high regard commenting that she had met him at a CIAM conference, 'I had recently returned from the seventh meeting of the congrés which had been at Bergamo, Italy. I had been very impressed by Le Corbusier", See RIBA Archive, F&D/4/1, "Le Corbusier and the relevance of his work in Chandigarh"', p. 3.

42 John Morrison Archive. *Jane Drew Biography*, unpublished manuscript, p. 6. This followed on from their work in the 1930s and was discussed at length in Elizabeth Denby, *Europe Re-Housed* (London, 1938).

43 10% of the worker's salary was paid in rent. See Gopal Krishan, *Inner Spaces – Outer Spaces of a Planned City.* (Chandigarh, 1999).

44 See S.Bahga and S. Bahga, *Le Corbusier and Pierre Jeanneret: Footprints in the sands of Indian Architecture* (New Delhi, 2000), p. 131.

45 Maxwell Fry, 'Problems of Chandigarh Architecture'. *MARG* 15 (1954): pp. 20–25, p. 21.

46 S. Sharma, *Corb's Capitol: A Journey through Chandigarh Architecture*. (Chandigarh, 2009). Sharma makes an interesting observation, if you don't know a person's address in Chandigarh, just tell the taxi driver their job title and he'll take you to the right sector.

47 See Christopher Gethin, 'Chandigarh: A Memorial to Arrogance', *Built Environment*, 2 (1973): pp. 291–94 and Madhu Sarin, *Planning and the Urban Poor: The Chandigarh Experience 1951–1975* (London, 1975).

48 In adopting a hierarchical approach to planning the Chandigarh scheme mirrored the New Delhi project and the Warrant of Precedence idea, see Anthony D King, *Colonial Urban Development: Culture, Social Power, and Environment* (London, 1976) and Nayantara Pothen, *Glittering Decades* (New Delhi, 2012).

49 Maxwell Fry 'Chandigarh: A new town for India'. *Town and Country Planning*, 21 (1953): pp. 217–21, p. 219.

50 Chandigarh Art Gallery and Museum, Letter from P. L. Varma to The Secretary, Secretaries Committee, Simla, 17 February 1950, *Selection of Architects and Town Planners for Chandigarh, Visit of P. N. Thapar and P. L Varma to Europe* file.

51 Chandigarh Art Gallery and Museum, Letter from P. L. Varma to The Secretary, Secretaries Committee, Simla, 27 February 1950, *Selection of Architects and Town Planners for Chandigarh, Visit of P. N. Thapar and P. L Varma to Europe* file.

52 Jane, Drew, 'Living: Sector-22', *MARG*, 15 (1961): pp. 22–25, p. 22.

53 Maxwell Fry, 'Chandigarh: A new town for India'., *Town and Country Planning*, 21 (1953): pp. 217–21, p. 219.

54 RIBA Archive, F&D/4/1, 'Planning and Development of Chandigarh', Town & Country Planning Summer School, Overseas Section, Cambridge, p. 4.

55 RIBA Archive, F&D/4/1, 'Le Corbusier and the relevance of his work in Chandigarh', p. 7.

56 Jane Drew, 'On the Chandigarh Scheme', *MARG*, 6 (1953): pp. 19–23, Maxwell Fry, 'Chandigarh: A New Town for India', *Town and Country Planning*, 21 (1953): pp. 217–21.

57 The Indian Government only had the finance to pay for half of the construction and land costs, circa £6m – the remaining £6m had to come from the private sale of land and buildings. It was therefore imperative that a good first impression was created.

58 Adams, *Recent Advances*, p. 298.

59 Ibid., p. 299.

60 See C. A. Perry, W. D. Heydecker et al, *Neighborhood and community planning: comprising three monographs. The neighborhood unit by C.A. Perry; Sunlight & daylight for urban areas, by W.D. Heydecker, in collaboration with E.P. Goodrich; Problems of planning unbuilt areas, by T. Adams, E.M. Bassett, R. Whitten*, (1929, New York).

61 Jane Drew (Ed), *The Architects' Year Book*, (1945, London).

62 Jaqueline Tyrwhitt, *Patrick Geddes in India*, (1947, London). Fry also dedicated his book, *Art in a Machine Age* to Tyrwhitt.

63 See H. V. Lanchester, 'Indian Cities and their Improvement', *Architectural Design and Construction* (June 1942): pp. 116–120.

64 Drew, 'Living: Sector-22'., p. 22.

65 RIBA Archive, F&D/4/1, p. 2. The photograph album on display at the 'Le Corbusier Centre' in Chandigarh also reveals a number of photographs taken of the villagers and existing settlements suggesting a curiosity and intrigue into the existing conditions.

66 Drew, 'Living: Sector-22', p. 22.

67 Fry, 'Chandigarh: The Capital of the Punjab'. *Journal of The Royal Institute of British Architects*, 62 (4 January 1955): pp. 87–94, p. 91.

68 RIBA Archive, F&D/30/1, Drew, 'Full Autobiography', p. 101.

69 Minnette de Silva in *Jane B Drew Architect: A Tribute from Colleagues and Friends for Her 75th Birthday 24th March 1986*, Sile Flower, Jean Macfarlane, and Ruth Plant, eds., (Bristol, 1986), p. 60.

70 RIBA Archive F&D/30/1, Drew, 'Full Autobiography', p. 101.

71 Kalia, *Chandigarh.*

72 Maxwell Fry, 'Max Fry – inspirations, friendships and achievements of a lifetime in the modern movement', *Building*, 229 (31 Oct 1975): pp. 52–58, p. 57.

73 RIBA Archive, F&D/4/2, 'India', 13 February 1983, p. 41.

74 Maxwell Fry, 'Harmony out of discord: Maxwell Fry speaks about his life and work, at the RIBA', *RIBA Journal*, 86 (1979): pp. 526–529, p. 528.

75 Maxwell Fry, 'Problems of Chandigarh Architecture', *MARG*, 15 no. 1 (1961): pp. 20–25, p. 20.

76 Ibid., p. 20.

77 See *The Architect and Building News*, (February 24 1960), p. 188

78 Fry, E. M, 'Chandigarh: The Capital of the Punjab', *Journal of The Royal Institute of British Architects,* 62(4 January 1955): pp. 87–94, p.91.

79 RIBA Archive, F&D/4/1, 'Planning and Development of Chandigarh', Town & Country Planning Summer School, Overseas Section, Cambridge, p. 10.

80 RIBA Archive, F&D/14/7, Fry 'An old man's epilogue', pp. 4–5.

81 RIBA Archive, F&D/25/3, Drew 'Reflections on My Life and Work', 1 March 1993, p. 4.

82 Drew, 'Living: Sector-22', p. 25.

83 RIBA Archive, F&D/4/2, Maxwell Fry. 'India', p. 33.

84 Fry, 'Problems of Chandigarh Architecture' p. 20.

85 Otto Koenigsberger, 'New Towns in India', *Town Planning Review*, 23 (July 1952). See also the extensive work undertaken by Rachel Lee, unpublished PhD thesis.

86 H. V. Lanchester, 'Architecture and Housing in India', *Architectural Design and Construction*, (May 1942), pp. 94–95 and Indian Cities and their Improvement, *Architectural Design and Construction*, (June 1942), pp. 116–120.

87 As quoted in Boesiger, W, *Le Corbusier the last works, vol. 8* (Zurich, 1970).

88 Each type relates to the Civil Service rank, plus the architects responsible for the design would add their initial, e.g. 13-D indicates it was designed by Drew.

89 RIBA Archive Drew, F&D/4/1, Drew, 'Le Corbusier and the relevance of his work in Chandigarh', p. 6.

90 A. F. Hare, 'Factory-made Housing for India', *Architect and Building News*, 197 (24 February 1950): pp. 217–8.

91 The intention was to prevent the lower castes from 'contaminating' the main rooms of the house. Various purification rituals would have to be performed as a result of any 'contamination'. Discrimination on the basis of caste was declared a sin by Gandhi and outlawed following partition, but the practice continued.

92 Joshi, K, *Documenting Chandigarh*, p. 78.

93 Fry, 'Problems of Chandigarh Architecture', p. 21.

94 Drew, 'Living: Sector 22', p. 23.

95 RIBA Archive, F&D/13/5, Fry, 'A Collection of texts by Maxwell Fry, 1958–1974'.

96 Fry, E. M, 'Chandigarh: The Capital of the Punjab', pp. 87–94.

97 See Sarin Madhu, *Urban Planning in the Third World: The Chandigarh Experience* (London: Mansell, 1982).

98 Chandigarh City Art Museum, 'Construction of godowns and staff quarters at Chandigarh', memorandum, circa February 1950, *Selection of Architects and Town Planners for Chandigarh, Visit of P. N. Thapar and P. L Varma to Europe* file, p. 101.

99 Aditiya Prakash, 'Mobile Shops in Chandigarh' in *Architects' Year Book 14*, Declan Kennedy and Margrit Kennedy (eds), (1974, London).

100 Although it has been described as 'curtain walling', it is in effect a large window per floor, designed to resemble curtain walling, which is 'hung' from above/fixed back to secondary structure and self supporting

101 Jane Drew, Edwin Maxwell Fry, and H. L Ford, *Village Housing in the Tropics: With Special Reference to West Africa* (London, 1947), p. 116. In turn this manual was heavily indebted to the Public Works Department building guides.

102 Jon Lang, *A Concise History of Modern Architecture in India* (Delhi, 2003), p. 24.

103 According to M N Sharma, he went to consult Fry over the lack of vehicular turning space at the entrance. Fry told him not to be concerned with such things, and instructed him to add a cantilevered canopy.

104 See for example, Christopher Wilk, 'The Healthy Body Culture', in *Modernism: Designing a New World*, ed. by Christopher Wilk (London, 2006). The notion of Health and Fitness was a major, if often overlooked, part of Le Corbusier's plan of Chandigarh – resulting in the Leisure Valley, fitness trails, the lake and extensive sports provision.

105 Linda Bryder, "'Wonderlands of Buttercup, Clover and Daisies': Tuberculosis and the Open-Air School Movement in Britain, 1907–39', in *In the Name of the Child: Health and Welfare, 1880–1940*, ed. by Roger Cooter (London: Routledge, 1992), p.76.

106 RIBA Archive, F&D/18/13, EMF, Ibadan, to JD, Chandigarh, 14th August 1953.

107 Drew developed an interest in healthcare from a young age, taking a particular interest in her father's surgical instruments business and her sister's career as a doctor.

108 See file entitled Museum at the Chandigarh Art Gallery and Museum.

109 See Fry and Drew's recollections in Russell Walden, ed., *The Open Hand: Essays on Le Corbusier* (Cambridge, Mass, 1977).

110 Maxwell Fry, 'Opening of High Court by Mr. Nehru', *Architect and Building News*, 207 (20 January 1955): p. 401.

111 RIBA Archive, F&D/31/1 Drew 'Full Autobiography', p. 139.

112 Le Corbusier, Oeuvre Complete, Volume V, 1946–52, p191

113 Charles Jencks, *Le Corbusier and the Tragic View of Architecture* (London, 1973), p. 157.

114 Maxwell Fry, 'Art in a Machine Age', (London, 1969), p. 61.

115 Ibid., p. 61.

116 Maxwell Fry, 'Problems of Chandigarh Architecture'. pp. 20–21, 25, p. 20

117 Maxwell Fry, 'Art in a Machine Age', (London, 1969), p. 137.

118 RIBA Archive, F&D/25/2, Jane Drew, October, 1989, 'Le Corbusier' / 'Corbusier's daily routine at Chandigarh', p. 21.

119 See Ernst Scheidegger, Maristella Casciato, Cerena Nievergelt, and Stanislaus Von Moos, *Chandigarh 1956: Le Corbusier, Pierren Jeanneret, Jane B. Drew, E. Maxwell Fry* (Zurich: Scheidegger & Spiess, 2010).

120 Ibid., p. 17.

121 Ibid., p. 17.

122 Ibid., p. 17.

123 Eulie Chodwhury in Sile Flower, Jean Macfarlane, and Ruth Plant, eds., *Jane B Drew Architect: A Tribute from Colleagues and Friends for Her 75th Birthday 24th March 1986* (Bristol, 1986), p. 74.

124 RIBA Archive, F&D/4/2, Fry 'India', 13 February 1983, p. 38.

125 Ibid., p. 38.

126 RIBA Archive, F&D/25/2, Drew, October, 1989, 'Le Corbusier' / 'Corbusier's daily routine at Chandigarh', p. 18.

127 Ibid., p. 19

128 Ibid., p. 4

129 RIBA Archive, F&D/25/3, Drew, 1 March, 1993, ' Jane Drew 'Reflections on My Life and Work', 1993', p. 4.

130 RIBA Archive, F&D/4/1, Drew, 'Le Corbusier and the relevance of his work in Chandigarh', p. 9–10.

131 RIBA Archive, F&D/25/2, Drew, October, 1989, 'Le Corbusier' / 'Corbusier's daily routine at Chandigarh', p. 24.

132 RIBA Archive, F&D/25/2, Drew, October, 1989, 'Le Corbusier' / 'Corbusier's daily routine at Chandigarh', p. 23.

133 Ibid., p. 18. Le Corbusier designed two houses and a 'Mill Owners Association' building in Ahmedabad .

134 RIBA Archive, F&D/4/2, Fry 'India', 13 February 1983, p. 5.

135 Maxwell Fry, 'Art in a Machine Age', (London, 1969), p. 49.

136 RIBA Archive, F&D/1/5, 'Transcript of an interview given by Fry and Drew on 24 November 1986', p. 5.

137 See V. Prakash, *Le Corbusier's Chandigarh: The Struggle for Modernity in Postcolonial India* (Seattle & London, 2002).

138 Lang, *A Concise History*, p. 68.

139 RIBA Archive F&D/6/4, Letter from Theo Crosby to Drew, 1 October 1953.

140 RIBA Archive F&D/6/4, Letter from Theo Crosby to Drew, 1 October 1953.

141 Such as M. N. Sharma and B. Doshi. By the time Jane had arrived on site there were 'eight qualified architects and about twenty-four draughtsmen', see Eulie Chowdhury, in *A Tribute from Colleagues and Friends for Her 75th Birthday 24th March 1986*, Sile Flower, Jean Macfarlane, and Ruth Plant, eds., *Jane B Drew Architect*: (Bristol, 1986), p. 73.

142 Interview with Mr. M. N Shama at his Sector-8 residence, 25 April 2012. See also Chowdhury, *Jane B Drew*, pp. 73–77.

143 Eulie Chowdhury, in *A Tribute from Colleagues and Friends for Her 75th Birthday 24th March 1986*, Sile Flower, Jean Macfarlane, and Ruth Plant, eds., *Jane B Drew Architect*: (Bristol, 1986), p. 74.

144 Interview with M. N Shama at the Le Corbusier Centre, 29 April May 2012.

145 RIBA Archive, F&D/4/2, 'Division on work within the office', Draft Letter from Jane Drew. It is not known to whom the letter was to be sent to, nor is it dated.

146 John Morrison Archive. *Jane Drew Biography*, unpublished manuscript, p. 6. Mr. M. N Shama also confirmed that they established an evening architecture school. Interview with Author, 25 April 2012, Chandigarh.

147 RIBA Archive, F&D/4/2, Fry 'India', p. 28.

148 RIBA Archive, F&D/1/1, Photocopy of letter from Jane Drew, date and recipient unknown, p. 3.

149 Ibid., p. 3.

150 Ibid., p. 3.

151 The total site of phase 1 was 8919 Acres (3609 Hectares) with an initial budget of £12,886,461 with around half that amount recouped from the sale of plots of land.

152 Maxwell Fry, 'Chandigarh: A new town for India', *Town and Country Planning*, 21 (May 1953): pp. 217–21, p. 221.

153 Norma Evenson, *Chandigarh*, p. 55.

154 The rules are carefully explained in, I. Bakshi, *Chandigarh Aesthetic Legislation: Documentation of Urban Controls in Chandigarh (1951–2001)*, (2002, Chandigarh).

155 RIBA Archive, F&D/1/2, Notes from Fry, p. 5.

156 RIBA Archive, F&D1/2, Letter from Fry to Nehru, 13 February 1954.

157 RIBA Archive, F&D1/2, Letter from Nehru to Fry, 15 February 1954.

158 see Achyut Kanvinde, in *A Tribute from Colleagues and Friends for Her 75th Birthday 24th March 1986*, Sile Flower, Jean Macfarlane, and Ruth Plant, eds., *Jane B Drew Architect:* (Bristol, 1986), p. 68.

159 RIBA Archive, F&D/4/2, Fry, 'India', p. 46.

160 RIBA Archive F&D/6/4, Letter from Theo Crosby to Drew, 8 February 1953.

161 RIBA Archive, F&D/4/2, Fry, 'India', 13 February 1983, p. 1.

162 RIBA Archive, F&D/6/3, Letter from Ove Arup to Drew, dated Sunday, Early May, 1954.

163 RIBA Archive, F&D/4/2, Fry, 'India', p. 1.

164 RIBA Archive, F&D/18/14, Letter from Fry to Drew, 13th April 1954.

165 UL Archive, D147/0A1, letter from Otto Koenigsberger to William Holford, 3 January 1963.

166 Part two of the course concerned with the design of various building types including residences, a small hospital for mining community and a village club with the final part of the course devoted to a thesis and 'no change in the usual procedure – except the choice of a tropical subject'. UL Archive, D147/LA27.

167 RIBA Archive, F&D/18/14, Letter from Fry to Drew, 19th May 1954.

168 RIBA Archive, F&D/18/14, Letter from Fry to Drew, 23rd May 1954.

169 RIBA Archive, F&D/18/14, Letter from Fry to Drew, 28th May 1954.

170 RIBA Archive, F&D/18/14, Letter from Fry to Drew, 28th May 1954.

171 RIBA Archive, F&D/18/14, Letter from Fry to Drew, 28th May 1954.

172 RIBA Archive, F&D/18/14, EMF, London, to JD, Ch. 30th June 1954.

173 RIBA Archive, F&D/18/14, EMF to JD, Ch. 6th July 1954.

174 RIBA Archive, F&D/1/1, Handing Over Report.

175 Correspondence from Mr. William Smail (Architect Partner at Wilson Mason from 1958) with Jackson, June 2012.

176 John Morrison Archive. *Jane Drew Biography*, unpublished manuscript, p. 1.

177 See British Petroleum Archives, 185155/10.

178 John Morrison Archive. *Jane Drew Biography*, unpublished manuscript, p. 1.

179 Jane Drew, 'Planning in Persian Oil Towns', *Town & Country Planning,* 28 (1960): pp. 38–42, p. 38.

180 John Morrison Archive. *Jane Drew Biography*, unpublished manuscript, p. 41.

181 Ibid., p. 41.

182 See sketch in 'Planning in Persian Oil Towns', *Town & Country Planning,* 28 (1960): pp. 38–42, p. 42.

183 Drew, 'Planning in Persian Oil Towns', pp. 38–42, 40.

184 John Morrison Archive. *Jane Drew Biography*, unpublished manuscript, p. 41.

185 Ibid., p. 41.

186 Ibid., p. 41.

187 Ibid., p. 41.

188 Ibid., p. 41.

189 RIBA Archive F&D/15/3 – Letters Drew to Fry, 1960s JD to EMF, London, 1 December 1960.

190 RIBA Archive F&D/15/3 – Letters Drew to Fry, 1960s JD to EMF, 21 February 1960.

191 RIBA Archive F&D/15/3 – Letters Drew to Fry, 1960s JD to EMF, London, 1 December 1960.

192 Thank you to Shireen Mahdavi for kindly sharing this information. Email to Jackson, 7th March 2012.

193 John Morrison Archive. *Jane Drew Biography*, unpublished manuscript, p. 41.

194 RIBA Archive, F&D/18/14, EMF, Calcutta, to JD, 8th September 1954.

195 RIBA Archive, F&D/18/14, EMF, Accra, to JD, Ch. 26th May 1954.

196 Maxwell Fry, and Jane Drew, *Tropical Architecture in the Humid Zone* (London, 1956), p. 137.

7

Humanism and Monumentality (A Post-war Compromise)

Following significant periods overseas, Fry and Drew began to consolidate their London practice in the 1950s and undertook projects that are perhaps the most overlooked of their lengthy careers. Fry and Drew returned home in 1953 and 1954 respectively, their international reputations at an all-time high following prestigious work implementing CIAM strategy with Le Corbusier and Pierre Jeanneret in Chandigarh. Drew took time out from India to participate in the ninth CIAM Congress at Aix-en-Provence, which was held in July 1953. She contributed to proceedings with a presentation of recent work undertaken at Chandigarh with Fry and Staff Architect N.S. Lamba,[1] which focused on the city's low-cost housing provision and illustrates her continued interest in a reductive Modernism that addressed the need for basic accommodation. CIAM 9 was the largest of the group's congresses, with a 500-strong delegation from 31 countries and 'observers numbering in the thousands'.[2] Amongst the group was the British architect Gordon Graham, who (more than 40 years later) recalled his first encounter with Drew:

> *The courtyard was full of chatter and conversation as individuals or small groups drank coffee in the shade of the plane trees, studied exhibition panels that had overflowed into the courtyard or began to arrange themselves into semi-formal discussion groups to get down to some serious work. Suddenly, I swear, the buzz of conversation and movement seemed to halt for a split second. The momentary silence, my memory tells me, coincided with the appearance on a first floor balcony of a figure adorned in a brightly coloured magnificent silk pyjama-suit and wearing the most enormous straw coolie-hat I had ever seen. The buzz of the background noise resumed as a wrought iron staircase down to ground level was elegantly negotiated with no apparent risk to the coolie-hat. Jane had arrived from Chandigarh for the Congress. It was an entrance of quite staggering star quality. I shall never ever forget it.*[3]

Drew would have been aware of the shifting architectural landscape of CIAM at Aix. The 1953 congress was to be the last attended by Le Corbusier and Walter Gropius as part of strategic attempts to 'hand over' CIAM to the next generation, as determined at the previous congress in Hoddesdon.[4] Indeed, the split between some of the younger members and the 'middle generation' became widely apparent in 1953, with the newly elected MARS members, Peter and Alison Smithson, disputing

CIAM discourse on the Functional City.[5] The Congress ended with a gathering on the rooftop of Le Corbusier's iconic Unité d'Habitation (1947–52) in Marseilles, a symbolic act, perhaps, that signalled a changing of the architectural guard.

As Chapter 3 has discussed, the MARS Group reached its ascendancy during the early post-war years, which led to Britain hosting CIAM conferences in 1947 at Bridgewater and in 1951 at Hoddesdon. Fry and Drew were significant in this regard, their work with the CIAM Summer School and the promotion of CIAM members' work in the *AYB* helping to publicise the group's architectural agenda. Following their return from Chandigarh they, no doubt, hoped to sustain this momentum. Yet it seems that, in the interim, British architectural discourse had moved on apace. Back in Britain, this shifting landscape – both on a domestic and at a wider CIAM level – led to a difficult period of readjustment for Fry and Drew. While their projects overseas had led to international recognition, a growing generational split of MARS members encouraged a negative reassessment of the architecture of Fry, Drew and Partners. The reassessment must have proved particularly difficult for Fry, who was irrevocably caught up in the interwar milieu and the initial development of the British Modern Movement. Fry had arrived home a year before Drew and experienced a difficult transition back to work in London. He sent doleful letters to Drew, remarking, 'O pet I wish you were here. I feel such a nothing. E. Nothing Esq CBE. What a joke'.[6] Drew, meanwhile, who had been just starting out in the 1930s, lost little momentum due to the war and, as Gordon Graham notes, her star was in the ascendancy. This distinction between the reputation of the practice and that of architects *within* the practice is an important one. Fry and Drew's individual responses to the shifting situation were increasingly divergent and this encouraged a divergence in their architectural approaches. Whilst Fry and Drew had always had different emphases in their work, in the 1960s their respective approaches – particularly in aesthetic terms – became increasingly apparent. These differences, as well as the similarities, are examined in this chapter; the period discussed is demarcated by Fry's return from Chandigarh in 1953 and his retirement in 1973, when the practice became Fry, Drew, Knight and Creamer.[7]

The integration of art and architecture was central to Drew and particularly Fry's developing ideas regarding a humanist architecture, which followed Fry's unsettled return from Chandigarh. Fry's personal take on this incorporated the ideas of close friends, such as Henry Morris, Herbert Read and Julian Huxley; Drew, meanwhile, built on her wartime network, particularly Kenneth Clark and Peter Gregory, at the Institute of Contemporary Arts. Centred around the ICA, Drew's involvement in the art scene led to an architectural approach that was distinct from that of Fry. Fry later wrote of Drew's 'rise to her fuller powers', noting how:

> *[she poured] her beautiful energies into fields of activity in South Persia I was never to tread, into a great hospital at Torbay of all the most humane, and into the creation of a new Institute of Contemporary Arts on the Mall that absorbed her body and soul until it could be said that the letters ICA were stamped indelibly across her breast.*[8]

Recent scholarship has highlighted the significance of the ICA for post-war culture in London,[9] although Drew is rarely mentioned in this context despite her deep commitment to the centre. She is seen as part of the institution rather than of the more progressive artistic groups that it supported. Yet Mark Crinson and Claire Zimmerman have recently highlighted the similarities between the work of British architects and critics – such as the Smithsons, James Stirling and Reyner Banham – and the 'neo-avant-garde' artists of the Independent Group that grew up around the ICA. Crinson and Zimmerman suggest that these young architects might also be considered as part of the neo-avant-garde, 'They felt nostalgia for the historic avant-garde, they heroicized it, and they felt compelled to return to its achievements. But they saw their own task in terms of the different challenges of the particular moment'.[10] Fry's ambiguous position in this avant-garde is demonstrated by his inclusion in the Smithson's 'Heroic Period of Modern Architecture', set out in 1965. His assignation to the history books suggests an exclusion from British post-war architectural culture, despite his status as a modernist pioneer. Although a generational (and therefore a presumed ideological) split between the avant-garde and the neo-avant-garde is often emphasized, it is useful to situate Fry and Drew's work in this milieu, thereby illustrating the links between groups during this period. By focusing on these formal and informal groupings – those established around the CIAM and the ICA – this chapter examines Fry and Drew's respective notions of architectural modernity during the 1950s and 1960s.

This context is widened to consider Fry and Drew continuing involvement in CIAM. Post-war, the CIAM sphere of influence had shifted to America following the emigration of many high-profile modernists, including J.L. Sert, Jacqueline Tyrwhitt and Sigfried Giedion. This group (led initially by Gropius) sought to continue their work, albeit with regional influences and, given their close relationships with these figures, Fry and Drew's position in this realignment deserves attention. With the rise of the younger generation, the implicit suggestion of Fry and Drew's fading relevance may be queried by the introduction of their activities in America. Indeed, their invitations to East Coast educational institutions suggest an alliance with American Modernism, a notion reinforced by later recognition from the American Institute of Architects, as both were elected as Honorary Fellows: Fry in 1963 and Drew in 1978.

The period discussed in this chapter is demarcated by Fry's return from Chandigarh in 1953 and his retirement in 1973, when the practice became Fry, Drew, Knight and Creamer.[11]

THE PARTNERSHIP AND THE LONDON SCENE

Fry and Drew's office was typical of the post-war internationalisation that contributed to the universalising of architectural culture and practice.[12] The couple's continuing involvement with the CIAM network guaranteed a supply of staff schooled in a modernist vein, to some extent perpetuating interwar ideas. For example, Hazen Sise, an employee of Fry's during the 1930s, returned to Canada to teach at McGill University. Sise's students, such as Ann Luke (BArch, 1948),[13] in turn made the journey to England to work for Fry and Drew. The couple's work overseas also contributed to the office's international outlook, as architects and students from around the world came to 63 Gloucester Place. A former Australian employee Graham Bligh later commented that the practice became a 'mini United Nations'.[14] Whilst foreign employees brought in new skills, the flow of staff often hampered the progress of projects, as one disgruntled client observed: 'There was a steady turnover of staff in the Drawing Office as, because of Mr. Fry's reputation, many young Architects stayed just long enough to claim that they had "worked for Maxwell Fry" ... Not one of the initial team remains'.[15]

The flow of staff was also, in part, a result of the increasing workload of the practice. In January 1950, 'The Office of Maxwell Fry and Jane Drew' (which had been established in 1946) officially became Fry, Drew and Partners in recognition of this growth. Exciting new commissions in Africa and Kuwait encouraged Fry and Drew to appoint five junior partners: John Cordwell, Kathleen Greenwell, John Shaw, Norman Starrett and S.E.A. Hounsell. Whilst many of their colleagues came and went, future Senior Partners Frank Knight and Norman Creamer were long-term employees, both joining the practice in 1947 following National Service in the Royal Air Force. Frank Knight replaced his friend and fellow AA student John Cordwell as a junior partner in 1951.[16] Little has since been written about Knight's work at the practice, yet he was evidently a core member of staff and worked closely with Drew from the 1950s onwards. Cordwell later said of Knight, 'He was the office. ... He was a nuts-and-bolts man. He wasn't a good designer, but he was meticulous about getting everything done properly'.[17] Cordwell's view is reinforced by Drew, who wrote to Fry during a visit to Iran, 'I am full of thoughts about the shining master plan. Frank not at all only on his fact finding'.[18] Knight's role as methodical, somewhat uninspired, technician – in contrast to Fry and Drew's charismatic personalities – has ensured his long-standing role in the practice has remained unacknowledged; a view that mirrors a wider tendency for architects to be perceived as lone designers, rather than as part of an extended team.

A couple of years later, new work necessitated another reorganisation. The partnership of Fry, Drew, Drake and Lasdun (FDDL) was formally established in 1952, with Lindsey Drake (1911–80) and Denys Lasdun (1914–2001) invited to run the office whilst Fry and Drew took up three-year posts as Site Architects in Chandigarh. The FDDL practice continued to operate out of the five-storey townhouse at 63 Gloucester Place, which could accommodate 'a surprising number of people'.[19] Drake and Lasdun shared the large, ground floor room overlooking the street, whilst Fry and Drew worked from individual offices on the second floor. In between, the first floor comprised two lively drawing offices, where the draughting work took place. The front office held up to seven staff and was run by Norman Creamer and Alec Redhouse, Creamer being Drake's Junior Partner, while Redhouse was Lasdun's. The rear office, meanwhile, undertook Fry and Drew's work, under their Junior Partner Frank Knight. The firm at first contracted and then expanded during the mid-1950s, upon Fry and then Drew's return from Chandigarh.[20] In line with the increasingly workload, new offices were found: a drawing office on the first floor of the mews at Gloucester Place; and a satellite office first at Seymour Place, later moving to a building on the corner of Albany Street facing Great Portland Street Station.[21] Staff moved around these offices and between the two partnerships, allowing ideas and practices to move with them.

The partnership has been labelled as a marriage of convenience – particularly by Fry and Lasdun – with Fry and Drew, and Drake and Lasdun, operating as separate partnerships under the umbrella practice.[22] Yet, whilst it is clear from their built works that the two partnerships held different architectural beliefs, it seems improbable that some cross-fertilization did not take place during almost a decade of co-operation. The generational split of post-war modernists was observed in microcosm at FDDL. Lasdun chose to continually draw attention to this difference in outlook, his 1960s' Brutalist aesthetic coming to the fore against what he perceived as the outdated model of MARS's interwar Modernism. A former Associate Partner, Derrick Lees, later wrote that 'it was recognised [amongst staff] that there was a spiritual split in the office'. Lees joined the practice in 1957 and worked under both principal partnerships. He highlighted Drake and Lasdun's focus on the 'sculptural content of schemes', which 'tended to be a bit black and white with the odd splash of colour', and observed that their design work was undertaken 'with great integrity and purpose. Tremendous concentration'. This contrasted with Fry's approach, which, he wrote, combined 'tremendous flair and innate spontainious [sic] ability with also a great sense of colour and fun'.[23] The distinction between Drake and Lasdun's repetitive, detailed work and the more artistic methods of Fry is clear, as is the contrast in their architectural styles. It is evident that Fry and Lasdun in particular were strong personalities, who each ploughed their own furrow. Indeed, some five years before the partnership was wound up, Fry wrote to Drew: 'I think we should revert to our old partnership as soon as we can. There is enough work and I have no feelings for the other two so that it seems wrong to go on'.[24]

However, this divide has been overstated. As Chapter 3 has touched upon, the work of Fry and Drew and of the Tecton partnership in the years immediately following World War Two shared a similar formalist approach in housing schemes such as Passfields and the Hallfield Estate. This pattern-making continued in Drake and Lasdun's work of the 1950s, with Alec Redhouse 'producing overlay after overlay of alternative elevations' for the flats at Green Park, suggesting the continuing influence of Lubetkin's façadism well into the 1950s.[25] The Hallfield Primary School, within the Estate and designed by Drake and Lasdun, is of particular interest as it opposed the standardized, functional approach of, in Ladsun's terms, a 'diagrammatic' architecture.[26] Their investigation into a humanistic architecture, enhanced by spaces of different character and form, is apparent in the free-flowing plan of the school, which the architects compared to a plant; the stem (circulation), linking the leaves (classrooms), flower (assembly hall) and seed pods (dining hall).[27] The scheme suggests a preoccupation with ideas and themes similar to those of Fry and Drew during this period, who were also investigating ways to keep the 'demon in the machine' at bay.[28]

Projects were also generously shared amongst the two partnerships. Drake and Drew were particularly adept at gaining new commissions and often entertained potential clients at Gloucester Place. Drew also wrote of the influence that Denys Lasdun had on her work and, whilst Fry became more concerned with organicism and symbolism, she moved nearer to the monumentalism of Lasdun's architecture. Significantly, Lees did not comment on Drew's design work, perhaps suggesting that her approach was more difficult to pigeonhole. FDDL was wound up at the end of 1959, with Lindsey Drake's retirement and Lasdun's establishment of his own practice. The loyalty of Knight and Creamer was rewarded with the beginning of the new practice in 1960: both were appointed as senior partners, along with J.R. Atkinson, who headed the West African office.[29]

Gloucester Place was a dynamic and creative place to work.[30] Fry and Drew ran a hospitable home and the welcoming atmosphere attracted fellow architects, engineers, writers, artists, and more besides. An employee Theo Crosby later wrote of 'Max and Jane':

> *They lived a full, noisy, and energetic life in the flat above the Gloucester Place office. There were modern paintings on the walls, and writers and artists came to visit. Among them was the beautiful, young Eduardo Paolozzi, who was persuaded to show his collection of slides and images, and gave the first of his Surrealist, disjointed lectures. He made us see the continuing vitality of the modern spirit, already weighed down by old men.*[31]

Paolozzi (1924–2005) was introduced to Fry and Drew by fellow artist Nigel Henderson at an exhibition of 'Kenneth King, Eduardo Paolozzi, William Turnbull' at the Hanover Gallery in early 1950 (Figure 7.1).[32] Paolozzi is well known as a member of the Independent Group, which met informally at the ICA from 1952 onwards. The Independent Group is known for its interest in American consumer culture and the Pop Art exhibited at 'This is Tomorrow', organised by Theo Crosby in 1956. Yet Fry's, and particularly Drew's, involvement in the ICA prior to and during the Independent Group years, illustrates an alternative aspect to the work of the neo-avant-gardes. Drew and Fry were tied to these people and places but not necessarily the same themes as the Independent Group; Drew had little interest in consumerism while Fry actively opposed it, and yet their friendship with Paolozzi was important in developing a method of combining art and architecture in meaningful dialogue. The association with Fry and Drew's employee Theo Crosby is also significant. The 'bountiful aesthetic' of the Independent Group was disseminated through the design of *Architectural Design*,[33] which Crosby co-edited, and emerged in his architectural work, to Fry's evident dismay. After a 1953 visit to Kumasi, Fry wrote to Drew, 'The new buildings of the Ashanti college work better than the first dormitory with a large and slightly brutal lavatory block of Theo's that does not entirely scale with the rest'.[34]

In 1950, Drew commissioned Paolozzi to work on the interiors of the Institute of Contemporary Arts' new headquarters at Dover Street. His contributions included wallpaper for the Members' Room and decoration of the bar for the opening; Nigel Henderson later recalled that Paolozzi created 'a kind of scarlet scheme round the

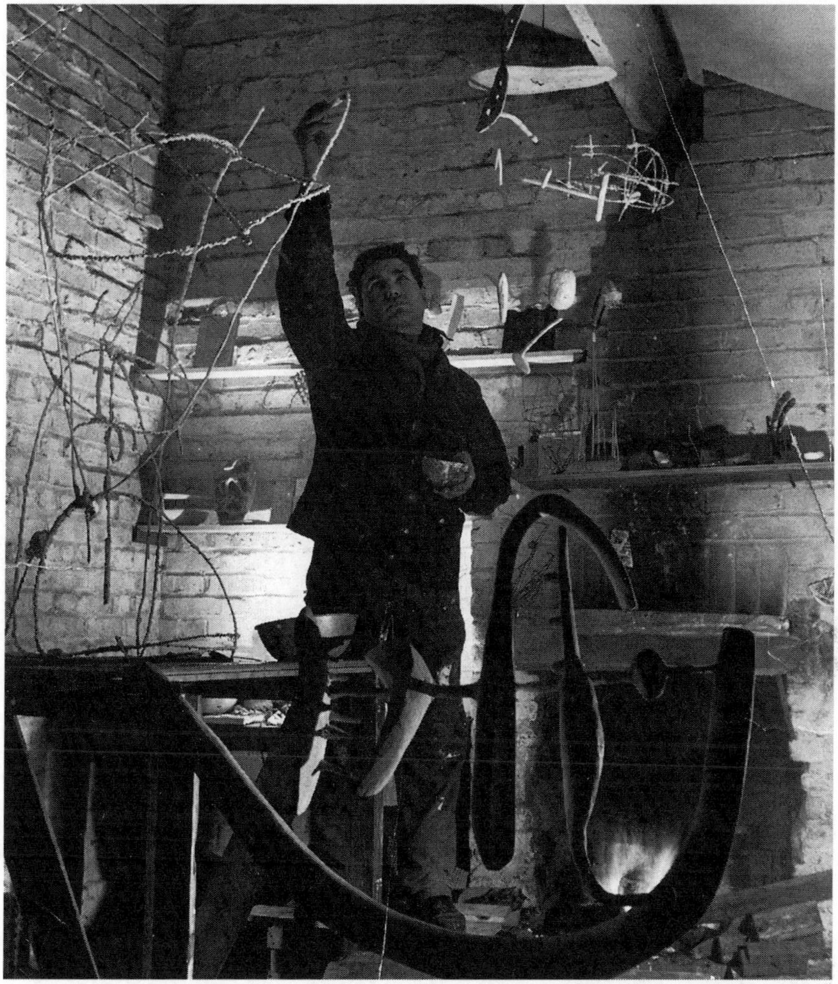

7.1 Eduardo Paolozzi in his studio

bar, with ... black spots' and 'bottles ... and the lovely glass, like bricks'.[35] Fry, Drew and Paolozzi shared common ground in their respect for Herbert Read's belief in the need for artists to create useful work for the benefit of society.[36] This view is embodied in Paolozzi's work in connection with Fry and Drew's unrealised block of flats at Whitefoot Lane in London. His design for a playground of concrete shapes, like giant building blocks, sought to provide a fantastical landscape for children in the urban environment.[37] A subsequent commission by Fry and Drew for a mural at Passfields Flats in Lewisham also went unrealised.[38] Although commissions were meagre in post-war London, Drew was quick to promote the young artist's work wherever possible. In 1952 she commissioned Paolozzi to design 'Collage Mural' for Gloucester Place, which was then hung in one of the first floor drawing offices. The collage of silkscreen fragments of geometric, overlapping forms – like his wallpaper designs of the same period – create unexpected associations and relationships in a rich connectedness that may be compared to CIAM's ideas regarding the 'core' of the city.[39]

Drew was also amongst the first architects to use wallpapers and fabrics by Hammer Prints, a company established in 1954 by Paolozzi with Nigel and Judith Henderson to exploit silkscreen techniques for the mass-production of artwork.[40] Drew perceived no distinction between their living and working quarters at Gloucester Place, and used Hammer Prints to decorate rooms throughout the building. A contemporary photograph of an office – perhaps Fry's room – shows this blend of work and home life, juxtaposing the practice's latest work pinned the wall with small sculptures and paintings (Figure 7.2). The couple's third floor dining room was dramatically papered with a Hammer Print to the ceiling, softened by curtains of the same design (Figure 7.3). A variation of this wallpaper was subsequently used to decorate the office ceiling of a structural engineer at

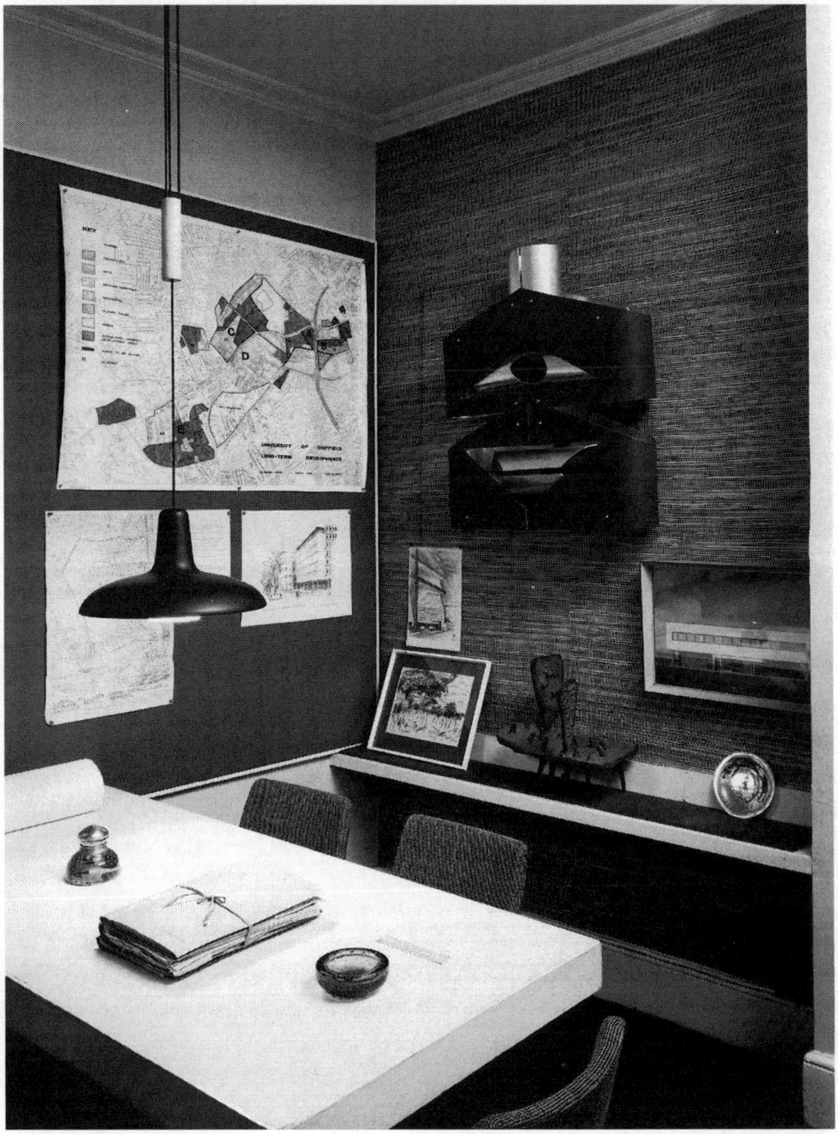

7.2 Meeting Room at 63 Gloucester Place, 1966

Ove Arup and Partners, Ronald Jenkins.[41] It is likely that Drew had introduced Jenkins to Paolozzi, as both worked with her on the Festival of Britain buildings. Indeed, Drew's commissions gave Paolozzi significant exposure amongst the tight-knit architectural community. In March 1952 the cover of the *Architectural Review* featured a textile by Paolozzi, designed originally for Drew's living room and adapted for use in F.R.S. Yorke's house at Wootton.[42] Inside, an article on 'Printed Textiles' followed the example set by the Festival of Britain, illustrating the relatively low cost methods of integrating artwork of 'the younger English designer' and architecture. Paolozzi's work for Drew, and for Eugene Rosenberg, is presented amongst others as an example of collaborative design.[43]

In their Tropical Architecture course, which had been established at the AA in 1954, Fry and Drew expounded the integration of art and architecture. Artists, including Paolozzi, instructed AA students on the merits of an artistic synthesis, an approach that Fry and Drew also sought to convey in their buildings overseas. In West Africa they had developed a language of sculptural concrete balusters and brise-soleil to provide sun shading and to add a decorative aspect to the otherwise undistinguished buildings. This used a single basic form with slight variations between each project to provide a cheap but instantly recognizable element – a Fry and Drew signature – to their architecture. Like Paolozzi's collages of the same period, the repeated elements suggest a pattern that might extend infinitely beyond the building (or canvas). As Diane Kirkpatrick notes, Fry and Drew 'were enchanted by the possibilities of the sort of never-ending pattern'.[44] The *Architectural Review*'s photography, and Gordon Cullen's accompanying artwork, of the concrete balusters at Adisadel College also emphasized this pattern-like quality and the primitivism of the sculptural forms (Figure 7.4).[45]

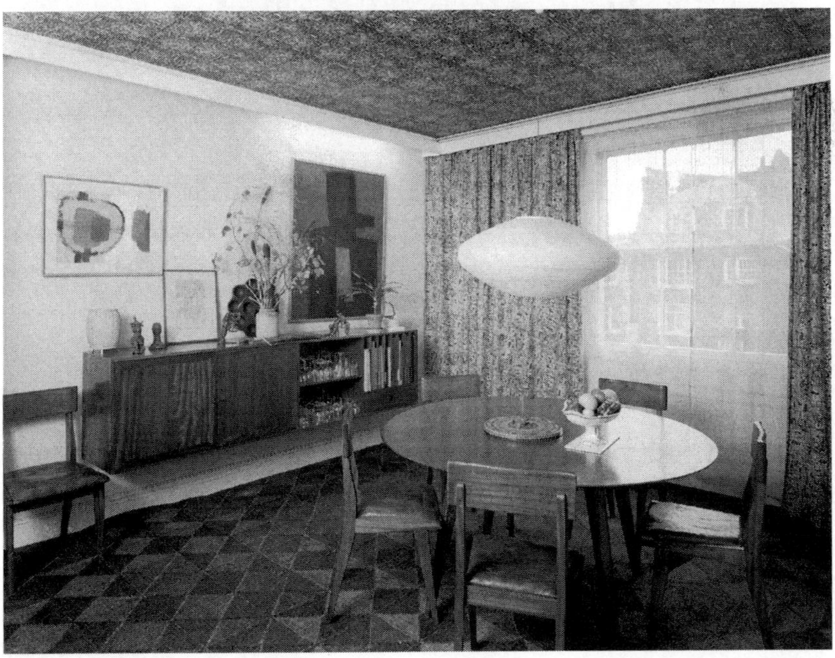

7.3 Dining Room at 63 Gloucester Place, 1966

This integration was also promulgated in the MARS Group's 'Turn Again' exhibition, which opened in July 1955 at the Royal Exchange in the heart of London. At Fry's suggestion,[46] the MARS Group staged 'Turn Again' in protest at the 'lowered architectural standards' of post-war building in London.[47] The exhibition focused on office development, which had increased dramatically in the early 1950s; in 1951 the London County Council permitted 1.7 million square feet of new office space, a figure that rose to 5.9 million in 1955.[48] The exhibition committee combined old and new generation MARS members, co-chaired by Fry and Drew, and selected photographs of new office work from New York, Rio de Janeiro and Milan.[49] In a passage that sounds remarkably like Fry, the exhibition catalogue asserted: '[T]he architect plays the tune – humanising the machine – collaborating with nature converting the arid city to a symphony of geometry

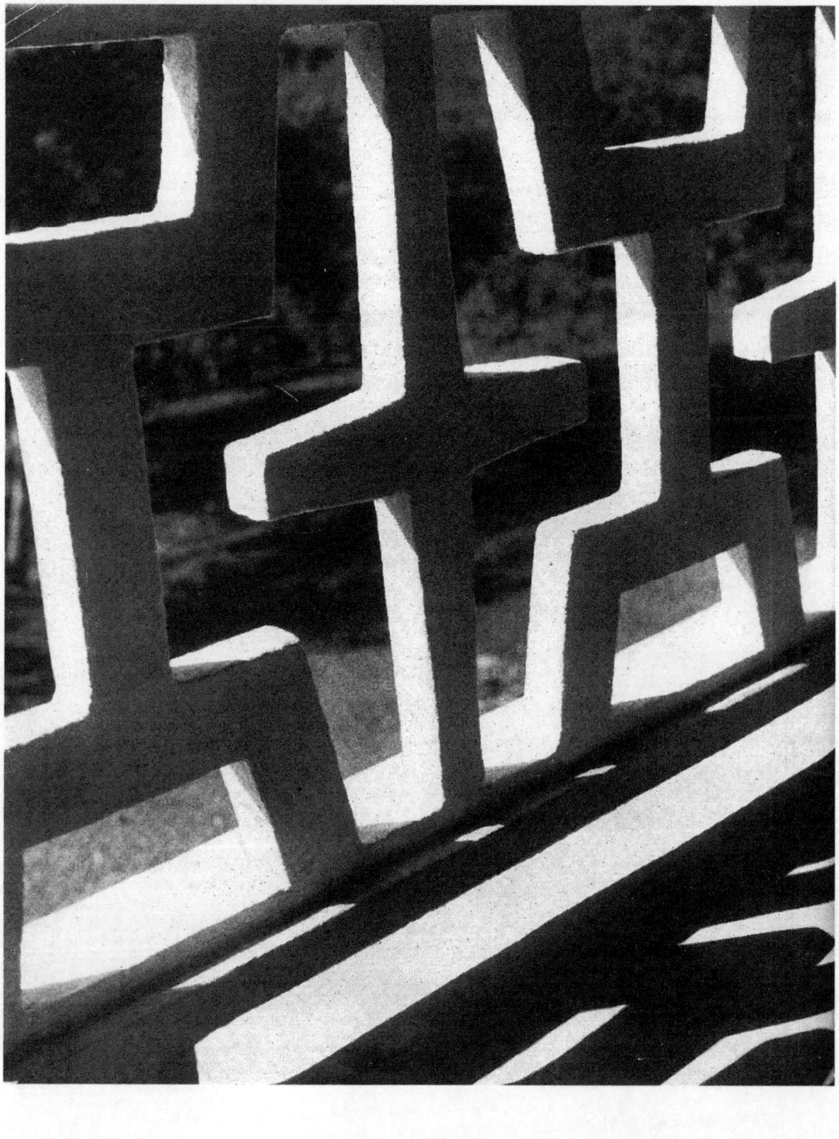

7.4 Sculptural balustrade, Adisadel College, Ghana, 1955

and natural form in which the scale of the common man is registered and his emotions find a place'.[50]

'Turn Again' signalled Fry and Drew's efforts to continue the immediate post-war impetus that had fostered so much activity amongst the MARS Group and CIAM. The exhibition was a comment on contemporary building in London, but for some MARS members 'Turn Again' was the continuation of accepted modernist practice, offering little innovation; Lasdun later commented that he referred to the exhibition as 'better late than never'.[51] However, the exhibition neatly encapsulates Fry and Drew's thinking during the mid-1950s. As a collaborative project between MARS architects and ICA-centred artists, the exhibition demonstrated how architecture might combine art and the machine. The exhibition was used as an opportunity to target the soulless 'machine technique' of contemporary office buildings, of which Fry declared, 'they horrify by their excellence in the thing they set out to be, which is a shining machine product, and some of these places have no colour even – the offices do pursue the machine technique to the end without artistry'.[52]

Tasked with the unenviable role of Exhibition Architect for the MARS Group, John Bicknell (1929–97) co-ordinated contributions that included large-scale typography and a mural by Edward Wright, a cartoon of the 'Whittington Cat' by Nicolas Bentley, and murals by both Eduardo Paolozzi and Victor Pasmore. Contemporary office furniture and fittings provided examples of good working environments, and were framed by typical MARS propaganda that hinted at the lingering difficulties (and frustrations) in post-war Britain (Figure 7.5). As the exhibition catalogue, designed by the editors of *Architectural Design* – Theo Crosby and Monica Pidgeon – proclaimed, 'This is the moment to Turn Again. When London rebuilds on the ashes of Hitler's fire, let us not be vanquished by our own past – the future lies within us'.[53]

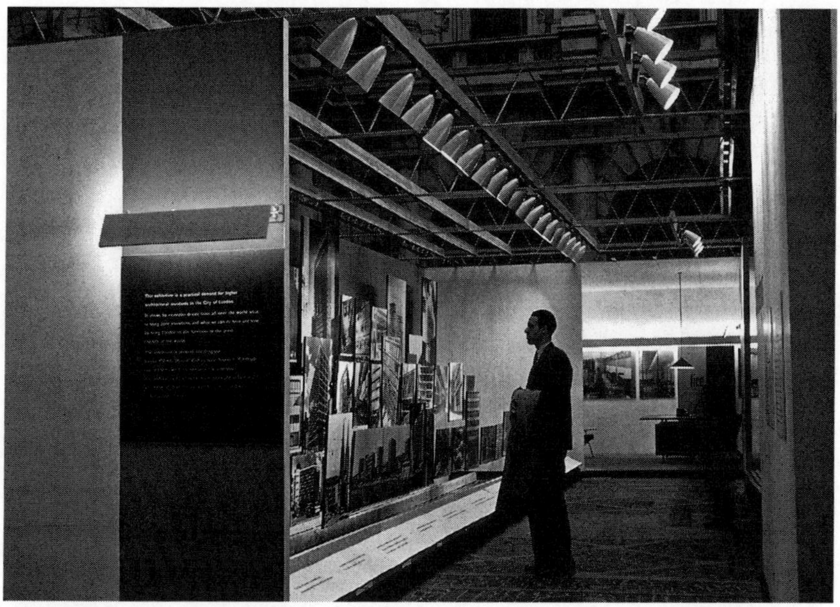

7.5 'Turn Again' Exhibition, 1955

PROPORTION VERSUS HUMANISM

It is evident that, post-war, Fry reconsidered his architectural outlook. Whilst he continued to follow Walter Gropius's 'synthetic vision' of the arts,[54] Fry became increasingly disillusioned with the 'solemn approach' taken by Gropius and instead looked to humanise his work.[55] This revisionism was part of a wider movement aiming to reconcile functionalism with humanist notions of symbolism and organicism.[56] As humanism had again come to the fore, post-war discussion focused initially on a Renaissance conception of humanism concerned with the use of geometry and proportional systems to invoke philosophical meaning in architecture.[57] Wittkower's writing on *Architectural Principles in the Age of Humanism* (1949) had set useful historical precedent that influenced modernists during the 1940s and 1950s.[58] At around the same time, Le Corbusier's had published *The Modulor* (1950), and both Wittkower and Le Corbusier attended the 1951 Milan Triennale, which considered 'De Divina Proportione'. Fry and Drew also contributed to proceedings, with their design for a small pavilion that investigated these divine proportions (Figure 7.6). Yet Fry remained unconvinced by Le Corbusier's Modulor as a universal system and later wrote of his inability to follow it, 'In his hands it is a chariot to heaven; but in others it can be a bus to a dusty terminus'.[59] However, he did experiment with proportional systems, demonstrated by his design for Wudil Teacher Training College. Planned as a series of related squares, the buildings turn inwards to provide shaded enclosures thus protecting inhabitants from the dry heat of Northern Nigeria.[60] Generally, however, Fry's approach to proportion was less than rigorous and he later highlighted the importance of an emotional, rather than a scientific, response to design.[61] Of his design for a Baha'-ist House of Worship (1955) in Kampala, for example, he wrote to Drew:

> *I have played on their magic number 9. The interior is 63" (7 nines) the outer enclosing wall 99 (11) the suspended ceilings 36' (4) the reading canopy 18 (2) the doors 9 (1) and by an accident the height of the dome from the ground is 63 (7). A good many dimensions are not to be mentioned, but some of these are in fact cardinal and may well bring proportion to the whole.*[62]

As Eva-Marie Neumann notes, much of the architectural community's widespread interest in proportion might be related to the development of industrialized building and systems of prefabrication.[63] While Drew had seized upon opportunities in industrial design, Fry to a cautious approach as, for him, the benefits of direct involvement in manufacturing were outweighed by the potential of a reduced role for architects and, moreover, 'the dehumanizing effects of uncontrolled industrialization'.[64] Indeed, for Fry, ideas of a proportional system were soon subsumed by a more comprehensive notion of humanist architecture. Fry enlarged upon his interwar work with educationalist Henry Morris (which was in turn based upon Geoffrey Scott's Renaissance sense of humanism) as Morris's ideas assumed new potency in connection with the post-war rebuilding programme. Morris argued that educational centres should be at

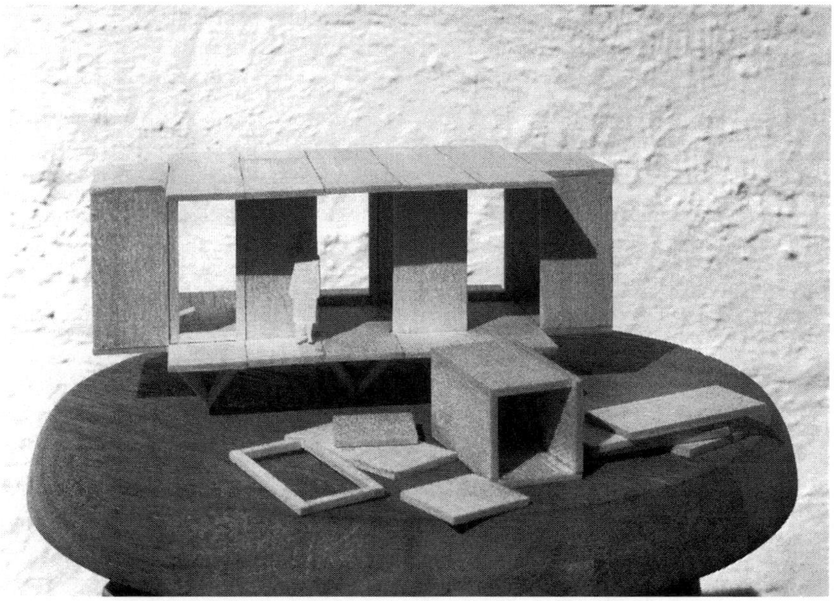

7.6 Model of a Pavilion for the Ninth Milan Triennale, 1951

the heart of planning New Towns and new communities. In an address of 1956 to the RIBA, Morris said:

> *We are living in a world dominated by applied science and technology ... Modern architecture, which is the result of new structural principles and materials with a mechanical logic of their own, is confronted with an imperative it must obey. This is, that, in addition to its practical utilitarian functions, modern architecture must nourish humanist values especially in its external service of expressing the significance of man's activities, of giving nobility to its environment, and ministering to his delight and appetite for beauty.*[65]

Morris's influential ideas were officially recognised and he was appointed to the Ministry of Town and Country Planning to advise on the cultural aspect of the New Towns. In 1957, with backing from the Government and the Welwyn Garden City Development Corporation, he established the Digswell Arts Trust. Drew, Herbert Read and Jack Pritchard all became trustees, and helped to promote the use of professional artists to create civic artwork for the benefit of society.

Fry and Drew's humanist approach sought to highlight the symbiotic relationship between the natural world and civilisation, and might be summarised as the following central concerns: direct application of artwork to the fabric of a building; use of hand-finished materials in contrast to a 'machine-made' structure; inclusion of humanist signifiers to educate and inspire the building's users; a basic use of geometry and proportion; careful integration with the landscape; and the creation of community, at project level and within the wider area. Fry and Drew's developing ideas of humanist architecture are manifest in the articles published by the *Architects' Year Book* (*AYB*) during this period. In 1953, the magazine included: 'Systems of Proportion' by Rudolf Wittkower; a text by Max Bill on the Bauhaus;

another from Leslie Martin on Gropius; and a contribution from Jacqueline Tyrwhitt on 'The Core and the City'.

In 1955, Fry and Drew were able to test out these ideas: the practice was commissioned to design a company headquarters for Pilkington Brothers at St Helens in Lancashire. As an historic glass manufacturer that promoted technological innovation, Pilkington Brothers provided the ultimate case study and Fry assumed the role of Project Architect to create, in his words, 'a monument to … industry' that expressed his belief in the progress of society.[66] A sizeable budget of £1 million (which would rise to almost £4 million upon completion) and enlightened clients – who saw themselves as patrons to the British art and design scene – enabled Fry to assemble a prestigious collective of artists to create an outstanding collection of post-war applied art to rival that of Basil Spence's Coventry Cathedral.[67] From the earliest stages, Fry envisaged the Pilkington Brothers' Headquarters as a synthesis of art and architecture taking its cue from Herbert Read's *The Grass Roots of Art* (1937). Read's volume of lectures are united by the central theme, in Read's words, 'that art is in some sense intimately related, not only to the social structure, but even by the very soil and landscape of a country'.[68] Fry wrote to Read after reading the recently re-published book in October 1955 and asked his friend to visit Gloucester Place to give advice on potential collaborations with artists.[69] No record of their meeting remains but, just a few months later, Fry issued a 'Memorandum on the Employment of [a] Sculptor' to the Pilkington building committee, which proposed the appointment of a sculptor or a painter to work in close connection with the architect. Fry's persuasive memorandum seals his intention in the post-war reassessment of Modernism: to create a synthesis of art and architecture in meaningful dialogue, which might give expression to a new era of craftsmanship, and do so against a growing tide of mechanisation and materialism. He pointed to the growing contemporary trend of using painters and sculptors to 'heighten the effects of architecture' and suggested that, by appointing an artist to 'humanise' the headquarters, Pilkington Brothers would lead the way in 'a somewhat fumbling and undecided, but none the less general movement'.[70] Fry suggested that Henry Moore (one of Read's favourite artists) be appointed to work with him, but this collaboration did not materialise.[71]

The building complex was intended by Pilkington Brothers to be 'considered as a work of art and a serious contribution' to architecture.[72] Situated on a green-field site of 16 acres, in a valley one mile from the town centre, the headquarters are bounded from the north-west to the north-east by the main route to Liverpool, by suburban property to the south-west and Pilkington's Ravenhead Glassworks to the south-east, which was then one of the company's five factories in St Helens (Figure 7.7). The natural contours of the site fall approximately 40 feet from the north-west boundary to the glasswork's cooling reservoir at the foot of the valley. Exploiting this change in level, the headquarters centres on a 12-storey office tower, which is visible on the route into St Helens from the west. The layout of four-storey lakeside and court blocks, with inward- and outward-looking offices, provide a variety of public and private spaces to foster a lively working environment (Figure 7.8).

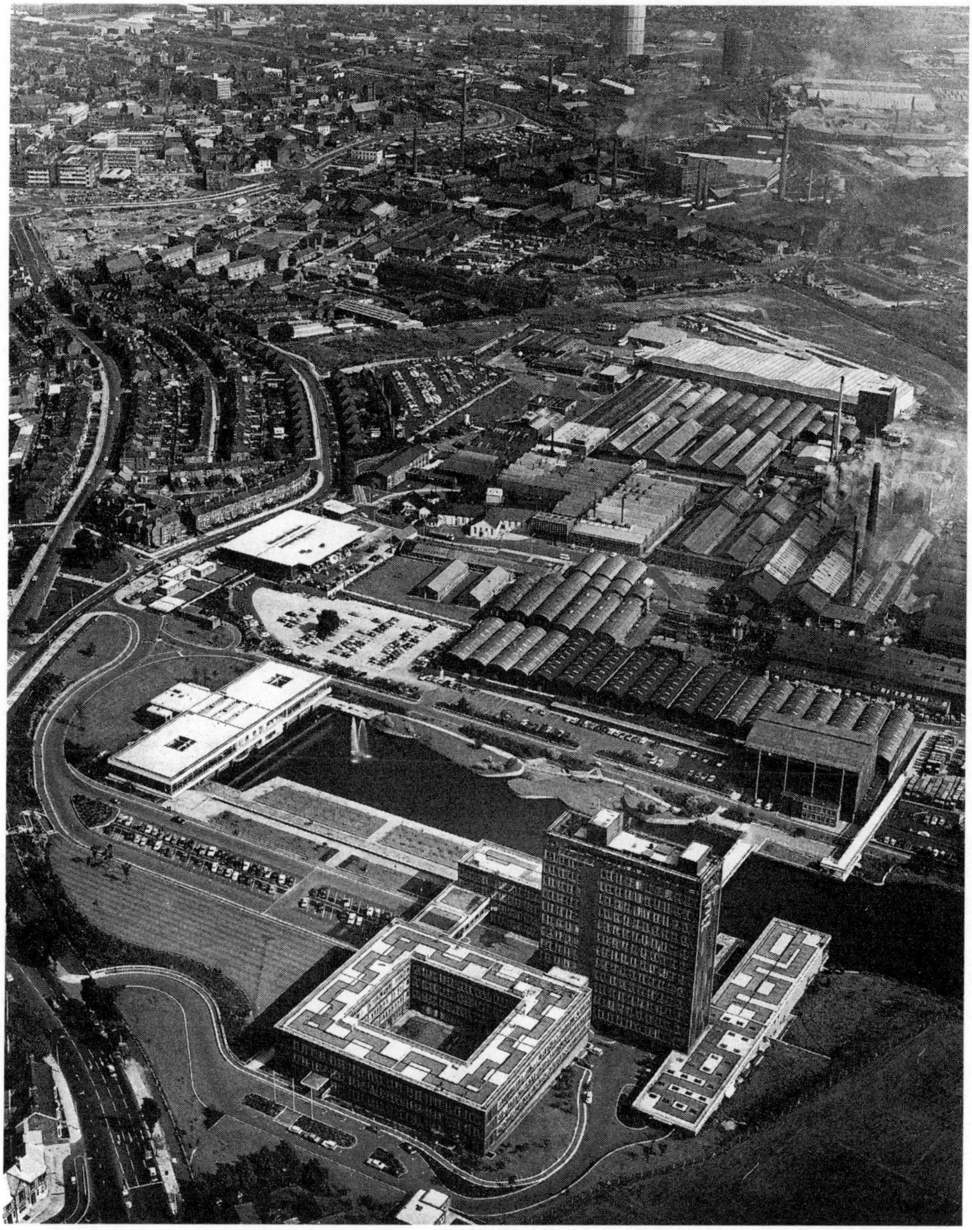

7.7 Aerial View from the West, Pilkington Brothers' Headquarters, St Helens, c. 1965

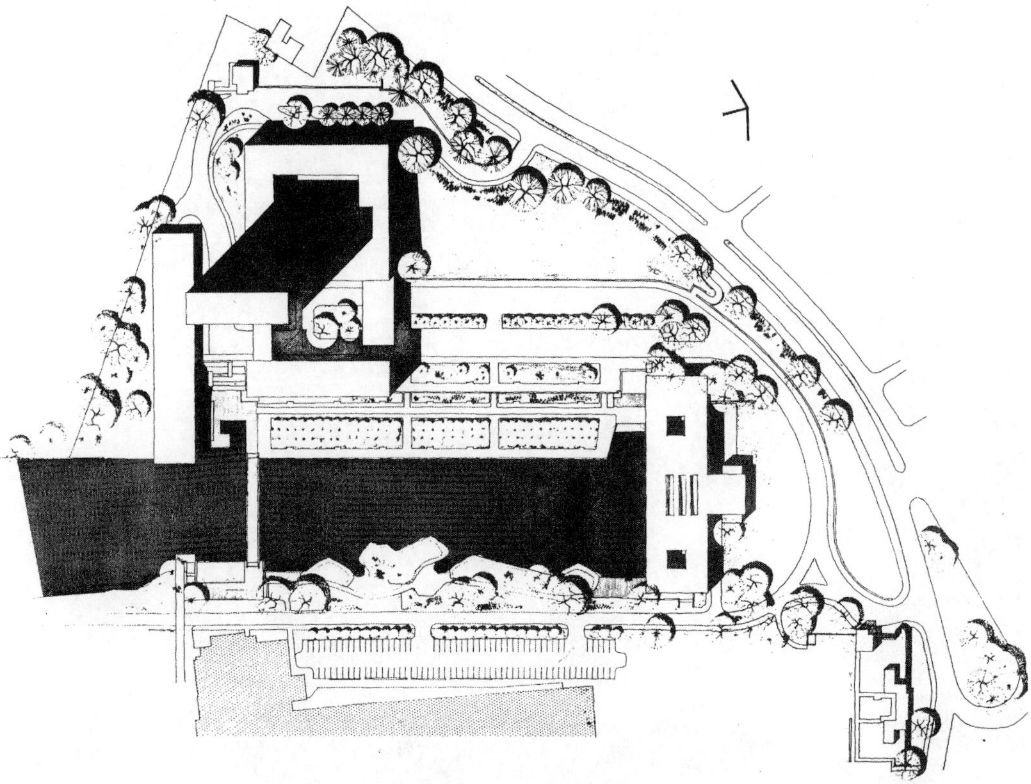

7.8 Site Plan, Pilkington Brothers' Headquarters, St Helens, 1958

Internally, accommodation provided a hierarchy of offices: senior staff housed in generous offices in the prestigious tower block; cellular offices for middle management around the perimeter of the court blocks; and innovative, open-plan style offices for administrative staff. The separate offices for Pilkington and for its subsidiary company, Fibreglass – previously located on different sites – were brought together to promote further a unified working community. In the lakeside court, healthcare facilities, including a medical centre with 'rehabilitation' services for staff injured at work, were combined with other staff amenities such as banking facilities, a hairdresser and a library.[73] This combination of workers' services and offices was an attempt by Fry to create a community 'core', resulting in a small-scale civic environment in which the workers might socialise.[74] This city-in-miniature also draws comparisons with campus design of New Universities during the 1960s, an area that Fry and Drew looked to go into.[75] A pavilion housing a 'museum of glass' and a staff canteen block completed the complex. Overlooking the reservoir –extended to form a sizeable lake – the two-storey canteen provided an open plan staff dining-hall on the first floor and smaller, private rooms for senior staff on the ground and first floors; internal courtyards were designed to bring natural light into the building's deep plan (Figure 7.9).

A deliberate reaction to the sealed-box Modernism of Mies van der Rohe's Seagram Building (1954–58) and Lever House (1951–52) by Skidmore, Owings and

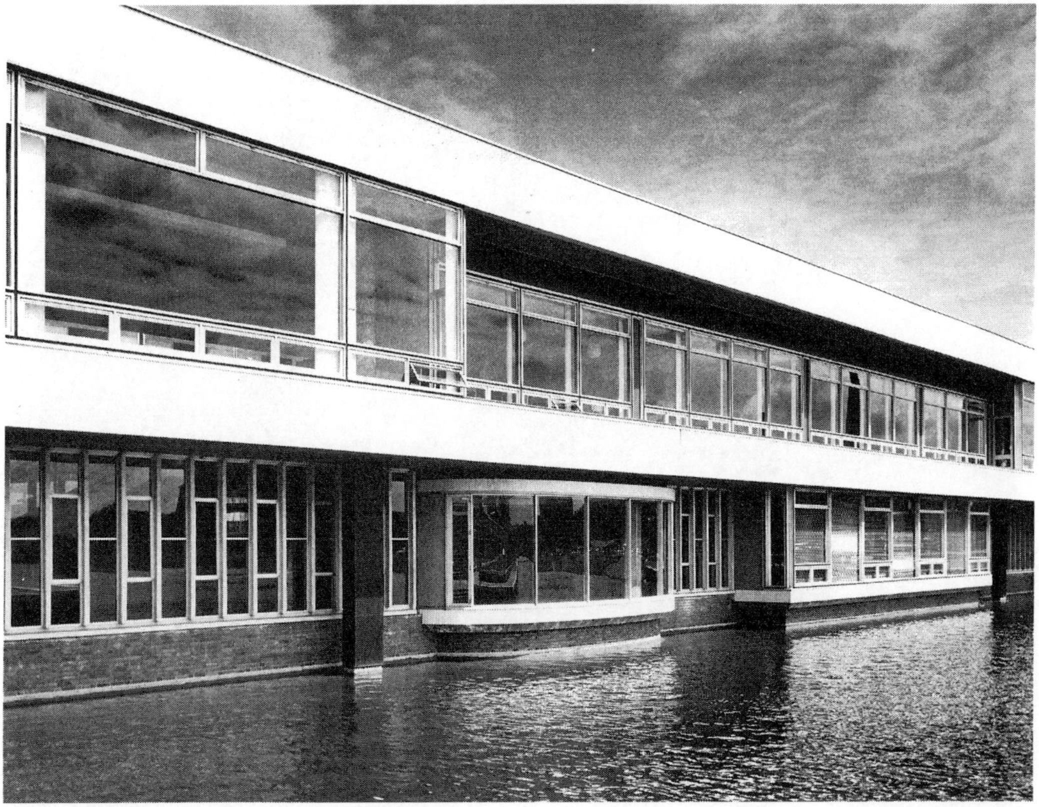

7.9 South Façade of the Canteen, Pilkington Brothers' Headquarters, St Helens, c. 1965

Merrill's Gordon Bunschaft and Natalie de Blois, Fry's intention to 'humanise the machine' is evident throughout. The tower and the court buildings are of pre-cast concrete frame, planned on a module of four feet six inches: the vertical elements are wrapped in polished Westmoreland slate, while the horizontal elements are clad with Pilkington-made 'Armourclad' spandrel panels. The natural colour variation of the slate contrasts with the highly reflective blue glass panels to provide colour and texture to the 'machine-made' structure. The north and south facades of the tower are clad entirely in 'Armourclad' panels in four shades of blue, while the east and west facades are infilled with floor to ceiling double-glazed 'Insulight' units with aluminium sash frames (Figure 7.10). The use of traditional sliding sash windows, manufactured at considerable expense, reinforces Fry's desire to enliven the building; in place of air-conditioning, the occupants are able to interact with their surroundings and have a degree of control over their environment, thereby humanising the façade of what might be an anonymous tower block. This contrast of high-tech and natural materials was particularly effective for showcasing the modernity of the Pilkington Brothers' product range and, by extension, the company.

The highpoint of the new headquarters was undoubtedly its artwork, used throughout the buildings to inspire and uplift the 1,500 office workers. Fry and the building committee members assembled a collective of 16 artists to undertake a total of 24 works.[76] Fry and Pilkington Brothers' sought to continue the spirit of

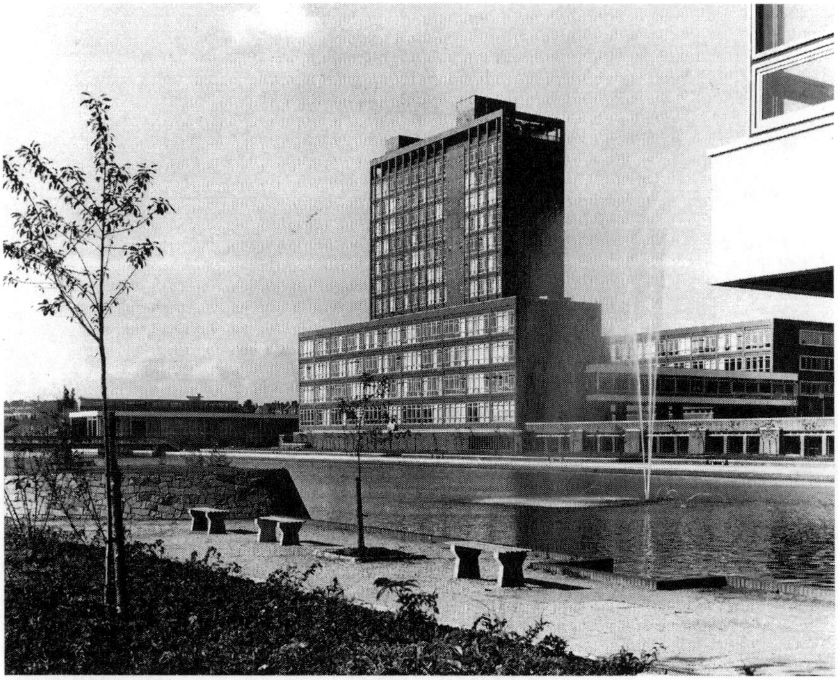

7.10 View of the Tower Complex from the North-East, Pilkington Brothers' Headquarters, St Helens, c. 1965

co-operation between art and industry fostered by the Festival of Britain with their commissions: the majority of the artists were graduates from the Royal College of Art and had participated in the Festival, thereby promoting the practical application of contemporary art. Commissions occupying principal locations in the canteen and the main entrance hall were awarded to high profile artists proposed by Fry: Victor Pasmore and Avinash Chandra.[77] Chandra's piece was amongst his series of large-scale, coloured glass murals undertaken for corporate clients during the 1960s.[78] The representation of fire, 'which lies at the heart of glassmaking', measures 37×9 feet and comprises laminates of coloured, clear and wired glass and plastic, in sweeping circular forms (Figure 7.11).[79] Meanwhile, Pasmore created three works for the canteen: a 60×10 feet mural of sculptural timber shapes on a green canvas background to the workers' canteen (Figure 7.12); a further mural painted and fired onto plate glass with a white Vitrolite background to the non-staff canteen; and a free-standing mural to the canteen lobby. Pasmore held similar views to Fry regarding the need to integrate art and architecture and was obviously a preferred choice, as Fry had proposed employment of the artist as early as August 1957.[80]

A variety of styles are evident. The Pilkington Directors evidently favoured decorative mirrors, which are used extensively in the directors' private rooms: Jennifer Simpson designed a mirror for the Senior Managers' Dining Room; for the Rainford Dining Room, John Drummond designed a work comprising three layers of silvered glass; Humphrey Spender used antique glass, glass mosaic and silvered bullions of glass to create a decorative mirror for the Roby Dining Room and, in a small conference room, he designed another piece comprising 299 pieces

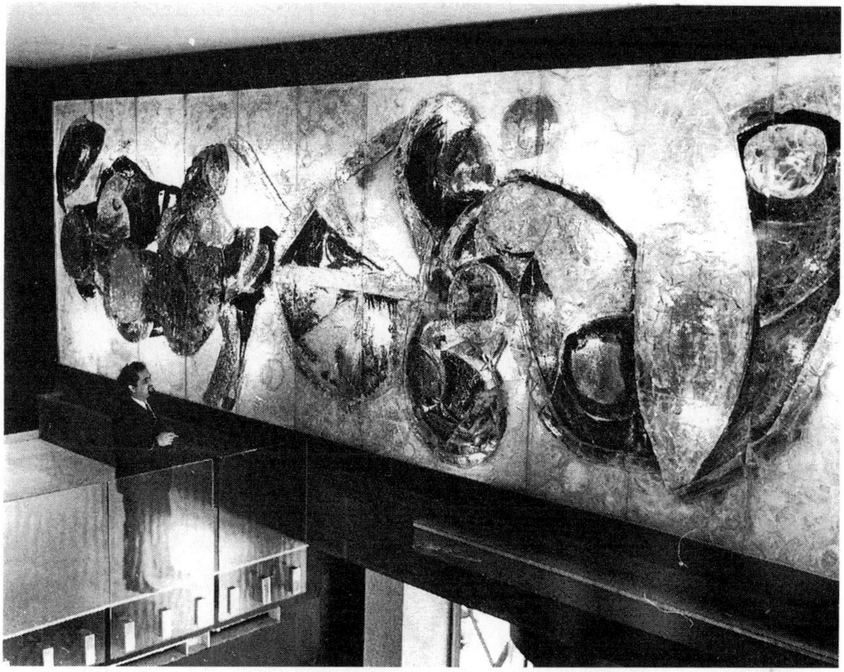

7.11 Avinash Chandra with his *Fire* mural, 1965

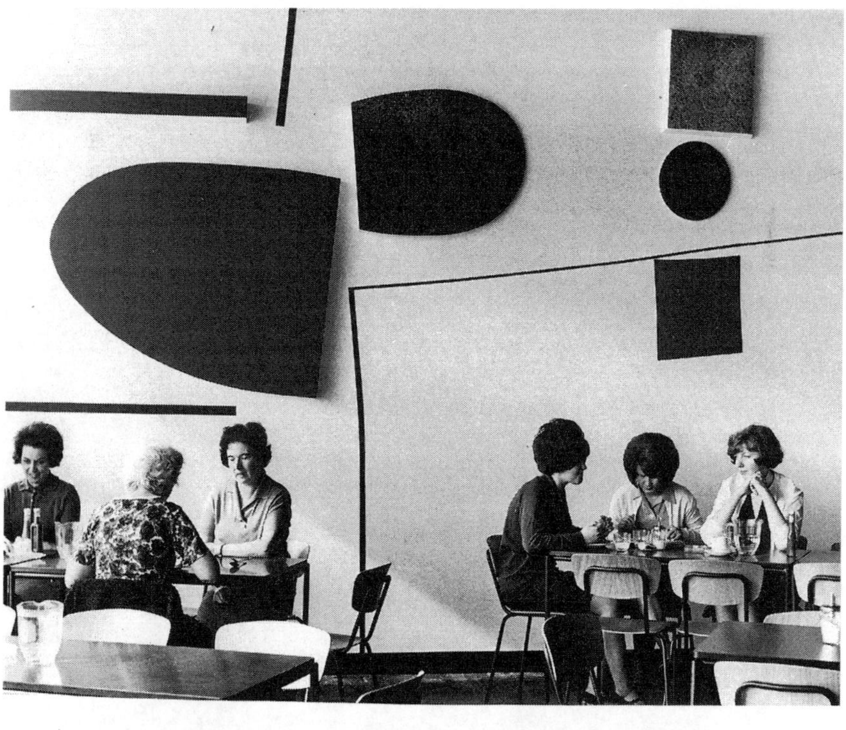

7.12 Victor Pasmore, *Untitled*, 1965

of mirrored glass;[81] John Hutton produced a mirror of blue-tinted glass, engraved with an ethereal depiction of the 'Three Graces' for the Windle Dining Room; for a small dining room in the tower, Edward Bawden created a brilliant-cut mirror; and, for the Directors' Large Dining Room, Professor Robert Goodden designed a topaz-tinted mirror, decorated with engraving, acid-etching and brilliant cut forms. Further internal decoration was provided by paintings in the office waiting rooms by William Kempster, which were incorporated into decorative units by James Gardner with mirrored maps illustrating the location of Pilkington's operations overseas; Edward Bawden painted a floor-to-ceiling image of glass-making processes for the Directors' Small Dining Room; and, for the entrance to the Fibreglass offices, in the court block, Paul Mount designed a relief sculpture fabricated from glass reinforced plastics.

Another group of artists experimented with techniques applied to the windows of the canteen block; these works foreshorten the view, reminding the onlooker of both the company's innovative future in the new buildings outside and the long tradition of skilled hand-craftsmanship in the artwork itself. Charles de Vic Carey designed four stained glass windows for the screen wall overlooking the lake; Don Foster used tinted glass and bullions to create a leaded light of geometric forms; Trevor Long created an abstract design; Barbara Jones produced a composition using brilliant cutting and partial silvering; J. Perez Roman designed two works using sand-blasting to tinted glass; and windows by John Hutton depict dancing figures that recall his West Screen at Coventry Cathedral (Figure 7.13). As the Pilkington brochure states, 'almost every possible effect with translucent glass is skilfully exploited'.[82]

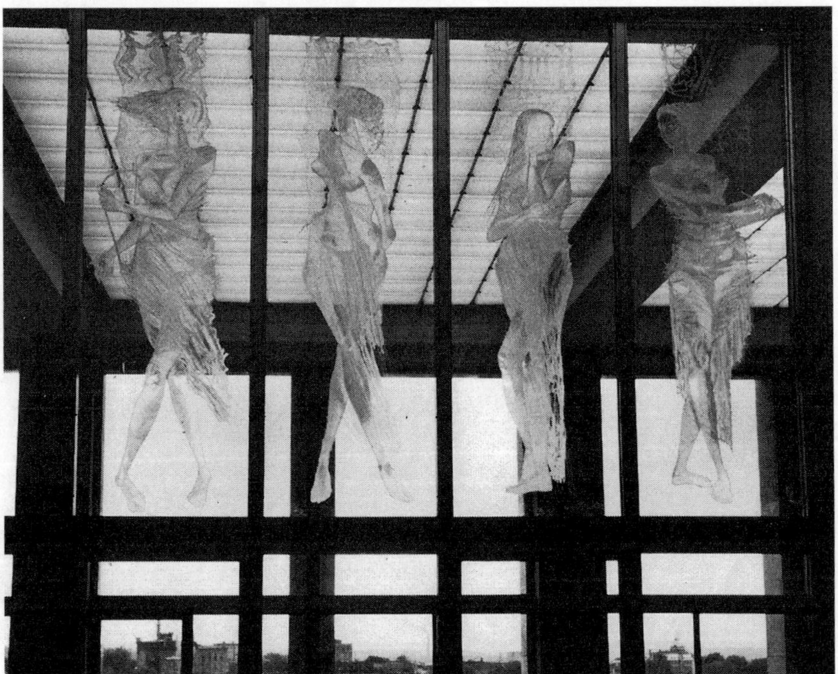

7.13 John Hutton, *Untitled*, 1965

The industrial folk museum was intended to reinforce the wider message of the glass-making tradition, and its integration of art and science. Fry believed that the exhibits would 'thrill' the Pilkington employees, fostering pride in their work, the company and in human ingenuity.[83] To develop the employees' understanding of their involvement in industrial innovation, Fry believed the museum should portray the long history of the industry, not just that of Pilkington Brothers. Fresh from his work at the 1958 Brussels World's Fair, which Fry and Drew had attended, James Gardner was selected by the building committee to fit out the museum 'shell'. Although Fry discussed his vision for the project with Gardner, there was some disparity between the architect and designer's ambitions and the museum suffered due to a lack of collaboration in its early stages. Fry's Miesian glass pavilion cantilevered over the lake had little relationship to its surroundings, as the exhibition turned its back on the full-length east window looking out over the water (Figure 7.14). Gardner later commented, 'Visitors didn't come here to view the lake, so I introduced a fibre glass curtain to soften, and structural panels to cut the direct sunlight'.[84] Nevertheless, Pilkington provided a sizeable budget of £10,000 to purchase rare museum pieces and they aimed to rival the collection held at the Corning Glass Centre.[85]

In 1961, the building committee appointed Drew as a consultant designer for office furniture and the interior design of key areas, such as the main reception.[86] Drew was experienced in designing prestigious interiors for international companies, such as a commission of 1959 for the interior layout, fixtures and fittings for Swan and MacLaren's for Shell Oil Company in Singapore. The 14-storey Shell office block was finished to the highest standards, and Drew combined new materials, including PVC wall covering and Formica-topped desks, with more traditional decorative elements, such as Japanese silk wallpapers.[87] Continuing close ties with British industry, at the Pilkington Headquarters furniture companies such as Kandya and Gordon Russell were approached, although budget cuts prohibited their eventual use.

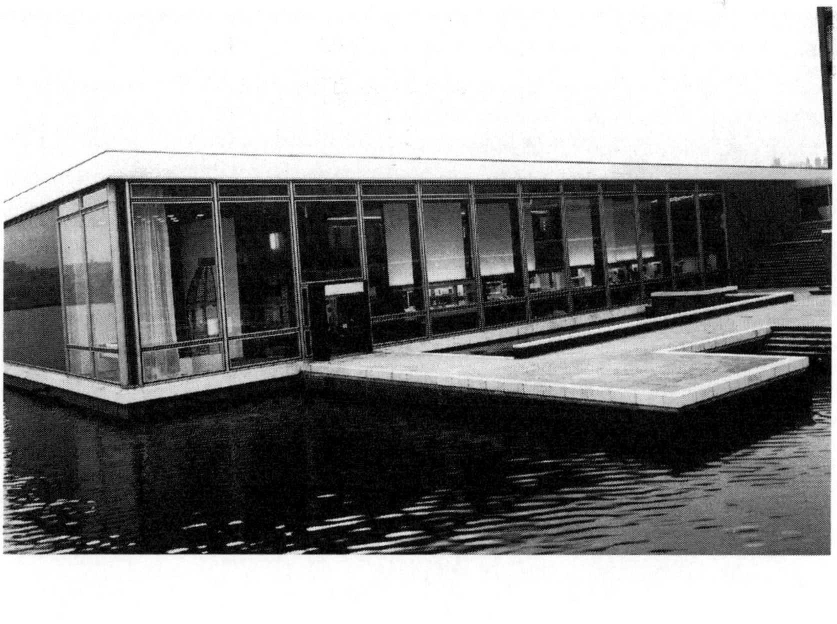

7.14 North Façade of the Museum, Pilkington Brothers' Headquarters, St Helens, c. 1965

The headquarters was clearly an exceptional commission. During the design process, from November 1956 to early in 1957, Fry took up a visiting professorship at the Harvard Graduate School of Design (GSD) – much to the despair of the Pilkington building committee. He was probably invited by the GSD Dean, Josep Lluís Sert, and the teaching post enabled both Fry and Sert to develop their work on the urban 'core' at smaller scale. They set design briefs to test out their ideas, asking the students to design 'A Headquarter Office and Welfare Group for a Large Industrial Corporation', which exactly followed the design brief set out by Pilkington Brothers.[88] Fry taught the module together with Ronald Gourley, Hideo Sasaki and Sert. This group also set a design problem for 'A Design Center for Harvard University',[89] with both projects investigating the impact of art and design on contemporary life. The student project was envisioned as a centre for 'the visual arts in action – as design for communication, as design for industry, as design for our environment'.[90] The project perhaps explored some of the ideas raised by the university's Committee for the Practice of Visual Arts, of which Sert was chairman, which at this time was looking to appoint an architect for such a centre, eventually leading to the commission of Le Corbusier's Carpenter Center (1959–63).[91]

The continuing influence of the core, is evidenced in a collection of readings assembled by Sert at this time to lead student discussion. He wrote of the core as the 'realm of the pedestrian' to enable the exchange of ideas, 'Man today observes, listens, and suffers – but he has no longer the means to be a participant'. The core is presented as a 'Centre of the Arts', an artefact that expresses 'the collective mind and spirit of the community, which humanises and gives meaning and form to the city itself'. Sert's 'Short Outline of the Core' includes a brief statement by Fry, which neatly summarises his humanist agenda, 'this idea of a city heart in harmony with the truth of times present, and with truth forever, is one in need of human contact at every stage of its inception'.[92]

During Fry's spell at the GSD, he may also have become aware of the 'corporate campus' of the Connecticut General Life Insurance Company Headquarters (1953–57), designed by Skidmore, Owings and Merrill's Gordon Bunschaft and Natalie de Blois.[93] The Connecticut headquarters opened shortly after Fry's professorship and it shares formal similarities with the Pilkington scheme; it comprises a three-storey curtain wall office block arranged around four internal courtyards, an executive wing, a cafeteria pavilion cantilevered over a reflecting pool, and extensive welfare and recreational facilities. Fry did not acknowledge any debt to the American 'corporate campus' but this is not surprising given the commercial interests of such international corporations, which were antithetical to his own belief in societal co-operation and unity with the natural environment. Instead, Fry's contemporary texts publicising the project emphasise the ideas of post-war civic rebuilding programmes and his continuing alignment with the city 'core' to create a multifunctional, vibrant headquarters.[94]

Contemporaneous with the Pilkington Brothers commission was Fry's work for a new Veterinary School and a Civil Engineering Building at Liverpool University. These educational projects demonstrate the pervasiveness of Fry's humanist architectural approach, which were undertaken on a much smaller budget than the Pilkington Headquarters. Writing in the early 1960s, Fry suggested that university buildings should 'nurture the soul, and this it does by investing the elements of composition with references to what lies very deeply bedded in both our animal nature and with what connects us with the universe'.[95] In the Civil Engineering Building (1955–60), for example, Fry made apparent the position of the students within the engineering tradition (and therefore the universe) through a huge decorative panel covering an entire façade of the six-storey, brick-clad tower (Figure 7.15).[96] Fry explained to his clients that the cast concrete panel of lettering placed directly above the main entrance, 'contains the names of the greatest civil and structural engineers as a potent reminder to students of the high standards of imagination and integrity upon which their profession is founded'.[97] Fry commissioned the Cornish artist Peter Lanyon to design a mural to the entrance hall. The piece, entitled *Conflict of Man with Tides and Sands*, represents the interaction of natural and human forces

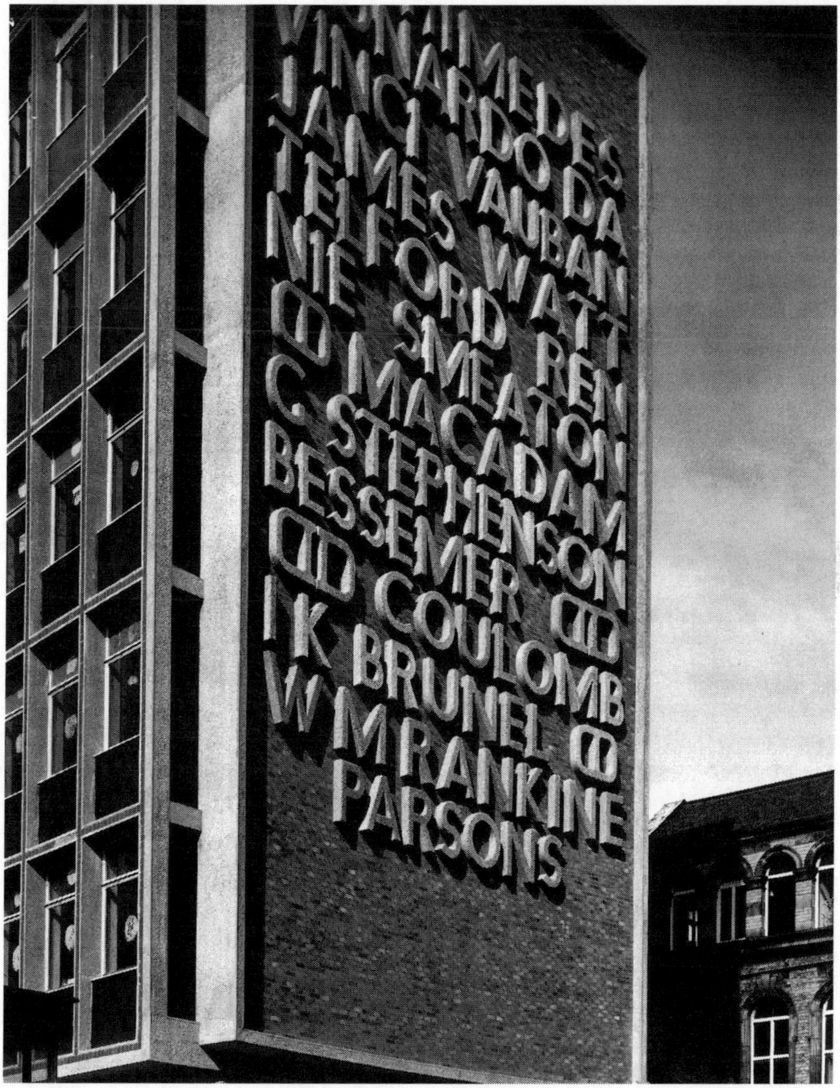

7.15 Civil Engineering Building, Liverpool University, 1960

and confronts students upon their entry to the building, encouraging a moment of metaphysical contemplation amidst the everyday (Figure 7.16). As with the Pilkington Brothers' Headquarters, the Civil Engineering Building was a humanised structure that combined a reinforced concrete frame with organic materials, including a skin of handmade brick with panels of blue tiles and Lancashire slate.

Across campus, Fry used the same palette of materials for the Veterinary School (1955–60, demolished 2012). The exposed concrete framework accentuated the factory-like function of the central block, which contained predominantly offices and laboratories. Ribs of concrete were extruded to form a projecting entrance canopy with blue tiles and, to the upper storey, this idea was repeated, with ribs wrapping around the rooftop plant room. These subtle protrusions ran down the otherwise flat front façade, creating a vertical contrast to the horizontality of the building (Figure 7.17). The concrete framework, hacked back to soften the finish, was infilled with metal-framed windows and flush brickwork, again emphasizing the straightforward design approach. To the roadside, the brickwork was embellished by panels of blue tiling again and slate, and two incised carvings by the artist Eric Peskett. His bas-reliefs of a bull and a horse alluded to the function of the building and provided a counterpoint to the stripped façades. The straightforward brickwork, factory-style windows and animal relief-work owed much to Harvard University's Biological Laboratories (1930) by Coolidge, Shepley, Bulfinch and Abbott, a building that Fry had probably studied during visits to Cambridge, Massachusetts, in 1944 or 1957. Fry's efforts to combine art and architecture at Liverpool were hampered by the Veterinary Building Committee,[98] and Peskett's two small carvings suffer by comparison to the incised sculpture of animals, birds and fish at the Harvard laboratory by Katherine Ward Lane, which are full of vitality (Figures 7.18 and 7.19).[99]

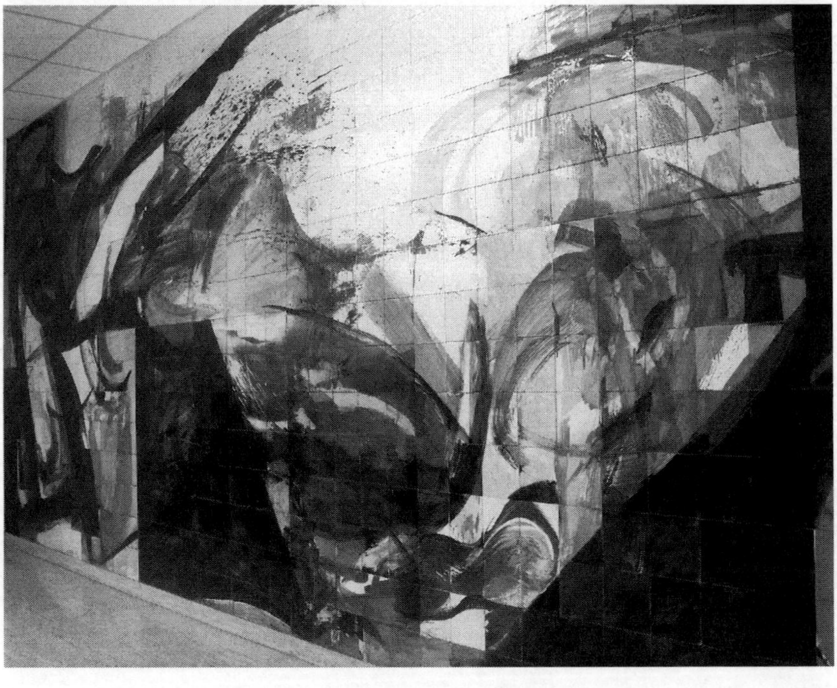

7.16 Peter Lanyon, *Conflict of Man with Tides and Sands*, 2013

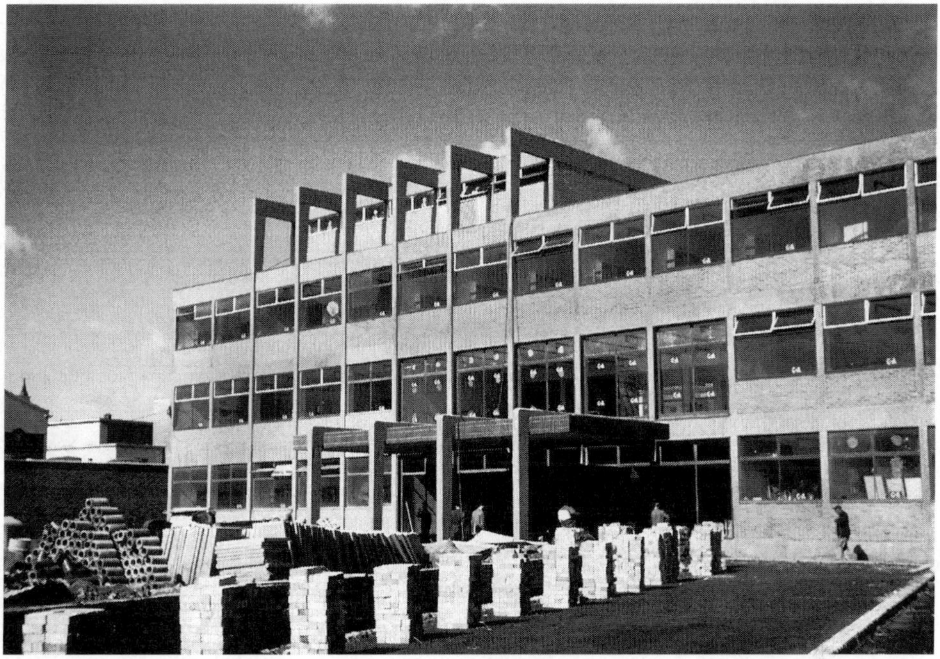

7.17 South Façade, Veterinary Building, Liverpool University, 1960

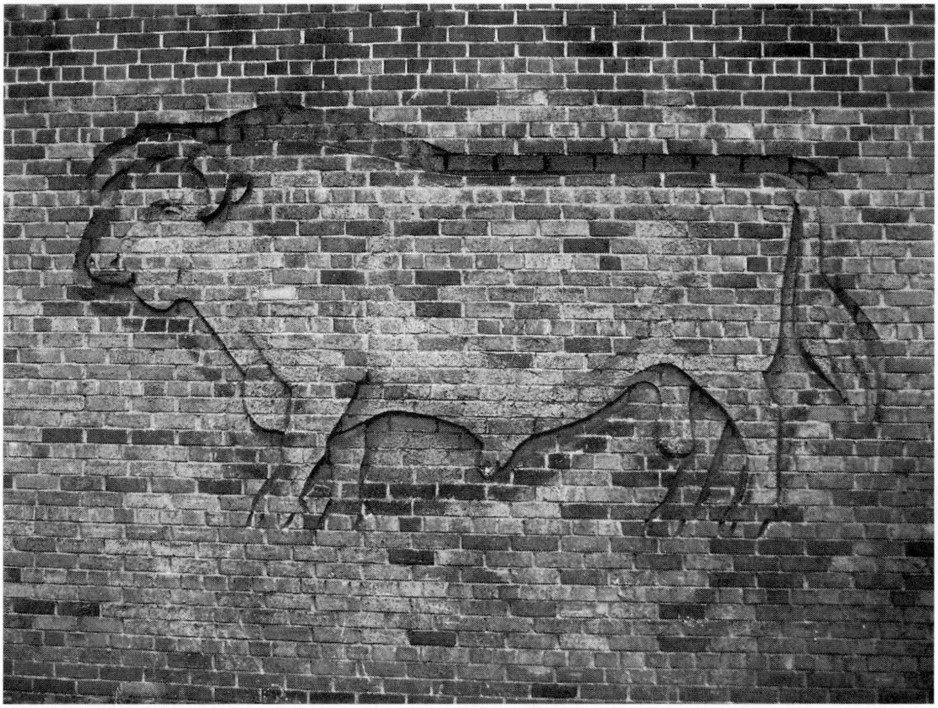

7.18 Bas-relief by Eric Peskett, Veterinary Building, Liverpool University, 2011

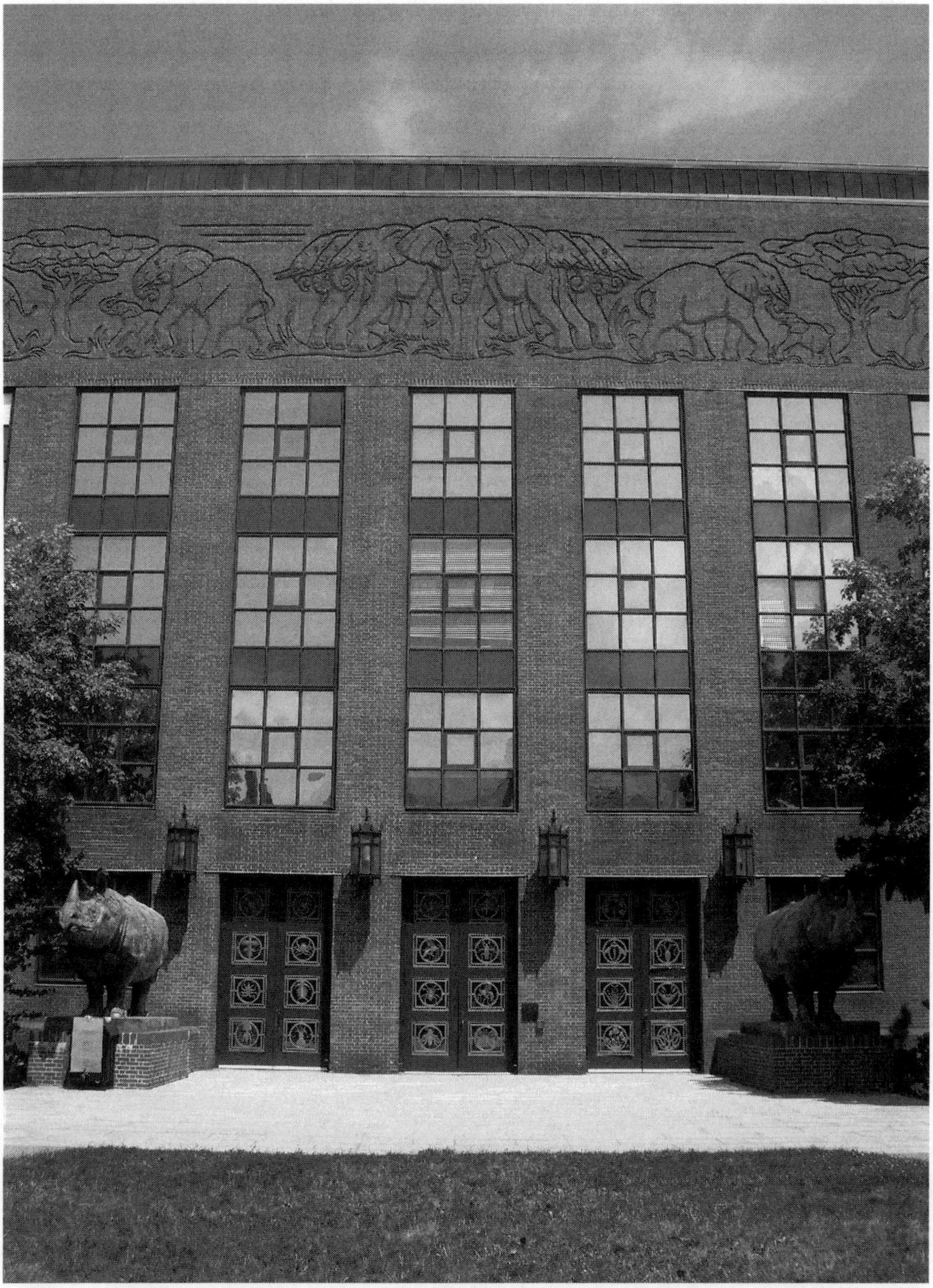

7.19 Bas-relief by Katherine Ward Lane, Biological Laboratories, Harvard University, 2013

Fry and Drew sought to build on their university work at Liverpool and at Ibadan with competition entries for Sheffield University campus (1953) and for Churchill College (1958), Cambridge, but both entries were unplaced. However, the Pilkington complex marked the first in a series of high-profile commissions in Britain for Fry, Drew and Partners. Offices for Gulf Oil Company, Dow Agro Chemicals and Rolls Royce followed, enabling the practice to establish itself as an expert in modern corporate architecture. Amongst this group is a Head Office for Wates building contractors (1963) at Norbury, in South London (Figure 7.20). The horizontal bands of handmade bricks and ceramic tiling give the building an interwar aesthetic that confirms Fry's involvement. The four-storey building uses similar formal devices as the Veterinary School, including the ubiquitous entrance canopy and an extruded stair tower. The offices are planned around a central courtyard furnished with an ornamental pool crowned by a sculpture, 'Girl with Doves', by David Wynne.

Like the Wates building, the Dow Agro Chemicals Office (1960) on the edge of King's Lynn is centred on a courtyard to provide a secluded open-air space for employees. The new offices mark the entrance to the 83-acre site, where the practice also designed a £1 million chemical factory. A (literally) sparkling pavilion of two storeys, the office building is planned on a wide grid of 30 feet to create a free plan. The building was designed for research and administration purposes. Cellular offices are distributed around the first floor perimeter, with circulation skirting around the central courtyard; the ground floor contains the sales department office, meeting rooms and the staff canteen. The generous entrance hall opens onto the courtyard, with a spiral staircase (an idea borrowed from the Regent Street Electricity Showrooms) leading up to a first floor waiting room with a large mural map depicting the company's worldwide concerns by Don Foster, who also contributed to the Pilkington Brothers' Headquarters.[100]

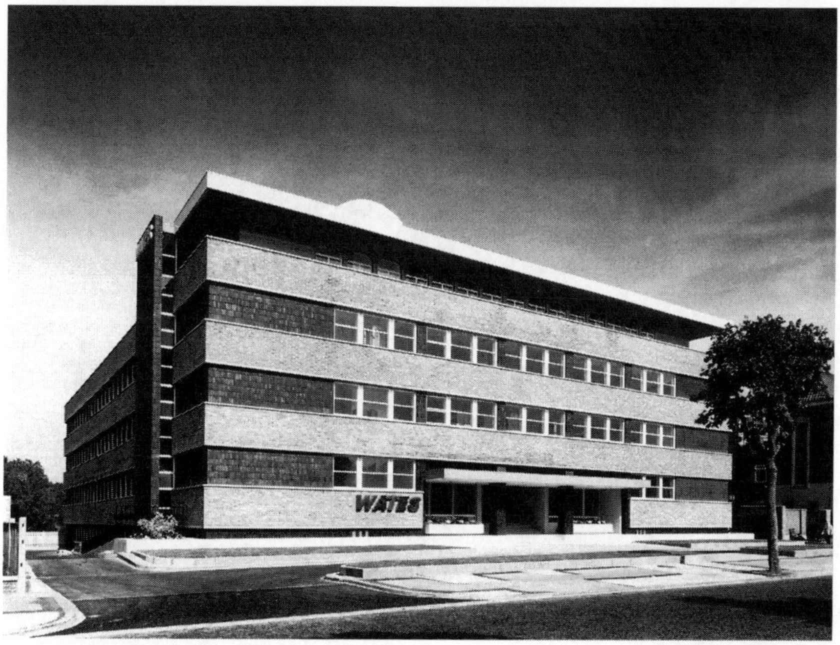

7.20 Wates Headquarters, Norbury, 1963

In elevation, the building is a knowing reference to Le Corbusier's Villa Savoye (1928–31), an idea that is enhanced by contemporary black and white photographs that highlight the building's form rather than texture. Closer scrutiny reveals the offices to be an essay in modern materiality. The first floor is clad in white-painted, vertical timber boarding that appears to float over a recessed ground floor faced in dark blue Staffordshire tiles. Free-standing steel columns are clad in pre-cast, stone-finished concrete and, typically for Fry, a projecting canopy demarcates the main entrance (Figure 7.21). The Dow offices present Le Corbusier's functional monument reworked as an example of humanist architecture. Thus Fry makes a statement about the end of the heroic period of architecture, following 'the close of a chapter' he had written of in the 1957 *AYB*.[101]

AMERICA

Like many European architects, Fry and Drew were drawn to the east coast of America. They visited the Gropius family in Massachusetts on several occasions and maintained their ties with the community of east coast CIAM émigrés. In 1944, Ise Gropius wrote of Fry and Drew's initial visit: 'Compared to the wind blowing from England everything here looks at the moment very conservative, so that the students here listening to Max and Jane got the impression of real "frontiers" people. … We wish we had more such visitors'.[102] In contrast to their reputations in Britain, the American view of Fry and Drew as progressive individuals appears to have remained and might be linked to the continuing importance of the CIAM in the US. Indeed, in 1954 Fry wrote to Drew of a recent trip to America by Lasdun: 'Denys is back and has just rung me up … He says our stock is still very high in the exalted circles of New York etc.'.[103]

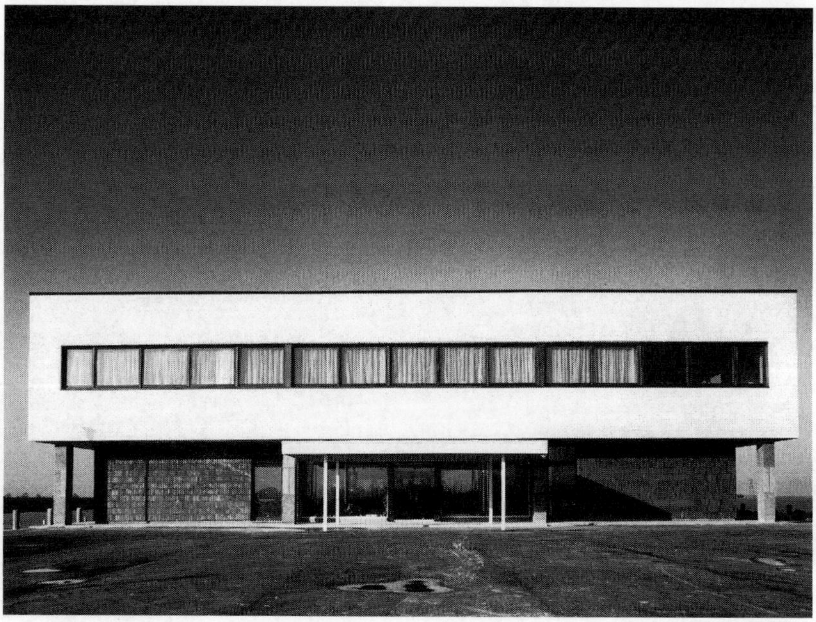

7.21 Dow Agrochemicals Office Building, King's Lynn, 1960

Following Fry's successful visiting professorship at Harvard University, Drew also benefitted from a teaching post and, from February to June 1961 she took up a visiting professorship at MIT.[104] Ostensibly the appointment was to provide an uninterrupted period of work on the manuscript for *Tropical Architecture in the Dry and Humid Zones* (1964), yet it also allowed Drew to undertake lucrative television and radio interviews, and she toured Canada giving lectures in Montreal, Toronto, Winnipeg and Vancouver.[105] Perhaps most important, however, was Drew's participation in the lively modernist community of her old CIAM friends; this group of émigré architects was, for Jane, the heart of post-war work in America.[106] She wrote to Fry back in England, 'I have seen everyone lunched with Gropius dined with Serge, drinks with Sert breakfast with Gidion [sic]. Have started my class taken part in a jury with Kahn it's all very stimulating and interesting and I am learning at quite a rate'.[107] The opportunity to learn was evidently important to Drew. She wrote repeatedly to Fry of learning new things and she revelled in the newness of North America, writing to Fry she admitted, 'I am becoming a bit seduced by America there is such scale and optimism and life'.[108] And later from British Columbia, 'I feel England is old and dingy'.[109]

Uncharacteristically, self-doubt crept into her work and she wrote, 'I wish I were more able myself though. Catelano [sic] who gives crits with me is a splendid man'.[110] Her relationship with Eduardo Catalano (1917–2010), the Argentinian-born architect and a professor at MIT, was an important one. They were perhaps already acquainted due to Catalano's teaching post at the Architectural Association, from 1945 to 1951. Together they established a Tropical Architecture course at MIT based upon the AA Tropical Architecture programme, which had been established in 1954 by Fry and Drew, but it was 'better', she thought.[111] Whilst disseminating her ideas on Tropical Architecture, Drew learned about concrete construction from Catalano. She wrote to Fry of Catalano's innovative methods of forming 'beautiful' parabolic vaults,[112] which he used in his own house at Raleigh, North Carolina (1954, demolished 2001).

Drew also became good friends with Sert and she admired his scheme for the Holyoke Centre at Harvard University (1962–67), designed with Gourley and Sasako. She wrote appreciatively to Fry that the 'use the filtered light of fibre glass and which control the little oddities of peculiar requirements that give life to a whole building within a strong form is pretty moving'.[113] Drew also lectured and participated in juries at Harvard and Yale, and often breakfasted with Giedion and Tyrwhitt. Through Drew's work at Yale University she became acquainted with Paul Rudolph and Louis Kahn, amongst others, and these associations seem particularly important to the development of her work.[114] The monumental approach of Kahn evidently left a lasting impression and she later spoke of her admiration of the forms created by Kahn.[115] Drew 'motored to Philadelphia' via New York to visit Kahn's studio, and there he showed her some ongoing work, including the Salk Institute (1959–66) in La Jolla, California, and the Richards Medical Research Laboratories (1957–65) in Philadelphia.[116] Indeed, the striking verticality of this project's service- and stair-towers, contrasted with the cellular laboratories, seems to foretell Drew and Frank Knight's hospital buildings in Mauritius. Moreover, the Operations Building of Drew's Open University campus (1969–77) uses Kahnian curved stair-towers to break up the monumentality of the long, low brick buildings.

Before leaving America, in a letter to Fry, Drew highlighted the buildings of Catalano, Sert and Kahn as particularly influential and spoke of her hopes of transferring these ideas into her own work.[117] Alongside a more monumental aesthetic, she also noticed small details that she was eager to try out: 'Last night I could not sleep. I was too excited. I was a great success in Winnipeg and spoilt. How I want to get back to work. June cannot come too fast. Little details I notice everywhere'.[118] Drew brought some of this inspirational milieu back to London; in May 1961, she appointed five 'MIT boys' to work at Gloucester Place.[119]

HOSPITALS AND OFFICES

Upon her return to London, Drew took over a project for Torbay Hospital (1961–70) in Devon from Norman Creamer – over 20 years after her competition-winning entry for a cottage hospital in Dawlish. The commission came via Dr Rex Cheverton at the Southwest Regional Health Authority, a medical officer with whom Drew had worked on an unbuilt design for University Hospital in Ibadan.[120] The Project Architect, Derrick Lees, later wrote that the 'basic principles of the scheme were established fairly quickly, in about a month (including research), but after that progress ticked along slowly mainly because of slowness on the part of the DHSS on agreeing precise schedules of accommodation'.[121] The design process ran from 1964, with work on site beginning in the autumn of 1966.

Drew designed a 300-bed extension to the existing Neo-Georgian hospital by Adams, Holden and Pearson, which included an operating wing, an outpatients' wing, a casualty department, X-ray rooms, a mortuary, a dispensary and a pathology department, plus a new boiler house and a new nurses' home with a school of nursing. The sloping site is skilfully exploited to enable the different floors containing the mortuary, casualty and the outpatients' department all to be accessed from ground level. A new connecting corridor link the new to the old, and brick and glazed tile skins were selected to 'marry' with the existing buildings.[122] In the outpatients' entrance hall, Arthur Goodwin created a mosaic mural, *The Tree of Life*, signifying the abundance of life. Despite the practice's typical palette of organic materials and integrated artwork, Torbay Hospital has a toughness and solidity of structure that distinguishes it as the first of Drew's monumental phase (Figures 7.22 and 7.23). The straightforward detailing continues her reductive Modernism, employed so successfully in Africa and India, and demonstrates her common-sense approach to architecture. Indeed, the project's budget limitations necessitated careful detailing in order to meet agreed standards of accommodation and (in contrast to Pilkington Brothers' Headquarters rising costs) the hospital was kept within budget; as Lees noted, the fine line between quality and economy was fundamental to Drew's approach to architectural design.[123]

The success of Torbay Hospital led to the practice undertaking a masterplan for the United Manchester Hospitals (1965–75). The sizeable project was intended to replace the existing hospital buildings, with the rationalisation of services and the construction of a new nine-storey hospital on an adjacent site situated to the south of the city's educational campuses (Figure 7.24). Despite undertaking detailed design work and setting up a regional office in Manchester, the project was cancelled in 1975, following the previous year's reorganization of the regional health services that signalled a change in priorities.[124] The hospital's unadorned, glazed curtain-wall façades show Frank Knight's involvement and a move away from the textural finish one associates with Fry's buildings (Figure 7.25).

The use of curtain-wall glazing with brickwork bookends had been employed earlier at the Rolls Royce Engineering Centre (c. 1960–65), built on the edge of Derby. Like the Pilkington Brothers' Headquarters, the project was built at a time

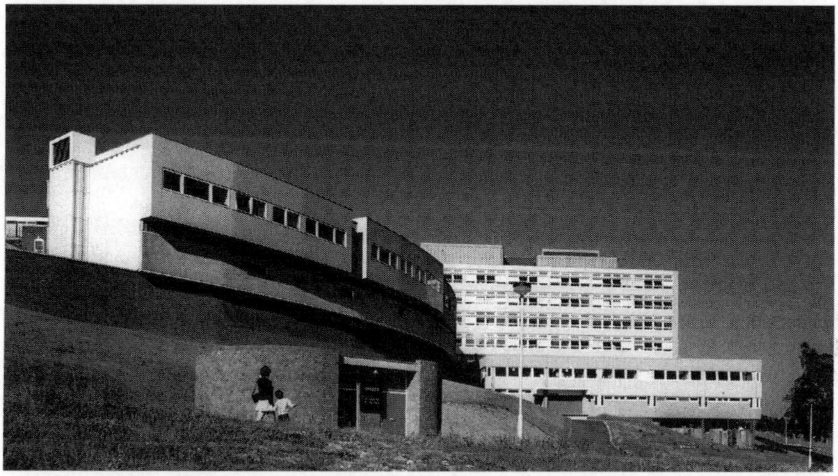

7.22 Theatre, pharmacy and pathology blocks, Torbay Hospital, Devon, 1970

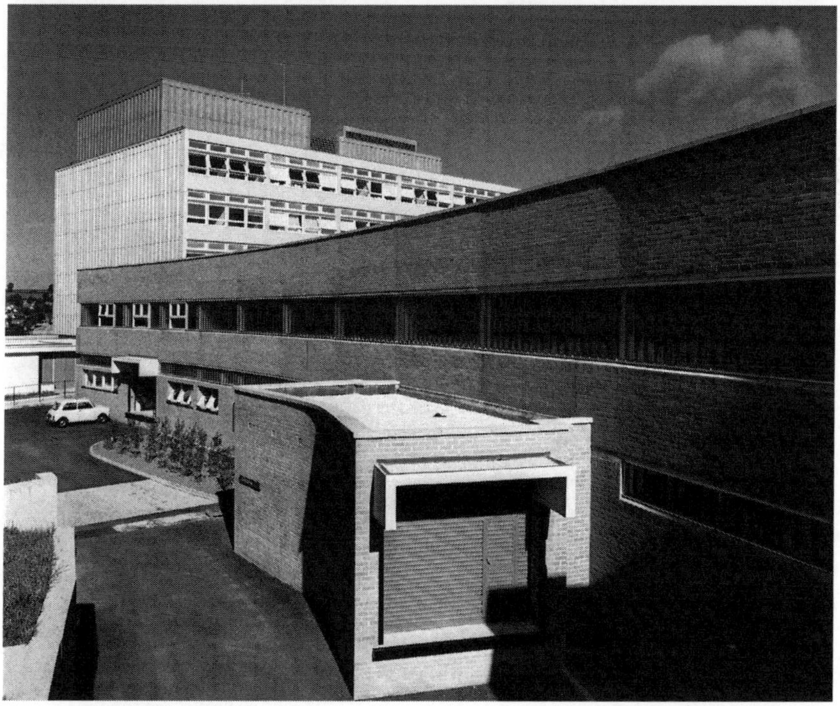

7.23 Theatre with pharmacy block to rear, Torbay Hospital, 1970

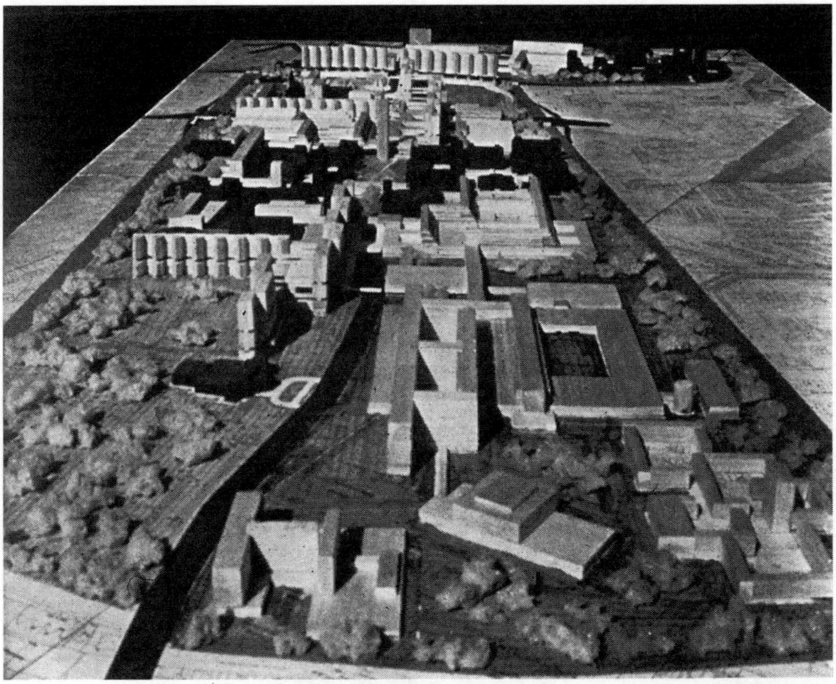

7.24 Model of United Manchester Hospitals Masterplan, c. 1965

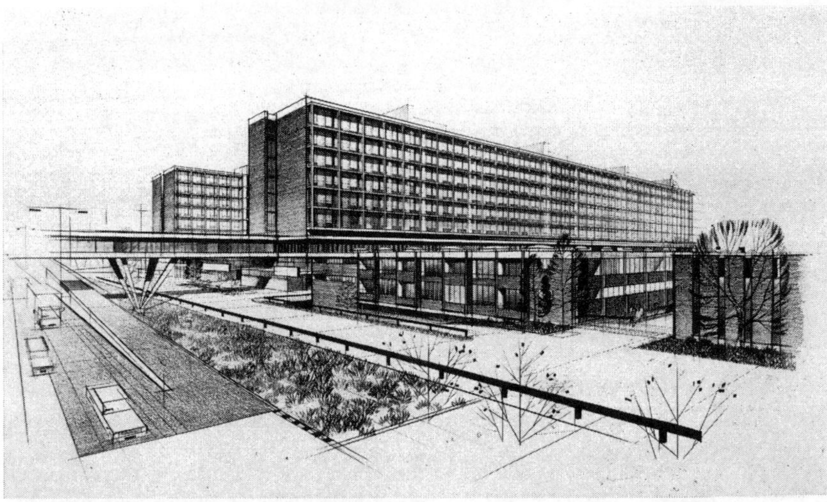

7.25 Perspective of wards, United Manchester Hospitals, c. 1965

when much of medium- and large-scale companies' work was still undertaken by hand, necessitating open-plan drawing offices for the company's Aero Engine Division. Yet, like Pilkington, a separate computer hall was also required to undertake specialist tasks. In the ultimate humanist statement, the module of the scheme was determined by the draughting tables and associated draughtsman's space, which measured nine-by-nine feet. Placing the workers at the heart of the building's design creates an interesting contrast to the otherwise high-tech, machine-oriented offices. The three-storey building was planned as a row of three

U-shaped courts, to allow phased occupancy as the construction progressed (Figure 7.26).[125] Quick to construct and easy to extend, the steel-frame building provided flexible offices with demountable internal partitions to allow for internal reconfiguration, as required by the occupants. Speed of construction and flexibility were again paramount for the client in the construction of an adjacent 8,000 square feet computer hall (Figure 7.27).[126] The north face of the curtain-walling was designed to be demountable for simple expansion of the two-storey hall.

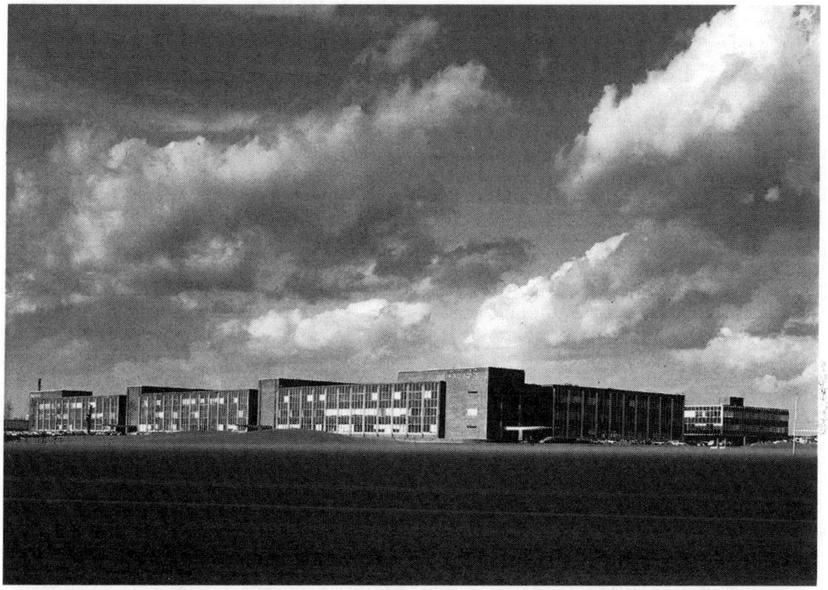

7.26 View of Rolls Royce Engineering Centre, Derby, c. 1965

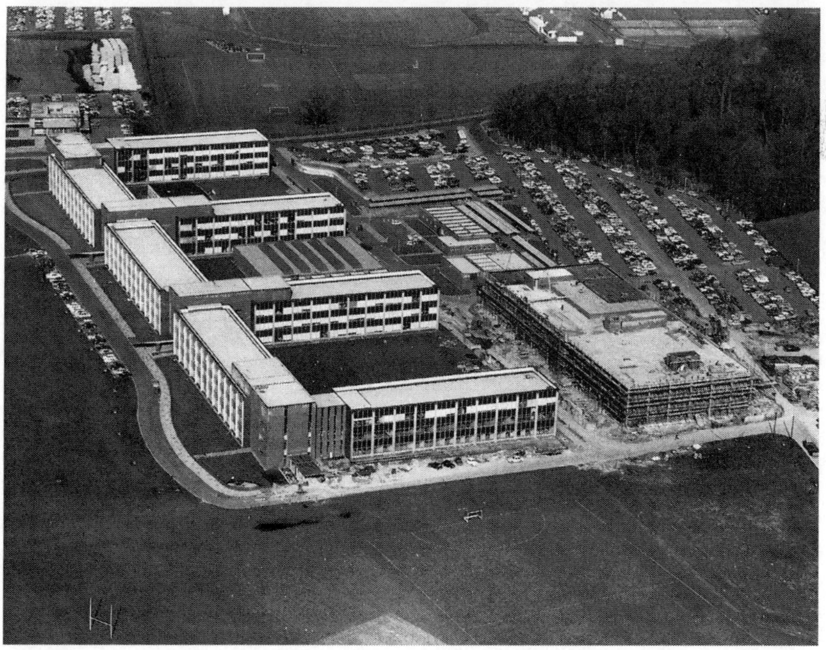

7.27 Computer Centre under construction (right), Rolls Royce Engineering Centre, Derby, c. 1965

The interiors of the office and computer hall were highly regulated: to counter the poor local air quality, the draughting offices were fitted with an advanced air filtration system; and the computer hall used a fully air-conditioned environment to ensure smooth-running of the machines. This high-tech interior led to the publication of articles, such as 'Designing for Computers' (1968) written by Frank Knight, illustrating Fry, Drew and Partners' absorption of the latest ideas in office design and their developing expertise in this specialized field.[127]

SPECULATIVE BUILDING AND DEPARTURE

Fry and Drew recognised the potential for architectural practices to collaborate with building contractors, and sought to capitalise on their long-standing professional relationships with developers, such as Wates. Around 1963, the practice was commissioned by a housing developer to design the Belfry System. The project for an industrialised housing system gave Fry an opportunity to return to his work in standardised housing from the early 1930s and continued Drew's work in low-cost housing. The Belfry System shares similarities with the hugely successful Span housing developments, designed by Fry's former employee Eric Lyons. Fry had written admiringly of Lyons's work as an example of good speculative building and, like Span, the Belfry System used a modular method to create new communities (Figure 7.28).

While Span houses catered for, in Fry's words, 'harried professional workers',[128] the Belfry System, was designed for working class occupants. Drew was lead designer for the scheme and explained her rationale thus: 'the council housing list lengthens and families are split apart for want of a place to live. Industrialised housing, in which everything from walls to plumbing is mass produced in the factory for rapid assembly at the site clearly offers part of the answer'.[129] Flexibility

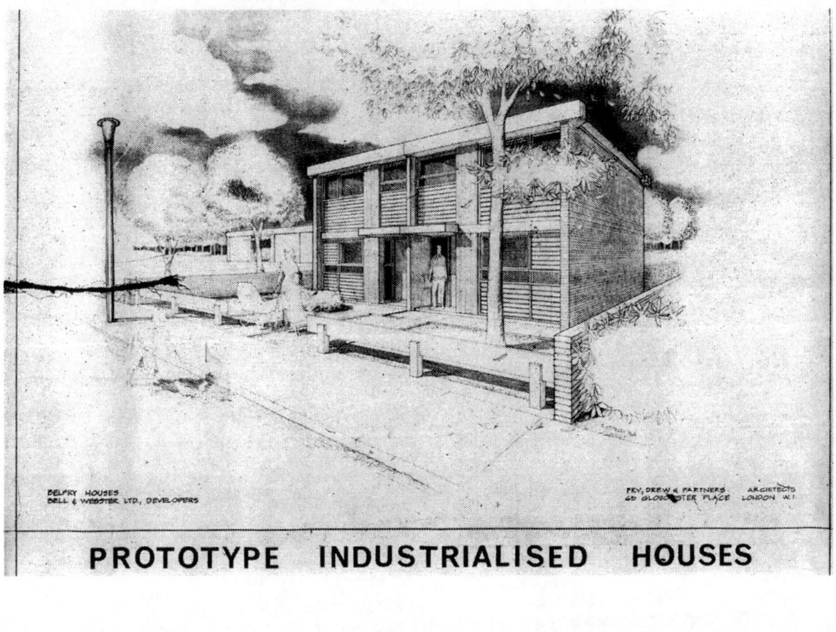

7.28 Semi-Detached Belfry Houses, drawing by R. Westbury, 1963

was the key aim in all aspects of the project. The system was intended to be adaptable in both construction and layout, to suit a range of site conditions and provide developments of varying density. Assembled from a kit of prefabricated parts, the units could be constructed using un-skilled labour to create semi-detached or terraced houses of one- to three-storeys, and blocks of flats of up to five storeys so as to give a maximum density of 30 units per acre. The overall aesthetic was also variable, with a range of component materials able to create a modern or more traditional aesthetic. Components included: a flat concrete roof or a low-pitched roof with timber trusses; metal or painted-timber windows and doors; and concrete or painted-timber cladding panels. The concrete spine walls gave a fixed depth of 27 feet to each dwelling, although the width might be varied by increments of one-foot to give variation in size and internal layout. With emphasis on meeting the needs of different occupants, the system fulfilled all the requirements set out by the 1961 Parker Morris report on space standards for housing and was one of the few systems approved by the National Building Agency at the time. The design was undertaken in close consultation with the Ministry of Housing.

In 1966, a development of 42 Belfry Houses was completed ahead of schedule at Hoddesdon in Hertfordshire for the Urban District Council. The promotional brochure tells of the 'Hoddesdon Story', illustrating the efficient process of erecting two-storey terraced housing with flat roofs (Figure 7.29). On day one, each of the pre-cast concrete separating walls were raised, on a poured-concrete foundation slab, and adjusted to fit on starter bolts; the following day, concrete beams and floor slabs were lowered into place, secured by steel ties and grouted in; next, plumbing and built-in services were positioned where practicable; on day four, roofs of six slabs of dense concrete were fitted, and finished with neoprene caps to the upstands; and, on the final day, pre-painted infill cladding panels, windows and doors were fitted, to give a water-tight building in just five days.[130]

In contrast to the Belfry System, Fry designed a genteel scheme for the building developer, Wates, at 2–7 Woodsford Square (1966–74) in Kensington. Woodsford Square is, in many ways, the continuation of Fry's interwar work, and the project's sales brochure shares many similarities with that of the unrealised Isokon 3 development. Highlighting the historic pedigree of the development, Woodsford Square was publicised as the 'rare event' of a new London square, following in the tradition of mid-seventeenth to mid-nineteenth century building of the 'urban counterpart' to the country manor set in park-like grounds.[131] Fry's post-war writing on speculative development illustrates his continued use of the interwar period's historicised Modernism and Trystan Edwards's writing on good and bad manners in architecture. One more, he argues that death of good, 'responsible' speculation occurred during the 1840s, when 'the rate of expansion increased enormously and the level of taste increasingly declined'.[132] The brochure mourns the loss of civilised squares of housing: 'Times and manners have changed and many a formerly graceful square has disappeared or become a packed car park'.[133]

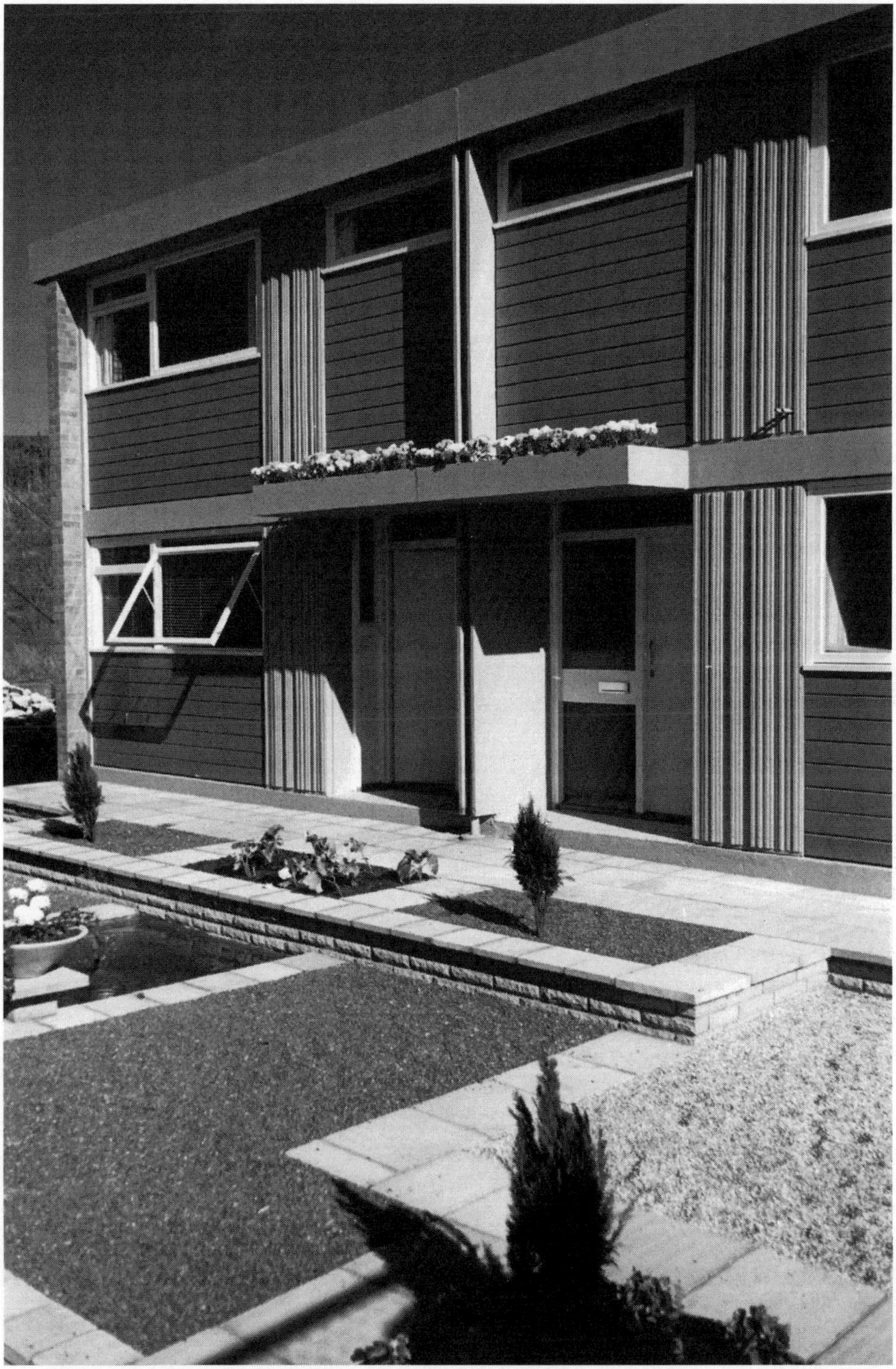

7.29 Terrace of Belfry Houses, Hoddesdon, c. 1966

Built on a seven-acre site (previously of five detached villas considered impractical to maintain), the development of terraced houses utilise traditional materials and construction of handmade load-bearing brick with timber floors and roofs. Emphasis was placed on quality of finish, with typical Fry details such as panels of deep blue ceramic tiles and painted softwood boarding to add interest to the brick façades (Figure 7.30). The four-storey houses offered 14 variations of plan, with each comprising five or six bedrooms and two bathrooms: a 'Type B' house with five bedrooms was marketed at a sizeable £24,500.

In the late 1960s, Fry also began work on two crematoria commissions for Coychurch Crematorium (1967–70) near Bridgend and for a chapel at Breakspear Crematorium (1971). The projects were amongst the last of his full-time work before his retirement in 1973 and, with Woodsford Square and Pilkington Brothers' Headquarters, the Coychurch Crematorium is amongst the most successful examples of Fry's post-war humanism. Fry's crematoria work was centred on a philosophy which he called an 'anatomy of mourning', developed during the 1960s and instigated by Fry, Drew and Partners' involvement with the Cremation Society.[134] Fry's ideas drew from his experience of active participation in a staff member's funeral at Chandigarh, which contrasted starkly with Western practices:

> *The body, wrapped simply in white cerements, was carried to the pyre by the river on a litter by six mourners. As the procession went forward I noticed that a man would leave it to tap one of the bearers on the shoulder and silently take his place; I did likewise and felt the burden in my turn.*[135]

At Coychurch, Fry sought to draw attention to 'the fact of death and the necessity for the full expiation of grief as a communal act' and created a ceremonial route through the grounds of the crematorium. The scheme uses a process of 'delay' and a heightening sense of urgency to engage mourners in the ceremony;

7.30 Woodsford Square, Holland Park, drawing by Fry, 1966

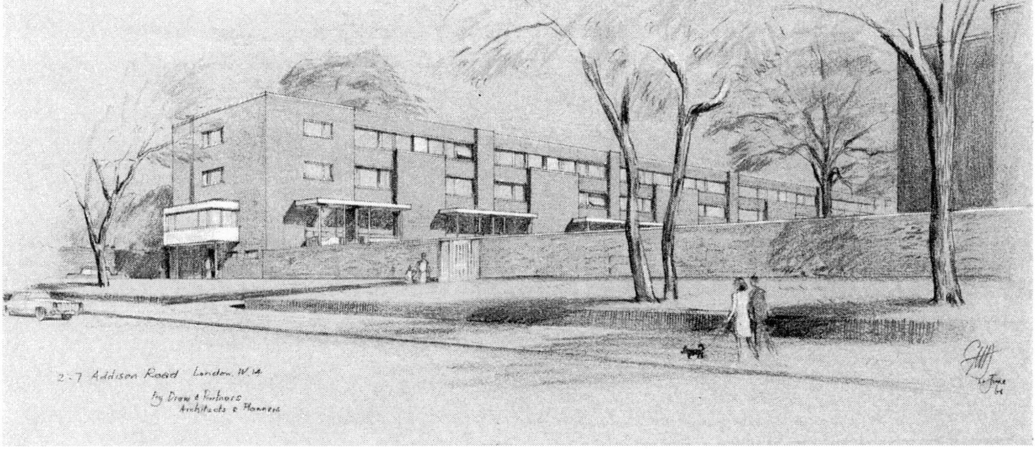

from the funerary stone monument at the entrance of the spacious landscaped site; down a gently sloping driveway, through a mass of trees and into view of the chapel; stopping at the porte-cochère and passing through the cloistered walk; and finally resting in the open space of the chapel, which inclines slightly towards the catafalque niche (Figure 7.31).[136] The building complex is constructed in Fry's typical blend of materials. Concrete is tempered by local stone, recycled from recently demolished old buildings, and by copper roofs. To the cloister, students from the nearby Swansea School of Art contributed stained glass windows and to the free-standing, drum-like Chapel of Remembrance, their tutor Timothy Lewis did the same. Like George Pace's nearby chapel (1957–59) for St Michael's College in Llandaff, Coychurch demonstrates a considerable debt to Le Corbusier's church at Ronchamp (1950–55). The crematorium is one of Fry's most sculptural buildings and Le Corbusier's influence is clearly apparent in the (since removed) concrete cowl sheltering a cross rising high over the chapel roof and the simplicity of the rendered interior walls.[137]

When compared with Drew's contemporaneous projects, the divergence between Fry and Drew's architectural aesthetic becomes apparent. Her work at Torbay Hospital, the Open University campus at Milton Keynes and series of buildings for the Mauritius Government all demonstrate the new direction of her work. Stemming perhaps from her professorship at MIT, Drew's work continued to develop throughout the 1970s. This work is beyond the scope of this book but sets the scene for further research.

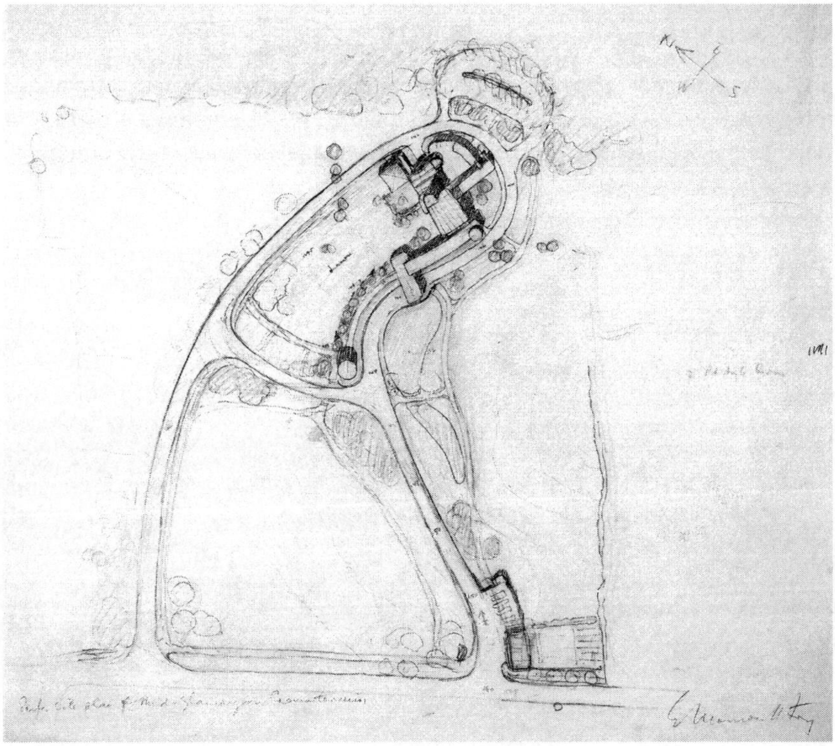

7.31 Site Plan of Coychurch Crematorium, near Bridgend, drawing by Fry, c. 1967

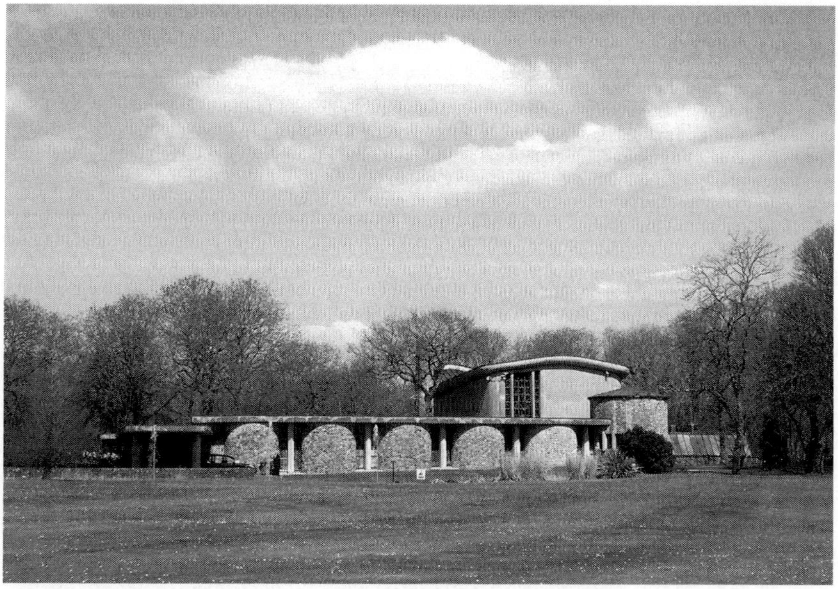

7.32 Coychurch Crematorium, near Bridgend, 2011

CONCLUSION

Ultimately, Fry resisted the major changes taking place in contemporary life, while Drew embraced them. The rise of consumer culture was anathema to Fry and he likened it to the first industrial age, which he – and much of the interwar architectural community – saw as the debasement of good manners in architecture and design. Unsurprisingly, Fry's best post-war work draws on historical precedent, such as Woodsford Square; or has a strong narrative on which to base his work, such as Coychurch Crematorium or the Pilkington Brothers' Headquarters. In contrast to Fry's despair of the second industrial revolution, Drew seized upon the possibilities offered by new synthetic materials. The post-war period shows a continuation of Drew's involvement in industrial design, in work such as the Belfry Houses. Despite concerns, the practice's development of expertise in office design, and its work in speculative building projects, illustrates Fry and Drew's ability to adapt to the rapid professionalization of architecture during the 1950s and particularly the 1960s.

Drew's MIT professorship in 1961 proved to be pivotal in the development of her architectural approach. The CIAM milieu in America gave fresh impetus to her work and she learned much from her colleagues. Indeed, Drew later chose to align herself with the internationalism of CIAM, rather than the British-based MARS Group.[138] The divergence in their careers is crystallized by Fry's award of the RIBA Gold Medal in 1964, which failed to, as he wrote, to 'provide the stimulous [sic] I still sought, [and] confirmed my dawning realisation that I was past my prime'.[139]

NOTES

1 Eric Mumford, *The CIAM Discourse on Urbanism, 1928–1960* (Cambridge, MA, 2002), p. 232.

2 Mumford, *CIAM Discourse on Urbanism*, p. 228.

3 Gordon Graham, testimonial at Jane Drew memorial event held at the Architectural Association, 13 November 1996. In the 1970s Graham and Drew served together on the RIBA Council.

4 Mumford, *CIAM Discourse on Urbanism*, p. 228.

5 Mumford, *CIAM Discourse on Urbanism*, p. 225.

6 RIBA Archive, F&D/18/14. Letter Fry to Drew, 13 April 1954.

7 With this reorganization, Robert Byng, Joseph Blackburn, Graeme Gibson and Maria Plant Zaccheo became Associates. See Hitchins, *Fry, Drew, Knight, Creamer*, p. 8.

8 RIBA Archive, F&D/13/2. E. Maxwell Fry, 'An Old Man's Epilogue', undated.

9 Mark Crinson and Claire Zimmerman (eds), *Neo-avant-garde and Postmodern Postwar Architecture in Britain and Beyond* (London and New Haven, CT, 2010); Anne Massey, *The Independent Group: Modernism and Mass Culture in Britain* (Manchester, 1995).

10 Mark Crinson and Claire Zimmerman, 'Introduction', in *Neo-avant-garde and Postmodern Postwar Architecture in Britain and Beyond*: pp. 7–25, p. 17.

11 RIBA Archive, F&D/24/1.

12 Ockman, *Architecture Culture 1943–1968*, p. 21.

13 Janice Anderson at the Canadian Women Artists History Initiative, Concordia University, correspondence with author, 22 January 2013.

14 Graham Bligh interviewed by John Macarthur, Robert Riddel and Janina Gosseye, 'Graham Bligh on working at Maxwell Fry's', Digital Archive of Queensland Architecture, 20 August 2012, https://www.youtube.com/watch?v=HlcVdcc0aPU. Accessed: 8 January 2013.

15 Although this comment refers to a period of nine years during the design and construction of Pilkington Brothers' Headquarters. See Pilkington Brothers' Archive, PB702 2/8. 'Report to Group Executive on Increased Cost of Prescot Road Offices. 3rd March 1964'.

16 John Cordwell also joined the practice in 1947 and was made a junior partner relatively quickly. He went to Ibadan University to work as Site Architect, before moving to Chicago in 1950. See Blum, *Oral History of John Donald Cordwell*.

17 Blum, *Oral History of John Donald Cordwell*, p. 85.

18 RIBA Archive, F&D/15/5. Letter Drew to Fry, undated.

19 UL Archive, Peter Dale Papers [uncatalogued]. Letter Derrick Lees to Peter Dale, 23 June 1993.

20 Fry wrote to Drew in anticipation of her return: 'There will have to be increases of secretarial and architect staff. I can't get typing done for ages. Lindsay [sic] fairly cut things down'. RIBA Archive, F&D/18/14. Letter Fry to Drew, 30 July 1954.

21 Lees erroneously refers to Albany Terrace. UL Archive, Peter Dale Papers [uncatalogued]. Letter Lees to Dale, 23 June 1993.

22 Denys Lasdun later commented that this was an administrative arrangement and 'there was never any, and there couldn't have been any design collaboration'. Lasdun cit. in Gold, *The Practice of Modernism*, p. 318 [11].

23 UL Archive, Peter Dale Papers [uncatalogued]. Letter Lees to Dale, 23 June 1993.

24 RIBA Archive, F&D/18/15. Letter Fry, Lake House, to Drew, 12 September 1955.

25 UL Archive, Peter Dale Papers [uncatalogued]. Letter Lees to Dale, 23 June 1993; For discussion of Lubetkin's approach, see Allan, *Berthold Lubetkin*, pp. 386–90.

26 William Curtis, *Denys Lasdun: Architecture, City, Landscape* (London, 1994), p. 46.

27 For an image of this sketch, see Curtis, *Denys Lasdun*, p. 42.

28 E. Maxwell Fry, 'Walter Gropius', *Architectural Review*, 117 (March 1955): pp. 155–7, p. 156.

29 Peter Bond and Robert Byng both became junior partners during this reorganisation.

30 The annual office party was legendary, with desks to the first floor drawing offices cleared away to form space for a band and the many guests; one year the ceiling was completely filled with balloons. UL Archive, Peter Dale Papers [uncatalogued]. Letter Lees to Dale, 23 June 1993.

31 Theo Crosby, 'Night Thoughts of a Faded Utopia', in *The Independent Group: Postwar Britain and the Aesthetics of Plenty*, ed. David Robbins (London and Cambridge, MA, 1990): pp. 197–9, p. 197.

32 Robin Spencer (ed.), *Eduardo Paolozzi: Writings and Interviews* (Oxford, 2000), p. 334.

33 Boyer, 'An Encounter with History', p. 137.

34 RIBA Archive, F&D/18/13. Letter Fry to Drew, 9 August 1953.

35 Nigel Henderson cit. in Robin Spencer (ed.), *Eduardo Paolozzi: Writings and Interviews* (Oxford, 2000), p. 79.

36 For Read's influence on Paolozzi's work, see Spencer (ed.), *Eduardo Paolozzi*, p. 2.

37 For an image of Paolozzi's playground, see Diane Kirkpatrick, *Eduardo Paolozzi* (London, 1970), p. 20.

38 Spencer, *Eduardo Paolozzi*, p. 334.

39 For an image of Paolozzi's collage, see Kirkpatrick, *Eduardo Paolozzi*, p. 41.

40 Spencer, *Eduardo Paolozzi*, p. 78.

41 Robin Spencer, 'Henderson + Paolozzi = Hammer Prints', in *Nigel Henderson & Eduardo Paolozzi: Hammer Prints Ltd, 1954–75*, (ed.) Michelle Cotton (Colchester, 2013), pp. 60–93, p. 64.

42 Lesley Jackson, 'Inside Out – Outside In: The guerrilla fabrics of Hammer Prints', in *Nigel Henderson & Eduardo Paolozzi*: pp. 94–109, p. 100.

43 Douglas Newton, 'Printed Textiles', *Architectural Review*, 111 (March 1952): pp. 190–5.

44 Diane Kirkpatrick correspondence with author, 13 February 2013.

45 'African Experiment', *Architectural Review*, 113 (May 1953): pp. 299–310.

46 RIBA Archive, Ove Arup Papers, ArO/1/4/13. 'Report on [MARS Group] ANNUAL GENERAL MEETING held on Monday October the 15th 1956 at the Architectural Association'.

47 MARS Group, *Turn Again: an exhibition presented by the MARS Group, Royal Exchange, 12–30 July 1955* (London, 1955).

48 Elain Harwood, 'White Light/White Heat: Rebuilding England's Provincial Towns and Cities in the Sixties', *Twentieth Century Architecture*, 6 (2002): pp. 57–70, p. 65.

49 For further discussion of the exhibition see Gold, *The Practice of Modernism*, pp. 89–93.

50 MARS Group cit. in Gold, *The Practice of Modernism*, p. 92.

51 Denys Lasdun cit. in Gold, *The Practice of Modernism*, p. 93.

52 Maxwell Fry cit. in Alison Smithson, Peter Smithson, Jane B. Drew, E. Maxwell Fry, 'Conversation on Brutalism', *Zodiac*, 4 (April 1959): pp. 73–81.

53 MARS Group, *Turn Again*.

54 Gilbert Herbert, *The Synthetic Vision of Walter Gropius* (Johannesburg, 1959).

55 RIBA Archive, F&D/13/4. E. Maxwell Fry, 'Maxwell Fry and the Modern Movement', undated.

56 Joan Ockman, 'Introduction', in *Architecture Culture 1943–1968: A Documentary Anthology*, (ed.) Joan Ockman (New York, 1993), pp. 13–24, p. 13.

57 Judi Loach, 'Le Corbusier and the Creative Use of Mathematics', *The British Journal for the History of Science*, 31/2 (June 1998): pp. 185–215, p. 211.

58 Henry Millon, 'Rudolf Wittkower, "Architectural Principles in the Age of Humanism": Its Influence on the Development and Interpretation of Modern Architecture', *Journal of the Society of Architectural Historians*, 31 (May 1972): pp. 83–91, p. 91.

59 Fry, *Art in a Machine Age*, p. 49.

60 For images of Wudil Teacher Training College, see 'Report of a Debate on the Motion 'that Systems of Proportion make good design easier and bad design more difficult', *RIBA Journal*, 64 (September 1957): pp. 456–63.

61 Fry, *Art in a Machine Age*, pp. 52–72.

62 RIBA Archive, F&D/18/15. Letter Fry to Drew, 9 October 1955.

63 Eva-Marie Neumann, 'Architectural Proportion in Britain 1945–1957', *Architectural History*, 39 (1996): pp. 197–221, p. 197.

64 Neumann, 'Architectural Proportion in Britain 1945–1957', p. 197.

65 UEA Archive, Pritchard Papers, PP/22/6/1/1. Henry Morris, 'Architecture, Humanism, and the Local Community', uncorrected proof for *RIBA Journal*, June 1956.

66 University of Victoria Archive, Herbert Read Papers. Letter Fry to Read, 9 October 1955.

67 For full discussion of the project, see: Jessica Holland and Iain Jackson, 'A Monument to Humanism: Pilkington Brothers' Headquarters (1955–65) by Fry, Drew & Partners', *Architectural History*, 56 (2013): pp. 347–90.

68 Herbert Read, *The Grass Roots of Art* (London, 1955), p. 13.

69 University of Victoria Archive, Herbert Read Papers. Letter Fry to Read, 9 October 1955.

70 Pilkington Brothers' Archive, PB702 2/8. Maxwell Fry, 'Pilkington Brothers Ltd: New Head Offices. Memorandum on the Employment of [a] Sculptor', 24 October 1956.

71 The memo was circulated amongst the committee and retains handwritten notes by the members, including a notable 'Oh no!' next to Fry's suggestion to employ Henry Moore as the associate artist.

72 FLL, Harvard University Archive. The GSD History Collection, Student Affairs – Student Work: An Inventory, CC165. 'Design Problem, issued November 21, 1956: A headquarter office for a large industrial corporation', 22 January 1957.

73 A medical examination room, a dental surgery, an optician, a doctor and a chiropodist were also on site. Pilkington Brothers' Archive, PB702 2/8. 'Minutes of Meeting held in the Research Conference Room 27th February, 1957'.

74 J.L. Sert, 'Centres of Community Life', in *The Heart of the City*: pp. 3–16.

75 See Stefan Muthesius, *The Postwar University: Utopianist Campus and College* (London and New Haven, CT: Yale University Press, 2000), pp. 94–202.

76 Other unsuccessful artists proposed by Fry include Terry Frost, Brian Winter, Patrick Heron, Roger Hilton, Alan Davie, Adrian Heath, Wilhelmina Barns-Graham, Alexander McKenzie, Ceri Richards, Sandra Blow, Peter Collingwood, David Gillespie, Robert Nicholson and Kenneth Armitage. Pilkington Brothers' Archive, PB547/1. Letter Maxwell Fry to Douglas Phelps, 24 November 1959; Letter Alan Hudson-Davies to Jane Drew, 11 December 1962; E. Maxwell Fry, 'Memorandum on choice of artist for decorative mirrors and wall panels. 26th August, 1963'; PB702 2/8. 'Minutes of Meeting held in the Head Office Conference Room on 15th August [1957]'.

77 Fry and Drew perhaps knew of Chandra from his first exhibition in England, hosted by the Royal India, Pakistan and Ceylon Society and held at the Commonwealth Institute (then the Imperial Institute) in 1957; a catalogue from the exhibition is present amongst the Pilkington papers. Pilkington Brothers' Archive, PB547/2 Artists Materials and Sundry Papers.

78 A mural for the Indian High Commission in Lagos (1962) and a Fibreglass mural for the Indian Tea Centre, Oxford Street, London (1964) were also undertaken.

79 Pilkington Brothers' Archive, PB157. 'Pilkington Head Office – Works of Art', undated.

80 Pilkington Brothers' Archive, PB702 2/8. 'Minutes of Meeting held in the Head Office Conference Room on 15th August [1957]'.

81 Spender discusses the processes involved in these pieces in detail; see National Life Stories, The British Library, C466/101/01–26. *Humphrey Spender*, Interviewed by Cathy Courtney, 1999–2002, pp. 374–5.

82 UL Archive, Robert Gardner-Medwin Papers, D688/4/2/7. *Pilkington Brothers Limited: The new head office* (St. Helens, 1965), p. 15.

83 James Gardner, *The ARTful Designer* (London, 1993), p. 269.

84 Gardner, *ARTful Designer*, p. 271.

85 Pilkington Brothers' Archive, PB702 2/8. 'Copy of Letter from Mr. James Gardner', 4 July 1963; PB547/1. Letter L.H.A. Pilkington to F.J. Hargreaves, 19 May 1959.

86 A fixed fee of £300 plus expenses was agreed for Drew's work. Pilkington Brothers' Archive, PB547/1. Letter Douglas Phelps to Drew, 27 July 1961.

87 'Shell Offices in Singapore', *Industrial Architecture*, (March/April 1961): pp. 130–1, p. 130.

88 FLL, Harvard University. The GSD History Collection, Student Affairs – Student Work: An Inventory, CC166. 'Problem II, Phase I', 21 November 1956.

89 FLL, Harvard University. The GSD History Collection, Student Affairs – Student Work: An Inventory, CA048 & CA137.

90 FLL, Harvard University. The GSD History Collection, Student Affairs – Student Work: An Inventory, CC165. 'Problem III, A Design Center for Harvard University', 21 November 1956.

91 Eduard F. Sekler, *Le Corbusier at Work: The Genesis of the Carpenter Center for the Visual Arts* (Cambridge, MA, 1978), p. 40.

92 FLL, Harvard University. HT166 .P76x. E. Maxwell Fry, 'CIAM 8', in *Proposed Readings: Session VI, Meeting Places*, (ed.) J.L. Sert, [1957].

93 Louise Mozingo asserts that the building 'redefined' the nature of corporate estate in America, as the architects created low-rise Modernist offices set in 280 acres of landscaped grounds within the Connecticut countryside, which she terms as a 'corporate campus'. Louise A. Mozingo, *Pastoral Capitalism: A History of Suburban Corporate Landscapes* (Cambridge, MA, 2011), p. 112.

94 UL Archive, Robert Gardner-Medwin Papers, D688/4/2/7. *Pilkington Brothers Limited: The new head office* (St. Helens, 1965).

95 E. Maxwell Fry, 'The Aesthetics of University Buildings', *Higher Education Quarterly*, 18 (1964): pp. 341–51, p. 344.

96 For further discussion of Fry's buildings for Liverpool University, see Iain Jackson, 'Post-War Modernism: Maxwell Fry's Buildings at the University of Liverpool', *Journal of Architecture*, 16 (October 2011): pp. 675–702.

97 UL Archive, Civil Engineering Building Sub-Committee, S167. E. Maxwell Fry, 'Engineering Building Committee', 31 May 1958.

98 The building committee rejected Fry's proposal to commission Elizabeth Frink to design a mural to the main entrance hall as they did not consider her work to be colourful enough. UL Archive, S715. 32nd Veterinary Building Committee Meeting Minutes, 16 March 1961.

99 As Bunting notes, a similar relief appeared soon after at the Mond Laboratory (1933) at Cambridge, with Eric Gill commissioned to carve a crocodile into the brickwork. Bainbridge Bunting, *Harvard: An Architectural History* (Cambridge, MA, 1998), p. 307 [24].

100 'Office Block and Chemical Works at King's Lynn', *Architect & Building News*, 219 (8 February 1961): pp. 185–90, p. 186.

101 Maxwell Fry, 'English Architecture from the 'thirties', *Architects' Year Book*, 8 (1957): pp. 53–6, p. 53.

102 UEA Archive, Pritchard Papers, PP/24/7/20. Letter Ise Gropius to Jack and Molly Pritchard, 7 March 1944.

103 RIBA Archive, F&D/18/14. Letter Fry to Drew, Chandigarh, 15 May 1954.

104 Drew returned to Boston as a visiting professor at Harvard University in 1970.

105 RIBA Archive, F&D/15/3. Letter Drew to Fry, 14 February 1961.

106 UL Archive, Peter Dale Papers [uncatalogued]. Jane Drew interview with Peter Dale, 1992.

107 RIBA Archive, F&D/15/3. Letter Drew to Fry, 8 February 1961.

108 RIBA Archive, F&D/15/3. Letter Drew to Fry, 9 March 1961.

109 RIBA Archive, F&D/15/3. Letter Drew to Fry, 29 March 1961.

110 RIBA Archive, F&D/15/3. Letter Drew to Fry, 9 March 1961.

111 RIBA Archive, F&D/15/3. Letter Drew to Fry, 19 July 1961.

112 RIBA Archive, F&D/15/3. Letter Drew to Fry, 10 July 1961.

113 RIBA Archive, F&D/15/3. Letter Drew to Fry, 19 July 1961.

114 Rudolph invited Drew to return to Yale as a Visiting Professor. RIBA Archive, F&D/15/3. Letter Drew to Fry, 30 May 1961.

115 UL Archive, Peter Dale Papers [uncatalogued]. Jane Drew interview with Peter Dale, 1992.

116 RIBA Archive, F&D/15/3. Letter Drew to Fry, 30 May 1961.

117 RIBA Archive, F&D/15/3. Letter Drew to Fry, 19 July 1961.

118 RIBA Archive, F&D/15/3. Letter Drew to Fry, 29 March 1961.

119 'I am now engaging M.I.T. boys for the office, I got a splendid boy this morning I enclose Grossmans v interesting work with underground roads and surface pedestrian ways please send it on with attached letter to Theo [Crosby]'. RIBA Archive, F&D/15/3. Letter Drew to Fry, 3 May 1961.

120 Flower et al, *Jane B. Drew*, p. 45.

121 UL Archive, Peter Dale Papers [uncatalogued]. Letter Lees to Dale, 23 June 1993.

122 Hitchins, *Fry, Drew, Knight, Creamer*, p. 16.

123 Lees moved to Exeter in September 1967 to establish a base office for supervision of the project's construction and noted: 'It was hoped that more work would materialise in that area but it did not'. The project was completed in Spring of 1970. UL Archive, Peter Dale Papers [uncatalogued]. Letter Lees to Dale, 23 June 1993.

124 Although work on a multi-purpose hospital building and a new geriatric unit did go ahead during the 1970s. See Hitchins, *Fry, Drew, Knight, Creamer*, pp. 50–1.

125 'Engineering Centre, Derby', *Architect & Building News*, 22 (13 April 1964): pp. 667–71, p. 668.

126 The computers were apparently due for delivery less than 12 months from the architects' instruction. RIBA Library Photographs Collection, P1985/23. 'Rolls Royce Computer Centre' typescript, undated.

127 Around 1958 the practice was also commissioned to fit out the shell of Portman House in London for Gulf Oil Company and Frank Knight later designed interiors for the Gulf Oil Computer Centre to the second floor. Frank Knight, 'Designing for Computers: Design Problems and Solutions by Frank S. Knight of Fry, Drew & Partners', *Building*, 215 (11 October 1968): pp. 87–91, p. 88.

128 E. Maxwell Fry, 'The Speculative Builder', *House and Garden*, 12 (June 1957): pp. 34–5, p. 35.

129 Jane Drew cit. in Barrie Fernandez, *Women in British Architecture*, unpaginated.

130 RIBA Library Photographs Collection, 5218. 'The Hoddesdon Story' Brochure, [1966].

131 RIBA Library Photographs Collection, 5222/17. 'Woodsford Square, Kensington' Brochure, [c.1974].

132 Fry, 'The Speculative Builder', p. 34.

133 RIBA Library Photographs Collection, 5222/17. 'Woodsford Square, Kensington' Brochure, [c.1974].

134 Fry, Drew and Peter Bond all gave lectures at meetings of the Cremation Society during the 1960s.

135 E. Maxwell Fry, *Art in a Machine Age* (London, 1969), p. 68.

136 Fry, *Art in a Machine Age*, p. 69.

137 As Hilary Grainger notes. For a full account of Fry's work at Coychurch, see Hilary Grainger, 'Maxwell Fry and The "Anatomy of Mourning": Coychurch Crematorium, Bridgend, Glamorgan, South Wales', in *Deathscapes: Spaces for Death, Dying, Mourning and Remembrance*, (eds) Avril Maddrell and James D. Sidaway (Farnham, 2010): pp. 243–61.

138 UL Archive, Peter Dale Papers [uncatalogued]. Jane Drew interview with Peter Dale, 1992.

139 RIBA Archive, F&D/13/2. E. Maxwell Fry, 'An Old Man's Epilogue', undated.

Conclusions

It is over 90 years since Fry first practised architecture, whilst Drew was still active and involved in the profession less than 20 years ago. Their lives seem very close to the present and yet they also span into the distant past beyond where living memory can reach them. Fry and Drew experienced some extraordinary times: theirs was an age of empires, unfathomable conflict, brutal poverty, social flux inherited from Victorian manufacturing, the emergence of the Welfare State, and its equally rapid demise. They worked during a time of extraordinary change, especially for architecture where so much of what was thought eternal, was discarded and dismissed. Politically, they witnessed the major transition that transformed Britain in the early-mid twentieth century, whilst inadvertently managing to benefit from many of the political decisions and hard-won liberations of overseas territories.

Fry's view of architecture as a gentlemanly pursuit put him at odds with it, and by the 1960s he seemed dismayed with the construction industry and the failure of architecture to deliver the high expectations he thought it capable of producing. Drew's quest for the 'sensible' shielded her from the knocks that offended Fry's sensibilities, her pragmatism and the passion she held for people, topping that of her love of art, seemed to give her work increased significance as the century wore on. Although her later work may be scorned by critics and ignored by historians she preferred to tackle the difficult commissions associated with social reform and healthcare rather than the photogenic and attention grabbing schemes that were on offer. Some architects follow the fashions, whereas Fry and Drew were stubbornly resilient to change and stuck to their preferred methods of a functionalist creed. It was a choice, or rather a principle, that offered clients a knowing outcome, but also limited the contributions they might have made to the later part of their careers had a more experimental and exploratory architecture continued to be pursued.

Recent scholarship has tended to focus on Fry's 1930s work or the African educational projects from the 1950s. Whilst these are certainly two of the most intriguing and fruitful periods of Fry and Drew's careers, one of the major objectives of this work has been to piece together the other various chapters of their lives with a view to providing greater insight into the architectural history of the twentieth century – for surely Fry and Drew must be positioned as instrumental in the century's major cultural leaps and twists. They are central to a particular and compelling story of modern architecture in the UK and its profusion throughout its (former) colonies. Modern architecture is neither a fixed entity nor a predefined and unproblematic term. It was a phenomenon that Fry would not blindly accept and he explored and grappled with it, initially through writing, attempting to decipher how it would fit within the English physical and cultural landscape. Fry's education equipped him with the ability to compose attractive façades and to comfortably handle the grandiose and the theatrical, but more importantly it gave him an international outlook and lessons in propaganda. It was these skills that he deployed whilst tentatively emerging as the UK's principal apologist of a modernism that was being explored in Germany. After completing some of the most recognisable and radical examples of concrete dwellings in the UK, he also guided the maturation of modernism (indebted to a rather dour Gropius inspired approach), incorporating a brick inflection and informal planning. For Fry, modern architecture was a means of bringing about social reform as well as eliminating wasteful and inefficient building techniques. Urging manufacturers to standardise their designs and seeing the nationalisation of land ownership as the only means of bringing about large-scale reorganisation of towns and cities he saw design, architecture and planning as being equal protagonists in a new reformation of society. Despite these grand visions, Fry steered a pragmatic course, dismissing extreme politics, dogma and grandiose sweeping statements. His architecture sought to express distinct functions with an equally distinct form; entrances were to be carefully composed and protected from the weather with canopies, facades were to be punctuated with a series of projecting and recessed elements such as balconies, window hoods and grills, and materials were used to highlight specific functions or roles (such as staircases). His major skill was managing to combine the interior planning arrangements with the demands of an attractive and delicately proportioned exterior.

After rejecting concrete as a facing material he developed a method that incorporated an exposed concrete structure with the wall infill completed in a buff or brown coloured masonry. The dwellings at Gower Street, Ladbroke Grove, Lewisham and Harlow all followed this approach, as well as some of the commercial, education and 'tropical' buildings. There were practical reasons, not least, exposed concrete finishes were particularly poor at the time in the UK. But also more principled motives behind Fry's decisions, such as his admiration of building craft and the resonance brickwork had with domestic tradition and 'instinctive' architecture. History was not to be abolished by modern architecture, it was not a radical new-beginning that Fry sought, but rather a patchwork approach to modernising the city.

There was undoubtedly a nostalgia expressed through these material choices, and despite his professed admiration for the machine, mechanical production and standardisation, Fry relied on 'traditional' construction procedures and availability of skills – perhaps awaiting the construction industry to catch-up with his ambitions for mechanised construction. It was, however, a system that would wear trait and rather dull – particularly when cheap materials, drab landscaping and a lack of artwork removed the durability and some of the joy of moving around and through those works.

Being part of the establishment, but also holding high currency with the avant-garde Fry positioned himself as a key 'taste maker', something he exploited with his connections in the printed media as well as in the social gatherings of architects, some of whom exclusively, but also shambolically, organised themselves as the MARS group. Despite these credentials, and his major achievements of building social housing (supplemented with nursery provision and carefully designed domesticity) in a modern manner, Fry realised that the photographs he enjoyed of a Frankfurt kitchen were very different to the reality of living with one. He could not ignore the practical limitations of his social housing experiments, but certainly took from them the importance of affordable rents, decent sanitary and kitchen facilities, balconies and heated access staircases.

Drew's early career could be surmised as being in resistance to the architectural profession. Finding work as a female architect was extremely difficult and although Drew did not recall any of her experiences on building sites, one can imagine that harassment was not uncommon. Furthermore being a single mother to twins and trying to establish a business cannot have been easy during 1930s Britain. Later, when Drew was in partnership with Fry and had accumulated extensive experience she was still subject to discrimination and treated as secondary to Fry. The Colonial Office docked the official allowance Drew should have been paid and she was subject to a degree of suspicion in West Africa that no male architects faced. Her design ability was frequently derided and her professional judgements were also undermined. Drew was not, however, defined by her gender. It was Drew who made many of the big decisions that influenced the practice and it was Drew who seized the opportunities and was prepared to take significant risks. Chandigarh would look very different today, for example, without her persuading both Fry and Le Corbusier to accept the commission. In India she was more readily accepted as an architect and exerted some considerable influence over Le Corbusier's design decisions. Drew was clearly a fighter and was more than capable of overcoming professional difficulties. Indeed, she flourished as an architect – employing and treating staff fairly and based on their ability rather than on social position or gender. She also relished the chance to surprise and shock, and became something of a much-loved matriarch.

Drew's joining the RIBA committees was, as she admitted, part of wanting to feel important but more than that, it was indicative of her wanting to contribute to something larger, to help shape her profession and society. Although Fry and Drew had very different temperaments they both had a desire to serve and to bring about change through the medium of planning and architecture. The ideas they wanted to implement in the late 1930s and early 1940s were clearly part of a larger motion that improved sanitation, healthcare, education provision, housing standards and design awareness, all contributing in a small way the changing political and social stratums of the countries in which they worked.

Drew's career needs to be examined using the objectives she set out for herself, such as developing a consultative and sociological approach to architecture. This can be considered a humanist-functionalist approach; if the building does not meet the specific requirements of its users then what use is it? Drew realised that in order to design effectively she would need to meet the people who would use and inhabit her designs. Interviews with inhabitants and clients were conducted at length and touring exhibitions arranged – perhaps at the expense of creating an aesthetic response to the façade. That said, when working as part of a larger team and taking the role of a project architect she delivered some delightful buildings, such as Prempeh College in Ghana and BP offices in Lagos. The housing she designed in Chandigarh is much loved by its residents, and her schools in the city have contributed to the educational provision that has delivered the high levels of literacy enjoyed in Chandigarh today.

Drew enjoyed being amongst people and rallying camaraderie –acting as a one-woman HR and PR department she brought vitality and vigour to the office, whereas Fry was the backroom-boy absorbed in his elevational treatments and literary prose – finding strength and solace through his drawing board. They formed a team that complemented each other's strengths and weaknesses; Fry the gentlemanly artiste and pacifier, and Drew something of an agitator and flamboyant instigator. Together they were able to successfully collaborate with a vast array of clientele, from one-off patrons through to the Colonial Office and large oil corporations. Politically their architecture and no doubt their personalities were considered appropriate, especially in Africa. They were sympathetic to the African cause and eager, albeit it in a paternalistic way, to bring about 'improvements', although they were sometimes reluctant to make any changes to the seductive landscape and villages they encountered. Fry considered Ibadan University to be the zenith of his career, yet the archival records reveal it to be a difficult and tortuous process of numerous arguments and bitter exchanges with the clients. The campus is renowned for the library building which is striking and impressive, frequently cited as a key work in the tropical building pantheon – yet Fry gave it a harsh critique and found it too decorative, lacking the sculpted quality of heavier hewn forms. The schools they designed in Ghana were seen not just as progressive educational tools, but part of the country's route to independence, and were to reflect this through progressive building design. Today, the schools are still, generally, in very good condition and seemingly effective teaching and community spaces. The schools and university were part of a neo-colonialist initiative, but have been re-appropriated; for all architectural meaning and intent is fluid, and political significance is imposed rather than entrenched – at Aburi school the fading and stained plaque referring to Arden-Clarke and signs stating that 'ladies do not sit on ironing boards' seem as Ghanaian as the tartan kilts of the school uniform.

Fry and Drew were able to appeal to both the conservative restraint of the Colonial Office and the nationalism of West Africa. They didn't see themselves as part of a late colonial endeavour, but as liberal agents able to intercede between Whitehall and the African village. Drew managed to reconcile designing a BP office in Lagos whilst celebrating freedom with the Nigerians on the eve of Independence. Fry and Drew may be raised as emblems and vectors of colonialism, but as individuals they followed the commissions and sympathised with the oppressed. Indeed, despite frequently working on behalf of government they preferred to retain their own private practice and resisted the temptation to join a progressive public sector, other than for a short spell in Africa, and even then they defiantly remained distinct from the PWD. Their independence was fuelled by a sense of adventure, as well as a fierce individualism that bordered on the anarchic and the desire to work with whomever they pleased, developing a design process that involved local residents and their 'devil-may-care' attitude to financial security.

Both Fry and Drew were able to court lasting friendships that invariably opened up new opportunities and collaborations and luck also certainly played its part in their success. The World Wars frustrated ambitions, but also reaped new prospects, and it was only Fry's chance posting to West Africa that enabled him and Drew to exploit the demand for architects and planners in that region. Furthermore, Drew's close friendship with Peter Gregory enabled the publication of *Village Housing in the Tropics*. Most publishers would have shied from such an obscure title, but Drew could be very persuasive and she turned what could have been a rather frustrating assignment into a publication that rapidly spread throughout the Empire, and had far greater significance than any of their planning schemes. The book was perfectly timed to receive enthusiastic interest from the Colonial Office who were just about to implement the £200m Development and Welfare grants. Fry and Drew fortuitously gathered an array of commissions throughout West Africa and with a progressive client were able to develop the PWD standard solutions into more radical proposals that were attempting to migrate from their colonial yoke. In India they became friends with the Prime Minister Nehru and as a result they reorganised the PWD structure to properly accommodate the Indian architects they had mentored, resulting in the legacy that continues to this day in Chandigarh.

Fry and Drew's attitude to the tropics did not ever stagnate and this is clearly revealed in their writing. The colonial attitude is entrenched in *Village Housing in the Tropics*, with its map of the world highlighting the British colonies, and many of the designs featured in the book are basic and inexpensive solutions – it was so novel for anyone to give any attention to the needs of African villagers. With the swing in political opinion towards self-governance and the use of architecture as a means to prepare for political independence, there is a reciprocal shift in the type of architecture being proposed, but also a distinct change in how the tropics are viewed. The linear transects of Cancer and Capricorn are replaced with other significant and notably 'scientific' modes of thinking about the tropics and factors such as rainfall, altitude and religion are now considered as important. The blanket approach of 1947 had been superseded with something far more specific and calibrated by the time *Tropical Architecture in the Humid Zone* was published in 1956. Fry and Drew had gained extensive experience not only in Africa but also in the Middle East and India within an intense ten-year period. They had begun to recognise that a 'one-size fits all' approach was not appropriate climatically or culturally. This married with other developments taking place at the time such as the UN Housing Missions and the emergence of building physics rather than 'rules of thumb' and historical precedent. Perhaps the greatest shift with regards to tropical dwelling was in the role of the architect which moved from that of overall designer to facilitator, with the building inhabitant building their own dwelling, frequently in a form originally condemned by Fry in the 1940s, but championed by him in the 1950s as something sacrosanct. It was not the failing of the architect to house the poor, but rather government decision not to pay for the housing shortfall. It was deemed more acceptable to let the labourer and farmworker build their own home and of course, the debate continues as more of the world's population resides in informal settlements in urban centres. Fry and Drew added to the tropical rhetoric with discussions on new cheaper building materials and air conditioning, recognising the limits of what they could achieve in terms of housing. They also wanted secure the latest commercial commissions that had begun to emerge as a result of increased international banking, oil discovery and political independence.

Fry and Drew were not averse to borrowing ideas from others when it suited. Indeed, Fry's early work is openly indebted to Taut and Van der Rohe. A.E.S Alcock was also preparing a publication on Village Housing in Africa at the same time they were and there was some controversy over ideas and ownership. Many of their proposals were rehashed PWD and Royal Engineer standards. Leo De Syllas and Gardner-Medwin's findings in the West Indies were readily absorbed into Fry and Drew's own designs for West Africa, and some of the Chandigarh housing was rather blatantly derived from others. Fry and Drew accumulated pertinent ideas from other architects engaged in innovative work, and readily incorporated them into their own designs – for it is one thing to have a good idea, and something different altogether to resolve, implement, build and effectively publicise one. Their writing certainly helped to promote their practice, and Drew used the AYB to stay abreast of the latest ideas and research not only in design, but also in art and new construction and material advances.

The most difficult period of Fry and Drew's career to research and interpret has undoubtedly been the most recent. Historians have tended to ignore this body of post-war work, preferring instead to look at James Stirling and the Smithsons, to name but three of the more dominant architects from that era. But, more than this, their later work remains difficult because in parts it lacks the glamour, charm and formal intrigue of their early UK work, equally, the quixotic orientalist allure found in their tropical buildings is also absent. The schemes are also generally of large, corporate and commercial buildings – which although wonderfully supplemented with artworks, murals and sculpture fail to develop the romantic functionalism into buildings that are formally and spatially enticing. The large headquarters for Pilkingtons' Brothers is a triumphal arrangement of programme, landscape and artistic endeavour; had the scale of the project been at that of the canteen or museum rather than of the 12 storey tower it may still be viewed with affection. A similar pattern emerges with Rolls Royce, which despite being a perfectly functional building, intriguingly designed around the module of drawing board dimensions, its bland exterior fails to deliver the humanist qualities that both Fry and Drew desired from their proposals. By this stage with Fry in his 60s and Drew taking a greater interest in writing and broadcasting, the practice had transformed from that of a creative studio into a large corporate firm with partners, junior partners and an increasingly large monthly salary demand to contend with. Fry slowly withdrew from the office, and but still delivered some highly successful projects such as the crematorium at Coychurch, South Wales. In many ways it was the ideal architectural commission; combining a clear spatial process endowed with memory, meaningful pauses and terminating with light streaming in from above in the chapel. The majestic materiality of stone and coloured glass is carefully set within the landscape and the overall result is very tranquil and appropriate. With the right client and an evocative programme Fry was more than capable of delivering a functionalist architecture rich with artwork and embedded narrative, along with a sense of journey and wonder.

The prolific output of Fry and Drew, coupled with their lectures, writing and collaborations has resulted in a fertile terrain that continues to offer rich pickings. Their chronology of work demonstrates that further work is still to be done, and if one extends this to their wider influence on the profession then surely a vast canon of material is opened up – some of which inevitably contradicts or undermines previous assumptions and seemingly steadfast interpretations. It is this changing stratum that has fuelled this study and kept the work fresh and interesting, as new ideas, buildings and tracts have been uncovered. Indeed, this has made the task of examining Fry and Drew's place and significance in the history of twentieth century architecture particularly challenging as it slides and distorts depending on which part of their careers one is examining. Their publications have made the largest single contribution, being held in high regard and frequently selling in vast quantities. Equally, their work is certainly appreciated by thousands who have never even heard of them in Chandigarh, throughout Ghana and Nigeria, as well as in certain buildings in the UK. A caretaker at Ladbroke Grove flats described the building in 2011 as 'looking a bit council', meaning they appeared to him more like social housing even though they are in fact private residences. In a way this sums up their work – Fry and Drew saw no distinction between the public and the private, if anything it was in the public and socially valuable projects that their work excelled and continues to have its greatest impact on residents, patients and students in numerous locations throughout the world today.

Chronology

Key events, Projects, Built works and corresponding reference where available.

1899
Fry born 'Edwin Maxwell Fry' at Liscard, Wallasey, Wirral.

1911
Drew born 'Joyce Beverly Drew' at Thornton Heath, Surrey.

1920–23
Fry studies at School of Architecture, Liverpool University.

1922
Fry works for Carrère and Hastings, New York (summer months).

1923
Fry is a finalist for the Rome Scholarship in Architecture and completes his BArch degree.

Fry joins the town-planning firm, Adams and Thompson.

Kemsley Village Plan, Kent.
Source: 'Kemsley Village, Sittingbourne, Kent, Adams, Thompson and Fry', *Builder*, 139 (10 October 1930): pp, 598–9.

1924–26
Fry works for Southern Railways.

Margate and Ramsgate Railway Stations, Kent. Attributed to Fry.
Source: 'New Stations at Ramsgate and Margate', *Builder*, 132 (14 January 1927): p. 87.

1925
Manchester Art Gallery, Competition entry (Second Premiated Design): Fry and G. Owen.
Source: 'Manchester Art Gallery', *Builder*, 128 (26 June 1925): pp. 969–75.

1926
Fry marries Ethel Speakman at Chelsea Registry Office.

1927
Fry becomes a Partner at Adams, Thompson & Fry.

Birmingham Civic Centre, Competition entry (Third Place):
Source: 'Birmingham Civic Centre Competition', *Architect and Building News*, 118 (19 August 1927): p. 298.

1928
Ridge End, Wentworth, Virginia Water, Surrey. Adams, Thompson & Fry.
Source: 'A Modern Country House', *Country Life*, 71 (19 March 1932): pp. 332–4. Now called 'Raven Morough' and much altered.

1929
Sittingbourne Village Club House, Kent. Adams, Thompson and Fry.
Source: Lionel B. Budden (ed.), *The Book of the Liverpool School of Architecture* (Liverpool, 1932): pl. xcv.

City of the Future Drawing (1929) 1931 Regional Plan of New York and its environs
Source: Thomas Adams, L Orton, and H Lewis, *Regional Plan of New York and Its Environs* (New York: Sage, 1931).
1929–34
Drew studies at the Architectural Association, London.

1930
9 Lansdown Crescent, Bath. Bathroom alterations and gas fire design. Fry (for Adams, Thompson and Fry).

Source:'New Bathroom: No. 9 Lansdown Crescent, Bath by E. Maxwell Fry', *Architects' Journal*, 71 (5 March 1930): pp. 377–8.

North-East Kent regional planning scheme. Adams, Thompson and Fry.

Ann Fry born (Fry and Ethel Fry's daughter).

1931

Venesta Exhibition Stand. Competition Entry, Fry (runner-up, won by Coates).

Eastbourne and District regional planning scheme. Adams, Thompson and Fry.

South East Sussex regional planning scheme. Adams, Thompson and Fry.

1932

Fry appointed Lecturer in architectural design at the Regent Street Polytechnic.

Recent Advances in Town Planning, by Adams Thompson and Fry published.

RIBA Headquarters Competition
Source: 'RIBA HQ schemes of note', *Architects' Journal*, 75 (11 May 1932): p. 625.

1933

Town Plan for Borough of Bexhill, East Sussex. Adams, Thompson and Fry.

De La Warr Pavilion, Bexhill-on-Sea, East Sussex. Competition Entry, Adams, Thompson and Fry.

Drew and James Alliston (1908–2000) marry at St Pauls, Thornton Heath.

Fry is a co-founder of the MARS Group with Wells Coates, P. Morton Shand, David Pleydell-Bouverie, H. de Cronin Hastings and John Gloag.

Westminster Electric Supply Corporation Limited Show Rooms, Victoria Street, London (demolished). Fry with photomurals by Hazen Sise.
Source: 'Working Details: 57–58. Shop-Front. Showrooms in Victoria Street, S.W.1. Adams, Thompson and Fry', *Architects' Journal*, 79 (31 May 1934): pp. 801–2.

1933–34

Sassoon House, St Mary's Road, Peckham, London. Fry (for Adams, Thompson and Fry) and Elizabeth Denby.

Source: 'Sassoon House Flats, Peckham', *Architects' Journal*, 79 (26 April 1934): pp. 611–15.

1933–37

Kensal House, Ladbroke Grove, Kensington, London. Fry and Elizabeth Denby (in association with Robert Atkinson, George Grey Wornum and C. H. James). Including a penthouse designed by Fry for Charles Kearley. Source: 'Kensal House', *Journal of the Royal Institute of British Architects*, 44 (20 March 1937): pp. 500–5.

1934

Exhibit of Modern Gas Apparatus, the Exhibition of Contemporary Industrial Design in the Home, Dorland Hall. Fry and Hazen Sise. Client: Gas, Light & Coke Company

Source: A.E. Frost, 'The Exhibition of Contemporary Industrial Design in the Home', *Architects' Journal*, 80 (1 November 1934): pp. 642–6.

1934

Isokon No. 2. Manchester. Client: Jack Pritchard.

1934–36

Little Winch, Chipperfield Common, Hertfordshire. Fry. Client: E. Butler.

Source: 'A House at Chipperfield Common, Buckinghamshire', *Journal of the Royal Institute of British Architects*, 43 (7 March 1936): pp. 483–6. See also Alan Powers, *Modern: The Modern Movement in Britain* (London, 2005), pp. 110–11 for recent photographs.

1934–37

Fry in partnership with Walter Gropius from November 1934.

1934–35

Isokon No. 3, St. Leonard's Hill, Windsor (unbuilt). Client: Jack Pritchard. For two blocks of flats, 110 dwellings in total. Gropius and Fry.

Source: *Architectural Review*, 'Cry Stop to Havoc', 77 (May 1935): pp. 188–92

1935

Glassware Gallery, the Exhibition of British Art in Industry, Royal Academy. Fry.

Source: Joseph Peter Thorp, 'R.A. Exhibition of British Art in Industry', *Architects' Journal*, 81(10 January 1935): pp. 45–8.

21 Queensmere Road, Wimbledon (demolished). Fry.
Source: Myles H. Wright (ed.), *Small Houses, £500–£2,500* (London, 1937), pp. 46–7.

White Lodge, Guildford Road, Bagshot, Surrey. Fry.
Source: Myles H. Wright (ed.), *Small Houses, £500–£2,500* (London, 1937), p. 46.

1935–36
Sun House, Frognal Way, Hampstead, London. Fry. Mural by Hans Feibusch, sculpture by Henry Ellison.
Source: 'House in Hampstead: E. Maxwell Fry, architect', *Architects' Journal*, 84 (13 August 1936): pp. 210–14.

1935–37
Papworth Tuberculosis Colony School, Cambridgeshire. Gropius and Fry (unbuilt).
Source: Harmut Probst and Christian Schädlich, *Walter Gropius. Band 2: Der Architekt und Pädagoge* (Berlin, 1986), p. 94.

1936
Miramonte, Warren Rise, Coombe, Kingston upon Thames, Surrey. Fry.
Source: 'House near Kingston, Surrey', *Architects' Journal*, 86 (18 November 1937): pp. 784–7.

66 Old Church Street Chelsea, London. Gropius and Fry. Client: Ben Levy and Constance Cummings. Much altered.
Source: 'Two houses in Church Street, Chelsea, designed by Walter Gropius and E. Maxwell Fry, by Mendelsohn and Chermayeff', *Architects' Journal*, 84 (24 December 1936): pp. 869–74.

Shepperton Film Studios, Gropius and Fry

Wood House, Upper Green Road, Shipbourne Green. Gropius and Fry.
Source: Alan Powers, *Modern: The Modern Movement in Britain* (London, 2005): pp. 128–9 for recent photographs.

Student Halls, Christ's College, Cambridge University (unbuilt). Gropius and Fry.
Source: 'The designs that Cambridge rejected: the Gropius-Fry scheme for Christ's College', *Architects' Journal*, 109 (3 February 1949): p. 116.

1936–38
Impington Village College, Cambridgeshire. Gropius and Fry.

Source: 'Impington village college, Cambridgeshire', *Architects' Journal*, 90 (21 December 1939): pp. 734–40.

1937

Mortimer Gall Electrical Centre, Cannon Street, London (demolished). Fry and Gropius.

Electricity Show Rooms, Regent Street, London. Fry and Gropius, with photomurals by Moholy-Nagy.

All Souls College, Oxford University (unbuilt).
Source: 'Drawings of the '30's', *Journal of the Royal Institute of British Architects*, 75 (March 1968): p. 112.

Doctor's house, Cliftonville, Kent. Drew (for Alliston Drew).
Source: 'A doctor's house at Cliftonville', *Architect &Building News*, 150 (28 May 1937): pp. 258–9.

Alliston House, Cambridge. Alliston Drew.

Putney Hospital, London. Drew with F. Halliburton Smith.

Competition entry for **Chester Royal Infirmary**, Cheshire (placed third, unexecuted). Alliston Drew.

Old Brightonians R.F.C. Pavilion. Alliston Drew.
Source: 'Old Brightonians R.F.C. Pavilion', *Architect & Building News*, 150 (21 May 1937): p. 231.

1937–39

Winning competition entry for a **Cottage Hospital**, Dawlish, Devon (unbuilt). Alliston Drew.

1938

65 Ladbroke Grove, Kensington and Chelsea, London. Fry. Penthouse flat designed by Rodney Myerscough-Walker.
Source: 'Flats in Ladbroke Grove', *Architects' Journal*, 88 (29 December 1938): pp. 1067–72.

Warham Ash, Breinton, Herefordshire. Fry. Client: Hudson-Davies family.

MARS Exhibition, Burlington Galleries, London. Fry, Moholy-Nagy.

Source: 'The Mars exhibition, new Burlington Galleries', *Architects' Journal*, 87 (20 January 1938): pp. 121–6.

St. Giles' Mount, Winchester, Hampshire. Alliston Drew.
Source: 'St. Giles' Mount Winchester : architect Jane B. Drew', *Architect & Building News*, 156 (25 November 1938): pp. 228–9.

Model Village and School, Kenya (unbuilt). Alliston Drew.

Competition entry for **St. George's Hospital** (unplaced). Hyde Park Corner, London. Alliston Drew.

Stand for Uxbridge Bricks, Building Trades Exhibition, Olympia, London. Fry.

1938–40
Cecil Residential Club. Gower Street, London. Hostel for 72 girls.
Source: 'Girls' hostel. Gower Street', *Architectural Review*, 87 (14 March 1940): p. 285.

1939
Fry assigned to 42 Royal Engineers: War Office Department of Fortifications and Works.

Drew and Alliston divorce.

Model Community Centre, Social Services Exhibit, British Pavilion, New York World's Fair. Fry.
Source: 'Community Centre', *Focus*, 4 (Summer 1939): pp. 25–9.

1940
Homerton College Nursery School, Hills Road, Cambridge.

Walton Yacht Works, Walton-on-Thames, Surrey. Drew.

1941
Fry and Ethel Speakman divorce (although separated since 1940).

1942
Fry stationed in Derby; Staff Captain in the Royal Engineers.

Fry and Drew marry on 25 April 1942 at Caxton Hall in Westminster.

Fry serves in West Africa, based in Accra, Ghana.

1943
Drew appointed as Consultant Architect to the Domestic Commercial Heat Services Committee, established by the British Commercial Gas Association

c. 1943
Extension to Factory, Harris & Sheldon, Birmingham. Trevor Dannatt for Jane B. Drew.

c. 1944
Drew: Member of the RIBA Housing Group with Elizabeth Denby, Jessica Albery and Judith Ledeboer

'Kitchen Planning' exhibition, Dorland Hall. Drew

1944
Fry appointed **Town Planning Advisor to the Resident Minister** and Drew **Chief of Staff**, West Africa. Plans produced for towns throughout West Africa, including, **Accra**, **Kumasi**, **Freetown**, **Bathurst**, **Lagos**.
Source: 'Town planning in West Africa', *Architects' Year Book*, 2 (1947): pp. 64–73.

Fine Building published by Faber.

1945
First edition of *Architects' Year Book* published by Paul Elek. Edited by Drew.

Architecture for Children published by Allen & Unwin.

Kitchen Planning: plans and suggestions based on joint research by the Gas industry and Jane B. Drew, published by the Gas Industry of Great Britain.

1946
Town-planning Scheme for Bathurst and the Kombo Area. Report prepared by Fry and Betty Benson (written in 1944, published in Accra in 1946).

'Britain Can Make It' Exhibition for Council of Industrial Design.

Alterations to **66 Old Church Street**, Chelsea. Drew.

c. 1947
Science School, Altofts, West Riding of Yorkshire (unbuilt). Fry.

1947
Town Plan and Deep Water Quay Planning Report, Freetown, Sierra Leone. Fry and K.W. Farms (Planning Assistant).

Asawasi Experimental Village, Kumasi, Ghana. In collaboration with A.E.S Alcock.

Village Housing in the Tropics published by Lund Humphries.

1948
Design Centre for British Rayon Industry, Upper Grosvenor Street, London. Source: 'Design centre for British Rayon Industry', *Architects' Journal*, 108, (23 December 1948): 575–578.

1949–51
Ahmadi General Hospital, Kuwait. Drew. Client: Kuwait Oil Company. Housing, clinics and social amenities also built at this time.

1949–60
University College of Ibadan, Nigeria. Campus masterplan, students residences, lecture theatres, administration offices, library, teaching rooms, Mosque, Catholic Church, bookshop, staff housing. Source: 'University College, Ibadan, Nigeria', *Architectural Design*, 25 (May 1955): 154–65.

1950
Fry, Drew & Partners established. Senior partners Fry, Drew; junior partners John Cordwell, Kathleen Greenwell, J.B. Shaw, Norman Starrett, S.E.A. Hounsell (Secretary).

1950–
Community Centre, Accra, Ghana. Funded by the United Africa Company. Mosaic mural to front façade by Kofi Antobam.
Source: 'Community Centre at Accra, Gold Coast', *Architect & Building News*, 197 (24 February 1950): pp. 185–7.

St. Francis College, Ho Hoe, Togoland (now Ghana).
Source: 'Recent educational buildings in the Gold Coast', *Architectural Review*, 113 (May 1953): pp. 301–10.

ICA Interiors, 17–18 Dover Street, London. Drew.

1950–51

Thameside Restaurant, the Waterloo Gate Bridges and Viewing Tower, Festival of Britain, South Bank, London. Drew.

1950–53

Amedzofe School, Ghana.
Source: *Architectural Design*, 25 (May 1955): p. 139.

Mfantsipim School, Cape Coast, Ghana.
Source: *Architectural Design*, 25 (May 1955): p. 167.

St. Andrew's College, Mampong, Ghana.
Source: 'Recent educational buildings in the Gold Coast', *Architectural Review*, 113 (May 1953): pp. 301–310.

1951

Festival of Britain, South Bank Exhibition (schools exhibition) Fry & Drew

The Ceylon Tea Centre, Lower Regent Street, London. Fry & Drew; Foyer design with mural by John Farleigh (1900–65)

Passfield, flats, Lewisham, London.
Source: *Architectural Review*, 109 (January 1951): pp. 7–15.

Harlow New Town, houses and flats (Tanys Dell and The Chantry).
Source: *Architect and Building News*, 204 (July 3 1953): pp. 12–15.

Adisadel College, Ghana
Source: *Architectural Design*, 25 (May 1955): p. 141.

1951–54

Chandigarh, Punjab, India. **Housing** in Sectors 4, 7, 10, 14, 16, 19, 22, 23, **Kiran Cinema, Government Printing Press, Nursery Schools, Boys and Girls Schools, Government Hostels** and **Colleges, Swimming Pool, Clinic, General Hospital, shops.** See Kiran Joshi, *Documenting Chandigarh* (Ahmedabad, 1999).

1952–59
Denys Lasdun and Lindsay Drake become partners in practice, renamed Fry, Drew, Drake & Lasdun. Lasdun and Drake act as office heads while Fry and Drew were engaged with Chandigarh.

1952–53
Prempeh College, Kumasi, Ghana.
Source: *Architectural Design*, 25 (May 1955): p. 143.

1953
Whitefoot Lane, Bromley, London

Sheffield University Campus (unplaced competition entry)

Mawuli School, Ho, Togo
Source: *Architectural Design*, 25 (May 1955): pp. 144–6.

Aburi School and College, near Accra, Ghana
Source: *Architectural Design*, 25 (May 1955): p. 149–53.

St. Joseph's Teacher Training College, Bechem, Ghana.

Presbyterian Teacher Training College, Amedzofe, Ghana.

1953–55
Opoku Ware School, Kumasi, Ghana.
Source: *Architectural Design*, 25 (May 1955): pp. 169–71.

1954
Apowa Roman Catholic Teachers' Training College and School, Takoradi, Ghana.
Source: *Architectural Design*, 25 (May 1955): pp. 147–8.

Town plan for Ummaidieh, Iran, Drew.

1954–?
Community Centre, Tarkwa, Ghana. Funded by the African Manganese Corporation.
Source: *Architectural Design*, 25 (May 1955): p. 174.

Falkland House, Bromley Road, Lewisham.
Source: *Architects' Journal*, 119 (11 February 1954): pp. 196–8.

1954–56

Co-operative Bank at Ibadan, Nigeria.
Source: *Architectural Design*, 25 (May 1955): p. 172.

1954–58

Oriental Insurance Building, Calcutta, India.

1954–59

Bank of West Africa warehouse and flats; village houses, Tema, Ghana. With Lindsey Drake.

1955

St. Andrew's College, Mampong, Ghana.
Source: *Architectural Design*, 25 (May 1955): p. 141.

Wesley Girls' School, Cape Coast, Ghana.
Source: *Architectural Design*, 25 (May 1955): p. 142.

St. Monica's School, Mampong, Ghana.
Source: *Architectural Design*, 25 (May 1955): p. 166.

Baha'-ist House of Worship, Kampala, Uganda (unbuilt). Fry. Nine-sided temple with dome.

Office Complex (unbuilt). Piccadilly, Manchester. Fry.

High Grade House for Tema, Ghana.
Source: *Architectural Design*, 25 (May 1955): p.173.

1955–58

Usk Street Housing Estate at Bethnal Green, London (with Denys Lasdun).

Leventis Store, Accra, Ghana.
Source: 'Leventis store, Accra' *Architectural Design*, 25 (May 1955): pp. 167–8.

1955–60

Veterinary Science Building, Liverpool University (demolished 2012).
Source: 'Veterinary building, Liverpool University', *Builder*, 202 (January 1962): pp. 113–18.

Civil Engineering Building, Liverpool University.

New Offices, Mining & Chemical Products Co., Folison Park, Berkshire. (unbuilt)

1955–65

Pilkington Brothers' Headquarters, St Helens.

Source: Jessica Holland and Iain Jackson, 'A Monument to Humanism: Pilkington Brothers' Headquarters (1955–65) by Fry, Drew and Partners', *Architectural History*, 56, (2013): 343–86; *Architect & Building News*, 226 (28 October 1964): pp. 817–26.

1956–57

Fry Visiting Professorship at Harvard Graduate School of Design.

1958

Teacher Training College, Wudil, Nigeria.

Shaw House, Singapore.

Churchill College, Cambridge Competition entry, unplaced.

Source: Mark Goldie, 'Corbusier comes to Cambridge: post-war Architecture and the competition to build Churchill College', (Cambridge, Churchill College, 2007).

Gulf Oil Computer Centre, Gulf House, Portman House, London. Knight.

Source: Frank Knight, 'Designing for Computers', *Building*, 215 (11 October 1968): pp. 87–92.

c. 1958

Women's Teacher Training College, Kano, Nigeria.

Broadcasting Building, Nigeria (exact location unknown).

1959

St. Patrick's Schools in Lagos, Nigeria.

Gach Saran new town for the Iranian Oil Exploration and Production Company, Iran. Drew and Knight.

Competition entry for **Churchill College, Cambridge University** (unplaced).

Interiors for **Shell Oil Company Offices**, Singapore by Swan and MacLaren. Drew.

Milan Triennale Project.

1959–60

Studio at Lake house, Sussex (Fry and Drew's country house - bought by Fry, Drew, Drew's father and Peter Gregory).

Source: 'Three studios designed by Fry, Drew & Partners', *Architectural Review*, 131 (February 1962): pp. 100–105.

1959–63

Houses of Parliament, Kuala Lumpur, Malaysia. Fry and Drew were advisors to the Project Architect Ivor Shipley, who subsequently worked in their London office. Source: Lai, *Building Merdeka*, pp. 80–93.

1960

Fry, Drew & Partners re-established. Senior partners Fry, Drew, Frank Knight, Norman Creamer, J.R. Atkinson (West Africa); partners Peter Bond, Robert Byng.

Secondary modern school, Okene, Nigeria.

British High Commission offices, Banjul, The Gambia.

Frew House, Accra, Ghana.

Broadbent House, near Hendon, Drew.
Source: H. Dalton Clifford, 'Topsy-Turvy by Design', *Country Life*, (14 April 1960): pp. 822–5.

BP Offices, Lagos, Nigeria.

Office building and factory for Dow Agrochemicals Ltd, King's Lynn, Norfolk.

Holy Cross School, Lagos, Nigeria.

Christ Cathedral Primary School, Lagos, Nigeria.

c. 1960s (exact date unknown)

Lionel Wendt Gallery, Colombo, Ceylon (Sri Lanka) Drew with Geoffrey Bawa.

White House Rural College, The Isle of Thorns, Chelwood Gate, East Sussex.

Mashhad development plan, Kitorassan, Iran.

Eleousa project, Cyprus.

Federal government offices, Petaling Jaya, Selangor, Malaysia.

University of Sheffield development plan.

Steel House, Eastcote, Harrow.

Sykes Chicks, offices.

Margaret Pyke Centre, Family Planning Association, London (built).

1961

Longmans Green house, Lagos, Nigeria.

Drew Visiting Professorship (Beamis Professor) at MIT, Boston (February to June).

Artist's studio added to a farmhouse, Hertingfordbury
Source: 'Three studios designed by Fry, Drew & Partners', *Architectural Review*, 131 (February 1962): pp. 100–105.

Steel House, Eastcote, Harrow.

1961–70

Extensions and School of Nursing, Torbay Hospital, Devon. Drew.

c. 1961

Artist's studio, near Paris, France
Source: 'Three studios designed by Fry, Drew & Partners', *Architectural Review*, 131 (February 1962): p. 100–5.

BOAC housing, Lagos, Nigeria.

1962

Paterson Simons & Co, Accra, Ghana (by Denys Lasdun?).

Temple Scheme, Abu Simbel, Egypt (unbuilt).

1963

Wates' Headquarters, London Road, Norbury, London. Central courtyard has an ornamental pool with 'Girl with Doves' sculpture by David Wynne. Substantially altered; two additional stories added during conversion to flats (c. 2003) by Stock Woolstencroft.

Fry elected Member of Town Planning Institute.

c. 1963–66

Industrialised **'Belfry' Housing**. Sites included a 42-house development at Hoddesdon, Hertfordshire (1966).

Source: 'Belfry system for housing in Castle Road, Hoddesdon', *Architects' Journal*, 143 (4 May 1966): 1149–56.

1964
Fry awarded the RIBA Gold Medal.

'An Exhibition of the Architecture of Maxwell Fry', Monks Hall Museum, Eccles (to coincide with Fry's 65th Birthday and the award of the RIBA Gold Medal).

Housing, Hatfield (Woods Avenue, Chantry Road, White Lion House).

ICA Interiors, 12 Carlton House Terrace, London.

Bank of West Africa and manager's house, Swedru, Ghana (with Lindsey Drake).

Ferdowsi University, IranDrew (unbuilt).

Pahlavi University, near Shiraz, Iran, Drew (unbuilt).

1964–65
Rolls Royce Engineering Centre and Computer Centre, Victory Road, Derby
Phased construction. Three U-shaped wings with computer centre in separate building.
Source: 'Engineering Centre, Derby', *Architect & Building News*, 225 (April 15 1964): pp. 667–71.

1965
Development of Rosia Bay, Gibraltar. For Gibraltar Study Group. Exhibited at the RA.

1967
Ahmadu Bello Stadium and Swimming Pool, Kaduna, Nigeria. Drew.

Hilton Hotel, Colombo, Sri Lanka. Drew and Geoffrey Bawa (unbuilt).

196?
Holiday apartments, Sandy Bay, Gibraltar. First of three projects for a development company (perhaps also Catalan Bay). Complex includes 138 flatlets, restaurant, bars, supermarket, launderette, and shops. Source: Hitchins, p. 144.

1965–70s

Masterplan for United Manchester Hospitals, Manchester (unbuilt).

1965

Chelwood House, Grosvenor Square, London. Fry.

1965–7

Kingston House, Bond Street, Hull.

1966

Woodsford Square, 2–7 Addison Road, Kensington, London. Fry. Development for Wates Construction.

Grove House School and Clinic for Deaf Children, Elmcourt Road, Lambeth, London. Drew.

Elm Court School for the Deaf, West Norwood, London. Drew.

City Hall Offices, Gibraltar, Fry, , (unbuilt).

1967–70

Coychurch Crematorium, near Bridgend, Wales. Fry.

1969

Anglo-Thai offices and workshops, Bangkok, Thailand.

1969–73

Extensions and School of Nursing, Torbay Hospital, Devon. Drew.

1969–77

Open University, Milton Keynes. Drew

Source: 'The Open University', *Architects Journal*, 153 (27 January 1971): 172–4.

1969–

Sir Seewoosagur Ramgoolam National Hospital, Pamplemousses, Mauritius. Drew and Knight.

1970

Drew visiting Professor at Harvard, (February to March).

The Chartered Bank, Bangkok, Thailand

1970–77

Yundum International Airport, The Gambia. Control tower and terminal buildings. New terminal building (opened 1997) adjacent to original by the Senegalese architect Pierre Goudiaby Atepa. Source: Hitchins, pp. 134–5.

1971

Chapel, Breakspear Crematorium, Ruislip, Middlesex. Fry.

1973

Fry, Drew, Knight & Creamer partnership established; Fry retires and becomes a consultant to firm Drew awarded Honorary Doctorate by the Open University.

Gestetner Factory, Stirling, Scotland. Steel-frame building, clad in 'Siporex' panels on a brick plinth. Situated below the Wallace Memorial. Altered and extended. Source: Hitchins, p. 10.

1974

Fry painting exhibition at the Drian Gallery, London.

1970s

Legislative Assembly and Government Centre, Port Louis, Mauritius. Fry.

Registrar General Building, Port Louis, Mauritius. Fry.

School for Mauritian, Oriental and African Studies, Le Reduit, Mauritius.

1977

Institute of Education, Le Reduit, Mauritius.

1977–81

Brikama College, Brikama, The Gambia.

1978
Right to Vote Exhibition, Westminster Hall, London. Drew.

Drew made Honarary Fellow of the American Institute of Architects.

1987
Fry died 3 September at Darlington Memorial Hospital and was buried at St Romald's Church, Romaldkirk, County Durham.

Drew awarded Hon. Doctorate from Newcastle University

1994
Drew awarded Honorary Doctorate of Architecture from Witwatersrand University, Johannesburg.

1996
Drew awarded DBE.

Drew died 27 July at Darlington Memorial Hospital and was buried at St Romald's Church, Romaldkirk, County Durham.

Bibliography

Adams, Thomas, L. Orton, and H. Lewis, *Regional Plan of New York and its Environs* (New York: Sage, 1931).

Adams, Thomas, F.L. Thompson, Edwin Maxwell Fry, and James Adams, *Recent Advances in Town Planning* (London: J & A Churchill, 1932).

Adams, W. Eric, *The New Town of Harlow* (Harlow: Harlow Development Corporation, 1954).

Adshead, Stanley Davenport, *Town Planning and Town Development* (London: Methuen & Co.: 1923).

Alcock, A.E.S., and Helga M. Richards, *How to Plan Your Village* (London: Longmans, 1953).

Alcock, A.E.S., and Helga M. Richards, *How to Build for Climate* (London: Longmans, 1960).

Alcock, A.E.S., 'Changes in the Use of Building Materials and Products for Low Cost Housing in the Tropics', *Planning Outlook*, 6 (Spring 1969): 38–45.

Alcock, A.E.S., 'Tema, a New Town in Gold Coast', *Town & Country Planning* (January 1955): 51–55.

Allan, John, *Berthold Lubetkin: Architecture and the Tradition of Progress* (London: Merrell, 1992).

Amarteifio, G.W., D.A.P. Butcher, and David Whitham, *Tema Manhean, a Study of Resettlement* (Accra: Ghana Universities Press, 1966).

Anker, Peder, *From Bauhaus to Ecohouse: A History of Ecological Design* (Baton Rouge, LA: Louisiana State University Press, 2010).

Ascot Gas Water Heaters Ltd, *Flats: Municipal and Private Enterprise* (London: Ascot Gas Water Heaters Ltd, 1938).

Asquith, Cyril, *Report of the Commission on Higher Education in the Colonies* (London: HMSO, 1945).

Atkinson, G. Anthony, 'British Architects in the Tropics', *Architectural Association Journal*, 69 (June 1953): 7–21.

Atkinson, Harriet, *The Festival of Britain: A Land and its People* (London: I.B. Tauris, 2012).

Bahga, Sarbjit, and Surinder Bahga, *Le Corbusier and Pierre Jeanneret: Footprints in the Sands of Indian Architecture* (New Delhi: Galgotia Publishing, 2000).

Banerjee, Tridib, 'U.S. Planning Expeditions to Postcolonial India', *Journal of the American Planning Association*, 75 (Spring 2009): 193–208.

Banham, Mary (ed.), *A Tonic to the Nation: The Festival of Britain 1951* (London: Thames and Hudson, 1976).

Bassett-Lowke, W.J., 'A Wonderful Experiment at Stuttgart', *DIA Quarterly*, 2 (December 1927): 8–10.

———, 'Recent Developments in Germany', *DIA Quarterly*, 6 (December 1928): 9–11.

Benton, Charlotte (ed.), *A Different World: Emigré Architects in Britain 1928–1958* (London: RIBA Heinz Gallery, 1995).

Bisson, Roderick F., *The Sandon Studios Society and the Arts* (Liverpool: Parry Books, 1965).

Blacklock, D.B., *An Empire Problem: The House and Village in the Tropics* (Liverpool and London: The University Press of Liverpool, Hodder & Stoughton 1932).

Blum, Betty J., *Oral History of John Donald Cordwell* (Chicago: Art Institute of Chicago, 1993).

Bourret, F.M., *The Gold Coast: A Survey of the Gold Coast and British Togoland 1919–1946* (London: Geoffrey Cumberlege, Oxford University Press, 1949).

Brassley, Paul, Jeremy Burchardt and Lynne Thompson (eds), *The English Countryside Between the Wars: Regeneration Or Decline?* (Woodbridge: The Boydell Press, 2006).

Brawne, Michael (ed.), *University Planning and Design: A Symposium* (London: Architectural Association, 1967).

Bryder, Linda, '"Wonderlands of Buttercup, Clover and Daisies": Tuberculosis and the Open-Air School Movement in Britain, 1907–39', in *In the Name of the Child: Health and Welfare, 1880-1940*, (ed.) Roger Cooter (London: Routledge, 1992).

Bullock, Nicholas, *Building the Post-War World: Modern Architecture and Reconstruction in Britain* (London: Routledge, 2002).

Bunting, Bainbridge, *Harvard: An Architectural History* (Cambridge, MA: Harvard University Press, 1998).

Burke, Catherine, *A Life in Education and Architecture: Mary Beaumont Medd* (Farnham: Ashgate, 2013).

Campbell, Louise, 'The MARS Group, 1933–1939', *RIBA Transactions*, 4/2 (1985): 68–79.

Cantacuzino, Sherban, *Wells Coates* (London: Gordon Fraser, 1978).

Chang, Jiat-Hwee, and Anthony D. King, 'Towards a Genealogy of Tropical Architecture: Historical Fragments of Power-Knowledge, Built Environment and Climate in the British Colonial Territories', *Singapore Journal of Tropical Geography*, 32 (November 2011): 283–300.

'The Chipperfield Case: Another House Rejected', *Architects' Journal*, 80 (20 September 1934): 406–8.

Clark, Jon, Margot Heinemann, David Margolis and Carole Snee (eds), *Culture and Crisis in Britain in the Thirties* (London: Lawrence and Wishart, 1979).

Cohen, J. L., *Scenes of the World to Come: European Architecture and the American Challenge 1893–1960* (Paris: Flammarion, 1995).

Colonial Office, 'Freetown: Canalisation of Streams and Surface Drainage (Typescript Correspondence, Annotated, and Memoranda Relating to Public Works in Freetown, between Acting Governor Cookson in Sierra Leone and the Colonial Office)' (London: National Archives, 1929).

Commission, Volta River Project Preparatory, *The Volta River Project* (London: Published for the Governments of the United Kingdom and of the Gold Coast by Her Majesty's Stationery Office, 1956).

Christopher Crouch, *Design Culture in Liverpool 1880–1914* (Liverpool: Liverpool University Press, 2002).

'Competition News: Cement Houses', *Architects' Journal*, 77 (8 February 1933): 197–8.

'The Contemporary Industrial Design Exhibition', *Architects' Journal*, 80 (1 November 1934): 644–5.

'Community Centre', *Focus*, 4 (Summer 1939): 25–9.

Cotton, Michelle (ed.), *Nigel Henderson & Eduardo Paolozzi: Hammer Prints Ltd, 1954–75* (Colchester: Firstsite, 2013).

Cox, Christopher William Machell, *Mass Education in African Society* (London: HMSO, 1944).

Creamer, Norman, 'The Work of Drake and Lasdun of Fry, Drew, Drake and Lasdun; Work in the Tropics', *Architectural Design*, 28 (February 1958): 55–74.

Crinson, Mark, *Modern Architecture and the End of Empire* (Aldershot: Ashgate, 2003).

Crinson, Mark, and Claire Zimmerman (eds), *Neo-avant-garde and Postmodern Postwar Architecture in Britain and Beyond* (London and New Haven, CT: Yale University Press, 2010).

Curtin, Philip D., 'Medical Knowledge and Urban Planning in Tropical Africa', *American Historical Review*, 90 (June 1985): 594–613.

Curtis, William, *Le Corbusier: Ideas and Forms* (London: Phaidon, 1986).

———, *Denys Lasdun: Architecture, City, Landscape* (London: Phaidon, 1994).

Dales, S. Phillips, 'Harlow', *Builder*, 173 (19 December 1947): 700–3.

Elizabeth Darling, '"The Star in the Profession she Invented for herself": A Brief Biography of Elizabeth Denby, Housing Consultant', *Planning Perspectives*, 20 (July 2005): 271–300.

———, '"The Scene in which the daily drama of personal life takes place": Towards the modern interior in early 1930s Britain', in *Designing the Modern Interior: From the Victorians to Today*, (eds) Penny Sparke, Anne Massey, Trevor Keeble and Brenda Martin (Oxford and New York: Berg, 2006): 95–105.

———, '"Into the world of conscious expression": Modernist revolutionaries at the Architectural association, 1933–39', in *Man-made Future: Planning, Education and Design in Mid-Twentieth Century Britain*, (ed.) Iain Boyd Whyte (London and New York: Routledge, 2007).

———, 'Kensal House: The Housing Consultant and the Housed', *Twentieth Century Architecture*, 8 (2007): 107–15.

———, *Re-Forming Britain: Narratives of Modernity before Reconstruction* (London: Routledge, 2007).

Datta, Partho, 'How Modern Planning Came to Calcutta', *Planning Perspectives*, 28 (2013): 139–47.

Dean, David, *The Thirties: Recalling the English Architectural Scene* (London: Trefoil, 1983).

Denby, Elizabeth, *Europe Re-Housed* (London: George Allen & Unwin, 1938).

Drew, Jane, *Kitchen Planning: A Brochure of New Plans and Suggestions for Labour-Saving Kitchens* (London: The Gas Industry, 1945).

———, 'Editor's Foreword', *Architects' Year Book*, 1 (1945): 5.

———, 'Housing', *Architects' Year Book*, 1 (1945): 43–54.

———, 'Domestic Appliances (Powered and Non-Powered)', in *Design '46: A Survey of British Industrial Design as Displayed in the 'Britain Can Make It' Exhibition Organised by the British Council of Industrial Design* (London: HMSO, 1946): 15–21.

———, 'West Africa', *Architectural Design* (April 1955): 137–39.

———, 'Planning in Persian Oil Towns', *Town & Country Planning*, 28 (January 1960): 38–42.

———, 'Living: Sector-22', *MARG*, 15 (1961): 22–25.

———, 'Indigenous Architecture: Architecture in the Tropics', *Perspecta*, 8 (1963): 57–58.

———, 'Le Corbusier a Personal Note', *Journal of the Town Planning Institute*, 51 (November 1965): 380–1.

Drew, Jane, and E. Maxwell Fry, *Architecture for Children* (London: George Allen & Unwin Ltd, 1945).

Drew, Jane, E. Maxwell Fry, and H.L. Ford, *Village Housing in the Tropics: With Special Reference to West Africa* (London: Lund Humphries, 1947).

Drew, Jane, E. Maxwell Fry, Alison Smithson, and Peter Smithson, 'Conversation on Brutalism', *Zodiac*, 4 (1960): 73–81.

Drew, Jane, and Margaret Garlake, *Drew, Jane. (2 of 9). National Story Collection: Architects' Lives*: British Library C467/03, 1995. (audio).

Dronamraju, Krishna R., *If I am to be Remembered: The Life and Work of Julian Huxley with Selected Correspondence* (London: Heinemann, 1993).

Dudley, Earl of, *Design of Dwellings: Report of the Design of Dwellings Sub-committee of the Central Housing Advisory Committee*, (London: HMSO, 1944).

Dumett, Raymond E., 'The Campaign against Malaria and the Expansion of Scientific Medical and Sanitary Services in British West Africa, 1898–1910', *African Historical Studies*, 1/2 (1968): 153–97.

Dunne, Jack, and Peter Richmond, *The World in One School: The History and Influence of the Liverpool School of Architecture 1894–2008* (Liverpool: Liverpool University Press, 2008).

Edwards, Arthur Trystan, *Good and Bad Manners in Architecture* (London: Philip Allan, 1924).

Esher, Lionel, *A Broken Wave: The Rebuilding of England, 1940–1980* (Harmondsworth: Penguin Books, 1981).

Evenson, Norma, *Chandigarh* (Berkeley and Los Angeles, University of California, 1966).

'Exhibition of Contemporary Furniture at the Mansard Gallery, W.1.', *Architects' Journal*, 83 (30 April 1936): 652–3.

Fair, Alistair, '"A New Image of the Living Theatre": The Genesis and Design of the Belgrade Theatre, Coventry, 1948–58', *Architectural History*, 54 (2011): 347–82.

Feibusch, Hans, *Mural Painting* (London: A & C Black, 1947).

'The Festival of Britain: Architectural Plans for South Bank Site', *Builder*, 177 (25 November 1949): 686–90.

'Flats in Ladbroke Grove', *Architects' Journal*, 88 (29 December 1938): 1067–72.

Flower, Sile, Jean Macfarlane, and Ruth Plant (eds), *Jane B. Drew Architect: A Tribute from Colleagues and Friends for Her 75th Birthday 24th March 1986* (Bristol: Bristol Centre for the Advancement of Architecture, 1986).

Folkers, Antoni, *Modern Architecture in Africa* (Amsterdam: SUN, 2010).

Ford, Boris (ed.), *Early 20th Century Britain: The Cambridge Cultural History* (Cambridge: Cambridge University Press, 1992).

Ford, Harry L., 'A New Town in Botswana', *Planner*, 61 (April 1975): 145–9.

Ford, Percy, *A Breviate of Parliamentary Papers, 1940–1954: War and Reconstruction* (Oxford: Blackwell, 1961).

Foyle, Arthur (ed.), *Conference on Tropical Architecture, 1953* (London: Organising Committee, Conference on Tropical Architecture, 1954).

Fraser, Valerie, *Building the New World: Studies in the Modern Architecture of Latin America 1930–1960* (London: Verso, 2000).

Fry, E. Maxwell, 'Modernism on Paisley', *Architects' Journal*, 67 (2 February 1928): 291–9.

———, 'The Work of the Schools: Liverpool', *Architects' Journal*, 72 (16 July 1930): 87–9.

———, 'Settings for Gas Fires', *Architects' Journal*, 72 (24 December 1930): 953.

———, 'Modern Cottages and Country Houses', *Architects' Journal*, 74 (8 July 1931): 54–6.

———, 'My Province is all Structure', *Architects' Journal*, 77 (12 April 1933): 492–3.

———, 'The Architect's Problem', *Architects' Journal*, 77 (22 June 1933): 844–6.

———, 'De-Slumming', *Architects' Journal* (15 March 1933): 366.

———, 'Housing Problems in 1933', *Architects' Journal*, 77 (8 February 1933): 208, 11.

———, 'Slum and the Land', *Architects' Journal*, 78 (26 October 1933): 550–1.

———, 'Rebuilding Britain'. *The Listener*, 255 (29 November 1933): viii.

———, 'An Urgent Question to the Manufacturers', *Architects' Journal*, 79 (5 April 1934): 499–500.

———, 'English Town Hall Architecture' *The Listener*, 275 (18 April 1934): 649.

———, 'London Housing an Itinerant Survey of Typical Schemes', *Architects' Journal*, 79 (17 May 1934): 712–29.

———, 'Mass-Production in Housing', *Architects' Journal*, 80 (1 November 1934): 629–30.

———, 'The Design of Dwellings', in *Design in Modern Life*, (ed.) John Gloag (London: George Allen & Unwin, 1934).

———, 'Design in the Countryside and the Town', in *Design in Modern Life*, (ed.) John Gloag (London: George Allen & Unwin, 1934).

———, 'Design and the Machine', *Architects' Journal*, 81 (24 January 1935): 157–8.

———, 'Town Planning in Northern Ireland' *The Listener*, 409 (11 November 1936): 886.

———, 'Can Belfast Be Reshaped?', *The Listener*, 410 (18 November 1936): 961.

———, 'Saving the Coastline's Beauty', *The Listener*, 412 (2 December 1936): 1040.

———, 'What Planning can do for Small Towns' *The Listener*, 414 (16 December 1936): 1147.

———,' Creative Housing', *The Listener*, 449 (18 August 1937): 358.

———, 'Town Planning', in *Circle: International Survey of Constructive Art*, (ed.) J.L. Martin, Ben Nicholson and N. Gabo (London: Faber & Faber, 1937).

———, 'Housing', *Architects' Journal*, 86 (9 December 1937): 947–8.

———, 'The New Britain Must Be Planned', *Picture Post*, 10 (1941): 16–20.

———, *Fine Building* (London: Faber & Faber, 1944).

———, 'Town Planning in West Africa', *African Affairs*, 45 (October 1946): 197–204.

———, 'Developing "The Most Beautiful Town in West Africa"', *West African Review*, June (1946): 625.

———, 'Town Planning in West Africa', *Architects' Year Book*, 2 (1947): 64–73.

———, 'Chandigarh: Birth of a Capital', *The Sunday Statesman* (6 May 1951): unpaginated.

———, 'African Experiment', *Architectural Review*, 113 (May 1953): 299–310.

———, 'Chandigarh: A New Town for India', *Town and Country Planning*, 21 (May 1953): 217–21.

———, 'Chandigarh: The Capital of the Punjab', *Journal of The Royal Institute of British Architects*, 62 (4 January 1955): 87–94.

———, 'The Architect's Dilemma – I', *The Listener*, 1355 (17 February 1955): 281.

———,'The Architect's Dilemma – II', *The Listener*, 1356 (24 February 1955): 331.

———, 'Opening of High Court by Mr. Nehru', *Architect and Building News*, 207 (20 January 1955): 401.

———, 'Walter Gropius', *The Architectural Review*, 117 (March 1955): 155–7.

———, 'The Speculative Builder', *House and Garden*, 12 (June 1957): 34–5.

———, 'A College in the Tropics: Ibadan', *Zodiac*, 2 (1958): 128–36.

———, 'Architects to the Rescue!', *RIBA Journal*, 66 (June 1959): 285–7.

———, 'Problems of Chandigarh Architecture', *MARG*, 15 (December 1961): 20–1, 25.

———, 'F.R.S. Yorke, 1906–1962', *Architectural Review*, 132 (October 1962): 279–80.

———, 'The Aesthetics of University Buildings', *Higher Education Quarterly*, 18 (August 1964): 341–51.

———, 'Le Corbusier: His Impact on Four Generations', *RIBA Journal* (October 1965): 497–500.

———, 'Housing and the Environment', *RIBA Journal*, 74 (August 1967): 331–3.

———, *Art in a Machine Age: A Critique of Contemporary Life through the Medium of Architecture* (London: Methuen, 1969).

———, 'Crematoria' *The Listener*, 2212 (19 August 1971): 239.

———, *Autobiographical Sketches* (London: Elek, 1975).

———, 'Max Fry: Inspirations, Friendships and Achievements of a Lifetime in the Modern Movement', *Building*, 229 (31 October 1975): 52–8.

———, 'Le Corbusier at Chandigarh', in *The Open Hand: Essays on Le Corbusier*, (ed.) R Walden (Cambridge, MA: MIT Press, 1977).

———, 'Harmony out of Discord: Maxwell Fry Speaks about his Life and Work, at the RIBA', *RIBA Journal*, 86 (December 1979): 526–9.

Fry, E. Maxwell and Betty Benson, '*Draft Townplanning Scheme for Bathurst and the Kombo Area*' (Accra: Office of the Townplanning Advisor to the Resident Minister, 1946).

Fry, E. Maxwell, and Jane Drew, *Tropical Architecture in the Humid Zone* (London: Batsford, 1956).

———, *Tropical Architecture in the Humid Zone* (London: Batsford, 1956).

———, *Tropical Architecture in the Dry and Humid Zones* (New York: Reinhold Publishing, 1964).

———, *Tropical Architecture in the Dry and Humid Zones* (New York: Reinhold Publishing, 1964).

Games, Stephen, 'Bobby Carter: An Appreciation', *RIBA Journal*, 89 (August 1982): 17.

John Gloag and E. Maxwell Fry, 'The Need for Planning Town and Countryside', *The Listener*, 232 (21 June 1933): 970.

Gardner, James, *The ARTful Designer* (London: James Gardner, 1993)

Gardner-Medwin, Robert, 'United Nations & Resettlement in the Far East', *Town Planning Review*, 22 (January 1952): 283–98.

Gardner-Medwin, Robert, 'The Decline of Architecture', *Higher Education Quarterly*, 10 (February 1956): 132–42.

Geddes, Patrick, *Patrick Geddes in India: [Extracts from His Official Reports on Indian Cities in 1915-19] Edited by J. Tyrwhitt, with an Introduction by L. Mumford, and a Preface by H.V. Lanchester* (London: Lund Humphries, 1947).

Gethin, Christopher, 'Chandigarh: A Memorial to Arrogance', *Built Environment*, 2 (May 1973): 291–4.

Gibberd, Frederick, 'Harlow', *Builder*, 173 (19 December 1947): 700–3.

Giedion, Sigfried, *Walter Gropius: Work and Teamwork* (London: Architectural Press, 1954).

Glass, Ruth (ed.), *The Social Background of a Plan: A Study of Middlesborough* (London: Routledge, 1948).

Gold, John R., 'The MARS Plans for London, 1933–1942: Plurality and Experimentation in the City Plans of the Early British Modern Movement', *Town Planning Review*, 66 (July 1995): 243–67.

———, *The Experience of Modernism: Modern Architects and the Future City, 1928–53* (London: E&F Spon, 1997).

———, *The Practice of Modernism: Modern Architects and Urban Transformation, 1954–1972* (London: Routledge, 2007)

———, '"A Very Serious Responsibility"? The MARS Group, Internationality and Relations with CIAM, 1933–39', *Architectural History*, 56 (2013): 249–75.

Goldie, Mark, *Corbusier comes to Cambridge: Post-war Architecture and the Competition to Build Churchill College* (Cambridge: Churchill College, 2007).

Gordon, Chris, and Kist Kilian, *Chandigarh: Forty Years after Le Corbusier* (Amsterdam: Architectura and Natura Press, 1992).

Gould, Jeremy, *Modern Houses in Britain, 1919–1939* (London: Society of Architectural Historians of Great Britain, 1977).

Grieve, Alistair, *Isokon* (London: Isokon Plus, 2004).

Gropius, Walter, *The New Architecture and the Bauhaus* (London: Faber and Faber, 1935).

Hare, A.F., 'Factory-made Housing for India', *Architect and Building News*, 197 (24 February 1950): pp. 217–18.

Hart, David, *The Volta River Project: A Case Study in Politics and Technology* (Edinburgh: Edinburgh University Press, 1980).

Hayward, Stephen, '"Good Design is Largely a Matter of Common Sense": Questioning the Meaning and Ownership of a Twentieth-Century Orthodoxy', *Journal of Design History*, 11/3 (1998): 217–33.

Herbert, Gilbert, *The Synthetic Vision of Walter Gropius* (Johannesburg: Witwatersrand University Press, 1959).

Hitchins, Stephen (ed.), *Fry, Drew, Knight, Creamer: Architecture* (London: Lund Humphries, 1978).

HMSO, 'Statement of Policy on Colonial Development and Welfare', (ed.) Colonial Office (London: HMSO, 1940).

———, Ministry of Education (Great Britain): 'Community Centres', (London: HMSO, 1945).

———, 'Report of the Commission on Higher Education in West Africa', (London: HMSO, 1945).

———, 'New Towns Act, 1946', (London: HMSO, 1946).

———, Advisory Committee on Education in the Colonies, 'Education for Citizenship in Africa', (London: HMSO, 1948).

———, 'Report of the United Kingdom Trade and Industrial Mission to Ghana', (London: H.M.S.O, 1959).

Holder, Julian, 'The Nation State or the United States? The Irresistible Kitchen of the British Ministry of Works, 1944 to 1951', in *Cold-War Kitchen: Americanization, Technology, and European Users*, (eds) Ruth Oldenziel and Karin Zachmann (Cambridge, MA: MIT Press, 2009): 235–58.

Holland, Jessica and Iain Jackson, 'A Monument to Humanism: Pilkington Brothers' Headquarters (1955–65) by Fry, Drew & Partners', *Architectural History*, 56 (2013): 347–90.

Home, Robert, 'Town Planning, Segregation and Indirect Rule in Colonial Nigeria', *Third World Planning Review*, 5 (May 1983).

———, 'Transferring British Planning Law to the Colonies', *Third World Planning Review*, 15 (November 1993): 397–410.

'A House at Chipperfield Common, Buckinghamshire', *Journal of the Royal Institute of British Architects*, 43 (7 March 1936): 483–6.

'House near Kingston, Surrey', *Architects' Journal*, 86 (18 November 1937): 784–7.

Hughes, Quentin, 'Before the Bauhaus: The Experiment at the Liverpool School of Architecture and Applied Arts', *Architectural History*, 25 (1982): 102–13.

Huxley, Elspeth, *Four Guineas: A Journey through West Africa* (London: Chatto and Windus, 1954).

Huxley, Julian, 'Did you hear that?', *The Listener*, 698 (28 May 1942): 682.

Ingleby, Richard, *Christopher Wood: An English Painter* (London: Allison and Busby, 1995).

Jackson, Anthony, 'The Politics of Architecture: English Architecture 1929–1951', *Society of Architectural Historians*, 24 (1965).

———, *The Politics of Architecture: A History of Modern Architecture in Britain* (London: Architectural Press, 1970).

Jackson, Frank, *Sir Raymond Unwin: Architect, Planner and Visionary* (London: A. Zwemmer, 1985).

Jackson, Iain, 'Maxwell Fry and Jane Drew's Early Housing and Neighbourhood Planning in Sector-22, Chandigarh', *Planning Perspectives*, 28 (2013): 1–26.

———, 'Tropical Architecture and West Indies: From Military Advances and Tropical Medicine, to Robert Gardner-Medwin and the Networks of Tropical Modernism', *The Journal of Architecture*, 18 (March 2013): 167–95.

Jackson, Iain, and Soumyen Bandyopadhyay, 'Authorship and Modernity in Chandigarh: The Gandhi Bhavan and the Kiran Cinema Designed by Pierre Jeanneret and Edwin Maxwell Fry', *The Journal of Architecture*, 14 (December 2009): 687–713.

Jencks, Charles, *Le Corbusier and the Tragic View of Architecture* (London: Allen Lane, 1973).

Jopp, K., and Development Secretariat Ghana, *Ghana's New Town and Harbour, Tema* (Accra: Ministry of Information, 1961).

Johnson, James, *The Influence of Tropical Climates on European Constitutions* (London, 1813).

Jordy, William H., *'Symbolic Essence' and Other Writings on Modern Architecture and American Culture* (New Haven, CT: Yale University Press, 2005).

Joshi, Kiran, *Documenting Chandigarh: The Indian Architecture of Pierre Jeanneret, Maxwell Fry & Jane Drew* (Ahmedabad: Mapin Publishing, 1999).

Kalia, Ravi, *Chandigarh: The Making of an Indian City* (New Delhi: Oxford University Press, 1987).

'Kensal House', *RIBA Journal*, 44 (20 March 1937): 500–5.

King, Anthony D, *Colonial Urban Development: Culture, Social Power, and Environment* (London: Routledge and Paul, 1976).

———, 'Exporting Planning: The Colonial and Neo-Colonial Experience', in *Shaping an Urban World*, (ed.) Gordon Cherry (London: Mansell, 1980).

Kirkpatrick, Diane, *Eduardo Paolozzi* (London: Studio Vista, 1970).

Koenigsberger, Otto, 'New Towns in India', *Town Planning Review*, 23 (July 1952): 95–132.

———, *Manual of Tropical Housing and Building* (Harlow: Longmans, 1974).

———, 'Obituary Denis C. Robinson (1921–1981)', *AAQ*, 13 (October 1981): 119.

Korn, Arthur, Maxwell Fry and Dennis Sharp, 'The M.A.R.S. Plan for London', *Perspecta*, 13/14 (1979): 163–73.

Krishan, Gopal *Inner Spaces - Outer Spaces of a Planned City*. (Chandigarh, Chandigarh College of Architecture, 1999).

Lai, Chee Kien, *Building Merdeka: Independence Architecture in Kuala Lumpur, 1957–1966* (Kuala Lumpur: Petronas, 2007).

Lagae, Johan, '"Montcassin, Montserrat Or ... An Alcazar?" Architecture, Propaganda and Everyday School Practices in the College Du Saint-Espirit in Bujumbura (Burundi)', in *Colonial Architecture and Urbanism in Africa*, (ed.) Fassil Demissie (Farnham: Ashgate, 2012).

Lanchester, H.V., 'Architecture and Housing in India', *Architectural Design and Construction* (May 1942): 94–5.

———, 'Indian Cities and their Improvement', *Architectural Design and Construction* (June 1942): 116–20.

Lang, Jon, *A Concise History of Modern Architecture in India* (Delhi: Permanent Black, 2003).

LeGates, R. and F. Stout (eds): *Early Urban Planning* (London: Routledge/Thoemmes Press, 1998).

Le Roux, Hannah, 'The Networks of Tropical Architecture', *Journal of Architecture*, 8 (Autumn 2003): 337–54.

———, 'Modern Architecture in Post-Colonial Ghana and Nigeria', *Architectural History*, 47 (2004): 361–92.

Lewis, David (ed.), *Architects' Year Book XI: The Pedestrian in the City* (London: Elek, 1965).

Lewis, L.J., 'Higher Education in the Overseas Territories 1948–58', *British Journal of Educational Studies*, 8 (1959): 3–21.

Liora, Bigon, *A History of Urban Planning in Two West Africa Colonial Capitals; Residential Segregation in British Lagos and French Dakar (1850–1930)*, (Lewiston, New York: Edwin Mellon Press, 2009).

Lipstadt, Hélène, 'Polemic and Parody in the Battle for British Modernism', *Oxford Art Journal*, 5/2 (1983): 22–30.

Liscombe, Rhodri Windsor, 'Modernism in Late Imperial British West Africa: The Work of Maxwell Fry and Jane Drew, 1946–56', *Journal of the Society of Architectural Historians*, 65 (2006): 118–215.

———, 'In-Dependence: Otto Koenigsberger and Modernist Urban Resettlement in India', *Planning Perspectives*, 21 (April 2006): 157–78.

Livsey, Tim, *'Suitable lodgings for students': modern space, colonial development and decolonization in Nigeria*. Urban History, Available on CJO 2014 doi:10.1017/S0963926813001065

Llewellyn, Mark, 'Producing and Experiencing Harlow: Neighbourhood Units and Narratives of New Town Life 1947–53', *Planning Perspectives*, 19 (2004): 155–74.

Loach, Judi, 'Le Corbusier and the Creative Use of Mathematics', *The British Journal for the History of Science*, 31/2 (June 1998): 185–215.

Macdonald, Susan (ed.), *Preserving Post-War Heritage: The Care and Conservation of Mid-Twentieth Century Architecture* (Donhead St Mary: Donhead Publishing, 2001).

Maddrell, Avril and James D. Sidaway (eds), *Deathscapes: Spaces for Death, Dying, Mourning and Remembrance* (Farnham: Ashgate, 2010).

MARS Group, *Turn Again: An Exhibition Presented by the MARS Group, Royal Exchange, 12–30 July 1955* (London: MARS Group, 1955).

Martin, J.L., Ben Nicholson, and N. Gabo (eds), *Circle: International Survey of Constructive Art*, (London: Faber and Faber, 1937).

Massey, Anne, *The Independent Group: Modernism and Mass Culture in Britain* (Manchester: Manchester University Press, 1995).

Matless, David, *Landscape and Englishness* (London: Reaktion Books, 1998).

May, Ernst, 'Kampala Town Planning', *Architects' Year Book*, 2 (1947): 59–63.

McMaskie, T.C., *Asante Identities: History and Modernity in an African Village, 1850–1950* (London: Edinburgh University Press, 2000).

Mellanby, Kenneth, *The Birth of Nigeria's University* (London: Methuen, 1958).

Millon, Henry, 'Rudolf Wittkower, "Architectural Principles in the Age of Humanism": Its Influence on the Development and Interpretation of Modern Architecture', *Journal of the Society of Architectural Historians*, 31 (May 1972): 83–91.

'"The Minimum Flat", Lawn Road flats, Hampstead, London; Architect: Wells Coates', *Building*, 9 (August 1934): 310–14.

Ministry of Health, *Housing Manual* (London: HMSO, 1949).

Moffett, Noel, 'Nigeria Today', *RIBA Journal*, 84 (June 1977): 244–55.

Moholy-Nagy: Paintings and Collages 1914–1946 (London: New London Gallery, 1961).

Morris, A.E.J., 'Chandigarh: The Plan Corb Tore Up?', *Built Environment Quarterly*, 1 (December 1975): 229–34.

Morris, Henry, *The Village College: Being a Memorandum on the Provision of Educational and Social Facilities for the Countryside, with Special Reference to Cambridgeshire* (Cambridge: Cambridge University Press, 1924).

Mozingo, Louise A., *Pastoral Capitalism: A History of Suburban Corporate Landscapes* (Cambridge, MA: MIT Press, 2011).

Mumford, Eric, *The CIAM Discourse on Urbanism, 1928–1960* (Cambridge, MA: MIT Press, 2002).

Muthesius, Stefan, *The Postwar University: Utopianist Campus and College* (London and New Haven, CT: Yale University Press, 2000).

Neumann, Eva-Marie, 'Architectural Proportion in Britain 1945–1957', *Architectural History*, 39 (1996): 197–221.

Newton, Douglas, 'Printed Textiles', *Architectural Review*, 111 (March 1952): 190–5.

Njoh, Ambe, *Planning Power: Town Planning and Social Control in Colonial Africa* (London: University College London, 2007).

———, *Urban Planning and Public Health in Africa: Historical, Theoretical and Practical Dimensions of a Continent's Water and Sanitation Problematic* (Farnham: Ashgate, 2012).

Ockman, Joan (ed.), *Architecture Culture 1943–1968: A Documentary Anthology* (New York: Columbia University Graduate School of Architecture, Planning, and Preservation, Rizzoli, 1993).

Okafor, Nduka, *The Development of Universities in Nigeria* (London: Longman Group, 1971).

Onokerhoraye, A.G., *Public Services in Nigerian Urban Areas* (Ibadan: Nigerian Institute of Social and Economic Research, 1982).

Osborn, Frederic James *New Towns: Their Origins, Achievements and Progress* (London: Leonard Hill, 1977).

Ossman, Laurie and Heather Ewing, *Carrère and Hastings: The Masterworks* (New York: Rizzoli, 2011).

Parker, Karen, 'Jane Drew, Architect and Practical Idealist', in *Women Designing: Redefining Design in Britain between the Wars*, (ed.) Jill Seddon and Suzette Worden (Brighton: University of Brighton, 1994): 136–40.

Parker, Karen, 'Jane Drew, Architect and Practical Idealist' (Brighton Polytechnic, 1988): Unpublished thesis

Parson, Kermit Carlyle (ed.), *The Writings of Clarence S. Stein* (Baltimore and London: John Hopkins University Press, 1998).

Payne, Alina, 'Rudolf Wittkower and Architectural Principles in the Age of Modernism', *Journal of the Society of Architectural Historians*, 53 (September 1994): 322–42.

Perera, Nihal, 'Contesting Visions: Hybridity, Liminality and Authorship of the Chandigarh Plan', *Planning Perspectives*, 19 (April 2004): 175–99.

Perry, C.A., W.D. Heydecker et al, *Neighborhood and Community Planning: Comprising Three Monographs. The Neighborhood Unit by C.A. Perry; Sunlight & Daylight for Urban Areas, by W.D. Heydecker, in Collaboration with E.P. Goodrich; Problems of Planning Unbuilt Areas, by T. Adams, E.M. Bassett, R. Whitten*, (New York, 1929).

Pevsner, Nikolaus, *Pioneers of the Modern Movement from William Morris to Walter Gropius* (London: Faber and Faber, 1936).

Phillips, Randal, 'A Modern Country House', *Country Life*, 71 (19 March 1932): 332–4.

Plageman, Nathan, 'Everybody Likes Saturday Night: A Social History of Popular Music and Masculinities in Urban Gold Coast/Ghana, c. 1900–1970' (Indiana University, 2008).

Pothen, Nayantara, *Glittering Delhi* (New Delhi: Penguin Books, 2012).

Powers, Alan, *Britain: Modern Architectures in History* (London: Reaktion Books, 2007).

———, 'Conservative Attitudes: Walter Gropius at Christ's College and Maxwell Fry at All Souls.' *Twentieth Century Architecture 11: Oxford and Cambridge Buildings*, (London, Twentieth Century Society, 2013).

Prakash, Aditya, 'Mobile Shops in Chandigarh', in *Architects' Year Book 14*, Declan Kennedy and Margrit Kennedy (eds) (London: Paul Elek, 1974).

Prakash, Vikramaditya, *Le Corbusier's Chandigarh: The Struggle for Modernity in Postcolonial India* (Seattle & London: University of Washington Press, 2002).

Pritchard, Jack, *View from a Long Chair* (London: Routledge, 1984).

Rand, Christopher, 'City on a Tilting Plan', *The New Yorker* (30 April 1955): 35–58.

'Rayon Industry Design Centre', *Architect and Building News*, 195 (28 January 1949): 76–9.

Read, Herbert, *Art and industry : The Principles of Industrial Design*, (London: Faber and Faber, 1934).

———, *The Grass Roots of Art* (London: Faber and Faber, 1955).

Rée, Harry, *Educator Extraordinary: The Life and Achievement of Henry Morris* (London: Longman, 1973).

Reidel, James, 'Walter Gropius: Letters to an Angel, 1927–35', *Journal of the Society of Architectural Historians*, 69/1 (March 2010): 88–107.

Reilly, C.H. (ed.), *The Liverpool Architectural Sketch Book: Being the Annual of the School of Architecture of the University of Liverpool* (London: Proprietors of the Architectural Review, 1920).

———, *Representative British Architects of the Present Day* (London: Batsford, 1931).

———, 'Criticism: Railways Stations, Terminal and Otherwise', *The Architectural Review*, 78 (October 1935): 147–9.

———, 'The Year's Work at Home', *Architects' Journal*, 85 (14 January 1937): 91–102.

Reith, John Charles Walsham, *Report of the New Towns Committee* (London: HMSO, 1946).

Richards, J. M., *An Introduction to Modern Architecture* (Harmondsworth: Penguin Books, 1940).

———, 'The Failure of the New Towns', *Architectural Review*, 114 (July 1953): 29–32.

———, 'Failure of the New Densities', *Architectural Review*, 114 (December 1953): 355–61.

Richards, J.M. (ed.), *New Buildings in the Commonwealth* (London: Architectural Press, 1961).

Richardson, Margaret, *66 Portland Place: The London Headquarters of the Royal Institute of British Architects* (London: RIBA Publications, 1984).

Richmond, Peter, *Marketing Modernisms: The Architecture and Influence of Charles Reilly* (Liverpool: Liverpool University Press, 2001).

Robbins, David (ed.), *The Independent Group: Postwar Britain and the Aesthetics of Plenty* (London and Cambridge, MA: MIT Press, 1990).

Robinson, Denis Curtis, 'Planning in the Commonwealth: Tema, the New Port of Ghana', *Journal of the Town Planning Institute*, 45 (March 1959): 90–3.

———, 'Tema, 1951–1961; Report on Ghana's New Town', (Accra, 1961).

Rovira, Josep, *José Luis Sert, 1901–1983* (London, Electa Architecture, 2003).

Royal Institute of British Architects, *Rebuilding Britain* (London: Lund Humphries 1943).

Said, Edward, *Orientalism: Western Conceptions of the Orient* (London: Routledge, 1978).

Saint, Andrew, *Towards a Social Architecture: The Role of School-Building in Post-War England* (London: Yale University Press, 1987).

Saler, Michael T., *The Avant-Garde in Interwar England* (Oxford: Oxford University Press, 1999).

Sarin, Madhu, *Urban Planning in the Third World: The Chandigarh Experience* (London: Mansell, 1982).

Scheidegger, Ernst, Maristella Casciato, Cerena Nievergelt, and Stanislaus Von Moos, *Chandigarh 1956: Le Corbusier, Pierre Jeanneret, Jane B. Drew, E. Maxwell Fry* (Zurich: Scheidegger & Spiess, 2010).

Schönfeld, Christiane (ed.), *Practicing Modernity: Female Creativity in the Weimar Republic* (Würzburg: Königshausen & Neumann, 2006).

Scott, Frederic Leslie, *Proposed Road Bridge at Charing Cross* (London: London County Council, 1931).

Sharma, Sangeet, *Corb's Capitol: A Journey through Chandigarh Architecture* (Chandigarh: Abhishek Publications, 2009).

Sharples, Joseph, Alan Powers, and M. Shippobottom, *Charles Reilly and the Liverpool School of Architecture 1904–1933* (Liverpool: Liverpool University Press, 1996).

Shoshkes, Ellen, *Jaqueline Tyrwhitt: A Transnational Life in Urban Planning and Design* (Farnham: Ashgate, 2013).

Simms, Barbara (ed.), *Eric Lyons and Span* (London: RIBA Publishing, 2006).

Simpson, W.J., *The Maintenance of Health in the Tropics* (London: London School of Tropical Medicine and John Bale, Sons & Danielsson, 1916).

Simon, A. P., *Manchester Made Over* (London: P.S. King and Son, 1936).

Spencer, Robin (ed.), *Eduardo Paolozzi: Writings and Interviews* (Oxford: Oxford University Press, 2000).

Stephenson, Flora, and Gordon Stephenson, *Community Centre* (London: Percy Lund, Humphries & Co., 1942).

Stockdale, Frank, Robert Gardner-Medwin, and Leo De Syllas, 'Recent Planning Developments in the Colonies', *RIBA Journal*, 55 (February 1948): 140–8.

Stonehouse, Roger, and Trevor Dannatt, *Trevor Dannatt Works and Words* (London: Blackdog, 2008).

'Sun House', *Architects' Journal*, 84 (13 August 1936): 210–14.

Sutherland, Alasdair C., 'Regional Planning Problems in Ashanti', *Planning Outlook Series 1*, 3 (1955): 16–28.

Swenarton, Mark, *Building the New Jerusalem*, (Bracknell: IHS BRE Press, 2008).

Taaffe, Peter and Tony Mulhearn, *Liverpool: A City that Dared to Fight* (London, Fortress Books, 1988).

Thompson, A.R., *Education and Development in Africa* (New York: St Martin's Press, 1981).

Thompson, F.L., *Site Planning in Practice: An Investigation of the Principles of Housing Estate Development* (London: Frowde, Hodder & Stoughton, 1923).

Tipple, Graham, *Extending Themselves: User-Initiated Transformations of Government-Built Housing in Developing Countries* (Liverpool: Liverpool University Press, 2000).

Tipple, Graham, Amole Bayo, David Korboe, and Helen Onyeacholem, 'House and Dwelling, Family and Household: Towards Defining Housing Untis in West African Cities', *Third World Planning Review*, 16 (November 1994): 429–50.

Towndrow, Frederic, 'Riba Hq Some Other Schemes of Note', *Architects' Journal*, 75 (11 May 1932): 625.

Tubbs, Ralph, *Living in Cities* (Harmondsworth: Penguin Books, 1942).

Tyrwhitt, Jaqueline, *Patrick Geddes in India* (London: Lund Humphries, 1947).

———, 'The Village Centre at the Exhibition on Low Cost Housing, Delhi, 1954', *Ekistics*, 52 (September–December 1985): 430.

Tyrwhitt, Jaqueline, J.L. Sert and E.N. Rogers (eds), *The Heart of the City: Towards the Humanization of Urban Life* (London: Lund Humphries, 1952).

Uduku, Ola, 'The Colonial Face of Educational Space', in *White Papers, Black Marks: Architecture, Race, Culture*, (ed.) Lesley Naa Norle Lokko (London: Athlone Press, 2000).

———, 'Modernist Architecture and "the Tropical" in West Africa: The Tropical Architecture Movement in West Africa, 1948–1970', *Habitat International*, 30 (September 2006): 396–411.

Unwin, Raymond, *Nothing Gained by Overcrowding! How the Garden City Type of Development May Benefit Both Owner and Occupier* (Westminster: P. S. King & Son, 1912).

Waters, Suzanne, 'In Search of Sir Gerald Barry, the man behind the Festival of Britain', *Twentieth Century Architecture*, 5 (2001): 39–46.

Watkins, Mark Hanna, 'The West African "Bush" School', *American Journal of Sociology*, 48 (May 1943): 666–75.

Waymark, Janet, *Thomas Mawson: Life, Gardens and Landscapes* (London: Frances Lincoln, 2009).

Whyte, William, 'The Englishness of English Architecture: Modernism and the Making of a National International Style, 1927–1957', *Journal of British Studies*, 48 (April 2009): 441–65.

Wilk, Christopher, 'The Healthy Body Culture', in *Modernism: Designing a New World*, (ed.) Christopher Wilk (London: V&A Publications, 2006).

Williams, Gareth, 'The Return of the Curve: Rodney Thomas, Architecture and Interior Design', *Journal of the Decorative Arts Society 1850 – the Present*, 19 (1995): 41–50.

Williams-Ellis, Clough, *Architect Errant* (London: Constable, 1971).

Wood, Alan G., 'Palm Trees and Iced Beer', *Journal of the Town Planning Institute*, 45 (March 1959): 117–19.

'Working Details: 57–58. Shop-Front. Showrooms in Victoria Street, S.W.1. Adams, Thompson and Fry', *Architects' Journal*, 79 (31 May 1934): 801–2.

Wright, Myles H. (ed.), *Small Houses, £500–£2,500*, (London: Architectural Press, 1937).

Yorke, F.R.S., and Colin Penn, *A Key to Modern Architecture* (London: Blackie and Son, 1939).

Index